CLASS No. 611.m2
BOOK No. 2877

This book is to be returned on or before the last date stamped below.

30 DEC. 1987
18. JAN. 1988
2 FEB 1988
15. JUL 1990

13 SEP 2002
19 OCT 2002
12 MAR 2003
13 APR 2004
13 OCT 2004

18 FEB 2005   T

DISPOSED
20.4
MAYBE NEWER INFO AVAILABLE

Human Anatomy

Human Anatomy

J. Robert McClintic, Ph.D.
California State University, Fresno, California

With 823 illustrations, 183 in full color

The C. V. Mosby Company
St. Louis · Toronto · London   1983

Editor: Diane Bowen

Editorial assistant: Tim Arnold

Manuscript editor: John Middleton

Book design: Nancy Steinmeyer

Cover design: Diane Beasley

Production: Linda Stalnaker, Judy Bamert

Cover: Study/Falling Man (Table Fig.)

Chrome-plated bronze-hinged figure on stainless steel base by Ernest Trova. (Courtesy of The Pace Gallery of New York.)

Photo by Dan Sindelar

Copyright © 1983 by The C.V. Mosby Company

All rights reserved. No part of this publication may be reproduced, stored in a retrieval system, or transmitted, in any form or by any means, electronic, mechanical, photocopying, recording, or otherwise, without prior written permission from the publisher.

Printed in the United States of America

The C.V. Mosby Company
11830 Westline Industrial Drive, St. Louis, Missouri 63141

Library of Congress Cataloging in Publication Data

McClintic, J. Robert, 1928-
   Human anatomy.

   Includes bibliographies and index.
   1. Anatomy, Human. I. Title. [DNLM:
1. Anatomy. QS 4 M127h]
QM23.2.M37   611   82-6351
ISBN 0-8016-3225-0   AACR2

C/VH/VH 9 8 7 6 5 4 3 2 1   05/D/640

A TRADITION OF PUBLISHING EXCELLENCE

To my wife, Peggy,

*whose contributions and constructive criticism
have improved the work*

and

to my daughters, Cathleen, Colleen, Marlene,

who have made the effort worthwhile

Preface

The study of the structure of the human body is regarded by many as a science of memorization of terms. That anatomy includes the learning of names cannot be denied, but the study of the body can be made more interesting by understanding how the body develops and functions, how it may suffer malfunctions or malformations, and how it can restore and maintain itself in the face of the many stresses it confronts. By integrating these concepts into a whole, we can gain greater appreciation of the exquisite purpose and organization of the body.

This book presents the body as a dynamic living organism. The first several chapters establish a foundation of terminology and relationships that will be built on in the rest of the book. General development and the cellular and tissue levels acquaint the reader with the basis for adult structure and function. That we do not present the same appearance or function throughout life is noted in an early chapter, and living anatomy is presented to create an appreciation of what most of us will continue to see throughout our lives.

Once into the consideration of body systems, the reader will find nearly each chapter or section to be introduced by a description of the specific development of that system or of the organs within it. Next, microscopic and gross anatomy are integrated with functional and pathological considerations. In those chapters dealing with body movement, the actual roles of muscles, joints, and bones are presented as they are utilized in the body and not merely as isolated organs.

Aids to the study of the body include correlated objectives, summaries, and questions. Material is summarized in tables. Readings are current and are for the most part chosen for ease of reading and comprehension.

The study of anatomy can only be enhanced by accurate and clear illustrations. For the superb artwork contained in this text, I thank the following artists:

The members of an illustrative team headed by David Mascaro, all of whom are located at the Medical College of Georgia, Augusta, Georgia. Team members include Mrs. Karen Waldo, Chief, Medical illustration section, and Mr. John Hagen and Mrs. Glenna Deutsch, staff illustrators.

Jeanne Robertson, illustrator, and Chris Hess, photographer, both of St. Louis, Missouri.

To Tom Manning, Diane Bowen, Sue Schapper, Tim Arnold, and Judi Wolken of The C.V. Mosby Company, I express my appreciation for their support and encouragement that ensured the completion of the project.

To my wife, Peggy, I extend my gratitude for her typing of and contributions to the manuscript.

Responsibility for errors in the book is mine.

J. Robert McClintic, Ph.D.

Contents

1 Introduction to the human body, 1

Objectives, *1*
The discipline of anatomy, *1*
How anatomy is studied, *1*
Some basic terminology, *2*
    Anatomical position, *2*
    Body regions, *2*
    Directional terms, *3*
    Planes of section, *4*
    Reference lines, *5*
    Body cavities, *7*
Summary, *8*
Questions, *8*
Readings, *8*

2 Organization, development, and basic tissues of the body, 9

Objectives, *9*
Atomic and molecular levels, *9*
Cellular level, *10*
    Structure of cells, *10*
    Cell replacement: mitosis, *18*
    Neoplasms, *18*
    Species replacement: meiosis, *19*
Development of a new individual, *20*
    First week, *20*
    Second week, *21*
    Third week, *22*
    Fourth week, *24*
    Fifth to eighth weeks, *25*
    Fetal membranes, *28*
Tissue level, *31*
    Basic tissue groups, *31*
    Epithelium, *31*
    Connective tissues, *42*
    Muscular tissues, *54*
    Nervous tissue, *55*
Organs, *56*
Systems, *56*
The whole organism, *63*
Summary, *64*
Questions, *65*
Readings, *65*

3 | *External appearance of the body, 66*

Objectives, *66*
Statement of purpose, *66*
External appearance of the body at different ages, *67*
Head, *72*
Neck, *74*
Upper appendage, *76*
    Shoulder and axilla, *76*
    Arm, *77*
    Forearm, *77*
    Wrist and hand, *77*
Thorax and abdomen, *78*
Lower appendage, *79*
    Hip, *79*
    Thigh, *80*
    Leg, *80*
    Ankle and foot, *80*
Summary, *81*
Questions, *81*
Readings, *81*

4 | *Integumentary system, 82*

Objectives, *82*
Development of the skin and its appendages, *82*
Gross features of the skin, *85*
Structure of the skin, *86*
    Epidermis, *86*
    Dermis (corium), *88*
    Hypodermis (subcutaneous layer), *88*
Blood and lymph vessels of the skin, *90*
Functions of the skin, *90*
Burns, *91*
Appendages of the skin, *92*
    Hair, *92*
    Nails, *93*
    Glands, *93*
Summary, *94*
Questions, *95*
Readings, *95*

5 | *The skeleton, 96*

Objectives, *96*
Tissues of the skeleton, *96*
    General remarks, *96*
    Development of bone (osteogenesis), *97*
    Blood vessels of bone, *100*
Bones of the skeleton, *101*
    General remarks, *101*
    Skull, *102*

Vertebral column, *118*
Thorax, *124*
Pectoral girdle, *126*
Upper appendage, *128*
Pelvic girdle, *134*
Lower appendage, *137*
Ossification of the skeleton, *142*
Fractures, *144*
Questions, *146*
Readings, *146*

6 | *Articulations (joints), 147*

Objectives, *147*
Classification of articulations (joints), *147*
    Fibrous joints, *147*
    Cartilaginous joints, *148*
    Synovial joints, *149*
Specific joints of the body, *152*
    Temporomandibular joint, *152*
    Joints of the vertebral column, *153*
    Joints of the thorax, *154*
    Joints of the shoulder girdle, *155*
    Shoulder joint, *156*
    Elbow joints, *157*
    Joints of the wrist and hand, *158*
    Sacroiliac joint, *159*
    Hip joint, *160*
    Knee joint, *163*
    Tibiofibular articulations, *164*
    Joints of the ankle and foot, *165*
Some selected disorders of joints and their associated structures, *166*
    Traumatic damage to a joint, *166*
    Inflammation of joints, *166*
    Inflammation of associated structures, *166*
Questions, *167*
Readings, *167*

7 | *Muscular tissue, 168*

Objectives, *168*
Development of muscular tissue, *168*
Types of muscular tissue, *169*
    Smooth muscle, *169*
    Striated muscle, *171*
Skeletal muscle, *173*
    Structure of the fibers, *173*
    Arrangement of fibers within a muscle, *174*
    Types of fibers, *174*
    Contractile mechanisms in skeletal muscle, *176*
    Sources of energy for contraction, *177*
    Properties of skeletal muscle, *177*

Connective tissue components of skeletal muscle, *178*
   Fasciae, *178*
   Organization of fibrous components of a muscle, *178*
   Tendons and tendon sheaths, *178*
Blood and nerve supply to skeletal muscle, *179*
Clinical conditions associated with skeletal muscle, *180*
Summary, *181*
Questions, *182*
Readings, *182*

8 | *The skeletal muscles, 183*

Objectives, *183*
Bones and muscles as body lever systems, *183*
How muscles are named, *185*
Muscles of the head, *188*
   Facial muscles, *188*
   Cranial muscles, *190*
Muscles of mastication, *193*
Muscles of the tongue, *194*
Muscles associated with the hyoid bone and larynx, *197*
Muscles of the neck, *199*
Muscles of the back, *205*
Muscles of the anterior and lateral abdominal walls, *209*
Muscles of the posterior abdominal wall, *210*
Muscles of respiration, *213*
Muscles operating the scapula and clavicle, *214*
Muscles moving the shoulder joint or humerus, *219*
   Flexors of the humerus, *219*
   Extensors of the humerus, *219*
   Abductors of the humerus, *220*
   Rotators of the humerus, *220*
Muscles moving the forearm, *222*
   Flexors of the forearm, *222*
   Extensors of the forearm, *222*
   Rotators of the forearm, *223*
Muscles moving the wrist and fingers, *224*
   Flexors of the wrist and fingers, *224*
   Extensors of the wrist and fingers, *226*
   Abduction and adduction of the wrist, *227*
Muscles of the thumb, *228*
Muscles within the hand, *230*
Muscles moving the hip and knee joints, *232*
Muscles moving the ankle and foot, *240*
Muscles within the foot, *246*
Readings, *247*

9 | *The nervous system, 248*

Objectives, *248*
Development of the nervous system, *248*
Organization of the nervous system, *250*

Central nervous system, *250*
Peripheral nervous system, *252*
Cells and tissues of the nervous system, *253*
Neurons, *253*
Glial cells, *259*
Synapse, *260*
Reflex arc, *261*
Nerves, *262*
Central nervous system, *263*
Brain, *263*
Spinal cord, *281*
Meninges, *286*
Ventricles and cerebrospinal fluid, *290*
Peripheral nervous system, *292*
Cranial nerves, *292*
Spinal nerves and plexuses, *300*
Autonomic nervous system, *303*
Summary, *306*
Questions, *308*
Readings, *308*

10 Organs of sensation, *309*

Objectives, *309*
Some characteristics of receptors, *309*
Classification of receptors, *310*
Cutaneous and membrane sensation, *310*
Location and structure of receptors, *310*
Pathways to the central nervous system, *312*
Visceral sensation, *314*
Location and structure of receptors, *314*
Pathways to the central nervous system, *314*
Muscle, tendon, and joint sensation, *315*
Location and structure of receptors, *315*
Pathways to the central nervous system, *316*
Eye, *317*
Development, *317*
Size, shape, and landmarks, *318*
Tunics of the eye, *318*
Visual pathways, *322*
Extrinsic eye muscles, *324*
Eyelids and lacrimal apparatus, *324*
Selected disorders of the eye and its associated structures, *325*
Ear, *328*
Development, *328*
Structure, *332*
Auditory and equilibrial pathways, *336*
Selected disorders of hearing and equilibrium, *336*
Sense of taste, *338*
Distribution and structure of taste buds, *338*
Taste pathways, *338*
Sense of smell, *340*

Location and structure of the olfactory epithelium, *340*
Olfactory pathways, *341*
Summary, *342*
Questions, *343*
Readings, *343*

11 | Circulatory system, *344*

Objectives, *344*
Development of the blood, *344*
Blood vascular system, *345*
    Blood, *345*
    Heart, *349*
    Development of blood and lymph vessels, *363*
    Structure of the blood vessels, *364*
    Major systemic arteries, *374*
    Major systemic veins, *382*
    Special features of the circulation, *386*
    Fetal circulation, *391*
Lymph vascular system, *392*
    Lymph, *392*
    Lymph vessels, *393*
    Lymph organs, *393*
Summary, *399*
Questions, *400*
Readings, *400*

12 | Respiratory system, *401*

Objectives, *401*
Development of the system, *401*
Organs of the system, *405*
Nose and nasal cavities, *406*
    Nose, *406*
    Nasal cavities, *407*
Pharynx, *408*
Larynx, *409*
    Cartilages, *409*
    Muscles, *410*
Trachea and bronchi, *411*
Lungs, *414*
    Relationships of the lungs to the thoracic cavity, *414*
    Gross anatomy, *414*
    Bronchioles, *414*
    Respiratory portion of the system, *416*
    Bronchopulmonary segments and lobules, *417*
    Blood supply of the lungs, *417*
    Nerves of the lungs, *418*
    Breathing, *418*
Summary, *419*
Questions, *420*
Readings, *420*

13 Digestive system, 421

Objectives, *421*
Development of the system, *422*
    Formation of the mouth and gut, *422*
    Disorders of formation of the gut, *424*
Organs of the system, *427*
Mouth and oral cavity, *427*
    Lips and cheeks, *427*
    Teeth and gums, *427*
    Tongue, *430*
    Salivary glands, *431*
Fauces, *433*
Pharynx, *433*
General plan of tissue layers in the alimentary tract, *433*
Esophagus, *434*
Abdomen, *434*
    Boundaries, *434*
    Peritoneum, *434*
    Mesenteries and omenta, *434*
Stomach, *436*
    Gross anatomy, *436*
    Microscopic anatomy, *437*
Small intestine, *438*
Large intestine, *439*
Rectum and anal canal, *440*
Liver, *441*
    Gross anatomy, *441*
    Blood supply, *441*
    Microscopic anatomy, *442*
Gallbladder and bile ducts, *444*
Pancreas, *446*
    Gross anatomy, *446*
    Microscopic anatomy, *446*
Digestion in the small intestine, *447*
Blood vessels of the system, *447*
Selected disorders of the system, *447*
Summary, *448*
Questions, *449*
Reading, *449*

14 Urinary system, 450

Objectives, *450*
Development of the system, *450*
Organs of the system, *452*
    Kidneys, *452*
    Physiology of the kidney, *457*
    Calyces, renal pelvis, and ureters, *458*
    Urinary bladder, *459*
    Urethra, *460*

Disorders of the system, *461*
Summary, *461*
Questions, *462*
Readings, *462*

15 Reproductive systems, *463*

Objectives, *463*
Development of the systems, *463*
Female reproductive system, *469*
    Ovaries, *469*
    Uterine tubes, *474*
    Uterus, *475*
    Vagina, *477*
    External genitalia and pelvic floor, *477*
    Mammary glands, *478*
    Endocrine relationships, *479*
    Disorders of the female reproductive system, *479*
Male reproductive system, *480*
    Scrotum and testes, *481*
    Ducts of the testes, *484*
    Accessory glands, *486*
    Penis and pelvic floor, *489*
    Endocrine relationships, *489*
Summary, *490*
Questions, *491*
Readings, *491*

16 Endocrine structures, *492*

Objectives, *492*
Development of the endocrine structures, *492*
Criteria for determining endocrine status, *493*
Pituitary gland (hypophysis), *495*
    Location, size, and structure, *495*
    Blood supply, *495*
    Cells of the pituitary gland, *496*
Thyroid gland, *500*
    Location, size, and structure, *500*
    Blood supply, *501*
    Thyroid follicles, *501*
Parathyroid glands, *502*
    Location, size, and structure, *502*
    Cells of the parathyroid glands, *502*
Adrenal glands, *503*
    Location, size, and structure, *503*
    Blood supply, *503*
    Adrenal cortex, *504*
    Adrenal medulla, *504*
Pineal gland, *505*
Review of endocrine structures described in connection with other systems, *505*
    Pancreas, *505*

　　　　　　Testes, *505*
　　　　　　Ovaries, *505*
　　　　　　Thymus, *505*
Gastrointestinal ''hormones,'' *506*
Prostaglandins, *506*
Disorders of the endocrine glands, *506*
　　　　　　Disorders of the pituitary gland, *506*
　　　　　　Disorders of the thyroid gland, *506*
　　　　　　Disorders of the parathyroid glands, *506*
　　　　　　Disorders of the pancreas, *506*
　　　　　　Disorders of the adrenal glands, *507*
　　　　　　Disorders of the gonads, *507*
Summary, *507*
Questions, *508*
Readings, *508*
General references, *508*

Glossary, *509*

1

Introduction to the human body

Objectives
The discipline of anatomy
How anatomy is studied
Some basic terminology
    Anatomical position
    Body regions
    Directional terms
    Planes of section
    Reference lines
    Body cavities
Summary
Questions
Readings

OBJECTIVES After studying this chapter, the reader should be able to:

Define the discipline of anatomy and some of the methods of studying anatomy.
Describe anatomical position and its importance in establishing directional terms and movements.
Name the major body regions.
List and define the directional terms appropriate to the human body.
Describe the planes of section appropriate to the human body.
Describe the major reference lines of the body, with emphasis on the abdominal lines, and the organs contained within the areas delineated.
Name the major body cavities and the organs housed within each cavity and its subdivisions.

The discipline of anatomy

The extraordinarily complex organism that is the human body has been studied for literally thousands of years. Interest in the human body is reflected in cave drawings of prehistoric times. The Egyptians gained some anatomical knowledge from their practice of embalming the dead, although dissection for purposes of study probably was not practiced. That removal of body organs *was* practiced is evidenced by the placing of viscera in separate receptacles during the embalming procedure. Amulets in the form of the heart, lungs, or trachea evidence some recognition of the importance of certain body organs as possessors of "magical" powers.

True dissection, at least of animals, may have occurred about 500 BC and led to generalized descriptions of internal organs. Indeed, the word **anatomy** (Gk. *ana*, up, + *tome*, cutting) is derived from the use of tools to dissect (L. *dissecare*, to cut up) animal and human bodies. Aristotle, regarded as the founder of comparative anatomy, Galen, and Avicenna are names of ancients associated with studies and publications in anatomy.

Real anatomical research began in the thirteenth and fourteenth centuries, and Mondino, da Vinci, and Vesalius should be recalled as people having published drawings or treatises on human anatomy.

How anatomy is studied

The earliest anatomists were of necessity dependent on their own eyes for observation of body structure revealed by their dissections. There were no microscopes available until the seventeenth century. Thus **gross anatomy**, or study of anatomy with the naked eye,

2 Human anatomy

was the forerunner of all other types of anatomy. **Surface anatomy** is a corollary of gross anatomy and is the study of the features of the living body that may be appreciated by *viewing*, or in some cases *palpating* (feeling), external and internal body structures from the outside. These procedures are used today by medical personnel.

With the development and improvement of the microscope, the body's cells and tissues could be visualized, and **cytology** (Gk. *cytos*, cell) and **histology** (Gk. *histos*, tissue, + *logos*, study of) developed. Modern instruments, such as electron and scanning microscopes, have enabled us to enlarge body parts many thousands or millions of times, thus revealing what is called the *ultrastructure* of cells and tissues. **Developmental anatomy**, or *embryology*, is concerned with prenatal (before birth) and postnatal (after birth) development and maturation of body structure and has led to an increased appreciation of disorders of development. **Teratology** (Gk. *teras*, monster, + *logos*) studies the mechanisms of development of *congenital* (born with) *malformations* and is a discipline that complements embryology.

Today, special techniques have been developed that enable medical personnel to visualize the structure of the living body for diagnostic purposes that may lead to surgical or other types of intervention. *X-ray* examinations, *endoscopy* (in which the interior of a hollow organ may be visualized by use of a special instrument), *radioactive isotopes*, and *computed tomography* (CT) are employed.

As can be seen, study of the structure of the human body has evolved to the point at which the smallest units of body structure may be examined. While consideration will be directed primarily at the gross morphology of the body, mention will be made of other aspects of anatomy whenever it is necessary for full understanding of body structure.

Some basic terminology

The discipline of anatomy of any living organism is of necessity involved with learning the names and relationships of the organisms' structures. The task of learning such names may be made easier if the student has a foundation from which to proceed. The following sections attempt to create such a foundation, and the material presented should be thoroughly learned, since it will be constantly employed in chapters to follow.

Anatomical position

So that there is a basic point of reference from which directions or motions may be said to originate, an arbitrary body position called **anatomical position** has been designated (Fig. 1-1). In this position, the body is in a standing posture, with head, eyes, and toes pointing forward, and with the upper appendages by the sides with the palms of the hands facing forward. All anatomical descriptions are based on this position, even though the body may be prone (lying on its face), supine (lying on its back), or on its side as it is viewed or dissected.

Body regions

In describing locations of body organs, such as muscles, blood vessels, and nerves, it is appropriate to be able to refer to rather gross regions of the body. Most of the terms employed are probably familiar. Nevertheless, students should review the body areas presented in Fig. 1-2 and incorporate any new or unfamiliar terms into their vocabulary. It may be noted that the body is divided into four major regions—head, neck, trunk, and appendages—and that each of these major regions is further subdivided. Attention is called to the terms *flank*, *loin*, *arm*, and *leg*, the latter two referring respectively to the body parts between shoulder and elbow, and knee and ankle, and *not* to the whole upper or lower appendage. Although flank and loin are terms

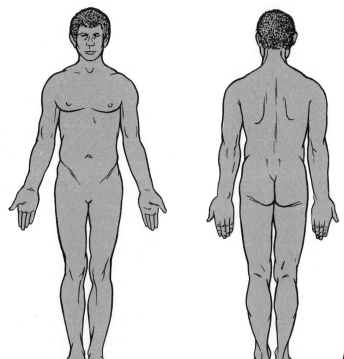

Fig. 1-1
Anatomical position.

most commonly heard in a butcher shop, medical personnel may speak of "flank pain," and we have all heard the expression "girding the loins."

Directional terms

Positions of body parts and relationships of one body part to another are given by **directional terms**. These also should be learned, since many specific body names are derived from or include directional terms. Remember that these terms assume that the body is in anatomical position.

A list of these terms follows:

anterior or *ventral* designates the "front" or belly side of the human body or a structure "in front of" an original point of reference. Anterior is perhaps more commonly used and really applies only to humans. In a four-footed animal anterior means toward the head; ventral refers to the belly side regardless of the posture of the organism. For example, the sternum is on the anterior thorax.

posterior or *dorsal* refers to the "back" of the body or a structure "behind" an original point of reference. The same cautions presented in the previous term apply here as well. The spine is on the posterior aspect of the body.

superior implies higher, toward the head, or something at a higher level on the body than an original point of reference. The prefix *supra-* is often used in naming body parts and implies *above*. The head is superior to the neck.

inferior describes lower, toward the feet, or something at a lower level than an original point of reference. *Infra-*, another prefix, implies *below*. The feet are inferior to the ankle.

medial implies a position closer to the body midline. A line running from the center of the forehead to a point between the feet defines the midline of the body. The nose is medial to the eyes.

lateral implies a position away from the body midline. The ears are on the lateral sides of the head.

external most frequently refers to something toward the body surface or toward the *outer* surface of a body organ. The term *superficial* is also used in the same sense. The skin covers the *external* body surface.

internal refers to a position away from the surface of the body or toward the *inner* aspect of a body organ. The term *deep* is used in the same sense. The pancreas is an *internal* abdominal organ.

proximal is used mainly in connection with the appendages and refers to a part closer to the attachment of that part to the midline of the body or closer to some specified point of reference. For example, the shoulder is *proximal* to the elbow, or the *proximal* convoluted tubule of the kidney is closest to the point of reference (in this case, the renal corpuscle).

distal implies a position further away from a point of attachment of an appendage or away from some specified reference point; it is the opposite of proximal. Thus the ankle is distal to the knee, and the kidney's distal convoluted tubule is further from the renal corpuscle.

palmar or *volar* refers to the palm (anterior) side of the hand.

plantar refers to the sole of the foot.

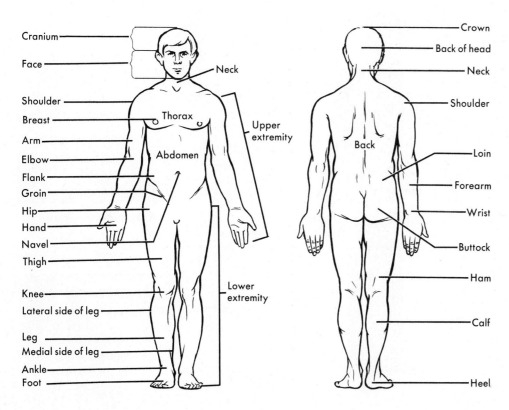

Fig. 1-2

General descriptive areas of body.

Planes of section

In many cases body structure may be further studied by cutting the body or its parts in various directions. These "directions" constitute *planes of section*. These are described by the following terms (Fig. 1-3):

midsagittal or *median plane* extends from anterior to posterior in the midline of the body and divides the body into *equal right and left halves*.

parasagittal or *sagittal plane* is parallel to the median plane but *not* creating equal right and left halves (that is, not in the midline).

coronal or *frontal plane* extends from side to side and divides the body into *front and back halves* (equal or not).

transverse or *horizontal plane* (section) or *cross-section* lies at right angles to the previous planes and usually cuts across the shortest dimension of a body part or organ. Often abbreviated *ts* or *xs*, this plane is commonly seen on labels of histological slides.

longitudinal plane or *section* follows the longest dimension of a part or organ and lies at right angles to the transverse plane. It is abbreviated *ls*.

oblique plane (diagonal or slanted) lies at more or less than a right angle to the previous planes.

Fig. 1-3 shows the orientations of these various planes and some of the shapes of the sections that would result.

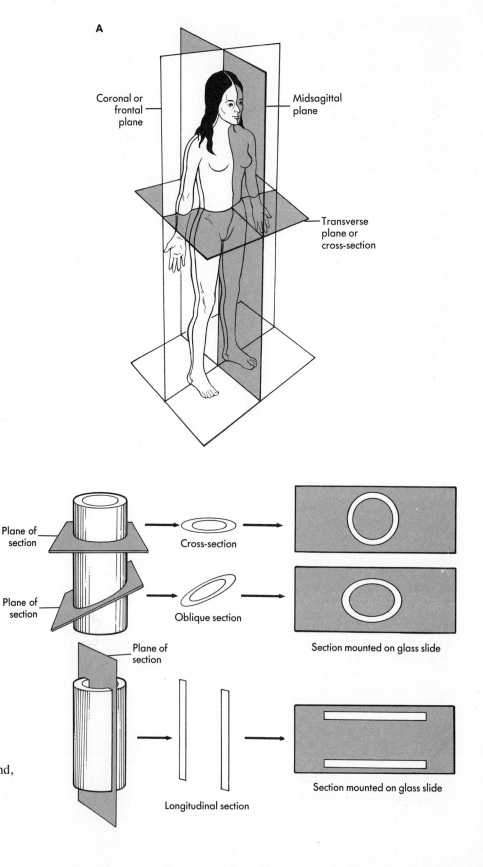

Fig. 1-3
Planes of section, **A,** of body and, **B,** of its organs.
After Ham.

Reference lines

Externally drawn *reference lines* form yet another method of locating superficial or deep body structures. *Midsternal* and *midclavicular* lines (Fig. 1-4) lie on the anterior chest; anterior and posterior *axillary* lines (Fig. 1-5) lie on the lateral chest; *vertebral* and *scapular* lines lie on the back (Fig. 1-6).

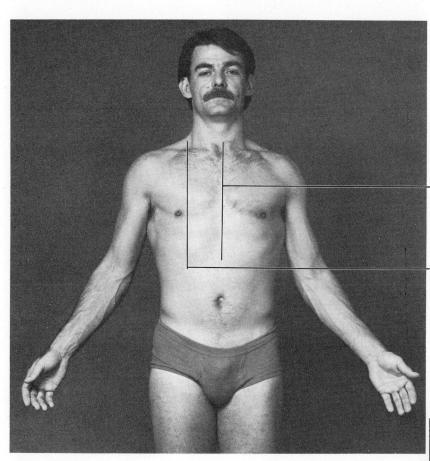

Fig. 1-4
Thoracic reference lines.

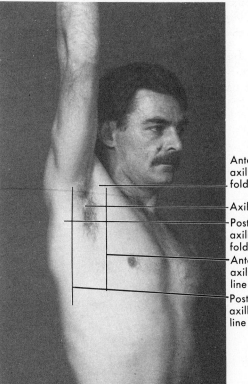

Fig. 1-5
Axillary reference lines.

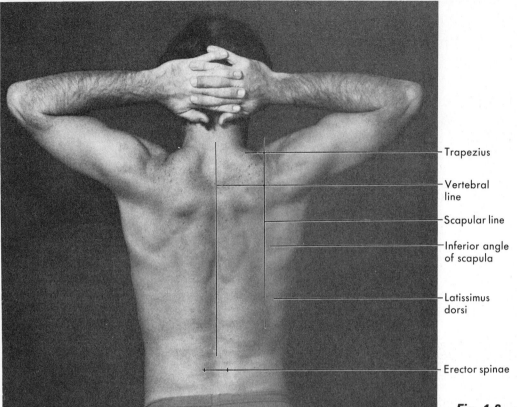

Fig. 1-6
Reference lines of back.

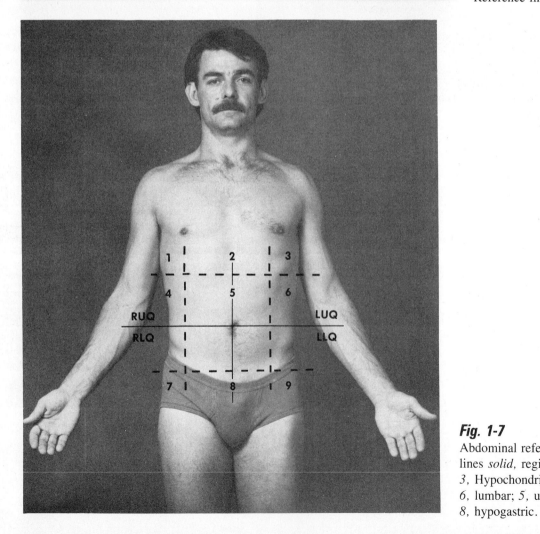

Fig. 1-7
Abdominal reference lines. Quadrant lines *solid*, region lines *broken*. *1* and *3*, Hypochondriac; *2*, epigastric; *4* and *6*, lumbar; *5*, umbilical; *7* and *9*, iliac; *8*, hypogastric.

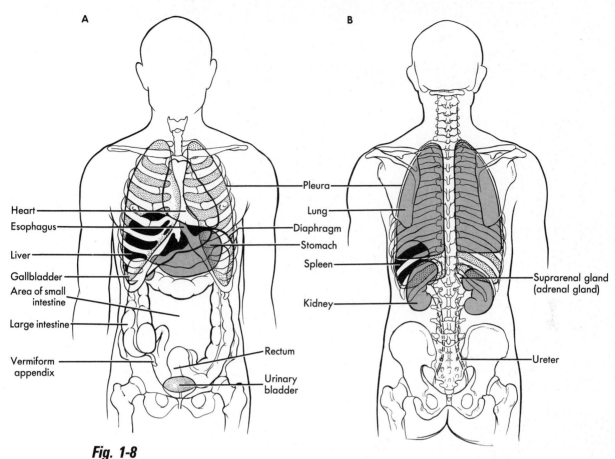

Fig. 1-8
Surface projections of some visceral organs. **A,** Anterior view. **B,** Posterior view.

On the abdomen, lines may be drawn through the umbilicus to create four *quadrants* (Fig. 1-7):

RUQ: right upper quadrant
RLQ: right lower quadrant
LUQ: left upper quadrant
LLQ: left lower quadrant

Other lines create nine smaller areas (see Fig. 1-7), each having its own name. The reader should be able to name the organs that lie in each of the quadrants or nine smaller areas (Fig. 1-8).

Body cavities

Most body organs are located within body cavities. Those cavities that do not communicate directly with the external body surface are called the "true" body cavities (Fig. 1-9). A *dorsal cavity* includes the *cranial* cavity

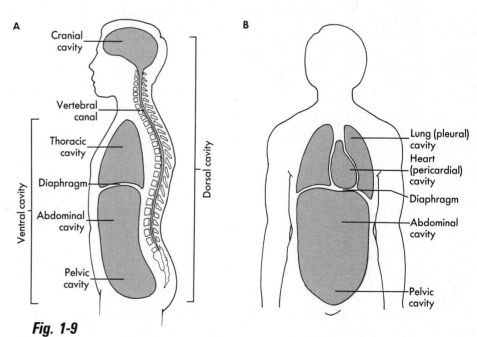

Fig. 1-9
True body cavities. **A,** Lateral view showing ventral and dorsal cavities. **B,** Anterior view showing divisions and subdivisions of ventral cavity.

and *vertebral canal,* housing the brain and spinal cord respectively. A **ventral cavity,** or more properly, ventral cavities, includes the *thoracic* cavity, subdivided into two pleural cavities housing the lungs, a pericardial cavity, housing the heart, and a potential space called the mediastinal cavity, filled with the great vessels of the heart and the trachea. The diaphragm separates the thoracic cavity from the abdominopelvic cavity. The **abdominopelvic cavity** is the other major subdivision of the ventral cavity and consists of a larger, superior, *abdominal* cavity and a smaller, inferior, *pelvic* cavity. No partition separates the abdominal and pelvic cavities.

Again, students should learn well the various terms presented in these sections and be able to use the words accurately to both describe positions and understand their meaning if they should appear as parts of other words used to designate specific body structures.

Summary

1. Anatomy is that discipline of science that deals with the organization and structure of a living organism.
2. Anatomy is studied in a variety of ways.
   a. Gross anatomy deals with structures visible to the unaided eye, usually revealed by dissection.
   b. Surface anatomy describes structures usually seen from outside the body.
   c. Cytology and histology deal with the microscopic anatomy of the body.
   d. Developmental anatomy is concerned with growth from conception onward and forms the basis for understanding mechanisms leading to malformation.
   e. Special techniques, including x-ray examination, endoscopy, and other techniques, enable visualization of living organs.
3. Anatomical position places the body in a standing posture with palms forward and serves as the reference point for directional terms and body movements.
4. General body regions describe gross areas of the body.
5. Directional terms provide a means of locating body parts and describing their relationships.
   a. Anterior indicates the front of the body; posterior indicates the back of the body.
   b. Superior indicates upward or toward the head; inferior indicates lower or toward the feet.
   c. Medial indicates closer to the body midline; lateral indicates further from the body midline.
   d. External indicates closer to the body surface; internal indicates away from the body surface.
   e. Proximal implies closer to a point of attachment; distal implies further from a point of attachment.
   f. Palmar refers to the palm of the hand; plantar refers to the sole of the foot.
6. Planes of section divide the body or its organs in different directions.
   a. Sagittal planes create right and left halves of the body.
   b. Coronal or frontal planes create anterior and posterior halves of the body.
   c. Transverse planes cut across the shortest dimension of a body part.
   d. Longitudinal planes follow the longest dimension of a body part.
   e. Oblique planes are diagonal or slanted cuts across the body or its parts.
7. Reference lines provide additional means of locating body features. Lines are described for the anterior, lateral, and posterior thorax and for the abdomen. Organs contained within the areas delineated by these lines are described or depicted.
8. True body cavities do not open directly on the body surface. The following are major cavities, subdivided as described:
   a. A dorsal cavity houses the brain (cranial cavity) and spinal cord (vertebral canal).
   b. Ventral cavities include the thoracic (subdivided into pleural, pericardial, and mediastinal cavities) and the abdominopelvic cavities.

Questions

1. What is anatomy, and how is it studied?
2. What is the anatomical position, and what is its usefulness?
3. What is the difference between?
   a. Loin and groin
   b. Flank and abdomen
   c. Arm and leg
4. Using terms of direction, describe the relative positions of each of the following:
   a. Elbow and hand
   b. Umbilicus and hip
   c. Thorax and head
   d. Skin and an abdominal organ
5. How does each of the following planes divide the body?
   a. Sagittal
   b. Coronal
   c. Transverse
   d. Oblique
6. What would be implied in the following sentence? The heart is located two thirds to the left of the midsternal line in the pericardial cavity.
7. What organ(s) is/are found in the following regions or cavities?
   a. Pelvic cavity
   b. Right upper quadrant
   c. Cranial cavity
   d. Pleural cavity
   e. Epigastric region
   f. Right iliac region

Readings

Gordon, R.: Image construction from x-ray, Sci. Am. **233:**56, Oct. 1975.

Kieffer, S.A., and Heitzman, E.R.: An atlas of cross-sectional anatomy, New York, 1979, Harper & Row, Publishers.

Miller, J.: The body in question, New York, 1978, Random House, Inc.

Nilsson, L.: Behold man, Boston, 1973, Little, Brown & Co.

Snell, R.S.: Atlas of clinical anatomy, Boston, 1978, Little, Brown & Co.

Yokochi, C.: Photographic anatomy of the human body, ed. 2, Baltimore, 1978, University Park Press.

2

Organization, development, and basic tissues of the body

Objectives

Atomic and molecular levels

Cellular level
  Structure of cells
  Cell replacement; mitosis
  Neoplasms
  Species replacement; meiosis

Development of a new individual
  First week
  Second week
  Third week
  Fourth week
  Fifth to eighth weeks
  Fetal membranes

Tissue level
  Basic tissue groups
  Epithelium
  Connective tissues
  Muscular tissues
  Nervous tissue

Organs

Systems

The whole organism

Summary

Questions

Readings

OBJECTIVES After studying this chapter, the reader should be able to:

State the various levels of organization in the body and give examples of each level.

Describe the structure of a generalized cell, including its ultrastructure.

Describe the stages of mitosis and meiosis, the events occurring in each stage, and the results of each process.

Outline the major developmental processes that occur in the first 8 weeks of human development.

Define a tissue, the four basic tissue groups, where they occur, the several types of tissues in each group, and where *they* occur.

List some special modifications that may occur within each tissue group that are functionally important to the tissue.

Describe the structure of glands in the body.

Define an organ and give examples.

Define a system, list the body systems, and name the major organs that compose each system.

Speculate on how systems are dependent on one another to ensure whole organism structure and function.

Atomic and molecular levels

Although a nuclear physicist might disagree, for our purposes it may be said that the fundamental building units of our body structure are its individual atoms. Of the over 100 atomic species or elements occurring in nature or synthesized by physicists, about 13 may be said to constitute the elements that form the basic body structure. These are carbon (C), hydrogen (H), oxygen (O), phosphorus (P), potassium (K, from L. *kalium*), iodine (I), nitrogen (N), sulfur (S), calcium (Ca), iron (Fe, from L. *ferrum*), magnesium (Mg), sodium (Na, from L. *natrium*), and chlorine (Cl). C, H, O, and N are commonly bound together to form larger units known as molecules or compounds. Some biologically important molecules and compounds formed by these atoms are the carbohydrates, commonly represented as CH_2O (many carbohydrates have the hydrogen and oxygen in a 2:1 ratio); lipids, represented as CHO because they are usually deficient in oxygen; proteins, represented as CHON because they always contain nitrogen in addition to the other elements; and the nucleic acids that form the hereditary basis of the body.

In our bodies the elements and molecules are organized in specific ways to confer that property we call life on a unit that forms the next level of body organization called the cell.

Cellular level

Structure of cells

The individual structural and functional units of the body are referred to as *cells.* Although individual types of cells vary widely in size, shape, and function, most possess certain components that enable them to sustain their own activity and to reproduce themselves. It is thus possible to speak of a *generalized animal cell* (Fig. 2-1) that illustrates the basic structure of cells in general.

Such a cell shows three main parts.

A cell or *plasma membrane* surrounds the cell and delimits it from its neighbors and the fluid environment surrounding the cell. It is almost certain that all cell membranes are not structured exactly alike on all body cells, but there is general agreement that two major types of substances compose the membrane. Lipids, usually in the form of phospholipids (a fat in combination with a phosphorus-containing compound), are believed to form a double (bimolecular) layer of molecules. At intervals in the double layer of lipid molecules are large proteins that may penetrate part or all the way through the lipid layers. The effect is somewhat like chips or logs floating in a shallow pond. Indeed, this idea of membrane structure has been described as "proteins in a sea of lipids."

In addition to serving as the limiting structure of a cell, the plasma membrane exerts a great deal of control over passage of materials across the membrane. The mere physical structure of the membrane can govern passage of

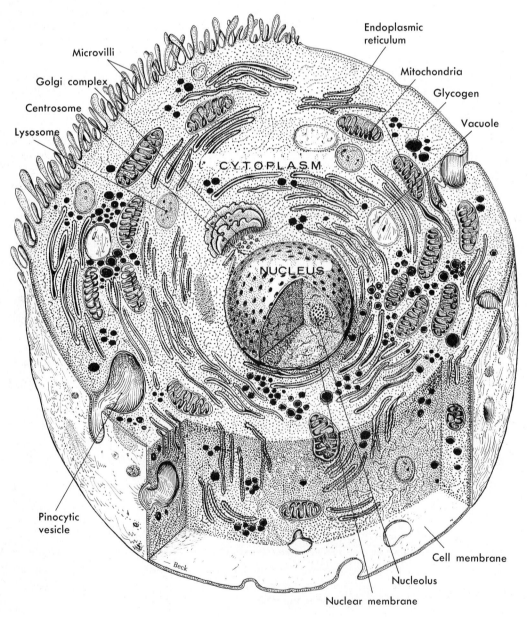

Fig. 2-1
Generalized cell as it might appear in a three-dimensional electron micrograph.

materials according to size, much in the manner that a sieve can separate materials by size. Since the protein components can become electrically charged by loss of electrons (e⁻) or hydrogen ions (H⁺), the entire membrane may bear an **electrical charge.** Substances may be attracted to the membrane if of opposite charge to the membrane or repelled if of like charge. Simple **thickness** of a membrane determines rate of passage of a substance in the sense that a thicker membrane takes longer to traverse. Since the membrane consists of much lipid, the ability of something to dissolve in the lipid, or its **lipid solubility,** may enhance its passage.

The membrane may become actively involved in moving materials across itself by providing **receptor sites** to which substances may bind before being taken into a cell. The effects of hormones, for example, on some cells but not others may be related to the presence of such sites on affected cells. Some molecules are passed across membranes by actually being transported on **carrier molecules** provided by the cell's metabolic machinery.

This discussion should emphasize the notion that the interior of a cell is not at the mercy of its environment but can to a large degree control not only what enters the cell but what may also leave the unit because of the ability of the cell membrane to be a selective structure.

The *cytoplasm* forms the bulk of the cell and contains a variety of formed, metabolically active structures designated as *organelles*. These organelles are often membrane-surrounded units that carry on specific types of metabolic reactions necessary for maintenance of cell activity and life itself. The following list of organelles emphasizes those that are most essential for cell survival:

endoplasmic reticulum (ER) (Fig. 2-2) is a system of interconnecting hollow tubular structures that permeate the entire cytoplasm and reach both the cell membrane and nucleus. It may serve as a device for distribution of materials throughout the cell, materials acquired either from the cell environment or manufactured within the cell itself. There are two varieties of reticulum, *smooth* and *rough*. The former has no structures attached to the outer walls of the reticulum and synthesizes complex lipids and carbohydrate-containing compounds; the latter has structures called ribosomes attached to the outer reticulum surface.

ribosomes (Fig. 2-3) are granules of nucleic acid and protein that may be found free in the cytoplasm or attached to the ER to form the rough reticulum. These structures are involved in protein synthesis, acting like a zipper to bring about the association of amino acids to form proteins.

Free ribosomes, that is, those not associated with ER, often form chains

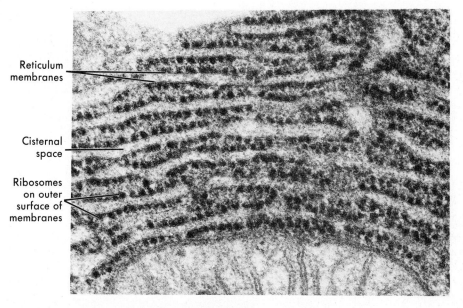

Fig. 2-2
Longitudinal sections of rough endoplasmic reticulum, with ribosomes on outer surfaces of membranes. Cisternal space is interior or cavity of reticulum tubules. (×88,000.)

From Anderson, W.A.D., and Kissane, J.M.: Pathology, vol. 1, ed. 7, St. Louis, 1977, The C.V. Mosby Co.

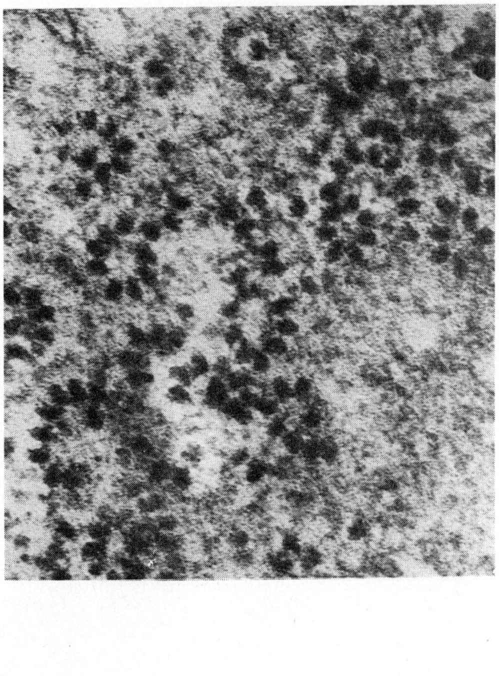

Fig. 2-3
Form of polyribosomes in cell cytoplasm. (×136,000.)

From Anderson, W.A.D., and Kissane, J.M.: Pathology, vol. 1, ed. 7, St. Louis, 1977, The C.V. Mosby Co.

called *polyribosomes* that are believed to be involved in the synthesis of proteins destined for use in the structure of the cell producing them. ER-associated ribosomes are postulated to be involved in the production of proteins that will be "exported" for use in or by cells that did not produce them. Enzymes and some hormones are examples of such exported proteins.

mitochondria (Fig. 2-4) are double-walled organelles that contain systems of enzymes leading to the formation of a compound known as adenosine triphosphate (ATP). ATP is the material used to energize and fuel physiological processes in the body, such as muscular contractions.

The outer wall of the organelle is smooth and appears to contain enzymes that degrade substances such as sugar and fatty acids. The inner wall is folded to form *cristae,* or shelflike partitions that house enzymes involved in ATP synthesis and carbon dioxide and water formation. Thus the unit sequentially breaks down molecules and uses the products of that breakdown to form the energy-rich ATP molecules.

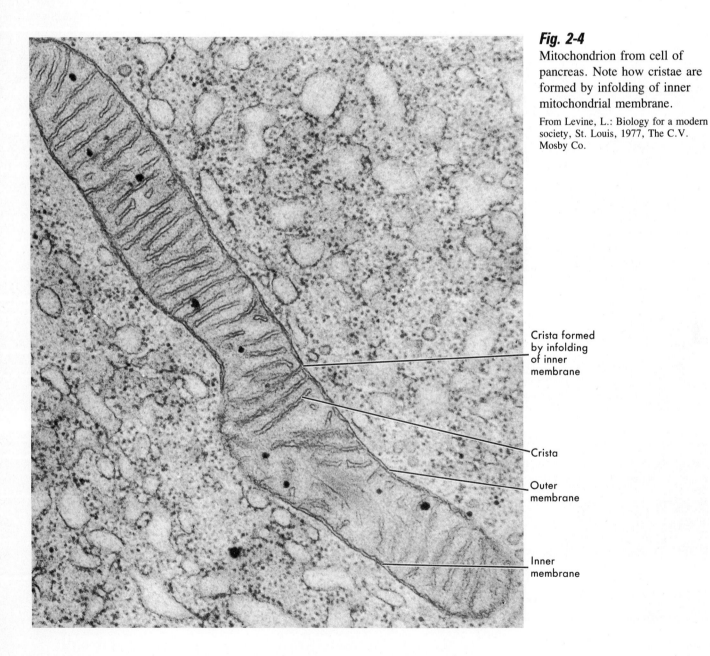

Fig. 2-4

Mitochondrion from cell of pancreas. Note how cristae are formed by infolding of inner mitochondrial membrane.

From Levine, L.: Biology for a modern society, St. Louis, 1977, The C.V. Mosby Co.

Golgi complex or ***body*** (Fig. 2-5) is a series of flattened channels (Golgi membranes) containing enlarged vesicles (Golgi vesicles) on the ends of the channels. It is restricted to one part of the cell, commonly lying between the nucleus and the free surface of the cell. It is concerned with producing secretions of certain cells (for example, mucus) and with placing membranes around enzymes or hormones that are intended for use elsewhere in the body (for example, digestive enzymes, insulin). An enzyme, or group of enzymes, surrounded by a Golgi-contributed membrane forms what is called a *secretion granule* and because of the membrane cannot act on the producing cell. Some synthetic activity (of lipids and glycoproteins, combinations of a carbohydrate and a protein) may also be carried out by the Golgi complex.

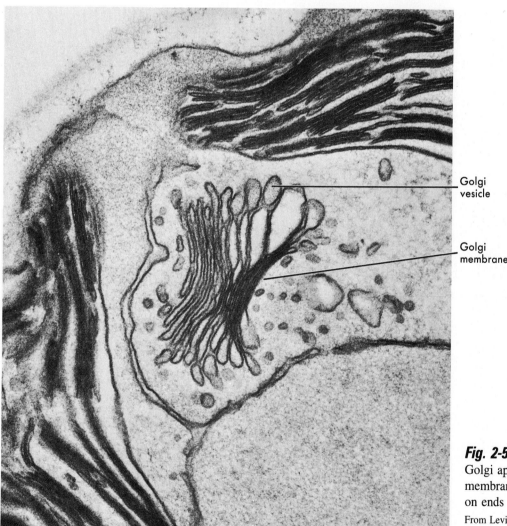

Fig. 2-5
Golgi apparatus. Note "stacked membranes" of apparatus and vesicles on ends of membranes.

From Levine, L.: Biology for a modern society, St. Louis, 1977, The C.V. Mosby Co.

lysosomes (Fig. 2-6) are membrane-surrounded sacs of powerful hydrolytic enzymes capable of breaking down large molecules into smaller ones by using water molecules. They may originate from cooperation of the ribosomes for enzyme synthesis, and Golgi complex, for their membranes, and their enzymes digest large molecules that enter the cell from its environment either as such or incorporated into microorganisms (viruses, bacteria). Thus, in addition to merely digesting large molecules and making the products available for cell use, lysosomes provide protection for the body.

White blood cells have numerous lysosomes that are important in the digestion of microorganisms (viruses, bacteria) that the cells have taken in by phagocytosis. Another aspect of lysosome activity is seen when cells are damaged and lysosome enzymes are released that digest the damaged cells to clear the area for replacement by new cells. This activity really involves self-destruction and is what gives the name "suicide packets" to lysosomes. If lysosomal enzymes are not synthesized properly, some large molecules accumulate in the cell instead of being digested, and various "storage diseases" arise. Tay-Sachs disease involves accumulation of a lipid in nerve cells, leading to mental deficiency; the glycogenoses are a result of accumulation of carbohydrates in liver and muscle cells with resulting abnormalities of function. Lysosomes are thus essential parts of the "digestive system" of a cell.

centrosomes *(central bodies)* are found in most cells and consist of a pair of *centrioles* (Fig. 2-7) and an area of differentiated cytoplasm called the *centrosphere*. Those cells possessing centrioles are capable of undergoing nuclear division for purposes of cell duplication or species perpetuation. Mature red blood cells have no centrioles; nerve cells, after about 3 to 5 years of age, either lose their centrioles or they no longer function. Thus nerve cells usually cannot replace losses by cell division.

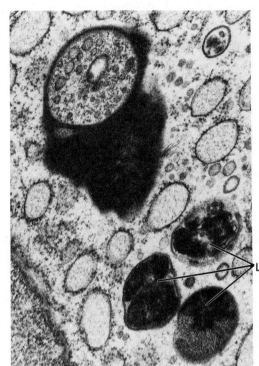

Fig. 2-6
Lysosomes, showing characteristic granular appearance. ($\times 27{,}200$.)

From Anderson, W.A.D., and Kissane, J.M.: Pathology, vol. 1, ed. 7, St. Louis, 1977, The C.V. Mosby Co.

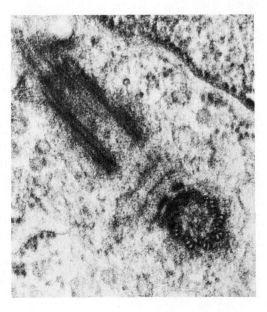

Fig. 2-7
Sections of centrioles. Nine sets of triplet microtubules compose the organelles.

From Anderson, W.A.D., and Kissane, J.M.: Pathology, vol. 1, ed. 7, St. Louis, 1977, The C.V. Mosby Co.

peroxisomes are membrane-surrounded structures containing oxidative enzymes capable of degrading hydrogen peroxide (H_2O_2) into water. H_2O_2 is a normal product of cellular metabolism and is also a chemical capable of killing the cell (it is used as an antiseptic). Thus peroxisomes protect the cell against buildup of H_2O_2.

Again, white blood cells are rich in peroxisomes because they produce more H_2O_2 than do other cells. In these cells the H_2O_2 aids in the destruction of engulfed microorganisms, and the cell itself is able to degrade the H_2O_2 to harmless products.

cilia or ***flagella*** (Fig. 2-8) (for example, sperm) that either enable movement of materials across the cell surface or movement of the whole cell through fluids are found in some cells. Both organelles have a structure similar to that of centrioles, from which they may be derived.

microtubules are tiny hollow tubular structures that appear to permeate the entire cytoplasm. They give strength to the cell by forming a *cytoskeleton* and may aid flow of materials through the cell. Microtubules form the spindle that appears at cell division and are found in flagella and cilia.

As this list of organelles in body cells is studied, it may be noted that the cell is provided with structures enabling it to survive, just as whole human bodies have systems that enable *them* to survive. The cell reticulum serves much the same purpose as the body's circulatory system, distributing materials; the lysosome acts as a digestive system; the mitochondrion might be compared to the liver, in which many energy-creating or liberating reactions occur. How many more comparisons between cell and organism can you make?

Cells also contain structures called ***inclusions*** (see Fig. 2-1). These are usually raw materials for cell use (glycogen, lipid droplets), products of cellular activity (secretion granules), or storage units (vacuoles). They have in common the fact that they are *not* metabolically active on their own.

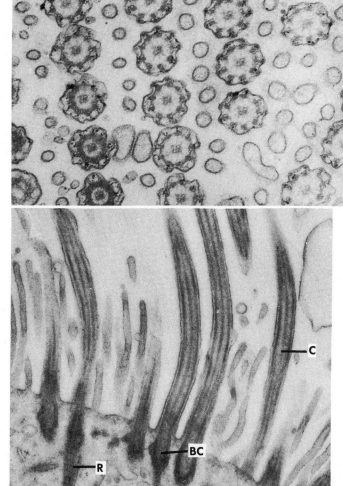

Fig. 2-8

A, Cross-sections of cilia from cat bronchiole epithelial cell. Note 9 + 2 arrangement of paired microtubules. ($\times 45,000$.)
B, Longitudinal sections of cilia from mouse tracheal epithelial cell. *BC,* Basal corpuscle; *C,* cilium; *R,* rootlet of basal body. ($\times 25,000$.)

From Bevelander, G., and Ramaley, J.A.: Essentials of histology, ed. 8, St. Louis, 1979, The C.V. Mosby Co.

Considered by many cytologists to be an organelle, the **nucleus** of a cell (Fig. 2-9) provides the ultimate direction of a cell's metabolic activity. Through its content of nucleic acids, the nucleus governs the rates and types of synthetic and destructive activity a cell exhibits. A two-layered nuclear *membrane*, possessing thin areas called *pores*, separates the nucleus from the cytoplasm; nucleic acid is usually appreciated as the dark-staining granular *chromatin material* in the *nuclear fluid*. One or more *nucleoli*, larger masses of nucleic acid, are usually seen within the nucleus. The nucleus is the "brain" of the cell, directing what the cell can do.

A particular type of nucleic acid designated *d*eoxyribo*n*ucleic *a*cid (DNA) forms the chromatin material. It is capable of duplicating itself in the process called **replication**. The nuclear DNA exists in the form of a double helix (Fig. 2-10). The building units of DNA are called **nucleotides** and consist of a nitrogenous base (adenine, guanine, cytosine, or thymine), a sugar (deoxyribose, $C_5H_{10}O_4$), and phosphoric acid

$$(-O-\overset{\overset{\displaystyle O}{\|}}{\underset{\underset{\displaystyle OH}{|}}{P}}-OH).$$

The helix resembles a twisted ladder, with the side rails composed of sugar-phosphate bonds between individual nucleotides, and the rungs of base-base bonds between the two strands. *Replication*, or duplication of the DNA, involves breaking the base-base links, allowing the separation of the two strands. Free nucleotides are attracted to the "rails," and each one will synthesize a new helix like the one from which it was separated. In pairing, only adenine-thymine and cytosine-guanine base pairing occurs to create a bond between the two original helixes or the new strands being synthesized.

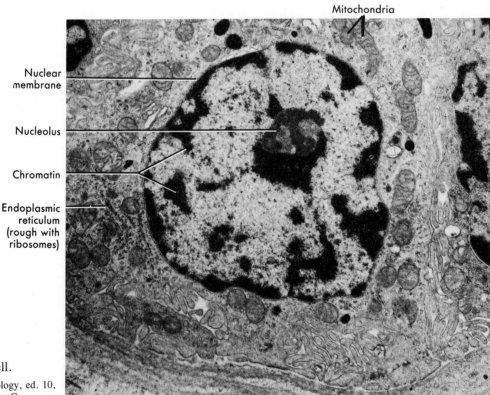

Fig. 2-9
Nucleus of human liver cell.

From Noland, G.B.: General biology, ed. 10, St. Louis, 1979, The C.V. Mosby Co.

Nuclear DNA also directs the synthesis of three kinds of *ribo*nucleic *a*cid (RNA) that differ from DNA in two important respects: the sugar is ribose ($C_5H_{10}O_5$), and the base uracil substitutes for thymine. RNA is synthesized in the nucleolus or is collected on it and then moves to the cytoplasm through nuclear pores. **Messenger RNA** (mRNA) carries "instructions" that tell the ribosome what the order of insertion of amino acids (the building blocks of proteins) into the protein will be. **Transfer RNA** (tRNA) picks up specific amino acids in the cytoplasm and brings them to the mRNA strand, where the ribosome inserts them at the proper place in the growing protein. **Ribosomal RNA** (rRNA) is synthesized, in organisms with a true nucleus, in the nucleolus and becomes incorporated into the ribosome. Ribosomes are about half rRNA and half protein and consist of a larger and smaller portion that associate when a protein is being synthesized and dissociate when the synthesis of the protein is completed. The ribosome is often thought of as a "workbench," complete with the machinery required to synthesize a protein according to the instructions carried on mRNA.

Production of RNA from DNA is called *transcription,* and the use of m-, t- , and rRNA to form a specific protein is called *translation* and **protein synthesis.**

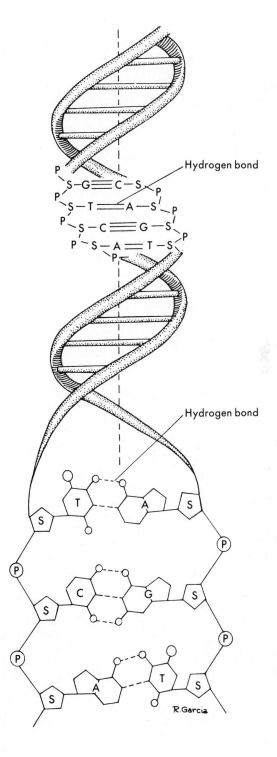

Fig. 2-10
DNA molecule. Three-way representation showing double helix, base pairing, and atomic arrangement in helixes. *A,* Adenine; *C,* cytosine; *G,* guanine; *T,* thymine; *S,* deoxyribose sugar; *P,* phosphate group.

Cell replacement: mitosis

Growth in body size is partly accounted for by increase in size of existing cells (hypertrophy) and partly by increase in numbers of cells (hyperplasia). Additionally, cells die, or are lost from the body (for example, skin), and must be replaced.

Mitosis is a type of cell division that results in the production of cells that are genetically, structurally, and functionally identical to the original cells. The process (Fig. 2-11), although a continuous one once initiated, is described as occurring in phases or stages, with each phase characterized by particular events as follows:

interphase (L. *inter*, between, + Gk. *phasis*, an appearance) is the stage in which the cell is not undergoing division and appears typical for its type. It is engaged in metabolic activities that are essential for its own survival, and, most importantly, there is exact duplication *(replication)* of the DNA of the nucleus and therefore of the entire genetic potential of the nucleus.

prophase (L. *pro*, before, + *phasis*) is the stage in which the centrioles migrate to opposite poles of the cell, and a spindle of tiny microtubules forms between the centrioles. The nuclear membrane disappears, and the chromatin material of the nucleus condenses to form visible *chromosomes*.

metaphase (Gk. *meta*, after or beyond, + *phasis*) is the stage in which the duplicated chromosomes align themselves on the equatorial plate of the cell. The equatorial plate lies midway between the centrioles across the midline of the cell and is an imaginary, not a real, line.

anaphase (Gk. *ana*, one of each, + *phasis*) is the stage in which the centromeres (the portions of the chromosomes to which the strands of DNA are attached and from which the DNA strands originate) divide, and, apparently attached to the spindle fibers by the centromere, the chromosomes separate. Since each centromere had attached to it two identical strands of DNA, and since the centromere divides in such a manner as to include one strand of DNA, each attached to the divided centromere, the products of the division are identical.

telophase (Gk. *telos*, end, + *phasis*) is the stage in which the nucleus reorganizes, with chromosomes returning to the form of granular chromatin material; the centrioles divide; and the cytoplasm "pinches" approximately in half (cytokinesis), following the equatorial plane of the cell.

Mitosis has a cyclical duration of 12 to 24 hours in those cells that have continual division activities, with the actual division of the cells occurring in about 1 hour. It has been estimated that some 500 million cells are replaced per day by mitosis, with a large percentage being skin, alimentary tract epithelial cells, and red blood cells.

Neoplasms

Cell division by mitosis is usually balanced to the needs of the body for cell replacement and repair. The mechanisms determining when and at what rate a cell divides are virtually unknown. Uncontrolled cell division constitutes a *neoplasm* (new growth) or a *cancer*.

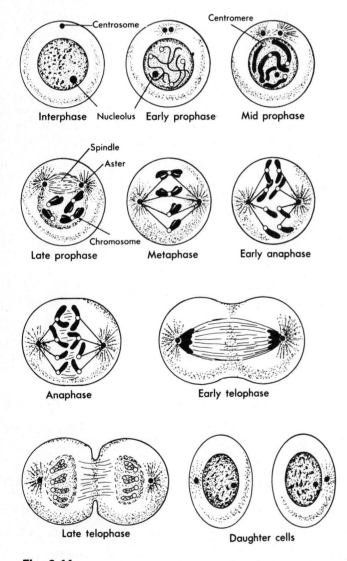

Fig. 2-11
Major events occurring in mitosis as depicted in cell having four chromosomes.

Cancer cells are not identical to the cells from which they are derived. In many cases, there is an increased frequency of division abnormalities, including deletions (losses of parts of chromosomes) and other malfunctions during the mitotic process. It is as if the process is being carried out at such a rate that the product is faulty in the genetic sense. Also, the chemicals on the surfaces of the cancer cells have been shown to differ from those of the parent cells. This fact *may* cause a response on the part of the body's immune system that destroys the abnormal cells. The threat posed by a cancer is due to mechanical effects of its enlargement (for example, a brain tumor) or in the "taking over" of a normal organ by cancer cells, ultimately destroying the organ's ability to carry out its normal functions.

Species replacement: meiosis

A second variety of cell division occurs in the production of sex cells (ova, sperm). It is called **meiosis** (Fig. 2-12). In this type of division, *haploid* (n) cells are produced that contain one half the chromosome number characteristic for the species. When egg and sperm unite to form the zygote (a fertilized egg) and the starting point of a new individual, chromosome number is restored to what is called the *diploid* number (2n). If this process did not occur, the next generation would contain twice the normal chromosome number. Such a situation is not compatible with human survival.

The process proceeds in two separate divisions that have the same stages as mitosis. Basically the same events occur as in mitotic division. The important difference occurs in first meiotic anaphase when chromosomes that are homologous (chromosomes in pairs, one from each parent, that have the same genes in the same order) are separated *without centromere division*. Thus one member of each pair passes to the daughter cells, *halving* the total chromosome number. Another difference lies in the fact that a second division, of mitotic nature, occurs after the first meiotic divisions. This division results in the final production of *four* meiotic cells from the original single parent cell. In sperm production four functional cells are produced, two carrying Y sex chromosomes and two containing X sex chromosomes. In ova production one functional cell containing an X sex chromosome is produced; the other three cells become *polar* bodies. In ova formation, cytokinesis is unequal so that *one* cell in effect survives with most of the cytoplasm.

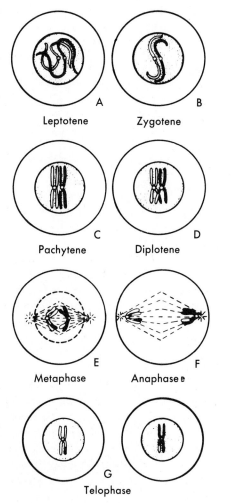
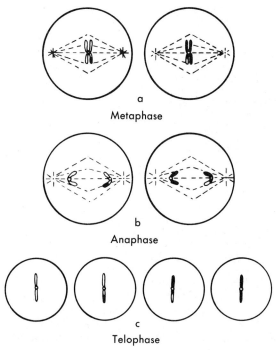

Fig. 2-12
Major events of meiosis. Only one of the 23 human chromosome pairs is shown. Chromosomes from one parent are shaded, those from other are not. *A* to *G,* First meiotic division; *a* to *c,* second meiotic division.

Development of a new individual

In a preceding section it was indicated that meiosis produces haploid sperm and ova. To create a new human individual, egg and sperm must unite to restore normal diploid chromosome number in the process of *fertilization.* This union is sometimes also called *conception.* At this time, a discussion of basic embryology to the eighth week is presented so that the reader can appreciate the origins of tissues and organs of the body.

First week (Fig. 2-13)

Fertilization not only restores normal chromosome number but also determines the genetic sex of the individual. An XY combination of the sex chromosomes results in a male individual; an XX combination, a female. Sperm penetration also initiates a series of mitotic divisions called **cleavage**. The ovum at fertilization is about the size of a period, and cleavage produces a solid ball of cells without much increase in size of the whole unit. The solid mass of cells is called a *morula,* and its individual cells are termed **blastomeres**. At this particular time, about 3 days after fertilization, each blastomere could, if separated from the others, give rise to an entire organism. It is this fact, plus the fact that the entire genetic machinery of each cell is still "turned on," that is the true basis for *cloning,* or the production of an organism that is *exactly* like another. To put it another way, there has been no differentiation of specific cells as to what each will ultimately form in the body.

Over the next 3 to 4 days of development, a **differentiation** of cells as to their specific potencies occurs, and a reorganization of the morula takes place, forming a **blastocyst.** A cavity, the **blastocoele,** develops in the morula. It enlarges and pushes the morula cells into an outer layer of cells, the **trophoblast,** and an **inner cell mass** attached to one side of the blastocyst. By this time, the blastocyst has lodged on the inner surface of the uterus. The divisions and reorganization described have nearly consumed the energy stores available in the zygote, and it becomes necessary for the blastocyst to embed in the glandular and vascular uterine wall to obtain nutrients for its further development.

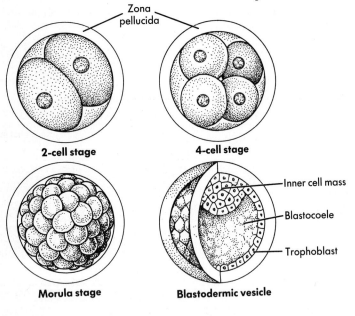

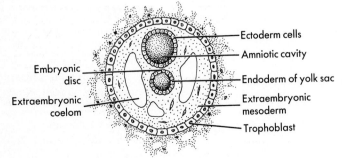

Fig. 2-13
Development during first week.

Second week

Two major events occur in the second week of development (Fig. 2-14). These are *implantation* and the *formation of a two-layered embryo.*

Implantation involves the blastocyst's "digesting" its way into the uterine wall. It may be that the trophoblast cells secrete enzymes that literally eat away the tissue of the uterine lining, permitting the blastocyst to "sink" into the wall. In any event the trophoblast is exposed to the secretions of the uterine glands and to nutrients in the blood vessels. The trophoblast soon develops extensions, or *villi,* that increase its surface area for absorbing the nutrients essential to continuation of development.

Within the inner cell mass, two cavities develop. One lies closest to the trophoblast and is known as the *amniotic cavity.* The cells lining it form a layer designated the *ectoderm.* The second cavity forms internally to the amniotic cavity and is the cavity of the *yolk sac.* Its lining of cells is termed the *endoderm.* Where the two layers contact one another between the two cavities is the *embryonic disc,* or *two-layered embryo,* where the embryo itself will develop.

As these changes are occurring, cells from the trophoblast are migrating into the blastocoele and differentiating to form the *extraembryonic mesoderm* (extra = outside of). This tissue becomes associated with the trophoblast to form the *chorion,* a fetal membrane forming a portion of the placenta. Cavities developing within this extraembryonic mesoderm are known as the extraembryonic coelom.

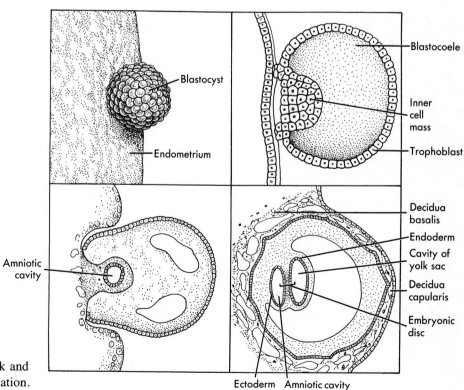

Fig. 2-14
Development during second week and implantation. See text for explanation.

Third week

The major events of the third week (Fig. 2-15) are the formation of a third layer of cells called the ***mesoderm*** and the formation of a ***three-layered embryo.*** The embryonic disc, if viewed from above, would present a leaflike shape with a narrow and wider portion. The narrowed end represents the *caudal* or tail end of the embryo, the broader portion the *cranial* or head end of the embryo. The midline of the disc is marked by a ***primitive streak*** composed of a mass of rapidly proliferating cells. Cells arising from the streak spread cranially and laterally and come to lie between the ectoderm and endoderm. These cells constitute the ***embryonic mesoderm*** and, with ecto- and endoderm, form the three ***germ layers*** from which all body structures will develop. Table 2-1 indicates the derivatives of these three germ layers.

Table 2-1. Derivatives of the germ layers

Ectoderm	Mesoderm (including mesenchyme)	Endoderm
1. Epidermis of skin, including Skin glands Hair and nails Lens of eye 2. Epithelium of Nasal cavities Sinuses Mouth, including Oral glands Enamel Sense organs Anal canal 3. Nervous tissues Hypophysis (neurohypophysis) Adrenal medulla	1. Muscle Skeletal, cardiac, smooth 2. Connective tissue including Cartilage Bone Blood Bone marrow Lymphoid tissue 3. Epithelium of Blood vessels Lymphatics Coelomic cavities Kidney and ureters Gonads and ducts Adrenal cortex Joint cavities	1. Epithelium of Pharynx Auditory tube Tonsils Thyroid Parathyroid Thymus Larynx Trachea Lungs Digestive tube and its glands Bladder Vagina and vestibule Urethra and glands 2. Hypophysis (adenohypophysis)

Adapted from McClintic, J.R.: Basic anatomy and physiology of the human body, ed 2. Copyright © 1980. Reprinted by permission of John Wiley & Sons, Inc.

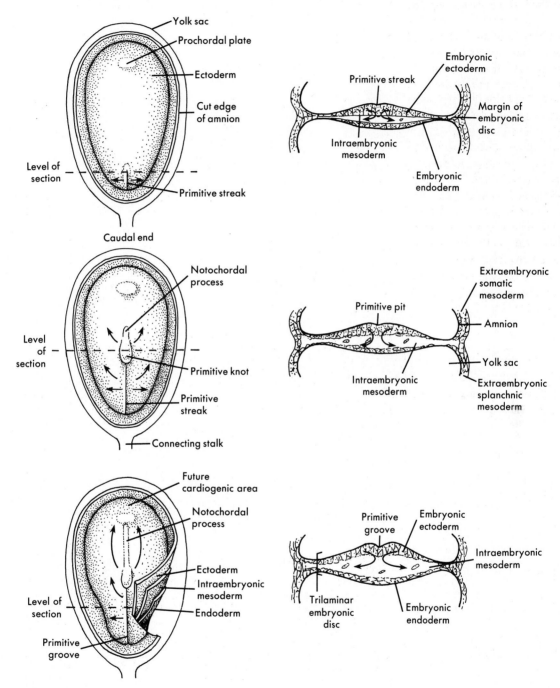

Fig. 2-15
Mesoderm formation during third week.
After Moore.

Fourth week

The fourth week (Fig. 2-16) sees the beginning of development of the nervous system, described in greater detail in Chapter 9. Also, the formation of *somites* occurs at this time. Somites are a series of, ultimately, 40 to 42 pairs of blocks of mesoderm that form either side of the developing nervous system. From this tissue will develop the connective tissue and muscular structures of the body.

During the fourth week the embryo begins to show features that suggest ultimate human form. A *head* is recognizable, a *heart* is forming and will soon begin to circulate blood through a rudimentary system of *blood vessels, arm* and *leg buds* are forming, and the *placenta* is assuming its form.

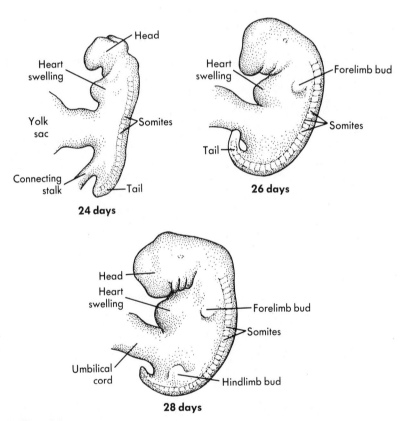

Fig. 2-16
Development in fourth week.

Fifth to eighth weeks

The period of the fifth to eighth weeks (Fig. 2-17) sees completion of the embryo in terms of the formation of the major body systems and the assumption of a clearly human form. At 8 weeks, the organism is known as a *fetus*. Finer details of the development of individual systems will be found in appropriate chapters later in the text. By 8 weeks, all major body systems are present or recognizable, although not all have assumed functional or mature status. Table 2-2 details the status of each body system by week or month.

It is of considerable interest to note the development of the limbs during this period of time (Fig. 2-18). Both upper and lower limbs originally project in similar directions from the body, with what will become the palm-of-the-hand sides and sole-of-the-foot sides facing the anterior (or ventral) midline of the body. The cranial edge of each limb will develop a radius bone (upper limb) or a tibia bone (lower limb). Caudally directed margins will develop an ulna (upper limb) or a fibula (lower limb). Further development sees the upper limbs rotating *laterally* about 90 degrees so that the radius becomes the lateral bone of the forearm, and the thumb projects laterally in anatomical position. The elbow points posteriorly. In the lower limb a *medial* rotation of about 90 degrees occurs, causing the fibula to lie laterally, the tibia medially, the knee to point anteriorly, and the great toe to assume a medial position in contrast to the lateral placement of the thumb.

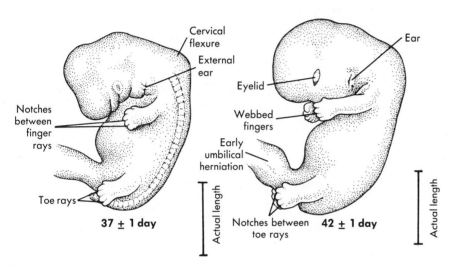

Fig. 2-17
Six-week embryos.

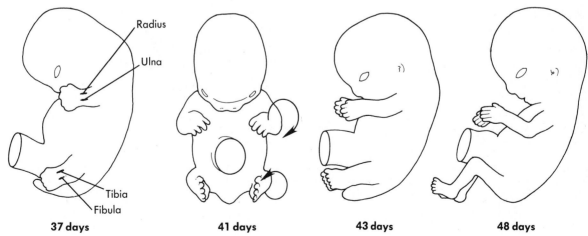

Fig. 2-18
Limb development and rotation.

Table 2-2. A reference table of correlated human development

Age in weeks	Size (C R) in mm	Body form	Mouth	Pharynx and derivatives	Digestive tube and glands	Respiratory system	Coelom and mesenteries
2.5	1.5	Embryonic disc flat. Primitive streak prominent. Neural groove indicated.	—	—	Gut not distinct from yolk sac.	—	Extra-embryonic coelom present. Embryonic coelom about to appear.
3.5	2.5	Neural groove deepens and closes (except ends). Somites 1 – 16± present. Cylindrical body constricting from yolk sac. Branchial arches 1 and 2 indicated.	Mandibular arch prominent. Stomodeum a definite pit. Oral membrane ruptures.	Pharynx broad and flat. Pharyngeal pouches forming. Thyroid indicated.	Fore- and hind-gut present. Yolk sac broadly attached at mid-gut. Liver bud present. Cloaca and cloacal membrane present.	Respiratory primordium appearing as a groove on floor of pharynx.	Embryonic coelom a U-shaped canal, with a large pericardial cavity. Septum transversum indicated. Mesenteries forming. Mesocardium atrophying.
4	5.0	Branchial arches completed. Flexed heart prominent. Yolk stalk slender. All somites present (40). Limb buds indicated. Eye and otocyst present. Body flexed; C-shape	Maxillary and mandibular processes prominent. Tongue primordia present. Rathke's pouch indicated.	Five pharyngeal pouches present. Pouches 1-4 have closing plates. Primary tympanic cavity indicated. Thyroid a stalked sac.	Esophagus short. Stomach spindle-shaped. Intestine a simple tube. Liver cords, ducts and gallbladder forming. Both pancreatic buds appear. Cloaca at height.	Trachea and paired lung buds become prominent. Laryngeal opening a simple slit.	Coelom still a continuous system of cavities. Dorsal mesentery a complete median curtain. Omental bursa indicated.
5	8.0	Nasal pits present. Tail prominent. Heart, liver, and mesonephros protuberant. Umbilical cord organizes.	Jaws outlined. Rathke's pouch a stalked sac.	Phar. pouches gain dors. and vent. diverticula. Thyroid bilobed. Thyro-glossal duct atrophies.	Tail-gut atrophies. Yolk stalk detaches. Intestine elongates into a loop. Caecum indicated.	Bronchial buds presage future lung lobes. Arytenoid swellings and epiglottis indicated.	Pleuro-pericardial and pleuro-peritoneal membranes forming. Ventral mesogastrium draws away from septum.
6	12.0	Upper jaw components prominent but separate. Lower jaw-halves fused. Head becomes dominant in size. Cervical flexure marked. External ear appearing. Limbs recognizable as such.	Lingual primordia fusing. Foramen caecum established. Labio-dental laminae appearing. Parotid and submaxillary buds indicated.	Thymic sacs, ultimobranchial sacs and solid parathyroids are conspicuous and ready to detach. Thyroid becomes solid and converts into plates.	Stomach rotating. Intestinal loop undergoes torsion. Hepatic lobes identifiable. Cloaca subdividing.	Definitive pulmonary lobes indicated. Bronchi sub-branching. Laryngeal cavity temporarily obliterated.	Pleuro-pericardial communications close. Mesentery expands as intestine forms loop.
7	17.0	Branchial arches lost. Cervical sinus obliterates. Face and neck forming. Digits indicated. Back straightens. Heart and liver determine shape of body ventrally. Tail regressing.	Lingual primordia merge into single tongue. Separate labial and dental laminae distinguishable. Jaws formed and begin to ossify. Palate folds present and separated by tongue.	Thymi elongating and losing lumina. Parathyroids become trabeculate and associate with thyroid. Ultimobranchial bodies fuse with thyroid. Thyroid becoming crescentic.	Stomach attaining final shape and position. Duodenum temporarily occluded. Intestinal loops herniate into cord. Rectum separates from bladder-urethra. Anal membrane ruptures. Dorsal and ventral pancreatic primordia fuse.	Larynx and epiglottis well outlined; orifice T-shaped. Laryngeal and tracheal cartilages foreshadowed. Conchae appearing. Primary choanae rupturing.	Pericardium extended by splitting from body wall. Mesentery expanding rapidly as intestine coils. Ligaments of liver prominent.
8	23.0	Nose flat; eyes far apart. Digits well formed. Growth of gut makes body evenly rotund. Head elevating. Fetal state attained.	Tongue muscles well differentiated. Earliest taste buds indicated. Rathke's pouch detaches from mouth. Sublingual gland appearing.	Auditory tube and tympanic cavity distinguishable. Sites of tonsil and its fossae indicated. Thymic halves unite and become solid. Thyroid follicles forming.	Small intestine coiling within cord. Intestinal villi developing. Liver very large in relative size.	Lung becoming gland-like by branching of bronchioles. Nostrils closed by epithelial plugs.	Pleuro-peritoneal communications close. Pericardium a voluminous sac. Diaphragm completed, including musculature. Diaphragm finishes its 'descent.'
10	40.0	Head erect. Limbs nicely modeled. Nail folds indicated. Umbilical hernia reduced.	Fungiform and vallate papillae differentiating. Lips separate from jaws. Enamel organs and dental papillae forming. Palate folds fusing.	Thymic epithelium transforming into reticulum and thymic corpuscles. Ultimobranchial bodies disappear as such.	Intestines withdraw from cord and assume characteristic positions. Anal canal formed. Pancreatic alveoli present.	Nasal passages partitioned by fusion of septum and palate. Nose cartilaginous. Laryngeal cavity re-opened; vocal folds appear.	Processus (saccus) vaginales forming. Intestine and its mesentery withdrawn from cord.

From Arey, L.B.: Developmental anatomy, ed. 7, Philadelphia, 1974, W.B. Saunders Co.

Urogenital system	Vascular system	Skeletal system	Muscular system	Integumentary system	Nervous system	Sense organs
Allantois present.	Blood islands appear on chorion and yolk sac. Cardiogenic plate reversing.	Head process (or notochordal plate) present.	—	Ectoderm a single layer.	Neural groove indicated.	—
All pronephric tubules formed. Pronephric duct growing caudad as a blind tube. Cloaca and cloacal membrane present.	Primitive blood cells and vessels present. Embryonic blood vessels a paired symmetrical system. Heart tubes fuse, bend S-shape and beat begins.	Mesodermal segments appearing (1 – 16±). Older somites begin to show sclerotomes. Notochord a cellular rod.	Mesodermal segments appearing (1 – 16±). Older somites show myotome plates.	—	Neural groove prominent; rapidly closing. Neural crest a continuous band.	Optic vesicle and auditory placode present. Acoustic ganglia appearing.
Pronephros degenerated. Pronephric (mesonephric) duct reaches cloaca. Mesonephric tubules differentiating rapidly. Metanephric bud pushes into secretory primordium.	Hemopoiesis on yolk sac. Paired aortae fuse. Aortic arches and cardinal veins completed. Dilated heart shows sinus, atrium, ventricle, and bulbus.	All somites present (40). Sclerotomes massed as primitive vertebrae about notochord.	All somites present (40).	—	Neural tube closed. Three primary vesicles of brain represented. Nerves and ganglia forming. Ependymal, mantle and marginal layers present.	Optic cup and lens pit forming. Auditory pit becomes closed, detached otocyst. Olfactory placodes arise and differentiate nerve cells.
Mesonephros reaches its caudal limit. Ureteric and pelvic primordia distinct. Genital ridge bulges.	Primitive vessels extend into head and limbs. Vitelline and umbilical veins transforming. Myocardium condensing. Cardiac septa appearing. Spleen indicated.	Condensations of mesenchyme presage many future bones.	Premuscle masses in head, trunk and limbs.	Epidermis gaining a second layer (periderm).	Five brain vesicles. Cerebral hemispheres bulging. Nerves and ganglia better represented. [Suprarenal cortex accumulating.]	Chorioid fissure prominent. Lens vesicle free. Vitreous anlage appearing. Octocyst elongates and buds endolymph duct. Olfactory pits deepen.
Cloaca subdividing. Pelvic anlage sprouts pole tubules. Sexless gonad and genital tubercle prominent. Müllerian duct appearing.	Hemopoiesis in liver. Aortic arches transforming. L. umbil. vein and d. venosus become important. Bulbus absorbed into right ventricle. Heart acquires its general definitive form.	First appearance of chondrification centers. Desmocranium.	Myotomes, fused into a continuous column, spread ventrad. Muscle segmentation largely lost.	Milk line present.	Three primary flexures of brain represented. Diencephalon large. Nerve plexuses present. Epiphysis recognizable. Sympathetic ganglia forming segmental masses. Meninges indicated.	Optic cup shows nervous and pigment layers. Lens vesicle thickens. Eyes set at 160°. Naso-lacrimal duct. Modeling of ext., mid., and int. ear under way. Vomero-nasal organ.
Mesonephros at height of its differentiation. Metanephric collecting tubules begin branching. Earliest metanephric secretory tubules differentiating. Bladder-urethra separates from rectum. Urethral membrane rupturing.	Cardinal veins transforming. Inf. vena cava outlined. Atrium, ventricle and bulbus partitioned. Cardiac valves present. Stem of pulm. vein absorbed into l. atrium. Spleen anlage prominent.	Chondrification more general. Chondrocranium.	Muscles differentiating rapidly throughout body and assuming final shapes and relations.	Mammary thickening lens-shaped.	Cerebral hemispheres becoming large. Corpus striatum and thalamus prominent. Infundibulum and Rathke's pouch in contact. Chorioid plexuses appearing. Suprarenal medulla begins invading cortex.	Chorioid fissure closes enclosing central artery. Nerve fibers invade optic stalk. Lens loses cavity by elongating lens fibers. Eyelids forming. Fibrous and vascular coats of eye indicated. Olfactory sacs open into mouth cavity.
Testis and ovary distinguishable as such. Müllerian ducts, nearing urogenital sinus, are ready to unite as utero-vaginal primordium. Genital ligaments indicated.	Main blood vessels assume final plan. Primitive lymph sacs present. Sinus venosus absorbed into right atrium. Atrio-ventricular bundle represented.	First indications of ossification.	Definitive muscles of trunk, limbs and head well represented and fetus capable of some movement.	Mammary primordium a globular thickening.	Cerebral cortex begins to acquire typical cells. Olfactory lobes visible. Dura and pia-arachnoid distinct. Chromaffin bodies appearing.	Eyes converging rapid. Ext., mid., and int. ear assuming final form. Taste buds indicated. External nares plugged.
Kidney able to secrete. Bladder expands as sac. Genital duct of opposite sex degenerating. Bulbo-urethral and vestibular glands appearing. Vaginal sacs forming.	Thoracic duct and peripheral lymphatics developed. Early lymph glands appearing. Enucleated red cells predominate in blood.	Ossification centers more common. Chondrocranium at its height.	Perineal muscles developing tardily.	Epidermis adds intermediate cells. Periderm cells prominent. Nail field indicated. Earliest hair follicles begin developing on face.	Spinal cord attains definitive internal structure.	Iris and ciliary body organizing. Eyelids fused. Lacrimal glands budding. Spiral organ begins differentiating.

Table 2-2. A reference table of correlated human development—cont'd

Age in weeks	Size (C R) in mm	Body form	Mouth	Pharynx and derivatives	Digestive tube and glands	Respiratory system	Coelom and mesenteries
12	56.0	Head still dominant. Nose gains bridge. Sex readily determined by external inspection.	Filiform and foliate papillae elevating. Tooth primordia form prominent cups. Cheeks represented. Palate fusion complete.	Tonsillar crypts begin to invaginate. Thymus forming medulla and becoming increasingly lymphoid. Thyroid attains typical structure.	Muscle layers of gut represented. Pancreatic islands appearing. Bile secreted.	Conchae prominent. Nasal glands forming. Lungs acquire definitive shape.	Omentum an expansive apron partly fused with dorsal body wall. Mesenteries free but exhibit typical relations. Coelomic extension into umbilical cord obliterated.
16	112.0	Face looks 'human.' Hair of head appearing. Muscles become spontaneously active. Body outgrowing head.	Hard and soft palates differentiating. Hypophysis acquiring definitive structure.	Lymphocytes accumulate in tonsils. Pharyngeal tonsil begins development.	Gastric and intestinal glands developing. Duodenum and colon affixing to body wall. Meconium collecting.	Accessory nasal sinuses developing. Tracheal glands appear. Mesoderm still abundant between pulmonary alveoli. Elastic fibers appearing in lungs.	Greater omentum fusing with transverse mesocolon and colon. Mesoduodenum and ascending and descending mesocolon attaching to body wall.
20-40 (5-10 mo)	160.0-350.0	Lanugo hair appears (5). Vernix caseosa collects (5). Body lean but better proportioned (6). Fetus lean, wrinkled, and red; eyelids reopen (7). Testes invading scrotum (8). Fat collecting, wrinkles smoothing, body rounding (8-10).	Enamel and dentine depositing (5). Lingual tonsil forming (5). Permanent tooth primordia indicated (6-8). Milk teeth unerupted at birth.	Tonsil structurally typical (5).	Lymph nodules and muscularis mucosae of gut present (5). Ascending colon becomes recognizable (6). Appendix lags behind caecum in growth (6). Deep esophageal glands indicated (7). Plicae circulares represented (8).	Nose begins ossifying (5). Nostrils reopen (6). Cuboidal pulmonary epithelium disappearing from alveoli (6). Pulmonary branching only two-thirds completed (10). Frontal and sphenoidal sinuses still very incomplete (10).	Mesenterial attachments completed (5). Vaginal sacs passing into scrotum (7-9).

Fetal membranes

There are several membranes or saclike structures that are formed during embryonic development. These include the chorion, amnion, allantois, yolk sac, and placenta.

Chorion. The chorion is derived from the trophoblast of the blastocyst and the extraembryonic mesoderm. It was pointed out previously that the trophoblast cells form extensions called villi that provide a route for exchange of materials between the trophoblast and the blood of the mother. Until about the eighth week of development, these villi cover the entire surface of the chorionic sac. As the sac grows, villi lying over the superficial portion of the developing organism become compressed, atrophy, and disappear. This area becomes known as the *smooth chorion (chorion laeve)*. The portion of the chorion beneath the embryo develops more villi that branch profusely to form the *villous chorion (chorion frondosum)*. This latter portion forms the fetal component of the placenta, to which blood is brought by the umbilical arteries and drained by umbilical veins. The villi formed are of two types: *anchoring,* which are not surrounded by a blood-filled cavity but are surrounded directly by uterine tissue; *free,* which lie in a blood-filled space in the uterine lining. The space is filled with maternal blood and the free villi are thus the exchange structures between mother and fetus.

Amnion. The amnion is derived from the layer of ectodermal cells that surrounded the original amniotic cavity. The cavity enlarges as development proceeds, gradually obliterating the space where the extraembryonic mesoderm was formed and eventually comes to surround the developing individual (Fig. 2-19). The ectodermal cells proliferate and form a membrane surrounding the cavity (the amnion itself). As the cavity enlarges, it is filled with amniotic fluid (formed initially from maternal blood) whose volume is increased in the last month or two of pregnancy by secretion of fetal urine into the cavity. The fluid and its amnion constitute the "bag of waters" in which the fetus floats.

Urogenital system	Vascular system	Skeletal system	Muscular system	Integumentary system	Nervous system	Sense organs
Uterine horns absorbed. External genitalia attain distinctive features. Meson. and rete tubules complete male duct. Prostate and seminal vesicle appearing. Hollow viscera gaining muscular walls.	Blood formation beginning in bone marrow. Blood vessels acquire accessory coats.	Notochord degenerating rapidly. Ossification spreading. Some bones well outlined.	Smooth muscle layers indicated in hollow viscera.	Epidermis three-layered. Corium and subcutaneous now distinct.	Brain attains its general structural features. Cord shows cervical and lumbar enlargements. Cauda equina and filum terminale appearing. Neuroglial types begin to differentiate.	Characteristic organization of eye attained. Retina becoming layered. Nasal septum and palate fusions completed.
Kidney attains typical shape and plan. Testis in position for later descent into scrotum. Uterus and vagina recognizable as such. Mesonephros involuted.	Blood formation active in spleen. Heart musculature much condensed.	Most bones distinctly indicated throughout body. Joint cavities appear.	Cardiac muscle appearing in earlier weeks, now much condensed. Muscular movements in utero can be detected.	Epidermis begins adding other layers. Body hair starts developing. Sweat glands appear. First sebaceous glands differentiating.	Hemispheres conceal much of brain. Cerebral lobes delimited. Corpora quadrigemina appear. Cerebellum assumes some prominence.	Eye, ear and nose grossly approach typical appearance. General sense organs differentiating.
Female urogenital sinus becoming a shallow vestibule (5). Vagina regains lumen (5). Uterine glands appear (7). Scrotum solid until sacs and testes descend (7-9). Kidney tubules cease forming at birth.	Blood formation increasing in bone marrow and decreasing in liver (5-10). Spleen acquires typical structure (7). Some fetal blood passages discontinue (10).	Carpal, tarsal and sternal bones ossify late, some after birth. Most epiphyseal centers appear after birth; many during adolescence.	Perinal muscles finish development (6).	Vernix caseosa seen (5). Epidermis cornifies (5). Nail plate begins (5). Hairs emerge (6). Mammary primordia budding (5); buds branch and hollow (8). Nail reaches finger tip (9). Lanugo hair prominent (7); sheds (10).	Commissures completed (5). Myelinization of cord begins (5). Cerebral cortex layered typically (6). Cerebral fissues and convolutions appearing rapidly (7). Myelinization of brain begins (10).	Nose and ear ossify (5). Vascular tunic of lens at height (7). Retinal layers completed and light perceptive (7). Taste sense present (8). Eyelids reopen (7-8). Mastoid cells unformed (10). Ear deaf at birth.

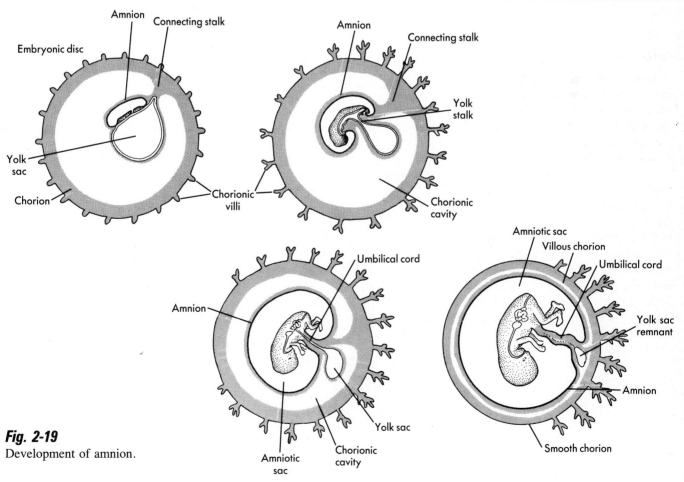

Fig. 2-19
Development of amnion.

The fluid provides several things:

- It permits symmetrical growth of the embryo and fetus, since they are "free floating" in the fluid, normally suffering no compression.
- It prevents the amnion from "sticking" to the embryo.
- It cushions the embryo from external or internal shocks or bumps.
- It aids in maintaining constant body temperature.
- It permits free movement of the fetus, aiding musculoskeletal development.

At birth the amnion ruptures and the fluid is lost via the vagina. It is possible to insert a needle through the mother's abdomen into the amniotic fluid to withdraw fluid for analysis. Cells are shed from the fetus into the fluid, and they can be cultured and examined for chromosome content. The procedure whereby the needle is placed in the fluid is called *amniocentesis*. Examination of withdrawn cells may demonstrate the presence of mutations that may be used as a guide to determine whether the pregnancy should be terminated.

Allantois. The allantois appears as an outpouching of the yolk sac at about 16 days of development. In some vertebrates (not the human), it serves as a reservoir for excretory products, functions as a respiratory chamber, or forms a part of the placenta. In humans it is richly vascularized, and its vessels contribute to the formation of the umbilical arteries and vein.

Yolk sac. The yolk sac is the innermost of the two cavities of the two-layered embryo. As development proceeds, the yolk sac is constricted in its center to form a gut tube, lying closest to the embryo, and a lower yolk sac remnant. By 12 weeks the sac has shrunk and becomes a solid mass of no functional significance. It does, however, serve important functions for the first month. These functions include the following:

- It transfers nutrients to the embryo during the second and third weeks.
- It forms blood cells up to the sixth week.
- It contributes to the formation of the epithelium of the respiratory system as it is incorporated into the primitive gut.
- It appears to be the earliest source of cells that migrate to the developing sex organs and that will ultimately become ova or sperm.

Placenta. The placenta (Fig. 2-20), as stated previously, functions as the organ for exchange of materials between mother and fetus. It is composed of a fetal (chorionic) component and a maternal (uterine, endometrial) component. In humans the placenta is usually disclike in shape and is composed of many subdivisions called cotyledons. Malformation of the placenta can interfere with the acquisition of nutrients essential to fetal development. It also produces several hormones essential for sustenance of the pregnancy (see Chapter 16).

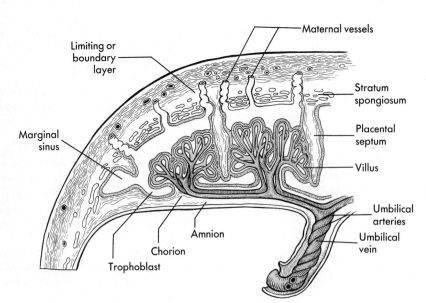

Fig. 2-20
Scheme of placental circulation.

Tissue level
Basic tissue groups

A tissue is defined as a group of cells that are similar in structure and function, together with all associated intercellular material. If we look at the whole body, it becomes apparent that there are only *four basic tissue groups* forming the body. To be sure, there are many different *types* or examples of tissues within each group. The four basic tissue groups are listed as follows:

epithelial tissues (epithelia) cover and line hollow organs and cavities and cover the external body surfaces. Epithelia can originate from all three germ layers.

connective tissues, except synovial membranes lining joint cavities, are tissues that are not found on body surfaces. They connect, support, protect, and in general hold body cells together into larger aggregates. This group usually includes the blood, and all its members are of mesodermal origin.

muscular tissues are characterized by their ability to contract or shorten and thus accomplish work in the body. Therefore movement of blood by heart action, movement of the body through space, and movement of materials through the gut will occur as a result of muscular activity. They too are mesodermal in origin.

nervous tissues are characterized by excitability and conductivity. They are responsible for the transmission of electrical disturbances called *nerve impulses* to muscle and glandular structures in the body and for the abilities to enjoy, store impressions of, and understand our lives. Ectoderm is the germ layer of origin for all but one type of cell in the nervous system.

Epithelium
General characteristics. Epithelia *cover* and *line* actual or potential free surfaces of the body. The surfaces of the skin and the digestive, circulatory, respiratory, excretory, and reproductive systems provide obvious examples of systems having hollow organs that are lined and covered by epithelium. The true body cavities (such as pleural and abdominal) are also lined with epithelium. Since they are found on body surfaces, epithelia must consist of **closely packed cells,** with minimal amounts of intercellular material between the cells if they are to form good barriers. Epithelia are **avascular,** that is, there are no blood vessels within the epithelium itself. The epithelial cells derive their nutrients from vessels in the connective tissue that underlies the epithelium. Again, vessels right at a surface would be exposed to constant "wear and tear." Nerves *are* found in some epithelia, such as the olfactory, retinal, or taste epithelia. In these cases the epithelium is usually protected by its location from "hard contact" with mechanical factors. A **basement membrane** is usually found beneath an epithelium and serves to attach the epithelium to the underlying connective tissue.

Glands are in most cases of epithelial origin although they may no longer retain a connection with the layer of origin.

The terms **endothelium** and **mesothelium** are applied, respectively, to the epithelial linings of the circulatory system (heart, blood vessels, lymph vessels), the ventral true cavities of the body (thoracic, abdominopelvic), and the outer epithelial coats of organs suspended within those cavities.

Types of epithelia. Most epithelia are named using two criteria:

Number of layers of cells in the epithelium

Shape of the cells in the epithelium

Those epithelia that cannot be named using these criteria are sometimes called *aberrant epithelia.*

Number of layers of cells. An epithelium having only one layer of cells is said to be a **simple** epithelium. One row of nuclei, all at or very near the same level and having the same shape, give evidence of a single cell layer. An epithelium having more than one layer of cells, as evidenced by more than one row of nuclei, is called a **stratified** epithelium. In such an epithelium the deeper lying cells tend to become more rounded or elongated, regardless of the shape of the surface layers of cells. Thus a stratified epithelium is *named by the shape of the topmost layer of cells.*

Shape of cells in the epithelium. Although there are many intergradations in shape, there are considered to be *three* basic shapes in which epithelial cells may occur.

squamous (flat) cells resemble a piece of tile, with length and width but very little height. Nuclei are compressed and follow the length and width of the cell, not the height. Thus the *nuclei are flattened parallel to the cell's surface* and may bulge at the surface.

cuboidal cells are shaped like a cube of sugar, with length, width, and height roughly equal. In such a shape the nucleus is free to assume its normal spherical shape and appears round, or nearly so, in section. Cuboidal cells are often found lining small tubes of the body (gland ducts, kidney) and in such a situation may appear pyramidal as their apexes are compressed to form the tube lining.

columnar cells are *tall*, with height greatly exceeding length and width. Again, nuclei are compressed, this time in a direction perpendicular to the surface.

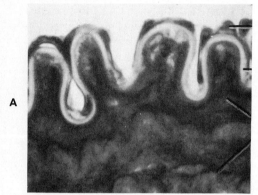
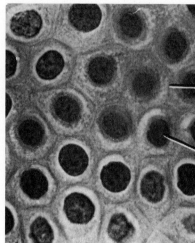
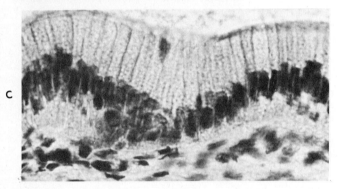

Fig. 2-21
Epithelia. **A,** Simple squamous epithelium (endothelium). **B,** Simple cuboidal (kidney) epithelium. **C,** Simple columnar (bile duct) epithelium. **D,** Stratified squamous (cheek) epithelium. **E,** Stratified cuboidal (sweat gland) epithelium. **F,** Stratified columnar (larynx) epithelium.

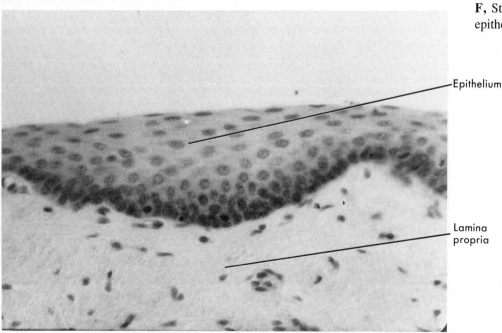

As stated previously, there are intermediate forms "in between" the three basic shapes. Perhaps the best clue as to how to name the cell shapes is to look at the nucleus. Any obvious flattening should, according to orientation, lead to diagnosis of squamous or columnar cells.

By combining these two criteria, six different types of epithelia may be defined (Fig. 2-21):

simple squamous epithelium consists of one row of cells whose nuclei are flattened parallel to the epithelial surface. This type of epithelium lines the lung alveoli and glomerular capsule of the kidney, covers the inner surface of the eardrum, and forms the *endothelium* and *mesothelium* described earlier. *Mesenchymal epithelium* lines certain cavities associated with the nervous system, such as the subdural and subarachnoid spaces

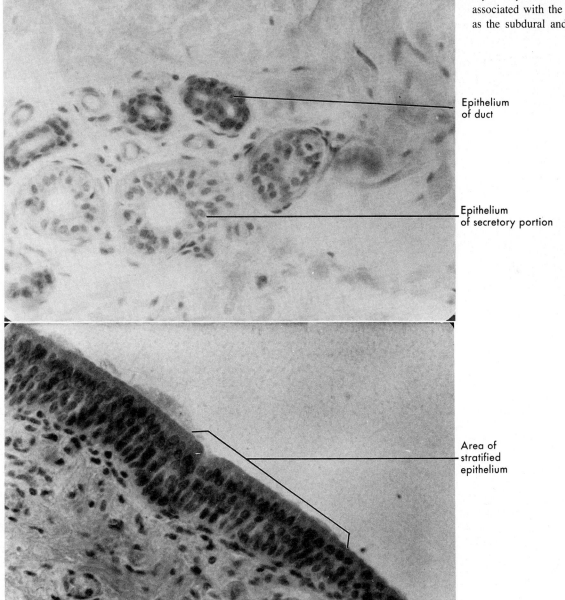

and the chambers of the eye and inner ear. It is identical in appearance to simple squamous epithelium seen elsewhere in the body.

Functionally, because of the thinness of a simple squamous epithelium, this type of tissue acts as an *exchange* epithelium, allowing passage of gas molecules (lungs) and dissolved or suspended material and fluids (capillaries generally and in the kidney particularly).

simple cuboidal epithelium consists of one row of cells whose nuclei are round, or nearly so. This indicates that the three dimensions of the cell are nearly equal (isodiametric) and that there is little or no compression of the nucleus. This type of epithelium covers the ovary (germinal epithelium), lines several of the kidney tubules, and lines the smaller ducts of many body glands.

In the kidney the epithelium is important in actively transporting solutes (dissolved materials) from the tubule cavity *(lumen)* to surrounding capillary beds in the process called *reabsorption*. The epithelium also transports solutes in the opposite direction (blood vessels to tubule lumen) in the process called *secretion*. This variety of epithelium is thus often very active metabolically.

simple columnar epithelium consists of a single layer of "tall" cells, whose nuclei are oval, perpendicular to the surface, and generally in the lower half of the cell. The lining of the alimentary tract from stomach to anal canal is simple columnar, as are the linings of medium-sized gland ducts and the gallbladder lining.

In the intestines this epithelium is very important in *absorption* of fluids and end products of the digestive process in the gut. It is as active in this process as is the cuboidal epithelium of the kidney.

stratified squamous epithelium has several cell layers with the top layer or two of cells flattened. Again, surface nuclei are flattened parallel to the epithelial surface. A look at Fig. 2-12 shows that deeper nuclei become round and sometimes oval at the basement membrane. This indicates that deeper lying cells may be rounded or columnar in shape. This type of epithelium is found where *protection* from environmental factors is essential. The layer is thick, and cells sloughed off from the surface are replaced by mitosis from the deepest layers of cells. The epidermis of the skin and the lining of the oral cavity, esophagus, anal canal, and vagina are of this type of epithelium. In the skin the surface layers of cells are filled with a waterproofing and toughening protein called *keratin;* thus the epithelium is said to be keratinized or cornified ("hardened" or "hornlike"). In the other areas mentioned previously, surface keratinization is not present or is very slight, and the epithelium is "soft" or noncornified.

stratified cuboidal epithelium is not widespread in the body. It usually has two layers of cells, with the top layer cuboidal, as evidenced by the round nuclei. The sweat glands of the skin have stratified cuboidal linings.

stratified columnar epithelium is also rare. Several rows of cells, with the top layer having perpendicularly oriented oval nuclei *in a row,* characterize this tissue. It occurs in the areas of the pharynx (throat) and larynx ("voice box"), where stratified squamous epithelium changes to pseudostratified epithelium.

Epithelia that cannot be accurately named as were those preceding constitute the *aberrant epithelia* (Fig. 2-22). Most, in addition to not "following the rules" of naming, have unique properties or structures that fit them for their tasks in the body as shown in the following:

pseudostratified epithelium appears at first glance to be a stratified tissue. It shows nuclei at many different levels; closer inspection shows that the *top nuclei are not in an even row,* and special techniques can demonstrate that although all cells touch the basement membrane beneath the epithelium, *not all reach the surface.* The epithelium is thus really simple but presents a false (pseudo) appearance of stratification. The greater part of the respiratory system and the ducts of the male reproductive system are lined with this type of epithelium.

In the respiratory system, pseudostratified epithelium commonly has cilia, and the epithelium contains one-celled mucus-secreting glands called goblet cells. The mucus traps particulate matter and the cilia move the mucus coat to the throat, thus cleansing incoming air of dust, pollen, and microorganisms that could damage the body. In the male reproductive system extensions of the surface of the cell (microvilli, see a later section in this chapter) provide increased surface area for channeling nutrients to sperm contained within the ducts.

transitional epithelium, or **uroepithelium,** is truly stratified, but a look at the top layers of cells shows them to be nonuniform in shape; some cells are flattened, some rectangular, others columnar or rounded. The tissue lacks an obvious basement membrane and is capable of being stretched without damage. When

A

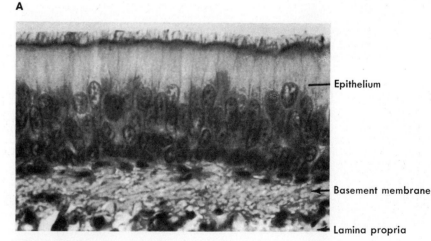

Fig. 2-22
"Aberrant" epithelia. **A,** Pseudostratified (trachea) epithelium.

stretched, the entire epithelium shows flattened cells and is often difficult to distinguish from stratified squamous epithelium. It lines the renal pelvis, ureters, urinary bladder, and a part of the urethra.

syncytial epithelium lacks membranes between nuclei, is regarded as a simple epithelium, and is basically one continuous multinucleated mass. It covers the villi of the placenta and serves as the epithelium for exchange of materials between mother and fetus.

germinal epithelium is the name properly applied to the stratified epithelium lining the seminiferous tubules of the testes. It produces secretions and sperm cells.

neuroepithelium is a highly specialized epithelial type found in the olfactory epithelium, taste buds, retina of the eye, and inner ear. Part of the epithelium consists of nerve cells, the rest of supporting cells. Details of this type of epithelium are provided in appropriate chapters to follow.

Some epithelial types are capable of changing to other types if irritants or stress-producing substances contact the epithelium. In the lungs the pseudostratified epithelium can become stratified squamous under, for example, irritants ("tars," nicotine) in cigarette smoke. In some cases squamous cell carcinomas (cancers) result.

Surface modifications of epithelia. Since one side of any epithelium has a free surface, there may be present on that free surface special modifications that enable the epithelium to carry out its functions more efficiently.

Cilia and flagella. Cilia (see Fig. 2-8) are multiple tiny hairlike projections arising from small granules (basal bodies) located in the apexes of certain ep-

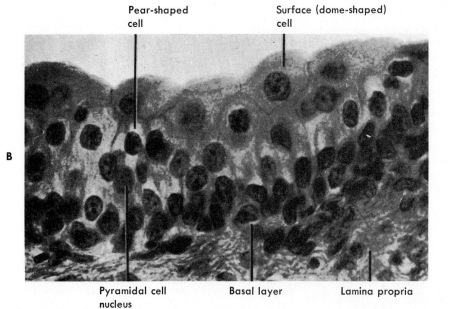

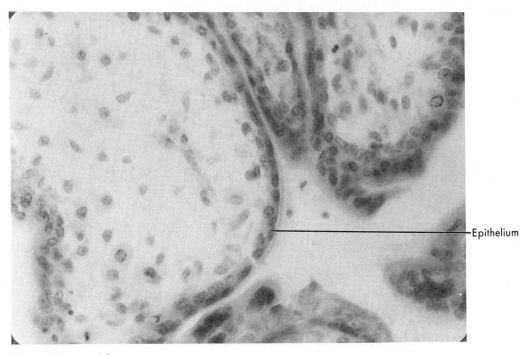

Fig. 2-22, cont'd
B, Transitional (bladder) epithelium. **C,** Syncytial (placental villus) epithelium.

Continued.

ithelial cells. The basal bodies represent derivatives of centrioles or centriole-like structures. Flagella are usually single and long processes, such as form the tails of spermatozoa. Both cilia and flagella have a similar structure consisting of nine pairs of peripheral microtubules around a central pair. Cilia are capable of waving or "beating" in a fashion so as to move materials across the ciliated surface of the epithelium. In the respiratory system, for example, a layer of sticky mucus is moved across the epithelium by the ciliary action. Trapped particulate matter is thus removed, and cleaning occurs. Cilia are also found in the uterine (fallopian) tubes of the female reproductive system and aid transport of ova (eggs) down the tube to the uterus. In a sperm cell the flagella confer independent movement through a fluid medium of the whole cell, a necessary requisite for fertilization of an egg as sperm move through the uterus and uterine tubes.

Pseudostratified and simple columnar epithelia commonly possess cilia.

Microvilli. Microvilli (Fig. 2-23) are small fingerlike projections of the apical surface of an epithelial cell. Their primary function is to greatly increase the surface area of the cell possessing them, and thus they are most common on those epithelia active in absorption and secretion of materials. For example, as stated earlier, the simple columnar epithelium of the small intestine and the simple cuboidal epithelium of certain tubules of the kidney have microvilli. Also, in the epididymis of the

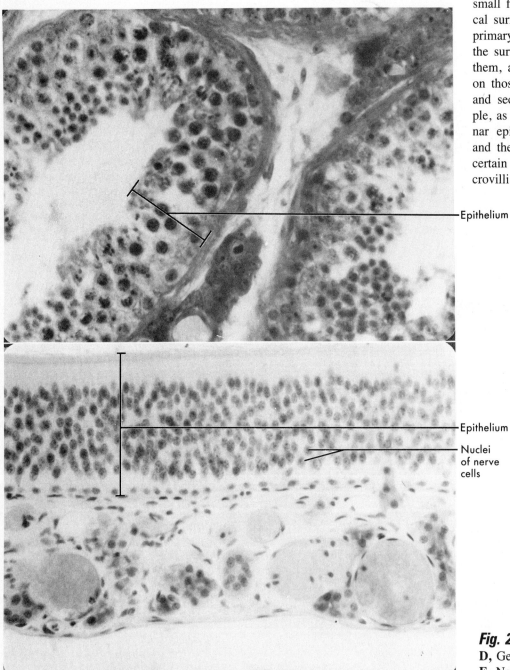

Fig. 2-22, cont'd
D, Germinal (testis) epithelium.
E, Neuroepithelium (olfactory organ).

male reproductive system, there are microvilli that channel nutrients to the sperm within the organ. Called "stereocilia" in older literature, these structures possess no ability to beat.

Cytoplasmic specializations at the free surface. Transitional epithelium possesses a unique ultrastructure at the apexes of the cells. There is a substructure of hexagonally arranged units consisting of what appear to be granular or short tubular units. A thickened membrane is created that may contribute to the ability of the surface cells to stretch as the bladder fills or to protect the epithelial cells from the acid urine within the bladder. The layer also prevents osmotic flow of water from the bloodstream into the hypertonic fluid (urine) in the bladder.

A layer of material that is produced by epithelial cells and that forms a separate layer that may be separated from the epithelial cells without harming them is called a *cuticle* (Fig. 2-24). The enamel layer of a tooth and the lens capsule are examples of cuticles in the human body. The "cuticles" of the nails are merely the surface epithelial cells of the epidermis that adhere to the nail and are pulled along as the nail grows.

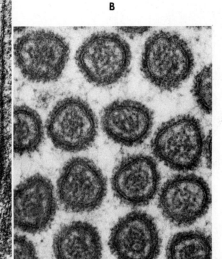

Fig. 2-23
A, Longitudinal sections of microvilli from small intestine epithelial cells. ($\times 88,000$.) Note trilaminar cell membranes and "core filaments."
B, Cross-sections of microvilli from small intestine epithelial cells. ($\times 145,000$.)

From Anderson, W.A.D., and Kissane, J.M.: Pathology, vol. 1, ed. 7, St. Louis, 1977, The C.V. Mosby Co.

Fig. 2-24
Example of cuticle (tooth enamel).

Modifications of intercellular surfaces. Several devices are employed to hold epithelial cells together and to seal their surface intercellular spaces so that free entry or exit of materials cannot occur (Fig. 2-25):

zonula occludens, or "tight junction," involves a very close approach of adjacent cell membranes to one another. This type of junction occurs at the free surface of the epithelial cells and is the main structure "sealing" the surface of the epithelium.

zonula adherens is a device holding cell membranes to one another. The area of involved cell membranes may be seen to be obviously separated; in the cytoplasm there is an area of diffusely staining material from which fibers radiate into the cytoplasm of the opposing cells.

macula adherens, or **desmosome,** is a more clearly defined holding device. It consists of a network of fine filaments embedded in a dense cytoplasmic matrix. In addition to holding cells together, the desmosome may contribute support to the cell in which it is found. Hemidesmosomes are often found on the basal ends of epithelial cells where they attach to a basement membrane and thus aid in affixing the epithelium to the underlying connective tissue.

Table 2-3 summarizes facts about epithelia.

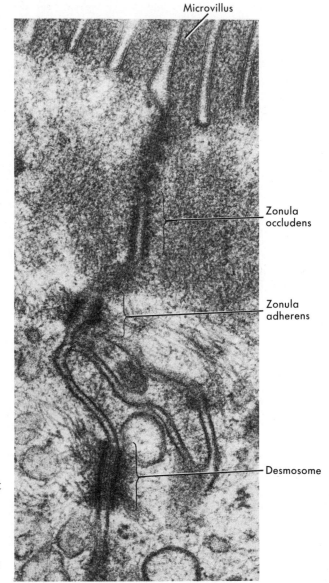

Fig. 2-25
Junctional complexes joining adjacent cells. Zonula occludens or "tight junction" seals cell surfaces. (×45,000.)

From Anderson, W.A.D., and Kissane, J.M.: Pathology, vol. 1, ed. 7, St. Louis, 1977, The C.V. Mosby Co.

Table 2-3. Summary of epithelial tissues

Type of epithelium	Characteristics	Typical locations	Surface modifications often present	Functions and comments
Simple squamous	One layer of flattened cells; nuclei parallel to free surface of epithelium	Glomerular capsule of kidney, endothelium, mesothelium	Microvilli of mesothelium	Exchange, because of thinness; lining of cavities not directly exposed to external environment, absorption
Simple cuboidal	One layer of cube-shaped cells; nuclei usually rounded	Most kidney tubules, lining of thyroid follicles, outer surface of ovary	Microvilli, in kidney tubules	Secretion, absorption; very active metabolically
Simple columnar	One layer of tall cells; nuclei compressed perpendicularly to free surface of epithelium	Inner lining of stomach, small and large intestines, gallbladder, larger ducts of exocrine glands, uterine tubes, bronchioles	Microvilli (in gut), cilia (in uterine tubes and bronchioles)	Secretion and absorption (gut), movement of material across surface if ciliated
Stratified squamous	Many layers of cells with top layer or two of cells flattened	Skin (cornified), vagina, mouth and esophagus, anal canal	None	Mechanically resistant tissue for protection
Stratified cuboidal	Several layers of cells with top layer cuboidal in shape	Sweat glands (secretory part and duct in dermis)	None	May be secretory as in sweat glands
Stratified columnar	Several layers of cells with top layer columnar in shape	Larynx, upper part of pharynx	Cilia	Commonly found as a transition form between stratified squamous and pseudostratified
Pseudostratified	Not all cells reach free surface, but all touch basement membrane	Nasal cavities, trachea and bronchi, vas deferens	Cilia, most commonly in respiratory system, microvilli (vas deferens)	Movement of materials across surface (if ciliated), secretion and absorption (with microvilli present)
Transitional	A stratified tissue with no uniform cell shape on free surface	Kidney pelvis, ureters, urinary bladder	"Condensed cytoplasm"	Capable of stretching and resisting acidic urine
Syncytial	A simple epithelium with *no* membranes between cells	Villi of placenta	Microvilli	Exchange, secretion, and absorption
Germinal	Multilayered, cells in various stages of maturation	Seminiferous tubules of the testes	None	Epithelium shows the several stages of sperm development
Neuroepithelium	Nerve cells form part of the epithelium and have supporting cells between the nerve cells	Taste buds, olfactory area, retina, cochlea	May have special modifications associated with their special sensory activities	Receive diverse stimuli and change them to nerve impulses

Adapted from McClintic, J.R.: Basic anatomy and physiology of the human body, ed. 2, 1980. Copyright © 1980. Reprinted by permission of John Wiley & Sons, Inc.

Epithelial membranes. The combination of an epithelium and its underlying connective tissue layer (often called the *lamina propria*) forms an *epithelial membrane.* Two types of such membranes are recognized in the body.

Serous membranes (Fig. 2-26) form the outer coats of visceral organs, the linings of the true body cavities, and omenta and mesenteries. The epithelial component of a serous membrane is a simple squamous epithelium that is moistened by a watery secretion. These moist membranes allow organs to move on one another and between an organ and the wall of its cavity (as in the lungs moving against the walls of the thorax).

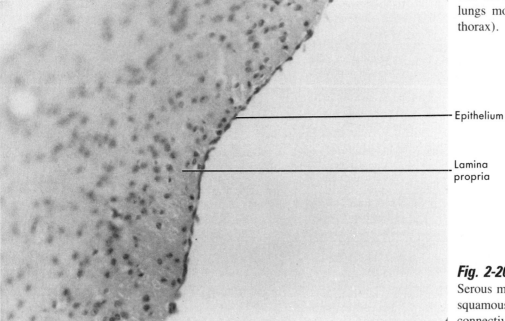

Fig. 2-26
Serous membrane, composed of simple squamous epithelium and underlying connective tissue (lamina propria).

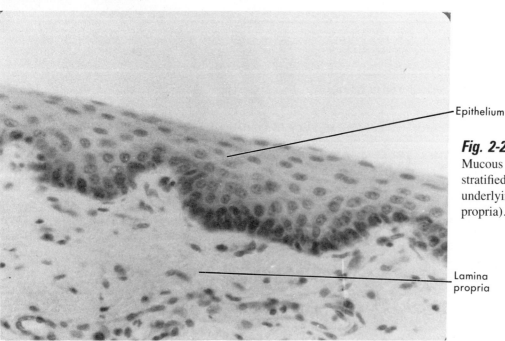

Fig. 2-27
Mucous membrane, here consisting of stratified squamous epithelium and underlying connective tissue (lamina propria).

Mucous membranes (Fig. 2-27) form the linings of hollow body viscera, such as the esophagus, stomach, intestines, uterus, vagina, urinary system, and the cavities that open on the body surface, such as oral, nasal, and anal cavities. The epithelial type varies, including stratified squamous, simple columnar, pseudostratified, or transitional, but is always moistened by mucus secreted by intraepithelial glands or glands in the underlying connective tissue.

Glands. Secretion forms one of the primary functions of epithelia. Glands are specialized epithelial cells or groups of epithelial cells that empty their secretions on the epithelial surface through ducts or into the bloodstream. The secretions are different from the fluids from which they are derived, suggesting active synthetic activity involving rough endoplasmic reticulum (ER) and the Golgi apparatus.

There are two major categories of glands found in the body. **Endocrine glands** originate from an epithelial surface but subsequently lose their connection with the epithelial surface and empty their secretions into blood vessels. **Exocrine glands** maintain a connection via ducts with an epithelial surface or consist of individual *intraepithelial* cells that secrete directly onto the epithelial surface. The *goblet cells* of the intestine and respiratory system are *unicellular intraepithelial* glands secreting onto the epithelial surface. All other glands are multicellular. Multicellular glands may contain **serous cells,** small granular cells secreting a watery fluid, or **mucous cells,** larger cells secreting viscous mucus; or they may contain both types of cells. If a gland contains only one duct and not a branching system of ducts, it is said to be *simple.* A *compound* gland has a branching series of ducts within it, although there may be only a single duct reaching the epithelial surface.

According to the *shape* of the secretory portion of the gland, further description of the gland may be made, for example, *tubular* or *saccular* (alveolar). If tubular, the secretory portion may be *coiled* or *branched.* Such terms are commonly included in naming the gland.

Exocrine glands may be further characterized as to the manner in which they produce their secretion. A **merocrine** gland produces a secretion in which no part of the producing cell is found as part of the secretion. An **apocrine** gland loses a portion of its cells in forming its secretion. A **holocrine** gland produces a cell as its "secretion" (for example, testis produces sperm) or an entire cell is "shed" into the secretion (for example, sebaceous or oil glands of the skin). Tables 2-4 and 2-5 summarize facts about exocrine glands.

Table 2-4. Summary of exocrine glands

Variety of gland*	Diagrammatic representation of gland†	Description	Examples
Unicellular		A one-celled intraepithelial mucus-secreting gland	Goblet cells of respiratory and digestive systems
Multicellular Simple tubular		One duct—secretory portion may be a straight, branched, or coiled tube	Sweat glands, gastric and intestinal glands
Simple alveolar or saccular		One duct—secretory portion saclike; gland may branch	Sebaceous glands
Compound tubular		Branched system of ducts in gland, secretory portions are tubular	Seminiferous tubules of testes, liver
Compound alveolar or saccular		Branched system of ducts in gland; secretory portions are saclike	Pancreas, certain salivary glands, mammary glands
Compound tubuloalveolar		Branched system of ducts in gland; some secretory portions are tubular, others are saccular	Certain salivary glands

Adapted from McClintic, J.R.: Basic anatomy and physiology of the human body, ed. 2. Copyright © 1980. Reprinted by permission of John Wiley & Sons, Inc.
*Simple implies only one duct serving the entire gland. Compound implies a series of branching ducts serving the secretory portions of the gland, even though only one duct may empty the secretion onto a surface.
†Heavily shaded portion outlines the duct(s).

Table 2-5. Manner of production of secretion by exocrine glands

Type of gland	Characteristics	Examples
Merocrine	Synthesized product independent of basic cell structure; secretion usually watery in nature	Pancreas, salivary glands
Apocrine	Some portion of cell issued as part of secretion; secretion rich in organic substances such as proteins and lipids	Mammary gland
Holocrine	Product of gland is a cell, or cell as a whole is shed in secretion	Testis, sebaceous glands

Adapted from McClintic, J.R.: Basic anatomy and physiology of the human body, ed. 2. Copyright © 1980. Reprinted by permission of John Wiley & Sons, Inc.

Connective tissues

General characteristics. With the exception of synovial membranes lining joints and the tissues forming bursae and tendon sheaths, connective tissues *do not have free surfaces*. They contain much *intercellular material* that is often *fibrous* (containing fibers) and in which *cells are widely spaced*. Most connective tissues are *vascular*. As a group, connective tissues support, protect, and hold cells into the larger units called *organs*.

Types of connective tissues. Classification of connective tissues is based on several criteria. If a tissue is unique to the embryo or fetus, it is termed an *embryonal* tissue. If it is a differentiated tissue, and is present as such after birth, it is called an *adult* tissue. Adult tissues are further subdivided according to *consistency of the intercellular material* (semiliquid, semisolid, or hard) and on the *type of fiber* predominating in the intercellular material.

A "classification tree" for connective tissues is presented in Fig. 2-28.

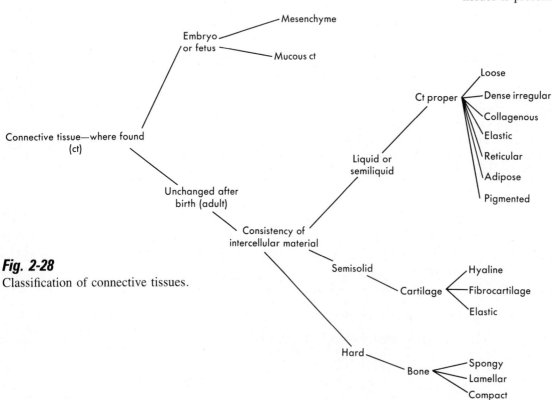

Fig. 2-28
Classification of connective tissues.

Embryonal connective tissue (Fig. 2-29). ***Mesenchyme*** is the primitive undifferentiated mesodermal tissue from which all other connective tissues, muscle, and blood arise. The cells are usually stellate (starlike) in shape, are ameboid, and usually form netlike arrangements. The cells lie in a nonfibrous, histologically homogeneous "ground substance" composed of water, inorganic substances, and protein polysaccharides (combinations of proteins and complex carbohydrates). Mesenchyme may be found in sections of vertebrate embryos under the external body covering and in those areas where bone and muscle will form.

Mucous connective tissue consists of flattened or elongated cells set in a mucoid (mucuslike) ground substance. In the ground substance are found fine fibrils composed of a protein known as *collagen*. Thus the fibrils are termed collagenous fibrils. They are strong but have little elasticity. This type of tissue is found in the umbilical cord (Wharton's jelly) and may also be found in

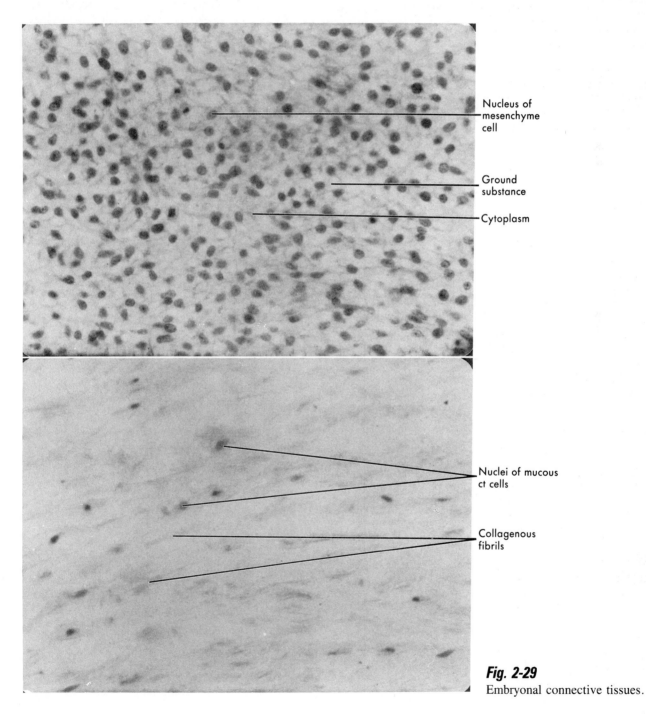

Fig. 2-29
Embryonal connective tissues.

the so-called sex skin of monkeys and in the combs of roosters.

Adult connective tissue

CONNECTIVE TISSUES PROPER. Connective tissues proper (Fig. 2-30) are that subdivision of connective tissues having a semifluid intercellular material and the fibroblast (fibrocyte) as the fiber-producing and tissue-maintaining cell. There are six types of connective tissue proper, separated on the basis of predominating fiber or special features:

loose (areolar) connective tissue contains all the structural elements that may be present in other members of the group. It is a nonoriented tissue; that is, the fibrillar elements run in a three-dimensional "feltwork." Some nine types of cells may be found in this tissue; the most common types will be mentioned here. *Reticular fibers*, regarded as antecedent or "immature" collagenous fibers, are one type of fiber in loose connective tissue. They require special stains (for example, silver) to demonstrate them and thus are not usually visible in slides of this tissue. *Collagenous fibers* are 1 to 12 μm in diameter, contain smaller units called *fibrils*, and are composed of a protein called *collagen*. Collagenous fibrils are in turn composed of *tropocollagen molecules* secreted by the fibroblasts. These molecules polymerize into the larger fibrillar units that then aggregate to form the fibers. Collagenous fibers are extremely strong (it takes 200 to 300 kg/cm^2 to break the fibers), but they have little elasticity. Collagenous fibers usually stain pink in slides of the tissue. *Elastic fibers* may also be 1 to 12 μm in diameter but are smaller in size than collagenous fibers in loose tissue. They are highly refractile, show no subunit of structure, branch freely, and are very elastic. They consist of a protein called *elastin*, also produced by fibroblasts. Only 20 to 30 kg/cm^2 can break these fibers, but they can stretch to about 150% of their original length and return. Loss of elastic fiber in the skin as we age contributes to the "sagging" of the skin in many individuals.

The cellular elements of loose connective tissue include the following as the most common types:

fibroblasts (fibrocytes) are connective tissue cells in which the cell cytoplasm is usually not visible in a typically stained slide but is flattened or spindle shaped. The large oval nucleus (the largest nucleus in the tissue) typically contains two or three prominent nucleoli. The cell gives rise to the fibers of the tissue, replaces fibers as they disappear (fibroblast), and maintains fibers once they are formed (fibrocyte).

histiocytes or ***macrophages*** are more or less fixed cells that can phagocytose (engulf) particulate matter. They "clean up" areas where inflammation has caused cell death and can engulf microorganisms that have found their way into the tissue. The cells are most easily recognized by their dark-staining nucleus that usually shows little internal structure, by an irregular cytoplasmic outline, and by the presence of irregular engulfed particles in the cytoplasm.

blood cells, because of their ability to pass through capillary walls *(diapedesis)*, may be found in loose connective tissue. *Lymphocytes, monocytes* (which may become macrophages), *basophils* (which may become tissue basophils or mast cells), and *eosinophils* (which may become tissue eosinophils) are present in the tissue. If the

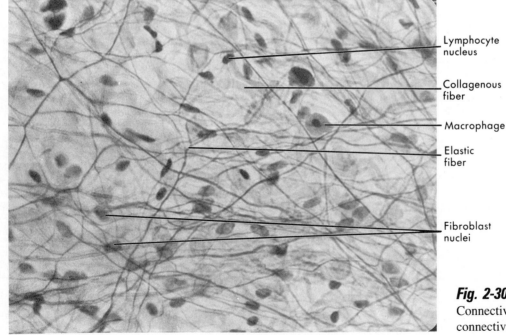

Fig. 2-30
Connective tissues proper. **A,** Loose connective tissue.

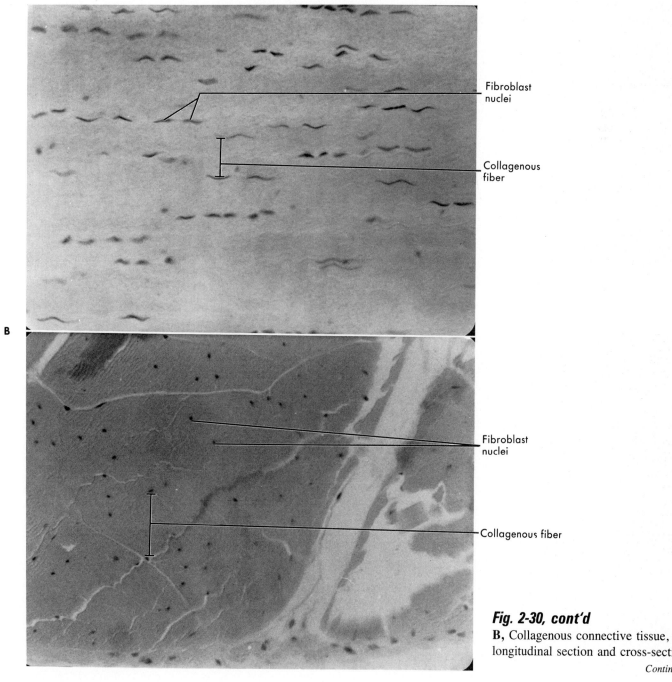

Fig. 2-30, cont'd
B, Collagenous connective tissue, longitudinal section and cross-section.

Continued.

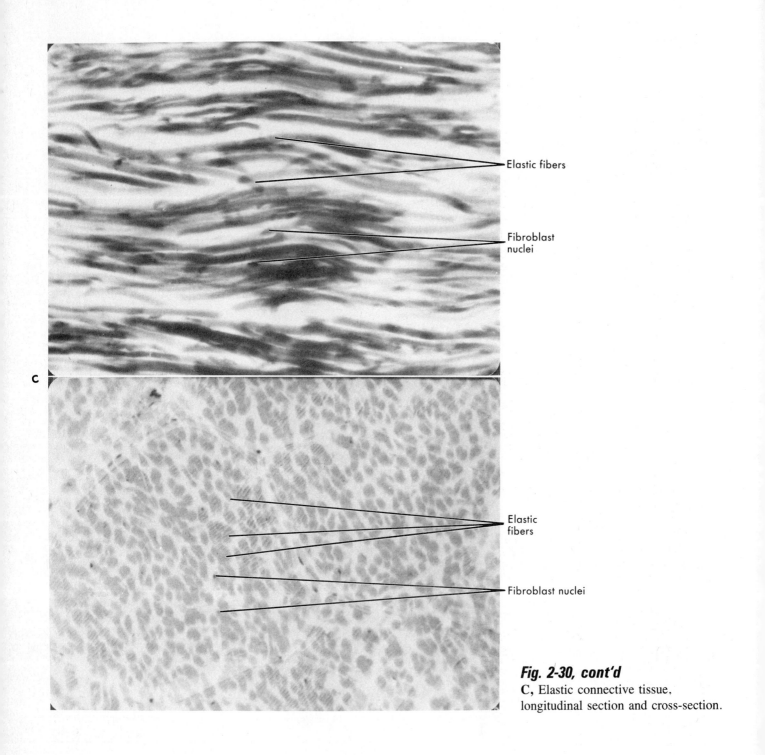

Fig. 2-30, cont'd
C, Elastic connective tissue, longitudinal section and cross-section.

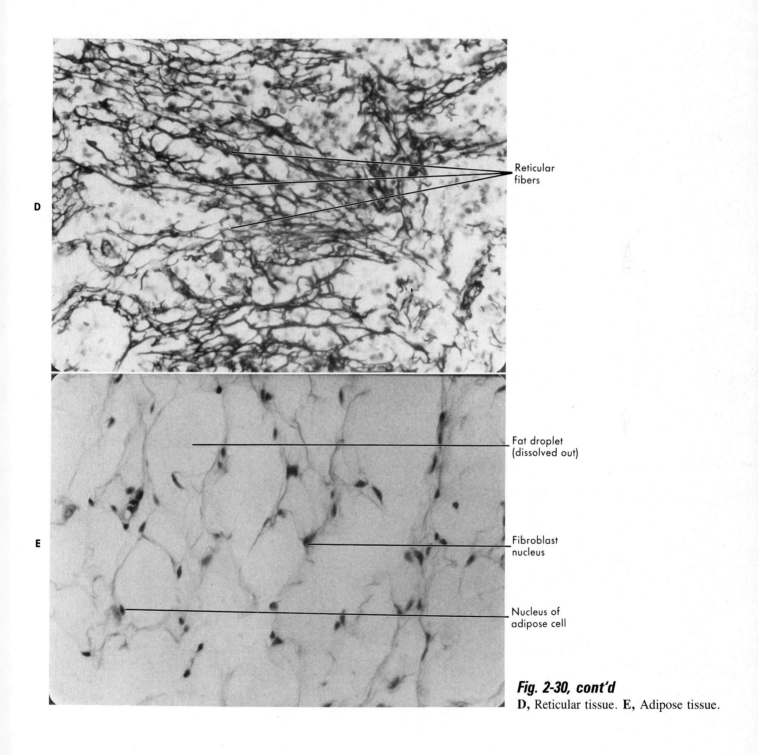

Fig. 2-30, cont'd
D, Reticular tissue. **E,** Adipose tissue.

tissue is inflamed, *neutrophils* may be found in quantity.

Lymphocytes, neutrophils, and monocytes are the best phagocytes and provide the greatest degree of protection. They also clean up products of cell death and aid in wound healing. Eosinophils appear to be involved in detoxifying foreign proteins (as in allergic states and in parasitic infections), whereas basophils may produce, as mast cells, chemicals that control blood flow (serotonin, a vasoconstrictor; histamine, a vasodilator) and change the viscosity of the ground substance in the tissue (heparin).

• • •

Loose connective tissue forms the subcutaneous layer of the skin (injections may be made into this layer) and occurs as a "packing material" around many body organs.

dense irregular connective tissue contains both collagenous and elastic fibers and the common cellular elements found in loose tissue. Fibers are larger and there is much less ground substance, so that the tissue resembles "compacted" loose connective tissue. This tissue forms the dermis of the skin and the lamina propria of tubular respiratory, digestive, and reproductive organs. The lamina propria is the connective tissue underlying the epithelium of the organ.

collagenous connective tissue is an oriented or regularly arranged tissue composed almost entirely of collagenous fibers, with flattened fibroblasts between fibers. It is extremely strong, forming the *tendons* that connect muscles to bones and many *ligaments* in the body. It also occurs as the sclera ("white") of the eye.

elastic connective tissue is also a regularly arranged or oriented tissue and consists almost entirely of elastic fibers with fibroblasts scattered in the spaces between fibers. The tissue forms the middle layer (media) of large arteries of the body (for example, aorta, pulmonary artery) and forms several of the ligaments of the spinal column.

reticular connective tissue is sometimes called a "special" connective tissue because a special silver stain is required to demonstrate the fibers. The small, irregular, and branching fibers form supporting networks for many internal body organs, including the liver, spleen, and lymph nodes. Muscle cells are held together by reticular fibers.

adipose tissue consists of large masses of adipose or fat cells. The mature fat cells have accumulated a large droplet of lipid that is usually dissolved by the alcohols used in preparing the slide. Thus the tissue contains many "holes." Adipose tissue serves protective or cushioning functions, insulates, and acts as a storage form of energy. The subcutaneous layer of the skin, the adipose capsule of the kidney, and the area around the coronary vessels of the heart are typical places to find the tissue.

pigmented connective tissue is loose tissue with many pigment cells in it. It is found in the iris of the eye.

CARTILAGES. Cartilages (Fig. 2-31) are much firmer tissues than the ones already described because they have a semisolid intercellular material called the **interterritorial matrix.** The typical cells, **chondrocytes,** lie in cavities or *lacunae* in the matrix. Chondrocytes may divide within a lacuna and form a **territorial matrix** between themselves; thus a "wall" is formed that divides the original lacuna into two compartments. A darker staining area of young matrix surrounds each lacuna and is designated as the **capsule.** The surface of the car-

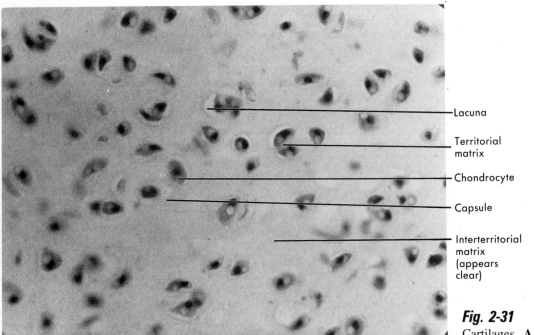

Fig. 2-31
Cartilages. **A,** Hyaline.

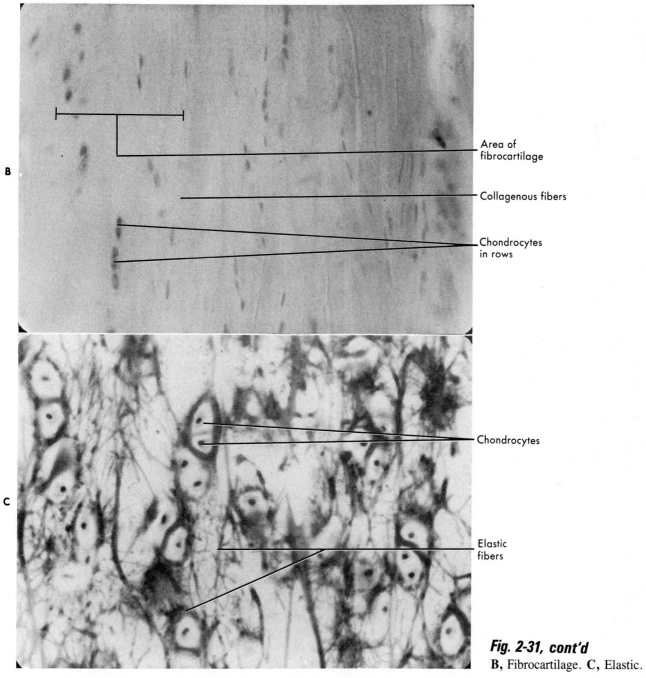

Fig. 2-31, cont'd
B, Fibrocartilage. **C,** Elastic.

tilage (except articular) cartilage in synovial joints) is covered by a membranous **perichondrium.** This membrane contains cells that can form new cartilage on the surface and increase the bulk of the tissue.

Three varieties of cartilage are recognized:

hyaline cartilage (gristle) is, in fresh section, a whitish, translucent, slightly compressible mass. The interterritorial matrix *appears* to lack fibers, but collagenous fibers may be demonstrated with special techniques. This type of cartilage forms a major portion of the embryonic and fetal skeleton, covers the ends of bones (articular cartilage) in synovial joints, forms the anterior nasal septum, and forms the costal, tracheal, and several laryngeal cartilages.

fibrocartilage is a very tough cartilage containing many visible collagenous fibers in its matrix. It forms the intervertebral discs of the spinal column and the pubic symphysis.

elastic cartilage is a flexible cartilage containing large amounts of elastic fibers. It forms the pinna of the outer ear and some of the nasal and laryngeal cartilages (for example, epiglottis).

Growth of cartilage, that is, increase in its mass, occurs by two methods. By **interstitial growth,** or growth from within, the formation of territorial matrix occurs. Obviously, little increase in mass can occur by this method, else the chondrocytes would be demolished. **Appositional growth,** or addition of new cartilage by perichondrial activity on the *surface* of the mass, accounts for most growth of cartilage.

If damaged, cartilage heals slowly, probably because of its poor blood supply.

FORMATION AND PHYSIOLOGY OF CONNECTIVE TISSUE PROPER AND CARTILAGE. Since the greater mass of a connective tissue proper or a cartilage consists of intercellular material, the function and integrity of the tissue depend on proper formation of the intercellular material. Ribosomes and the Golgi bodies of fibroblasts are instrumental in formation of the proteins that polymerize into the fibers of connective tissues proper. In cartilage, chondrocytes perform a similar function.

Both groups of tissue are greatly influenced by growth hormone (from the pituitary gland) for setting the rate of growth of the tissues. In hyaline cartilage that will eventually become bone, parathyroid hormone (from the parathyroid glands) and calcitonin (from the thyroid gland) influence the removal and deposition (respectively) of inorganic material in the tissue. Vitamins *A* and *C* are essential for formation of fibers and ground substance, whereas vitamin *D* ensures absorption of inorganic calcium and phosphate from the gut for eventual formation of bone.

If collagenous fibers are produced in excessive quantities because of a genetic abnormality of their production, *scleroderma* may result. Particularly evident in skin and blood vessels, the collagenous fibers render such tissues tough and inflexible. *Achondroplasia* (chondrodystrophy) is a condition in which cartilage matrix is not properly formed, and the long bones of the skeleton are misshapen and short. It is also a genetic disorder.

BONE. Bone is a connective tissue with an intercellular material that is *hard* because of the deposition of inorganic material in the matrix by the process of *calcification.* The typical cells, **osteocytes,** lie in **lacunae** in the matrix. Tiny canals, or **canaliculi,** connect neighboring lacunae for transport of nutrients and waste removal.

There are several types of bone distinguished on the basis of *arrangement* and not on the basis of any chemical differences. All types consist of about 65% inorganic matter, chiefly $[Ca_3(PO_4)_2 \cdot Ca(OH)_2]$ and 35% organic matter, that is, cells and *bone collagen* fibers.

Cancellous or *spongy* bone (Fig. 2-32) consists of interlacing plates or bars of bony material, leaving many spaces between the bone, hence the spongy name and appearance. Spongy bone forms the interior of the ends of long bones and the interior of flat and irregularly shaped bones (skull, vertebrae).

Compact bone (Fig. 2-33) is ivorylike in its density, forms the shafts of the long bones of the body, and consists of longitudinally oriented cylindrical subunits of structure called **osteons** or **haversian systems.** A cross-section of an osteon shows a centrally placed *haversian canal* containing blood vessels, concentric layers of bony tissue called *lamellae* (singular, lamella), *lacunae* containing *osteocytes,* and the *canaliculi* connecting lacunae with one another and with the haversian canal. The blood vessels of the haversian canals are derived from blood vessels out-

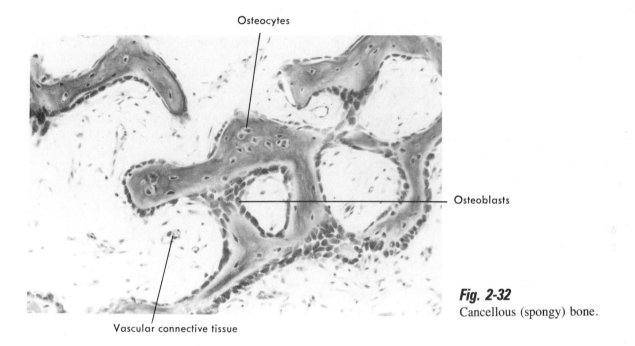

Fig. 2-32
Cancellous (spongy) bone.

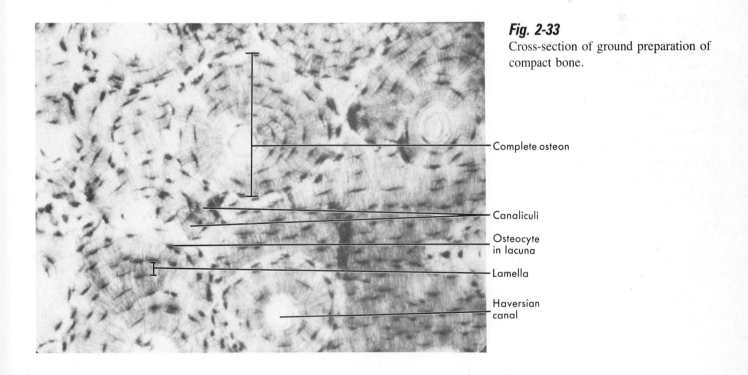

Fig. 2-33
Cross-section of ground preparation of compact bone.

side the bone that enter the bone through **Volkmann's canals** (Fig. 2-34). The blood vessels also reach the bone interior (spongy bone or the marrow cavity of long bones), and thus blood cells produced in the *bone marrow* may be delivered to the circulation.

A type of compact bone sometimes called **dense** or **lamellar** bone lacks osteons and forms the *outer circumferential lamellae* covering the outer surfaces of all bones. *Inner circumferential lamellae* of dense bone line the walls of the marrow cavity of long bones.

A fibrous **periosteum** covers the dense bone on all except articular surfaces. A much thinner **endosteum** lines the marrow cavity of long bones. Cells in these membranes can become bone-forming cells to heal fractures and deposit new bone on surfaces.

Bone is constantly remodeled, especially during youth, and sometimes incomplete osteons in the compact bone may be seen. Such osteons are called **interstitial lamellae.**

Two additional types of cells other than osteocytes are found in bony tissue. **Osteoblasts,** usually found in rows on surfaces of spongy and dense bone, are *bone-forming cells*. They calcify newly forming bone as one of their duties. **Osteoclasts** are giant (90 to 100 μm), multinucleated cells found on bone surfaces. They are *bone-destroying cells*. Thus bone remodeling occurs by the activity of these two cell types.

Table 2-6 summarizes facts about the connective tissues.

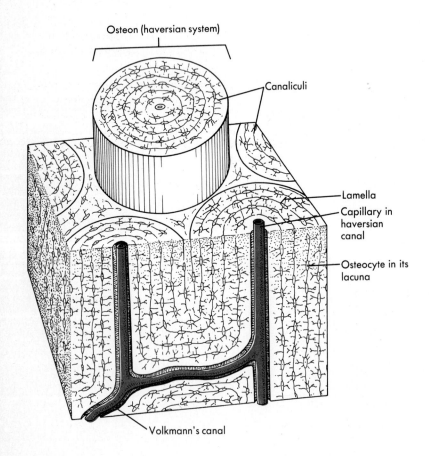

Fig. 2-34
Diagram of longitudinal section of bone to show Volkmann's canals.

Table 2-6. Summary of connective tissues (ct)

Adult, embryonic, or fetal tissue	Consistency of intercellular material (icm) and tissue group	Cell characteristic of the tissue	Tissue name	Other cells common in the tissue	Characteristics	Typical locations
Embryonic and fetal	Fluid	Mesenchyme (mesodermal)	Mesenchyme	None	Stellate cells in homogeneous icm	Under skin of embryo; around blood vessels
	Fluid, with collagenous fibrils	Modified mesenchyme	Mucous ct (Wharton's jelly)	None	Flattened cells in fibrillar icm	Umbilical cord
Adult	Fluid—connective tissue proper	Fibroblast	Loose (areolar) ct	Macrophage, white blood cells, fat cells	Feltwork of collagenous and elastic fibers	Around organs, hypodermis of skin, mesenteries
			Dense irregular ct	As above	Feltwork of large collagenous and elastic fibers, very compact tissue	Dermis of skin, submucosa of gut
			Collagenous (fibrous) ct	None	Oriented dense mass of collagenous fibers; fibroblasts flattened between fibers	Ligaments, tendons, sclera of the eye
			Elastic ct	None	Branching network of elastic fibers	Middle layer of aorta, some ligaments of spinal column
			Reticular ct	Reticular cell	Fine, irregular branching fibers	Framework of spleen, liver, lymph nodes
			Adipose tissue	Fat-storing fibroblasts	Large cell with stored fat droplet	Hypodermis of skin, around organs
			Pigmented ct	Melanocytes	Loose ct with many pigment-laden melanocytes in it	Iris and chorioid of eye
	Semisolid—cartilages	Chondrocyte	Hyaline cartilage	None	Matrix *appears* devoid of fibers; chondrocytes in lacunae	Embryonic skeleton, costal and some nasal cartilages, articular surfaces
			Fibrocartilage	None	Visible collagenous fibers in matrix	Intervertebral discs, symphyses
			Elastic cartilage	None	Visible elastic fibers in matrix	Epiglottis, pinna
	Hard—bone	Osteocyte	Spongy or cancellous	None *in* bony tissue	Joined plates and bars of tissue	Bone ends and interiors
			Lamellar bone	As above	Layers of dense bony tissue	Outer surfaces of all bones
			Compact bone	As above	Osteons present	Interior of shafts of long bones

Muscular tissues

Muscular tissues (Fig. 2-35) are the contractile tissues of the body. Because of their ability to shorten, they are able to do work. Details of the structure of muscular tissues will be found in later chapters. Here, it should be pointed out that there are three types of muscular tissues found in our bodies:

skeletal (striated) muscle forms the so-called muscular system. It attaches to the skeleton or connective tissue components of the body and moves the skeleton to achieve motion of body parts. It is also responsible for facial expression, breathing, and speech. It is a *voluntary* type of muscle, a term indicating that it normally depends on nervous stimulation to contract.

cardiac muscle occurs only in the heart. Contraction of this type of muscle causes circulation of blood throughout the body. It is an *involuntary* type of muscle, not dependent on outside nervous stimulation for its contraction.

smooth (visceral) muscle is found as a major component of the walls of many hollow body organs. In areas such as the

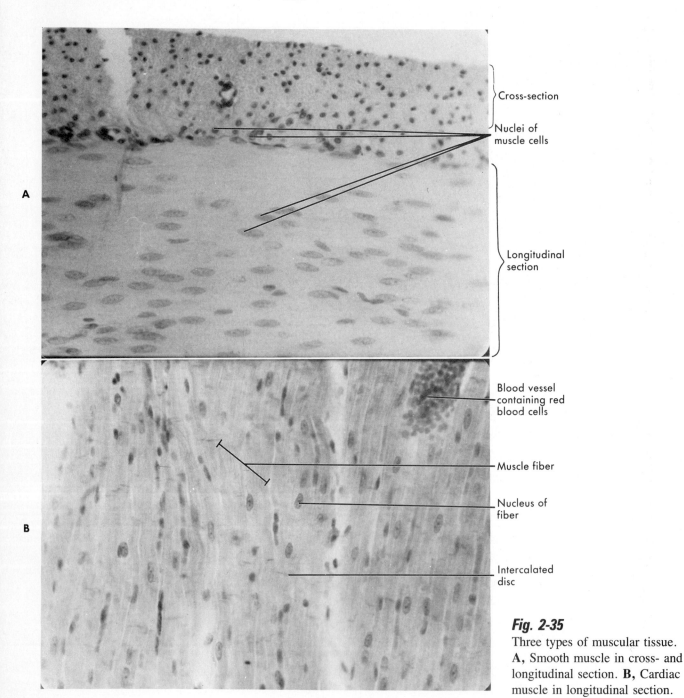

Fig. 2-35
Three types of muscular tissue. **A,** Smooth muscle in cross- and longitudinal section. **B,** Cardiac muscle in longitudinal section.

digestive, urinary, and reproductive systems, contraction of the muscle propels materials through the organ. In blood vessels and respiratory organs, smooth muscle is primarily circularly oriented, and contraction causes a change in the diameter of the organ. Thus change in blood pressure or volume of the organ may be achieved.

Smooth muscle found in the areas mentioned is called *unitary smooth muscle* and is primarily involuntary; its contraction is not due to external nerves but may be modified by nerves. In areas such as the iris of the eye, contraction *does* depend on nerves, and the muscle is called *multiunit smooth muscle*. Reflex control of light entering the eye is the result.

Nervous tissue

Nervous tissue (Fig. 2-36) is the highly *excitable* and *conductile* tissue of the body. Again, more detailed descriptions of this tissue will be found in later chapters. Nerve cells are typically provided with long cytoplasmic extensions or *processes*. These processes enable nerve impulses to be conducted to areas of the nervous system for integration and analysis and to be carried to the body areas that respond to give movement or secretory activity.

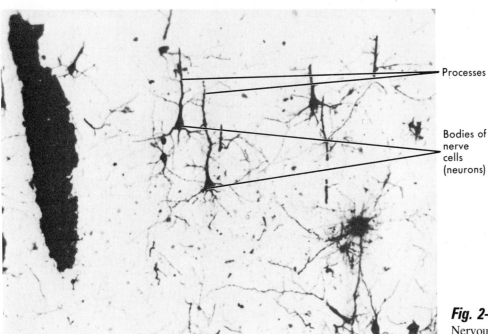

Fig. 2-35, cont'd
C, Skeletal muscle in longitudinal section.

Fig. 2-36
Nervous tissue of cerebrum.

Organs

Two or more types of tissue organized in a specific way constitute an *organ*. Additionally, individual body organs have specific tasks to perform in the body. Space precludes a complete listing of all body organs and their tasks. Several examples should serve to illustrate the organ concept:

The heart is composed of (at least) epithelial, connective, and muscular tissues and has the task of pumping blood through the body.

The stomach contains members of all four tissue groups, stores eaten foods, and carries out partial digestion of those foods.

The kidney contains epithelium and connective and muscular elements; it is responsible for the elaboration of urine and control of blood composition.

Systems

Two or more organs are joined together to form a *system* that has a broader function or process to perform than the organs composing the system.

According to the authority cited, the number of systems forming the body may be from 11 to as many as 13. The following list of systems and the major organs composing each reflects a functional as well as an anatomical organization:

urinary system (Fig. 2-37), including the kidneys, ureters, urinary bladder, and urethra, provides for regulation of blood composition and elaboration of urine.

endocrine system (Fig. 2-38) consists of several glandular structures producing chemicals known as *hormones*. These hormones are secreted into the bloodstream and have the potential to influence all body cells. Processes such as growth, metabolism, and reproduction are controlled (in part) by hormones.

circulatory system (Fig. 2-39), with the heart, blood vessels, and blood, interconnects all other systems, providing nutrients and controlling chemicals and removing products of cellular activity.

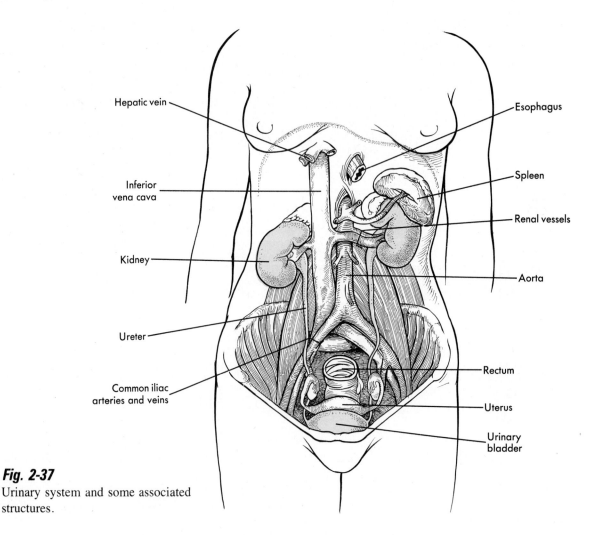

Fig. 2-37
Urinary system and some associated structures.

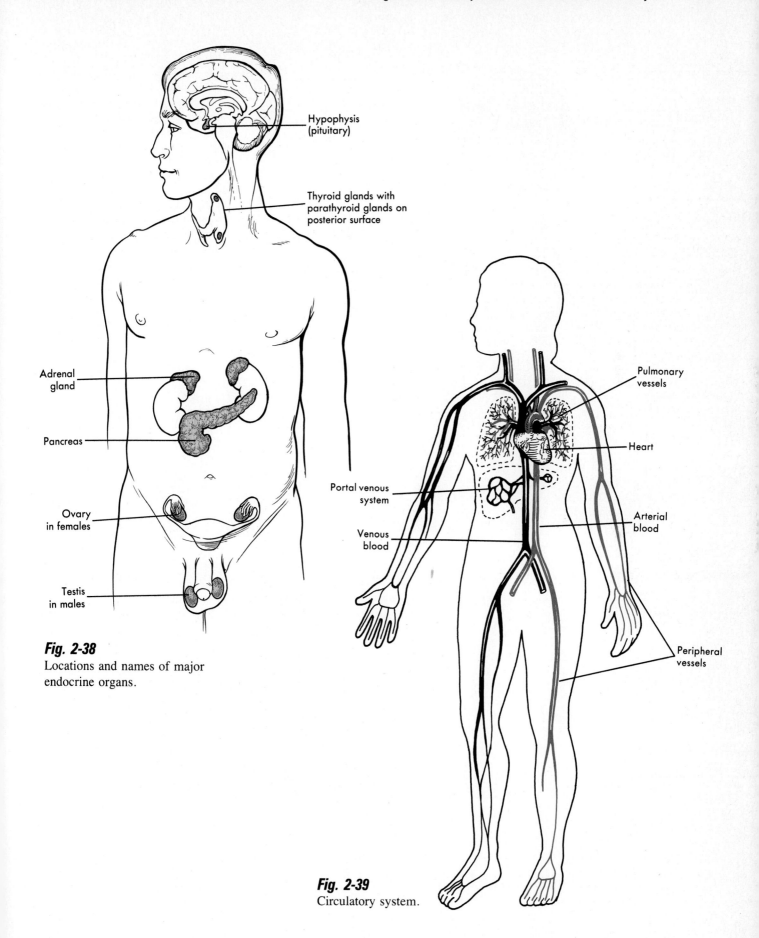

Fig. 2-38
Locations and names of major endocrine organs.

Fig. 2-39
Circulatory system.

respiratory system (Fig. 2-40), including the nasal cavities, larynx, trachea, and lungs, along with the muscles of respiration, provides for acquisition of oxygen and elimination of carbon dioxide.

nervous system (Fig. 2-41) consists of the *central nervous system,* which includes the brain and spinal cord, and the *peripheral nervous system,* which includes all nerves, ganglions, plexuses, and other nervous tissue outside the central nervous system. This system is one of the major controlling systems of the body.

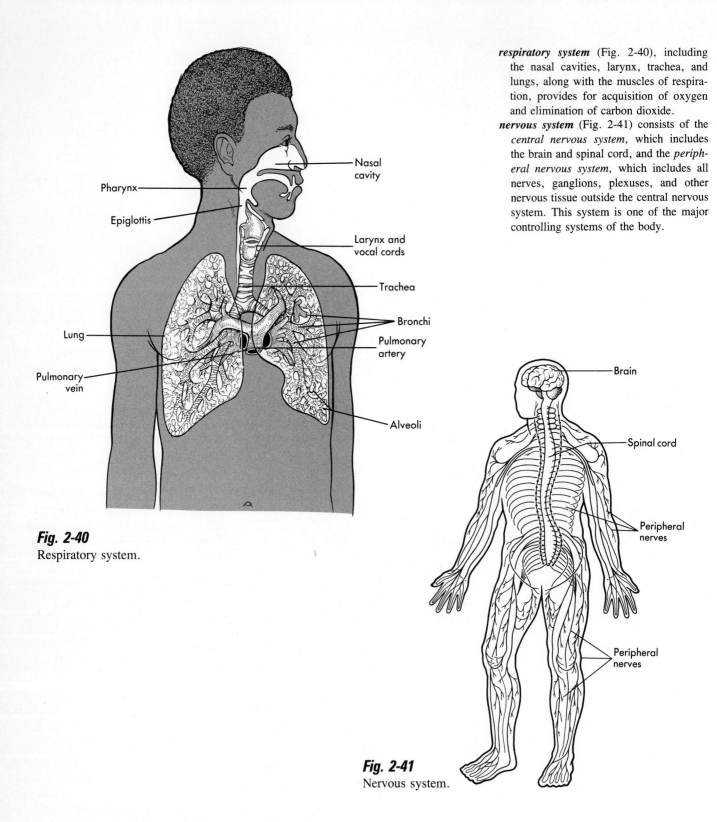

Fig. 2-40
Respiratory system.

Fig. 2-41
Nervous system.

digestive system (Fig. 2-42) includes such organs as the mouth, esophagus, stomach, intestines, liver, and pancreas and is responsible for the processes of digestion, absorption, and elimination of food residues.

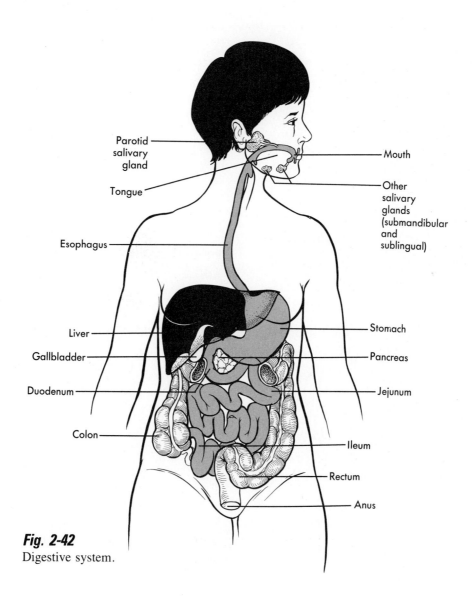

Fig. 2-42
Digestive system.

integumentary system (Fig. 2-43) includes the skin and its derivatives such as hair, nails, and skin glands and is a major protective and temperature-regulating system for the body.

reticuloendothelial (RE) system (immune system) consists of free and fixed cells in the body that are capable of engulfing (phagocytosis) foreign chemicals, microorganisms, and products of cellular destruction and that produce antibodies to create immunity to certain diseases. The cells are scattered throughout the body and form a *functional* system.

lymphatic system (Fig. 2-44) consists of vessels (lymphatics), a fluid (lymph), and lymph organs such as spleen, tonsils, and lymph nodes. A source of RE and certain white blood cells, the system also returns tissue fluids to the circulation.

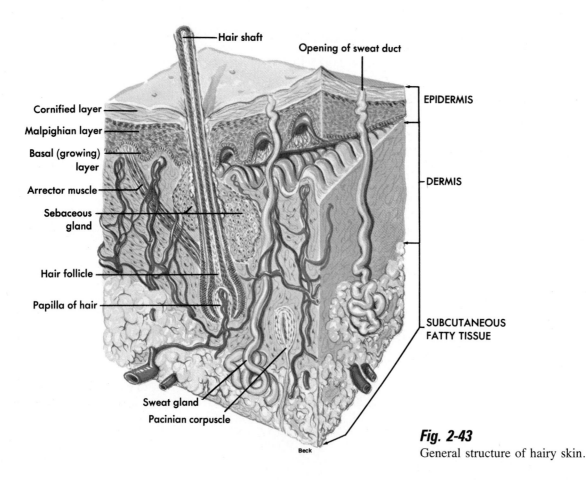

Fig. 2-43

General structure of hairy skin.

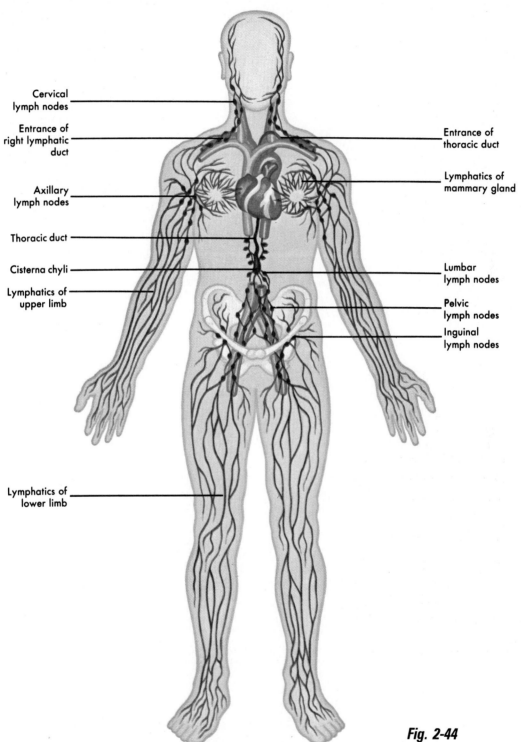

Fig. 2-44
Some organs of lymphatic system.

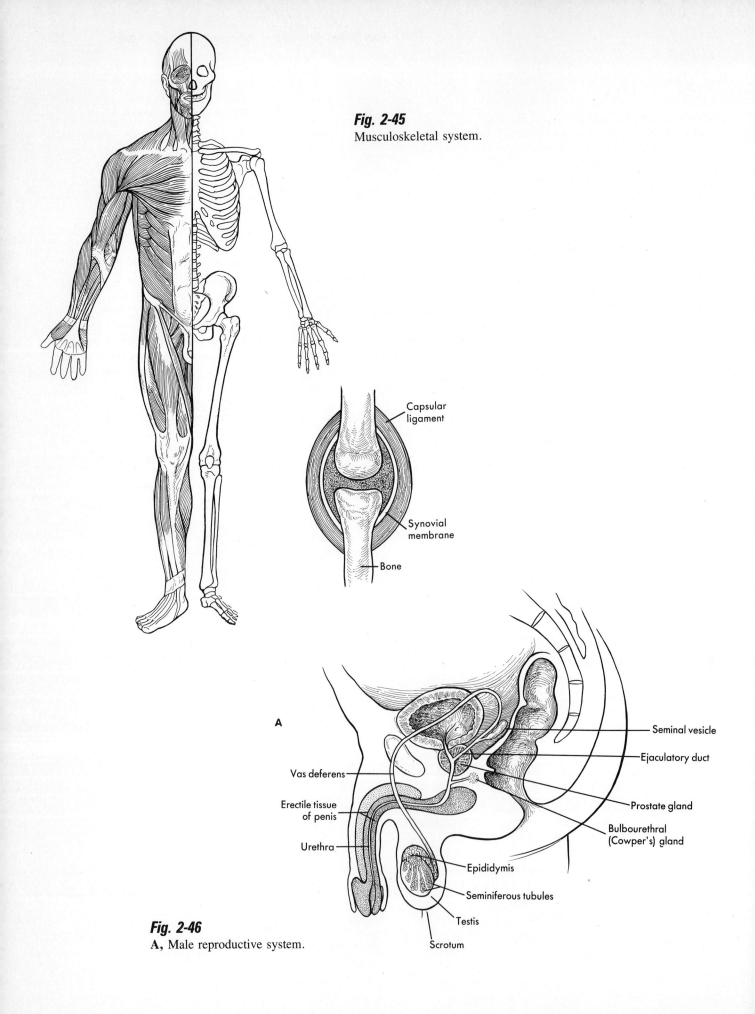

Fig. 2-45 Musculoskeletal system.

Fig. 2-46 A, Male reproductive system.

muscular and *skeletal systems* (Fig. 2-45) work together to support, protect, and move the body. A tissue called *bone marrow* is found in many bones and is a major source of blood cells.

reproductive systems (Fig. 2-46) in the male and female include many diverse organs designed to provide sex cells (sperm and ova) for perpetuation of the species and development and nourishment of offspring. *Hormones* essential to growth and development are produced by the testes, the ovaries, and, in the case of pregnancy, the placenta.

The whole organism

The body systems collectively form the human organism. All work together to achieve optimal body function, and although we may be able to function without certain organs (for example, a lung, one kidney), normal function depends on interdependence. Some systems or organs are vital to survival (heart, brain), and if they fail, so does the organism as a whole. The term *homeostasis* aptly summarizes the major aim of the cooperating systems of the body as they act to control body functions within limits that ensure survival. Homeostasis is a state of dynamic equilibrium in which basic structure and function remain essentially constant. We can appreciate the necessity of maintaining temperature, pH, water content, and other factors so that normal function can continue. Examples of the operation of homeostasis will be mentioned in later chapters.

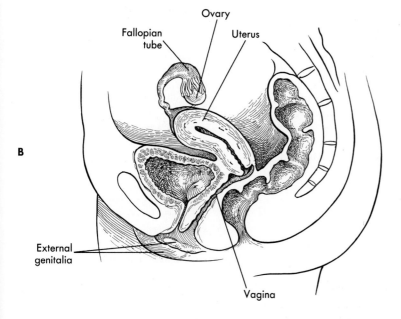

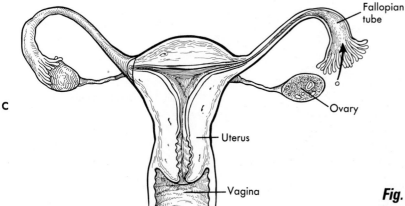

Fig. 2-46, cont'd
B and **C**, Female reproductive systems.

Summary

1. There are several levels of organization in the body; the first two include the following:
   a. The atomic and molecular levels consist of chemical elements and their organization into molecules and compounds that form the basis of body structure. Thirteen elements may be considered most important.
   b. The cellular level consists of complex molecules organized into membrane systems, composed of protein and lipid that limit the cell and control passage of substances; the cytoplasm with its organelles (ER that distributes materials, ribosomes that manufacture protein, mitochondria that produce ATP, Golgi complex that packages products, lysosomes that digest molecules, central body involved in cell division, peroxisomes that metabolize H_2O_2, and cilia and flagellae that move cells or substances); and inclusions (glycogen, lipid, crystals, vacuoles). The nucleus controls cellular activity by protein synthesis. It contains DNA and RNA that determine the type and amount of proteins synthesized.
2. Two processes are involved in production of cells for repair and replacement and for species perpetuation.
   a. Mitosis occurs in five stages and results in the production of two cells genetically like the parent cell. Examples of its occurrence include replacement of skin cells and red blood cells.
   b. Uncontrolled mitosis results in the development of neoplasms (cancers).
   c. Meiosis is a process occurring in the production of sex cells (ova, sperm). It produces cells containing one half the normal species chromosome number, and the four resulting cells show a different genetic potency than does the parent cell.
3. Development of a new human organism begins with fertilization of an ovum by a sperm. Further events of importance may be considered on a weekly basis.
   a. During the first week cleavage of the cells of the zygote produces a morula that develops into a blastocyst. In the blastocyst, a trophoblast and an inner cell mass are formed.
   b. During the second week the blastocyst embeds in the uterine wall and develops amniotic and yolk sac cavities. The two layers of cells between the cavities are the ectoderm and the endoderm and mark the site of embryo development.
   c. The third week sees the development of a third germ layer called the mesoderm.
   d. During the fourth week somites form, and the embryo begins to assume human form.
   e. From the fifth to eighth weeks all major body systems form, and the organism is called a fetus.
   f. Fetal membranes form: the chorion becomes part of the placenta; the amnion contains the amniotic fluid in which the fetus floats; the allantois contributes to the formation of the umbilical cord; the yolk sac forms blood and other cells.
4. The tissue level is composed of cells similar in structure and function and their associated intercellular material. Four tissue groups compose the body.
   a. Epithelia cover and line body organs.
   b. Connective tissues connect, support, and protect.
   c. Muscular tissues contract or shorten and do work.
   d. Nervous tissues carry bioelectrical impulses that cause body activity to occur.
5. Epithelia have the following characteristics:
   a. They consist of closely packed cells, with minimal amounts of intercellular material.
   b. They are not supplied with blood vessels *in* the epithelium (avascular).
   c. They may contain nerve cells (sensory epithelium).
   d. They usually possess a basal or basement membrane.
   e. They give rise to a variety of glandular structures.
   f. They are usually named according to number of layers of cells (one layer, simple; more than one layer, stratified), shape of cells, (squamous or flat, cuboidal, columnar, or tall), or other special features.
6. Types of epithelia include the following:
   a. Simple squamous epithelium consists of one layer of flattened cells; it forms endothelium (vascular system linings) and mesothelium (linings of "true" body cavities). It is the exchange epithelium of the body.
   b. Simple cuboidal epithelium consists of one layer of cube-shaped cells; it is common in kidney tubules, where absorption and secretion are carried out.
   c. Simple columnar epithelium consists of one layer of elongated cells; it typically forms the lining of the alimentary tract and is an absorbing epithelium.
   d. Stratified squamous epithelium consists of several layers of cells with the top layer or two composed of flattened cells.
   e. Stratified cuboidal epithelium consists of several layers of cells with the top layer cube shaped.
   f. Stratified columnar epithelium consists of several layers of cells with the top layer of columnar cells.
   g. Pseudostratified epithelium consists of cells that all reach the basal membrane but do not all reach the surface. It is common in respiratory and male reproductive systems, where something is moved by ciliary action.
   h. Transitional epithelium is a stratified tissue in which the top layer of cells is varied in shape. It is found in the urinary system.
   i. Syncytial epithelium (placenta), germinal epithelium (testes), and neuroepithelium (special sense organs) are specialized epithelia.
7. Epithelia may have special modifications of or at their surfaces.
   a. Cilia and flagella are structures associated with movement of substances or the whole cell.
   b. Microvilli are minute cytoplasmic extensions of the cell surface to increase surface area.
   c. Transitional epithelium possesses a surface sheet of tubular units that may permit stretching (urinary bladder).

d. Several devices hold epithelial cells together or seal their free surfaces.
8. A combination of an epithelium and its underlying connective tissue forms an epithelial membrane. Serous membranes line true body cavities and cover internal body organs. They are moistened by a watery (serous) secretion. Mucous membranes line body cavities opening on the body surface and are moistened by mucus.
9. Glands originate from epithelial surfaces.
   a. Endocrine glands have lost their epithelial connection and place their product(s) into the bloodstream.
   b. Exocrine glands use ducts to empty their products onto an epithelial surface. They are classified according to number of ducts, shape of secretory portion, and method of production of secretion.
10. Connective tissues are characterized by the following:
    a. They usually do not occur on surfaces.
    b. They have abundant intercellular material and scattered cells.
    c. They are usually fibrous.
    d. They are usually vascular.
    e. They connect, support, and protect body parts.
    f. They are classified by consistency of intercellular material and by type of fiber predominating in the tissue.
11. There are many varieties of connective tissue (ct).
    a. Mesenchyme and mucous ct are found only in the embryo or fetus.
    b. Ct proper has a more or less fluid intercellular material and includes loose (areolar) ct collagenous ct, elastic ct, irregular ct, reticular ct, and adipose tissue.
    c. Cartilages are tougher cts because of a semisolid intercellular material and occur in hyaline, fibrous, and elastic varieties.
    d. Formation of ct requires cells to produce their fibers, vitamins (A, C, D), and hormones (growth, parathyroid hormone, calcitonin).
    e. Bone has been calcified and occurs in spongy and compact varieties. The latter may or may not contain osteons (haversian systems).
12. Muscular tissue has the following characteristics:
    a. It is contractile.
    b. It occurs in three varieties, termed skeletal (striated), cardiac, and smooth (visceral).
13. Nervous tissue is characterized by the following:
    a. It is conductive and highly excitable.
    b. It contains typically elongated cells to carry new impulses throughout the body.
14. Organs consist of two or more tissues organized for a specific job.
15. Systems consist of two or more organs carrying out a body process. This text lists 13 systems, which are presented on pp. 56-63.
16. Systems form interdependent parts of the whole organism.

Questions

1. List the various levels of body organization, characterizing each as to structure and tasks.
2. If you had to "invent" a generalized cell, what organelles and other structures would you provide to ensure that the unit could survive? Justify each choice.
3. Compare and contrast mitosis and meiosis as to nature and number of cells resulting from each process.
4. What events characterize the stages of cell division?
5. Define a tissue and list the tissue groups forming the body.
6. Contrast epithelia and connective tissues for general characteristics.
7. For each of the following statements, name the tissue described and give a body location and function for the tissue.
   a. A one-layered tissue consisting of cells having little height
   b. A multilayered tissue permitting distention of the organ it lines
   c. A tissue appearing to be multilayered but actually not
   d. A multilayered, primarily mechanically resistant tissue
   e. A tissue serving as a "packing material"
   f. The strongest connective tissue proper
   g. A tissue forming an internal framework for many body organs
   h. A connective tissue that protects and serves as a lever
   i. A tissue specialized for fat storage
8. What would be implied by the following statements or words?
   a. Simple, branched tubular, merocrine gland
   b. Compound tubuloalveolar gland
   c. Histiocyte
   d. Osteoclast
   e. Periosteum
   f. Interstitial growth
   g. Microvilli
   h. "Tight junction"
9. What are the major "weekly developments" in the first 2 months of embryonic life?

Readings

Allison, A.: Lysosomes and disease, Sci. Am. **217**:62, Nov. 1967.
Bloom, W., and Fawcett, D.W.: A textbook of histology, ed. 10, Philadelphia, 1975, W.B. Saunders Co.
DeRobertis, E.D.P., and DeRobertis, E.M.F., Jr.: Cell and molecular biology, Philadelphia, 1980, W.B. Saunders Co.
Fawcett, D.W.: An atlas of fine structure: the cell, Philadelphia, 1966, W.B. Saunders Co.
Ham, A.W., and Cormack, D.H.: Histology, ed. 8, Philadelphia, 1979, J.B. Lippincott Co.
Hopkins, C.R.: Structure and function of cells, Philadelphia, 1978, W.B. Saunders Co.
Kessel, R.C., and Kardon, R.H.: Tissues and organs: a test-atlas of scanning electron microscopy, San Francisco, 1979, W.H. Freeman & Co., Publishers.
Moore, K.L.: The developing human, Philadelphia, 1973, W.B. Saunders Co.
Neutra, M., and LeBlond, C.P.: The Golgi apparatus, Sci. Am. **220**:100, Feb. 1969.
Nomura, M.: Ribosomes, Sci. Am. **221**:28, Oct. 1969.
Reith, E.J., and Ross, M.H.: Atlas of descriptive histology, ed. 3, New York, 1979, Harper & Row, Publishers.
Sanborn, E.B.: Light and electron microscopy of cells and tissues New York, 1972, Academic Press, Inc.

3
External appearance of the body

Objectives
Statement of purpose
External appearance of the body at different ages
Head
Neck
Upper appendage
   Shoulder and axilla
   Arm
   Forearm
   Wrist and hand
Thorax and abdomen
Lower appendage
   Hip
   Thigh
   Leg
   Ankle and foot
Summary
Questions
Readings

OBJECTIVES After studying this chapter, the reader should be able to:

Explain the practical importance of surface anatomy as learning and its significance to those who deal with the human body.

Describe how the external appearance of the body changes with age and nutritional state.

Indicate the significant differences in external appearance of mature male and female bodies.

Point out the major bony, muscular, vascular, and nervous landmarks on the head, neck, trunk, upper appendages, and lower appendages.

Outline the triangles of the neck and the important anatomical features contained within each triangle.

Comment on the use of superficial blood vessels as pressure points to control bleeding.

Statement of purpose

Having considered the "building blocks" of the body and its development, it is perhaps good to conduct an inspection of the external appearance of the body at different ages. This inspection will emphasize that changes occur with age and will also underline those body structures that can be palpated (felt) through the skin surface. Medical personnel and those trained in first aid may use such body landmarks in their own right as guides to the location of deeper lying structures or as devices (for example, pressure points) to stem bleeding or locate pain sources.

In describing the various body regions, the reader may find it appropriate to review Fig. 1-2 to refresh the memory as to the names of the body regions and should also remember that terms reflecting direction assume the body is in anatomical position. In general there are five major body regions to be described, some having subdivisions:

   Head
   Neck
   Upper appendage
      Shoulder
      Axilla
      Arm
      Forearm
      Wrist and hand
   Thorax and abdomen
   Lower appendage
      Hip
      Thigh
      Leg
      Ankle and foot

Skeletal, muscular, nervous, and vascular landmarks will be described for these several areas.

External appearance of the body at different ages

It is a statement of obvious fact that external body appearance depends on age, sex, and state of nutrition. Premature infants (Fig. 3-1) bear the appearance of fetuses because they are born before they have accumulated the typical thick subcutaneous layer of adipose tissue characteristic of full-term fetuses (Fig. 3-2). The skin of the premature infant thus appears to hang loosely on the body. The full-term infant has a chubby appearance because of overall deposition of fat, with concentrations of fat in the cheeks (buccal fat pad), abdomen, and buttocks. As growth occurs, fat distribution tends to become more even over the body (assuming normal nutrition), and the older child usually presents the appearance shown in Fig. 3-3. At the time of puberty and

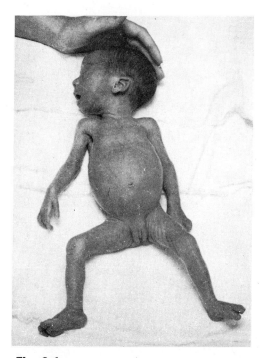

Fig. 3-1
Premature fetus. Note paucity of skin fat and "looseness" of skin.

From Ingalls A.J., and Salerno, M.C.: Maternal and child health nursing, ed. 4, 1979, The C.V. Mosby Co.

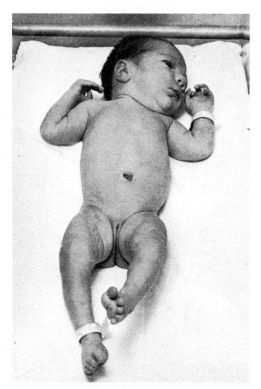

Fig. 3-2
Full-term infant. Note smooth body contours resulting from well-developed subcutaneous fat layer.

From Ingalls, A.J., and Salerno, M.C.: Maternal and child health nursing, ed. 4, 1979, The C.V. Mosby Co.

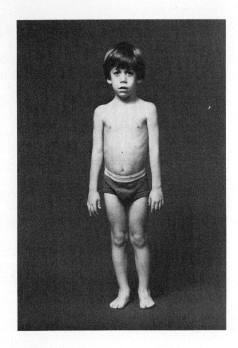

Fig. 3-3
Body development in 6-year-old boy. Body contours exhibit "sexless" appearance.

adolescence (Fig. 3-4), sex differences become apparent in terms of fat distribution on the body. The female shows accumulation of adipose tissue in particular regions of the body, including breasts, shoulders, buttocks, inner and outer sides of the thighs, lower abdomen, and over the pubic area. The adult female body is shown in Fig. 3-5. Other than the obvious genital differences, the male body, beginning at adolescence, tends to show a more angular appearance, as the fat pads are thinner and more evenly distributed over the body (unless there is gross overweight). Muscular development is greater in the male, and the adult male (Fig. 3-6) usually shows a greater upper body development than the female. With maturity also comes a different hair distribution over the body. Males are, in general, hairier than females, at least in the visibility of the hair. Particularly evident is the sex distribution of

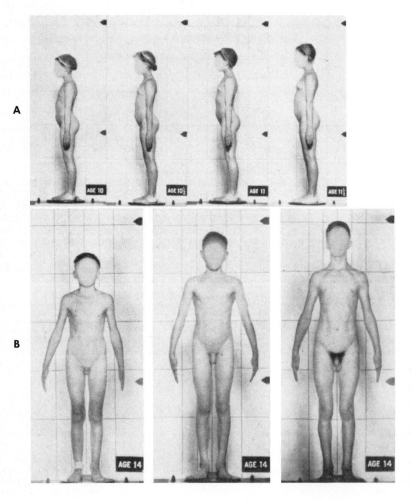

Fig. 3-4
Body development in adolescents. **A,** Female development in same individual taken at 6-month intervals. **B,** Male development in three different individuals of same age to show degrees of development and normal progression toward maturity.

3 External appearance of the body 69

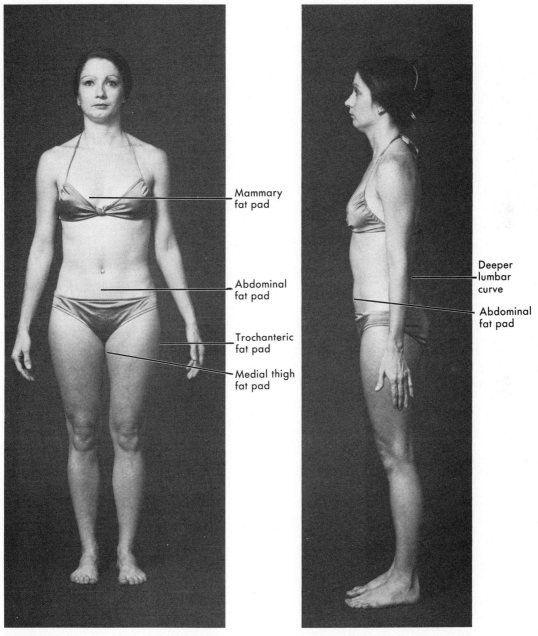

Fig. 3-5
Adult female.

70 Human anatomy

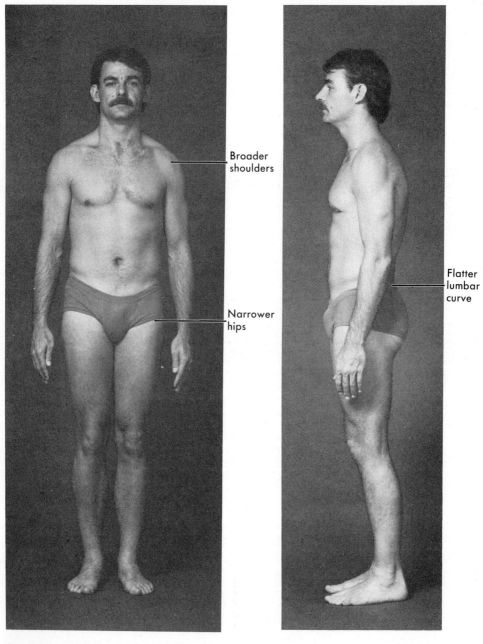

Fig. 3-6
Adult male.

facial, pubic, and abdominal hair. In the female, pubic hair is usually in the form of an inverted delta (∇) that does not extend far up the abdomen. Male pubic and abdominal hair is diamond shaped in distribution (◇), usually extending to the umbilicus or higher. Axillary hair is usually about equally abundant in both sexes but is commonly removed by females.

In middle age there is a tendency to overweight, as food consumption remains the same or increases, and activity levels decrease. Old age (Fig. 3-7) is commonly associated with loss of fat and loss of elastic fibers of the skin. The skin again tends to hang loosely on the body.

If body appearance over the whole life span is considered, there seems to be a "return to fetal status" as to skin folds and teeth (in many individuals).

Fig. 3-7
Aged human. Facial muscles tend to wrinkle skin at right angles to their lines of pull. Loss of elastic fibers and fat in skin accentuate these wrinkles.
From Kaluger, G., and Kaluger, M.F.: Human development, ed. 2, 1979, St. Louis, The C.V. Mosby Co.

Head

The upper or *cranial* portion of the head (Fig. 3-8) presents an outline determined by the bony structures of the skull. Outlines may be modified by length and style of the hair, but very little or no fat is found in the skin of the scalp, the skin lying almost directly on the bones. The *frontal eminences* are slight rounded prominences to either side of the midline of the upper forehead. *Parietal eminences* also occur on either side of the midline about two thirds of the way posterior on the upper cranium. Continuing posteriorly, the *external occipital protuberance* may be felt on the lower posterior aspect of the cranium. On the lateral aspect of the skull, the *external auditory meatus* (mē-ā′-tus) marks the opening of the ear. The *zygomatic arch* may be felt as a ridge of bone extending from the meatus to the "cheek." Behind and below the external ear's pinna (auricle), the *mastoid process* may be appreciated. In front of and about level with the upper third of the pinna, the pulsations of the *superficial temporal artery* may be felt. The vessel is easily compressed against the underlying bone, and pressure here is often used to control scalp bleeding. Clenching the jaws causes the *temporalis* muscle to bulge on the lateral skull above the zygomatic arch. Simultaneously, the *masseter* bulges on the lateral side of the mandible (lower jaw).

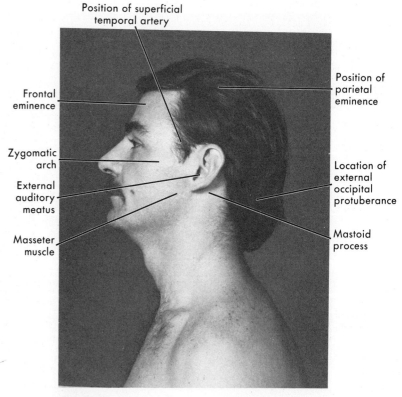

Fig. 3-8
Some anatomical landmarks on cranium.

On the face (Fig. 3-9), bony landmarks are somewhat obscured by muscle and fat. Just superior to the medial margins of the *orbits* (housing the eyeballs) are the *superciliary ridges*. They are usually larger in males than in females (sometimes giving a "beetle-browed" appearance) and form one of several criteria used to determine the sex of skeletons. Internal and deep (inside the frontal bones) to the ridges are the *frontal sinuses*, cavities within the bone. "Colds" or "sinus problems" may lead to inflammation of the sinuses' linings and cause headaches. The *zygomatic* bones form the "cheek." Between the cheek and orbit, the *infraorbital nerve* exits from the skull to supply the skin of the midface, anterior teeth, and nose with sensory fibers. The infiltration of an anesthetic in this area may be carried out for purposes of oral surgery or tooth repair. At the superior midorbital margin the *supraorbital nerve* exits from the skull to supply the respective (left or right) anterior quadrant of the scalp with sensory nerves. In the infant the *anterior fontanel* ("soft spot") may be seen as a slight depression above the forehead in the midline. It is an area not yet filled in with bone. Observation of the fontanel may indicate increased intracranial pressure (bulging) or dehydration (greatly depressed). The fontanel normally bulges when the infant cries. It usually closes by 18 months of age. The lower face shows the *chin, lips, nasolabial fold* (from the nose to the corners of the mouth), and *philtrum*, the broad medial vertical groove running between the upper lip and the nose. Disappearance of the nasolabial fold may result from nerve damage to muscles of that area. On the mandible, pulsations of the *external maxillary artery* may be appreciated about an inch anterior to mandibular angle. The artery carries blood to the side of the face, and pressure on it at the mandible may control facial bleeding.

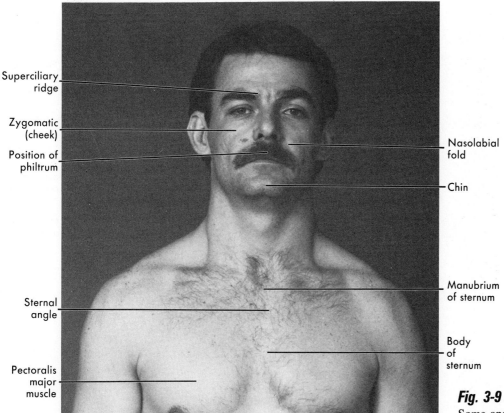

Fig. 3-9
Some anatomical landmarks on face and thorax.

Neck

The bony parts of the neck (Fig. 3-10) are well covered by muscle and fascia (connective tissue planes between muscle layers) and are difficult to palpate. In the *posterior midline* of the neck, at its base, a protuberance may be felt. It is the **spinous process of the seventh cervical vertebra (vertebra prominens)**. Knowing the number of this vertebra makes it possible to "count" the spinous processes, up or down, of the other vertebral spines to determine vertebral levels. In the *anterior midline* of the neck the **body of the hyoid bone** may be felt on a level with the lower margin of the mandible (lower jaw). Inferior to the hyoid body is the prominent **thyroid cartilage** of the larynx, or "Adam's apple." The **cricoid cartilage** lies just inferior to the "apple." Below the cricoid the **tracheal rings** may be appreciated. The **thyroid gland** may be felt on the lateral sides of lower larynx and upper trachea as soft masses beneath the fingers. If the head is turned to the sides, the **sternocleidomastoideus muscles** are seen to stand out on either side of the neck. Placing the fingers just anterior to these muscles will detect pulsations of the **common carotid arteries**. This is *not* a good pressure point because the arteries cannot be compressed against a bony part. The **external jugular vein** lies alongside the artery and commonly bulges when a person is angry or lifting heavy objects. The **trapezius muscles** extend from the midshoulder up the posterior sides of the neck to the posterior skull.

The muscles of the neck and the bones and cartilages outline a series of **triangles** (Fig. 3-11) containing important vascular and nervous structures of particular interest to surgeons operating on the neck.

The **posterior cervical triangle** lies between the *trapezius* muscle posteriorly, the *sternocleidomastoideus* muscle anteriorly, and is delimited by the *clavicle* (collarbone) at its base. The **omohyoideus muscle**, not palpable,

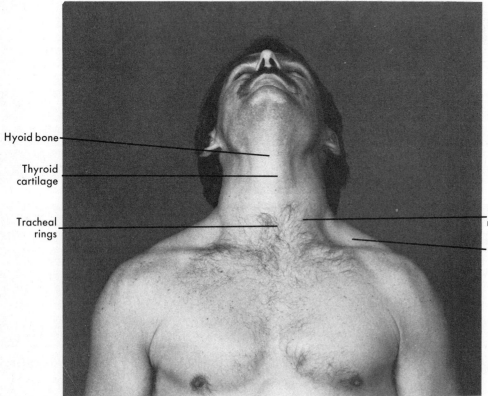

Fig. 3-10
Some anatomical landmarks of neck.

crosses the triangle about a third of the way up, dividing it into an upper, larger *occipital* triangle and a lower, smaller *omoclavicular* triangle. The occipital triangle contains the *eleventh cranial (accessory) nerve* on its way to innervate the trapezius. The omoclavicular triangle contains the roots of the *subclavian artery and vein* and the large nerve trunks of the *brachial plexus* on their way to innervate each upper appendage. The space behind the clavicle is termed the supraclavicular notch, in which pulsations of the subclavian artery may be felt. This is another poor pressure point because of the difficulty in compressing the vessel against a bony prominence.

The **anterior triangle** is bounded by the *anterior* midline of the neck, *posteriorly* by the sternocleidomastoideus muscle, and *superiorly* by the lower margin of the mandible. It is subdivided into four smaller triangles.

The *inferior carotid (muscular) triangle* lies between the anterior midline of the neck and the omohyoideus muscle. It contains the thyroid gland, larynx, upper trachea, base of the common carotid artery, lower part of the internal jugular vein, and the first portion of the tenth (vagus) cranial nerve.

The *superior carotid triangle* is bounded by omohyoideus, sternocleidomastoideus, and posterior digastricus muscles, the latter running nearly horizontally between the hyoid bone and an area near the mastoid process. The bifurcation (division) of the common carotid artery into an internal (brain) and external (face) branch occurs in this triangle.

The *submandibular (digastric) triangle* is bordered by the anterior digastricus muscle (between hyoid and "posterior chin" surface) and the stylohyoideus muscle (alongside the posterior digastric) and the mandibular body. The seventh (facial) cranial nerve and the parotid and submandibular salivary glands lie in this triangle.

The *suprahyoid (submental) triangle* is formed by the anterior digastricus muscle, neck midline, and hyoid body. Several large cervical lymph nodes and the anterior jugular vein are found in this triangle.

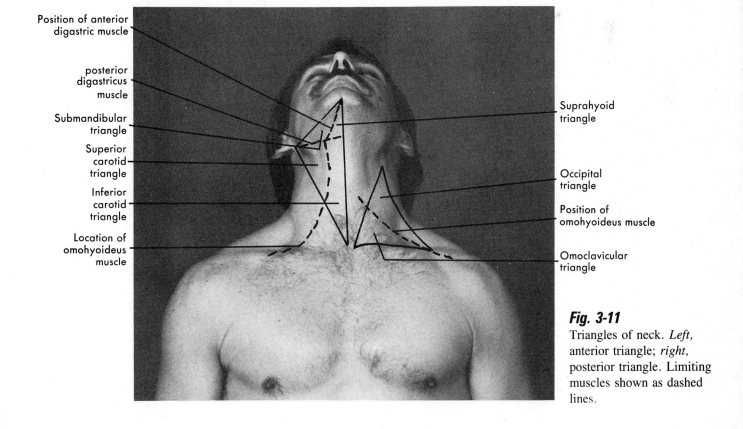

Fig. 3-11
Triangles of neck. *Left*, anterior triangle; *right*, posterior triangle. Limiting muscles shown as dashed lines.

Upper appendage
(Fig. 3-12)

Shoulder and axilla

The *clavicles* (collarbones) and *scapulae* (shoulder blades) are the bones of the *pectoral girdle* that attach the upper appendage to the trunk. The clavicle is subcutaneous throughout most of its length and is easily palpated. The spine of the scapula is palpable as a nearly horizontal ridge on the upper back. Tracing the clavicle laterally brings us to the **acromioclavicular joint** as a small elevation on the "point" of the shoulder. The joint is formed by the lateral end of the clavicle and the acromion process of the scapula. The latter is a prominent projection from the scapular spine that "arches" over the superior end of the humerus. The joint forms the *only* attachment of the pectoral girdle to the trunk and is one factor that permits the great range of movement the upper appendage shows. By moving laterally and inferiorly from the joint, we may feel the **greater tuberosity** *(tubercle)* of the humerus. The **coracoid process** of the scapula may be felt about an inch below and slightly medial to the outer end of the clavicle. The **deltoideus** muscle gives the rounded contour to the shoulder and has *anterior, medial,* and *posterior* *heads*. By raising the arm directly sideward, we may feel the three heads of this muscle.

In the **axilla** *(armpit),* the anterior limit (fold) is formed by the **pectoralis major muscle** and the posterior limit by the **latissimus dorsi muscle.** Near the center of the axilla, several **lymph nodes** may be felt as rounded structures. Gentle pressure in the same area discloses the cordlike nerves of the **brachial plexus** as they pass to the upper appendage. It is also possible to feel the pulsations of the **axillary artery** next to the medial biceps tendon. It is derived from the subclavian artery and becomes the **brachial artery** of the arm.

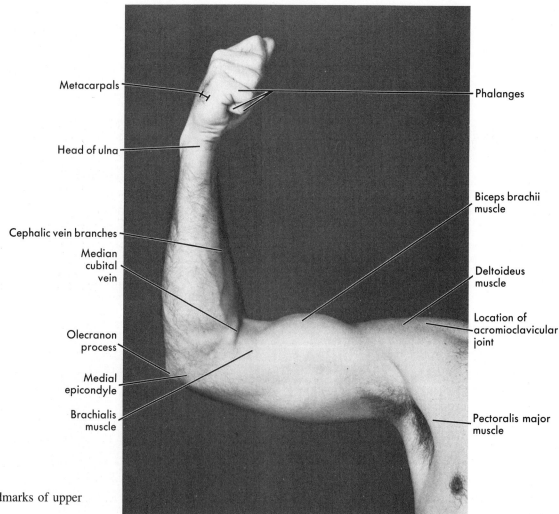

Fig. 3-12
Some anatomical landmarks of upper limb.

Arm

The anterior aspect of the arm is covered by the superficial **biceps brachii muscle** and the deep **brachialis muscle**. The separation of the two heads of the biceps may be appreciated by pushing a finger between them about two thirds of the way up the muscle. The brachialis muscle shows as bulges on either side of the biceps tendon just above the elbow joint. The posterior, lateral, and medial aspects of the arm are occupied by the three heads of the **triceps brachii muscle**. By forcibly straightening (extending) the elbow, we may feel the three heads of this muscle. On the medial side of the distal end of the arm, the **medial epicondyle** of the humerus forms a sharp projection. Proceeding posteriorly from the epicondyle brings us to the **olecranon process,** actually part of the forearm (ulna). In the groove between these two features is the ulnar nerve, subcutaneous at this point. A blow to this nerve (the "crazy" or "funny bone") causes a tingling sensation in the hand. The **lateral epicondyle** of the humerus forms a smaller projection on the lateral side of the distal end of the humerus. The brachial artery lies covered by the muscles of the anterior arm, but a tourniquet placed around the arm may compress both muscles and artery against the humerus to control bleeding. Veins are not prominent on the arm, usually being obscured by the subcutaneous adipose tissue.

Forearm

The forearm is composed of two bones, the medially placed **ulna** and the laterally placed **radius**. The previously mentioned olecranon process of the ulna is the only feature easily palpated; other bony features are covered by muscle. The anterior aspect of the forearm contains muscles flexing the wrist and digits. Proceeding anteriorly and medially from the **styloid process of the radius** that is most lateral at the wrist, we encounter the tendon of the *abductor pollicis longus* muscle that pulls the thumb to the side. A depression medial to this tendon contains the *radial artery,* one of the two branches (the other is the *ulnar artery*) formed from the brachial artery after it crosses the elbow. The pulse is commonly taken here. Next, and particularly obvious if the wrist is bent (flexed), is the tendon of the *flexor carpi radialis muscle.* Next is the tendon of the *palmaris longus muscle,* and on the medial side, the tendon of the *flexor carpi ulnaris muscle.* The posterior aspect of the forearm is covered by muscles that extend the wrist and digits. The **head of the ulna** forms a prominent feature at the wrist on the medial side.

A network of superficial veins may be seen on the anterior forearm. While there are many connecting veins, the vessels form a laterally positioned trunk, the **cephalic vein,** and a medially positioned trunk, the **basilic vein.** Across the anterior aspect of the elbow, the two trunks are connected by the *median (ante) cubital vein,* a common site of venipuncture to obtain blood for testing purposes or for transfusion of blood.

Wrist and hand (Fig. 3-13)

The wrist is composed of eight **carpal bones.** Of these eight, the lateral and anterior prominences that may be felt are produced by the **scaphoid** and

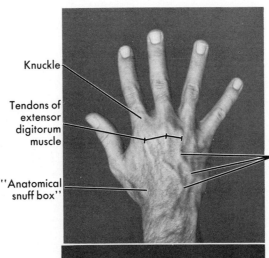

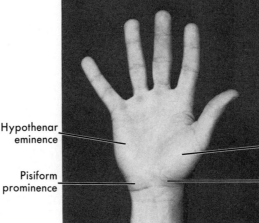

Fig. 3-13
Some anatomical landmarks of hand and wrist.

trapezoid bones. A prominent anterior and medial elevation is produced by the **pisiform bone**. The **metacarpals** are best appreciated on the posterior aspect of the hand as five bones (one for each digit) that end in the "knuckles." The tendons that stand out on the posterior hand when fingers are extended belong to the extensor digitorum muscles. Fourteen *phalanges* form the digits, three in each finger and two in the thumb (a person has five *digits:* four fingers and a thumb—*not* "five fingers" on each hand).

On the anterior and lateral aspect of the hand, a fleshy prominence at the base of the thumb is the **thenar eminence**—it is formed by the **opponens** *pollicis* and *flexor pollicis brevis muscles* operating the thumb. The **hypothenar eminence** lies on the palm side of the hand at the base of the little finger. It is formed by the muscles operating the little finger.

On the posterior aspect of the hand, the "anatomical snuff box" is a depression that appears when the thumb is pulled to the side. Snuff was placed in the depression and then sniffed into the nose.

On the posterior side of the hand, the veins of the **dorsal venous network** may be seen. They drain into the cephalic and basilic trunks mentioned previously.

Thorax and abdomen
(Fig. 3-14)

The thorax or chest is composed of the **thoracic vertebrae**, the *ribs*, and the *sternum (breastbone)*. Anteriorly, the **manubrium** of the sternum, with its superiorly placed **jugular notch**, may be felt in the midline. The manubrium joins the **body of the sternum** at the **sternal angle**, a prominent transversely oriented ridge about 2 inches below the notch. The second rib attaches at the angle. The **xiphoid process of the sternum** projects downward from the sternal body into the upper abdomen. The ribs and their **costal cartilages** (which connect the first 10 pairs of ribs to the sternum, directly or indirectly) are easily felt on the front and sides of the thorax. The anterior superficial muscle of the thorax is the **pectoralis major**.

The posterior thorax or "back" is marked medially by a furrow in which the **spinous processes** of the thoracic vertebrae may be felt. The margins of

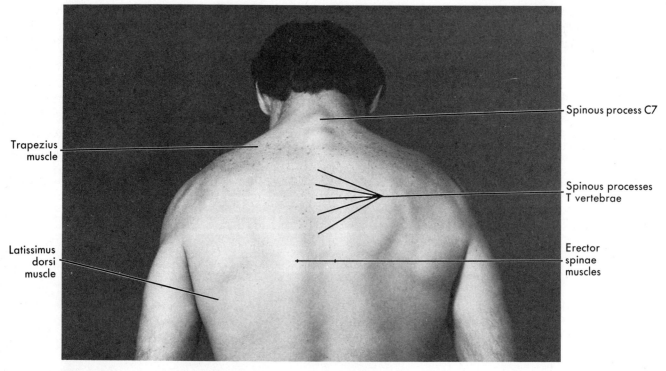

Fig. 3-14
Some anatomical landmarks of back.

the furrow are formed by the masses of muscle constituting the ***erector spinae*** that keep the spine extended against the force of gravity when we sit or stand. The major muscles of the back are the superiorly placed trapezius and inferiorly placed latissimus dorsi.

The abdomen is generally considered to be the portion of the trunk not contained by bony structures. Anteriorly and laterally, the abdomen is delimited by muscles, including the ***rectus abdominis muscle*** in the midline, and the ***external oblique, internal oblique,*** and ***transverse abdominal muscles*** to the sides. The contour of the abdomen is determined primarily by the "tone" of these muscles and the amount of adipose tissue in the skin and internal structures. Posteriorly, the abdomen is delimited by the lumbar vertebrae and the muscles of the lower back. The floor of the abdomen is a muscular sheet that is penetrated by openings for the digestive, reproductive, and urinary systems.

The reader would do well at this time to review the *reference lines* presented in Chapter 1 and correlate them with the structures discussed in this section.

Lower appendage
(Fig. 3-15)

Hip

Anatomically speaking, there is really no such area as the "hip." People may speak of the "hip bones" in referring to the iliac crests, but measurements of the "hip" are actually taken across the greater trochanters of the femurs (the "thigh bones").

What is referred to when a person says "hip" are the bones of the *pelvic girdle,* the ***os coxae,*** which connect the lower appendages to the trunk. The ***anterior superior iliac spine*** is the most

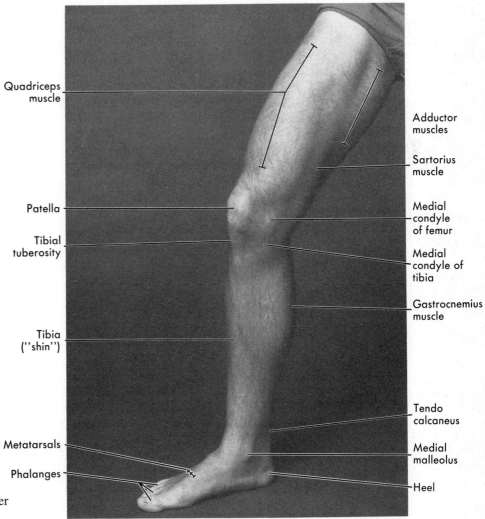

Fig. 3-15
Some anatomical landmarks of lower limb.

anterior point of the bones, and posteriorly, the *iliac crests* may be felt in all but the most obese persons. The *pubic crest* is in the anterior midline at the base of the abdomen. Posteriorly, the *ischial tuberosities* are covered by the *gluteus maximus* muscles of the buttocks.

In the fold between abdomen and thigh (the inguinal area) lie the femoral artery, vein, and nerve. This is often used as a pressure point to control bleeding, although it may be difficult to fully compress the artery.

Thigh

The bone of the thigh is the *femur*. Its *greater trochanter* may be felt on the lateral aspect of the upper thigh. Anteriorly, the thigh is covered by the *quadriceps muscle,* a four-headed muscle. Medially, the *adductor muscles* are found and posteriorly are the *hamstrings,* a series of three muscles. Distally, the *medial* and *lateral condyles* and *epicondyles* of the femur are palpable, and the *patella* (kneecap) covers the knee joint anteriorly. In lean individuals the *sartorius* (tailor's muscle) is sometimes visible, running from the anterior superior iliac spine to the medial tibial head.

In the *popliteal space* on the posterior aspect of the knee, the pulsations of the *popliteal artery* may sometimes be felt.

Leg

The leg is composed of two bones, the medially placed *tibia* and the laterally placed *fibula.* Proximally, the *medial condyle* of the tibia may be felt, and laterally, the head of the fibula forms an obvious prominence. Distally, the *medial malleolus* of the tibia forms a large bulge at the ankle. Commonly referred to as part of the ankle, the malleolus is part of the leg. Laterally, the *lateral malleolus* of the fibula forms another distal bulge.

The tibia is subcutaneous on its anterior aspect, forming the "shin." Just lateral to the skin is the large *tibialis anterior muscle,* important in running and maintaining the longitudinal arch of the foot. Posteriorly, the calf muscle, or *gastrocnemius,* and the *soleus muscle* beneath it, continue into the *tendo calcaneus* (Achilles tendon) to the heel. These muscles are essential in walking and running, and cutting the tendon results in inability to perambulate.

The *saphenous veins* form a network on the anterior and medial aspects of the leg.

Ankle and foot (see Fig. 3-15)

The foot, unlike the hand, is set at about a right angle to the leg to provide a base of support for the erect body. As such, the foot takes the weight of the body and is often subject to disorders (for example, flat feet) that may lead to pain and locomotion disabilities. The *calcaneus* ("heel bone") is one of seven *tarsal* bones that compose the ankle. It affords attachment to the muscles essential for locomotion. The talus actually forms the ankle joint with the cuplike depression formed by the distal ends of tibia and fibula. The remaining tarsals are largely responsible for the formation of the arches of the foot that give "springiness" to our locomotion. Five *metatarsals,* numbered from the big toe side, connect the tarsals to the bones of the toes. As in the hand, 14 phalanges form the foot's digits, 3 in each of the four outer toes, 2 in the great toe.

As the popliteal artery comes onto the leg, it gives off three branches, the *anterior tibial, posterior tibial,* and *peroneal* arteries. The anterior tibial artery continues onto the upper surface of the foot as the *dorsalis pedis* artery. This vessel is commonly palpated to determine if sufficient blood is reaching the lower extremity. A *difference* in pulses on the two sides of the body may indicate a process that is narrowing the vessel on a side.

The surface anatomy landmarks presented in this chapter must be considered to be only the more obvious ones. The reader is encouraged to add to these lists any other landmarks observed, since good observational powers will only increase appreciation of body structure.

Summary

1. Surface anatomy provides a means of learning certain bony, muscular, vascular, and nervous features of the body. Clues as to the location, and in some cases state of health, of internal organs may be gained from surface anatomy.
2. Several major divisions are made in body regions: with some having subdivisions. These are as follows:
   a. Head
   b. Neck
   c. Upper appendage
      (1) Shoulder and axilla
      (2) Arm
      (3) Forearm
      (4) Wrist and hand
   d. Thorax and abdomen
   e. Lower appendage
      (1) Hip
      (2) Thigh
      (3) Leg
      (4) Ankle and foot
3. External body appearance depends on age, sex, and state of nutrition.
4. The major bony, muscular, vascular, and nervous landmarks are presented for the body areas previously listed. Attention to the photographs in this section is probably the best way to learn quickly these major landmarks.

Questions

1. Describe where to locate the following arteries (a to e) and veins (f to i):
   a. Dorsalis pedis
   b. Superficial temporal
   c. Radial
   d. Common carotid
   e. Subclavian
   f. Saphenous
   g. Internal jugular
   h. Medial cubital
   i. Cephalic
2. Forcible clenching of the teeth causes two muscles on the skull to bulge. What are they?
3. Name three bony prominences that may be felt at the elbow region.
4. What are three bony structures that may be palpated at the ankle?
5. Around what bony feature of the femur are "hip measurements" taken?
6. Name the most obvious muscular structures of the following:
   a. Upper arm
   b. Calf
   d. Chest
   d. Upper back
   e. Thigh
7. What do fontanels represent, and of what practical importance are they to "the owner" and medical personnel?
8. What is the "crazy bone"?

Readings

Lockhart, R.D.: Living anatomy, London, 1970, Faber & Faber, Ltd.
Royce, J.: Surface anatomy, Philadelphia, 1965, F.A. Davis Co.

4

Integumentary system

Objectives
Development of the skin and its appendages
Gross features of the skin
Structure of the skin
   Epidermis
   Dermis (corium)
   Hypodermis (subcutaneous layer)
Blood and lymph vessels of the skin
Functions of the skin
Burns
Appendages of the skin
   Hair
   Nails
   Glands
Summary
Questions
Readings

OBJECTIVES After studying this chapter, the reader should be able to:

Give a general outline of the development of the skin and its appendages.
Describe the gross features of the skin.
Describe the structure of the epidermis, dermis, and hypodermis, the functions of the skin, and what happens when it is removed, as in a burn.
Outline the pattern of blood and lymphatic vessels in the skin.
Describe the structure of hairs, nails, and glands of the skin.

Development of the skin and its appendages

The skin consists of a superficial epithelial layer derived from the ectodermal germ layer of the embryo and a deeper layer of mesodermally derived connective tissue. The epithelial layer is designated as the *epidermis,* whereas the connective tissue layer is the *dermis.* Not considered as a part of the true skin is the mesodermally derived *hypodermis (subcutaneous layer)* that is usually heavily infiltrated with adipose tissue.

The epidermis begins its development at 4 weeks (Fig. 4-1) as a single layer of squamous to cuboidal cells that undergoes proliferation to form, by 7 weeks, a surface layer of nucleated flat cells termed the *periderm* and a layer of cuboidal cells that form a *germinal* or *basal layer.* The periderm cells are constantly replaced by mitosis of the basal or germinal layer and become *keratinized,* or filled with a protein that toughens and waterproofs the surface cells. Shedding (desquamation) of the periderm cells creates part of the whitish cheeselike material *(vernix caseosa)* that covers the fetal external body surface. Periderm production ceases by 21 weeks and the surface cells become keratinized to form the **stratum corneum.** By about 11 weeks, the basal layer has, through mitosis, produced several layers of cells that constitute an intermediate layer. The **intermediate layer** will differentiate into the three middle layers of the adult epidermis *(strata spinosum, granulosum, lucidum)* that are present at birth.

As these changes are occurring, the mesenchyme of the dermis is differentiating to form *fibroblasts.* By about 11 weeks, both elastic and collagenous fibers are being formed. Downgrowths *(epidermal ridges)* of basal epidermal cells are pushing into the dermis, and between the ridges are found the *dermal papillae.* Capillary loops important in control of body temperature develop in some papillae, and sensory corpuscles for touch develop in others.

The *patterns* of dermal and epidermal ridges are genetically determined and form the characteristic "prints" of fingers, hands, and feet that are often used for identification purposes. By 17 weeks the characteristic pattern is "set."

Another interesting development occurring in the third month of development is the differentiation of cells that are part of the nervous system into *melanoblasts* that enter the epidermis and become **melanocytes** (see Fig. 4-1). A brown to black pigment designated *melanin* is produced by these cells and initially remains within the melanocytes. After birth the pigment granules are distributed to the epidermal cells. In dark-skinned races greater amounts of melanin are produced and are to be found in all layers of the epidermis. In light-skinned races there is less pigment produced, and it is usually confined to the lower few layers of epidermal cells. All human races have about the same numbers of melanocytes in their skins, pigmentation depending only on amount and shade of pigment produced.

Hairs begin developing at about 12 weeks (Fig. 4-2) as solid downgrowths of epidermal basal cells into the dermis. This downgrowth is termed the

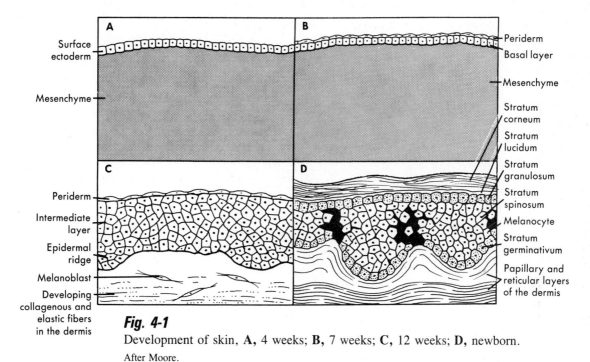

Fig. 4-1
Development of skin, **A**, 4 weeks; **B**, 7 weeks; **C**, 12 weeks; **D**, newborn.
After Moore.

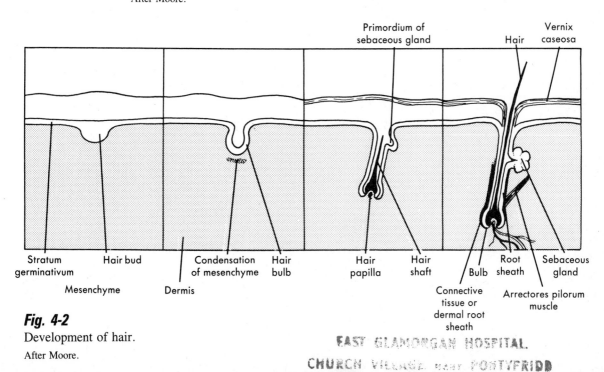

Fig. 4-2
Development of hair.
After Moore.

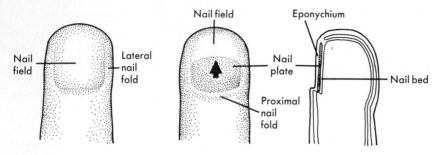

Fig. 4-3
Development of nails.
After Moore.

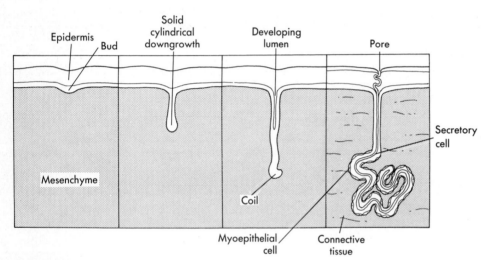

Fig. 4-4
Development of sweat gland.

Fig. 4-5
Development of mammary gland.
After Moore.

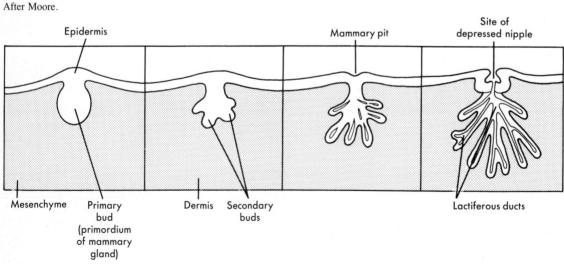

hair bud. It soon develops an enlarged end called the **hair bulb.** Into the bulb pushes a vascular dermal structure called the **hair papilla.** Nutrients to sustain the formation of hairs are thus ensured. The central basal cells form the **germinal matrix** that will give rise to the hair itself, whereas peripheral cells form the **epithelial root sheath** of the developing *hair follicle.* Dermal derivatives form the external **connective tissue sheath** of the follicle. As cells of the germinal matrix proliferate, they become keratinized at the surface of the hair to form the *cuticle* of the hair. The centrally placed cells do not keratinize and form the *cortex* and *medulla* of the hair.

The first hairs formed are very fine and are called **lanugo** hairs. Coarser hairs replace these and are called **vellus** hairs. These hairs remain after birth except in axillary and pubic regions, where they are replaced at puberty by very coarse **terminal hairs** (in the male, such hairs also appear on the face and chest).

Nails develop (Fig. 4-3) beginning at about 10 weeks. A **nail field** develops on the posterior aspect of each digit. Laterally and proximally the fields are surrounded by **nail folds.** Cells from the proximal fold grow over the field and become keratinized to form the nail itself.

Glandular structures, including **sweat** and **sebaceous glands** arise (Fig. 4-4) as downgrowths of the epidermis and outgrowths of the root sheath of a hair follicle respectively. In the sweat gland the downgrowth coils and develops a central cavity, or lumen. Sebaceous glands do not develop lumina but are filled with cells that degenerate to form the oily *sebum* that helps coat the hair and skin surfaces.

Mammary glands are modified sweat glands that form (Fig. 4-5) in a manner similar to sweat glands. The bud *branches* rather than coils to form the secretory and duct portions of the gland.

Gross features of the skin

An adult has an average skin *surface area* of about 1.75 m^2, and the skin weighs about 11 kg. Being composed of epithelial and connective tissue, it qualifies as an organ, the largest of the body. **Flexion creases** are present or appear, as on the palm of the hand, where the skin folds during movement. **Flexion lines** occur where the skin folds but to a lesser degree and are seen on the back of the hand or over the backs of the finger joints. **Friction ridges,** such as form the "fingerprints" and "toe prints," give added ability to grasp small objects. The elasticity of the dermis gives **cleavage lines** to the skin (Fig. 4-6). The skin, if cut, draws away from these lines; surgical incisions involving the skin should parallel these lines to ensure the most rapid healing and the least cosmetic abnormality.

Skin *color* depends on several factors. The **melanin** mentioned previously gives varying shades of brown or black. **Carotene** is a yellowish pigment found in the upper epidermis. It is somewhat more abundant in the skin of the Oriental races. Mixtures of these two pigments give shades of red. The **amount of blood** and its **degree of oxygenation** influence skin color. The skin appears flushed when increased amounts of blood are circulating and may appear bluish *(cyanotic)* if oxygenation of the blood is low (note the bluish veins on the anterior forearm of white persons). With liver disease the skin may appear yellowish *(jaundice)* as bilirubin accumulates in the bloodstream instead of being excreted into the intestine. Lastly, *tanning* of the skin, involving increased production of melanin in response to ultraviolet radiation, tends to protect the skin from harm. However, excessive exposure to sunlight or ultraviolet lamps dehydrates the skin and toughens it and can cause loss of elastic fibers. A deep tan looks great but can age the skin prematurely.

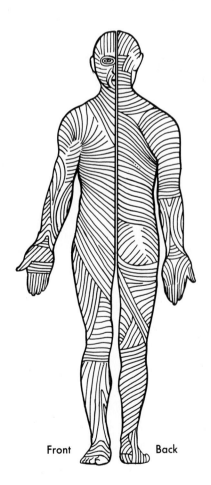

Fig. 4-6
Cleavage lines of skin.

Structure of the skin

Histologically, the skin is classed as **thick** or **thin**. These terms refer not only to total depth of the skin but also to structural differences. Thick skin will be described first, and then differences between thick and thin types will be noted.

Epidermis

The thickest skin (5 to 6 mm) is found on the palm side of the hand and fingers and on the soles of the feet. Of this total thickness, the epidermis constitutes 1 to 1.5 mm. The epidermis in these areas has 50 to 60 layers of cells arranged in five clearly defined layers, or *strata*. From the deepest layer outward, they are as follows (Fig. 4-7):

stratum basale *(synonyms include stratum cylindricum,* named for the shape of the cells, or *stratum germinativum,* named for the fact that the greatest rate of mitosis and cell replacement occurs here) is composed of a single row of cuboidal to cylindrical cells whose lower ends have processes that fit into "pockets" in the basal lamina to help hold the epidermis to the underlying dermis.

stratum spinosum ("prickle-cell layer") consists of eight to ten layers of polygonal cells that appear to bear short spines that attach to similar structures on adjacent cells. These processes are actually short cytoplasmic extensions that are affixed to one another by desmosomes. Thus the cells are held to one another. These two layers (basale and spinosum) are sometimes called the malpighian layer.

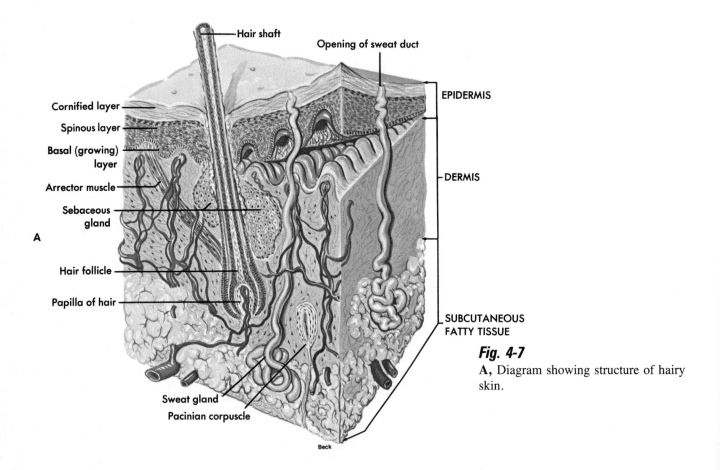

Fig. 4-7

A, Diagram showing structure of hairy skin.

stratum granulosum ("granular cell layer") is composed of two to five layers of rhombic cells containing dark-staining granules of *keratohyaline*. The cells are starting to degenerate, and nuclei are degenerate or missing. The keratohyaline may represent the initial stage in the keratinization or cornification process that toughens and waterproofs the surface cells.

stratum lucidum ("clear layer"), which is about as thick as the granulosum, contains nonnucleated, flattened, clear cells containing a substance called *eleidin*. This material is regarded as the next step in the keratinization process.

stratum corneum ("horny" or "cornified layer") contains 25 to 30 layers of dead, scalelike keratinized cells. These cells are shed singly or in groups from the surface and form the so-called *stratum disjunctum* (not a separate layer). Forty to fifty pounds of cells will be shed from the corneum in a lifetime.

Skin on the rest of the body, called *thin skin*, has a total depth of 1 to 3 mm, of which the epidermis constitutes 0.07 to 0.12 mm. The thinnest skin is found on the eyelids. Epidermis of thin skin has 20 to 30 layers of cells, with spinosum and corneum showing the greatest decrease in numbers of layers. The granulosum is reduced and contains scattered cell granules, whereas a lucidum is usually not visible.

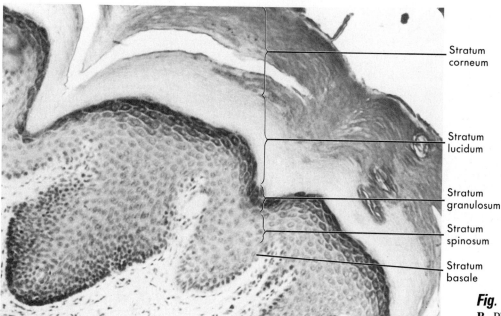

Fig. 4-7, cont'd
B, Photomicrograph of epidermis of thick skin.

Dermis (corium)

The dermis may be characterized as a layer of vascular, dense, irregularly arranged connective tissue. It averages 3 to 4 mm in depth in thick skin and 1 to 2 mm in thin skin. The upper fifth or sixth of the dermis projects its papillae into the epidermis and is known as the *papillary layer.* In some of these papillae are found *Meissner's corpuscles* (Fig. 4-8) serving the sense of touch. Also, the elastic and collagenous fibers are somewhat smaller here, and the layer looks a bit "looser" in arrangement. The remainder of the dermis is designated as the *reticular layer.* Fiber bundles here are coarser, and many blood vessels and sweat glands are found, as are the *pacinian corpuscles* (Fig. 4-9) serving the sense of pressure. The latter resemble an onion, containing lamellae of connective tissue around a centrally placed nerve fiber. *Krause's corpuscles* (cold) and *end organs of Ruffini* (warmth), depicted in Fig. 4-10, are additional sensory structures found in the dermis.

Hypodermis (subcutaneous layer)

Not usually considered a part of the skin, the *hypodermis* consists of loose connective tissue typically containing large amounts of adipose tissue *(panniculus adiposus)*. Pacinian corpuscles, sweat glands, and hair roots may be found in this layer. On the abdomen, thickness of this layer may reach 3 cm or more because of storage of fat. The hypodermis is loose enough to accommodate significant volumes of injected fluid (subcutaneous injection). The looseness of the tissue also permits the skin to move when joints change position. On the elbow, for example, the tissue is extremely loose to permit arm flexion.

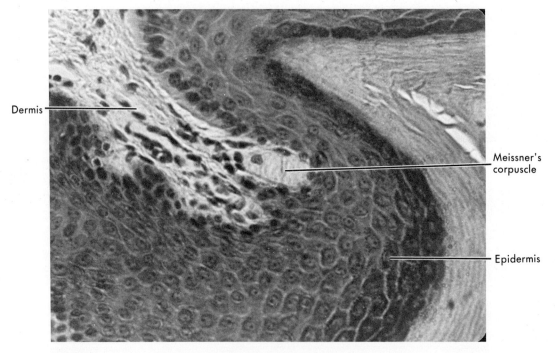

Fig. 4-8
Meissner's *(touch)* corpuscle in skin.

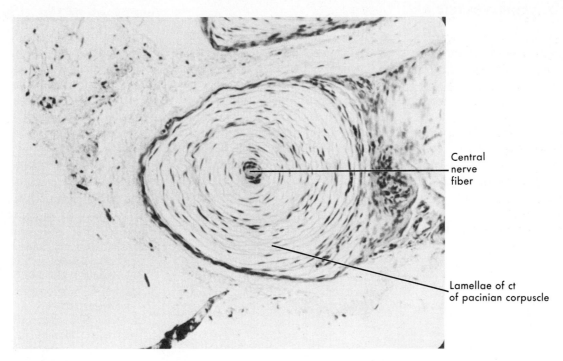

Fig. 4-9
Cross-section of pacinian *(pressure)* corpuscle in subcutaneous layer of skin.

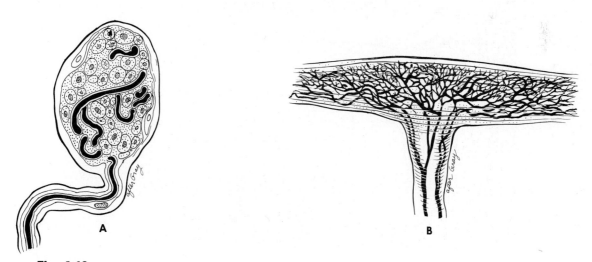

Fig. 4-10
Diagrammatic representations of, **A,** Krause's *(cold)* and, **B,** Ruffini's *(warmth)* corpuscles.

Blood and lymph vessels of the skin (Fig. 4-11)

The larger arteries that supply the skin are found in the subcutaneous layer. From the deeper side of this *sinuous network*, vessels supply the fat cells and tissue of the subcutaneous layer. Superficially directed arteries form a *subpapillary plexus* in the upper part of the reticular layer of the dermis. From here, *capillary loops* extend into the papillae of the papillary layer. Veins draining these capillary loops also form a subpapillary plexus and then pass to the sinuous venous network in the subcutaneous layer. *Arteriovenous shunts* (see Fig. 4-11) connect the arteries and veins at the capillary loops. The arrangement of vessels permits more blood (when the blood is flowing through the loops) or less blood (when flowing through the shunts) to enter the upper dermis. Thus loss of heat through the skin surface may be controlled according to diameter of vessels and whether the loops are filled.

The skin is particularly rich in lymphatic vessels. They originate as blind-ended tubes or networks in the dermal papillae and form a deeper network at the dermis-hypodermis junction. From this deeper network, larger lymphatic vessels follow the course of the skin's arteries and veins.

Functions of the skin

As the external covering of the body, the skin *protects* the internal structures from assault by the microorganisms, radiation, and other threats that abound in our environment. The skin also guards against the loss of vital internal body constituents such as water, salts, and organic materials.

Several devices are employed to give this protection:

- A *surface film* formed from the water secreted by sweat glands and from the breakdown of sebaceous cells that contribute lipids, amino acids, and small protein molecules has a pH of 4 to 6.8 (acidic) and effectively retards the growth of microorganisms and fungi on the skin surface.
- The *keratinized surface cell layers* act as a barrier to chemical and microorganism penetration and prevent loss of internal constituents.
- At the junction of keratinized and nonkeratinized layers, a *layer of ions* tends to repel charged substances approaching it from either side.

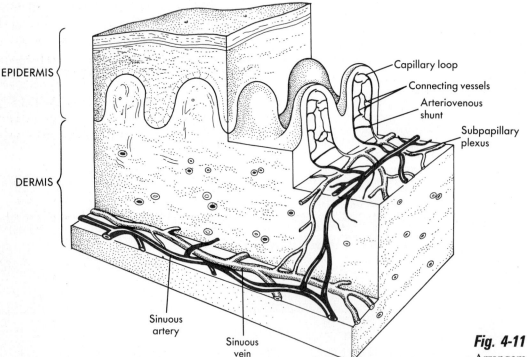

Fig. 4-11
Arrangement of blood vessels in skin.

Temperature regulation is another function in which the skin is involved. A layer of adipose tissue *insulates* the body. The capillary loops, when filled with blood, permit radiation of heat *from* the skin if blood temperature is higher than that of the environment. *Evaporation* of eccrine sweat (mostly water) carries heat with it and becomes the predominant route of cutaneous heat loss when environmental temperature is higher than that of the body.

Absorption of certain materials can occur through the skin. Oxygen and carbon dioxide pass easily through the skin; the skin "breathes." A substance (keratolytic) that can dissolve keratin can more easily pass through the epidermis—aspirin is such a substance and may be found in ointments that may be applied for "aches and pains." Liniments also often contain keratolytics and may also cause vasodilation of dermal blood vessels, leading to the feelings of warmth that may help alleviate pain or soreness.

Excretion of sodium chloride (salt) and urea takes place via sweat glands. Although not great in the short term, heavy sweating can deplete the body of salt to the point at which muscles may cramp, thus the suggestion that salt pills be available to those working hard in a hot environment.

Because of the sensory corpuscles present in the skin, ***reception of stimuli*** may be mentioned as a final skin function.

Burns

It is only necessary to have encountered a burn patient to appreciate the protective function played by the skin. About 1 million people are burned each year, and approximately 1% will die from their burns. Infants and children are especially liable to burns, as from pulling pans of boiling water off a stove. Removal of skin by a burn leads to infection and loss of body fluids and protein. A burn is described according to the extent of the burn and by the degree of destruction of skin layers. The "rule of nines" (Fig. 4-12) is used to calculate the area of skin destroyed, and the area is expressed as a percentage of the total body surface. Depth, or which layers are destroyed, is expressed by "degrees."

A first-degree burn damages only the epidermis, although dermal blood vessels may be affected. A sunburn, with reddening of the skin as capillary beds dilate, is a good example.

A second-degree burn destroys much of the epidermis but leaves *some* epidermal remnants. Regrowth from these remnants is thus possible. Blisters are common and pain is often severe, since skin nerves are irritated by the products of cellular destruction.

A third-degree burn reaches to and through the dermis, often exposing bone and muscle. *No* epidermal remnants are present. No feeling is present because of destruction of nerves. Treating this type of burn involves skin grafts to provide epidermal cells to cover denuded areas.

Burn treatment must be directed toward preventing infection, maintaining fluid and electrolyte balance to offset losses, and encouraging food intake in a patient who is not usually interested in eating.

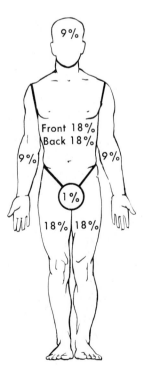

Fig. 4-12
Rule of nines used to calculate extent of burns.

Appendages of the skin
Hair

Development of hair has been described in a previous section of this chapter. The mature hair (Fig. 4-13) lies in a *follicle* derived from both the epidermal downgrowth and the dermis. That part of the hair that lies immediately above the dermal papilla is termed the **hair root**. The **hair shaft** lies within the follicle and also forms the visible portion of the hair. Within the root and lower shaft of the hair, an inner core of more or less clear cells forms the *medulla* of the hair. Several layers of cells containing pigment granules constitute the *cortex* of the hair. Flat, overlapped, and cornified cells form the *cuticle* of the hair. The peripheral layers of the root cells form the **inner root sheath** consisting of three layers: A *cuticle;* *Huxley's layer* (the cells contain red-staining granules of trichohyalin); and *Henle's layer* (a layer of flattened cells with elongated nuclei). The **outer root sheath** is a continuation of the skin's spinous and basal layers. A **connective tissue sheath** is composed of a *vitreous membrane,* a *middle layer,* and an *outer layer* continuous with the subcutaneous or dermal tissues. Hairs are set at an angle into the skin and in the obtuse angle is found a bundle of smooth muscle, the **arrector pili**. This muscle pulls the hair into a more vertical position, increasing insulating capacity in hairy animals. In the human, "gooseflesh" results as a mound of skin is pushed up when the hair is straightened.

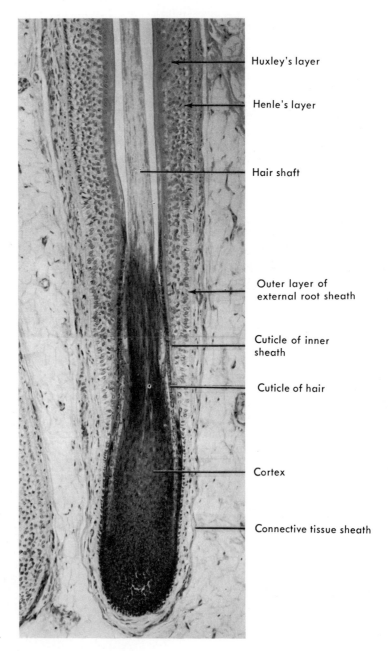

Fig. 4-13
Structure of hairs and hair follicles.

Nails

Mature nails (Fig. 4-14) possess a proximal *root* lying beneath the proximal nail fold, a *body* overlying the main portion of the *nail bed,* and a *free edge* that projects from the finger. The *eponychium (cuticle)* is epidermal tissue that often clings to the nail as it grows, whereas the *hyponychium (quick)* is similar tissue under the free edge of the nail. The edges of the nail lie in *nail grooves* that are bordered by *nail folds.* Nails are curved longitudinally and transversely, a device that increases their strength and minimizes their bending backward as objects are grasped or picked up.

Glands

Sweat glands (Fig. 4-15) are of two general types. The "ordinary" or *eccrine sweat glands* are simple (one duct) coiled tubular structures. They produce, under nervous stimulation in response to heat, a watery secretion that cools the body surface as it evaporates. These glands are distributed on all parts of the skin except the lips, glans penis, and nail bed. In the armpit and around the anus and genital area, *apocrine sweat glands* are found. They are larger than the eccrine type and produce a secretion that contains parts of the secretory cells. The secretion is thus rich in organic material and, when acted on by bacteria, may break down into odoriferous substances. Apocrine sweat glands respond to nerve stimulation during excitement or aroused emotions rather than to heat. Ducts of sweat glands are lined with cells as they pass through the dermis but are mere hollow unlined channels through the epidermis.

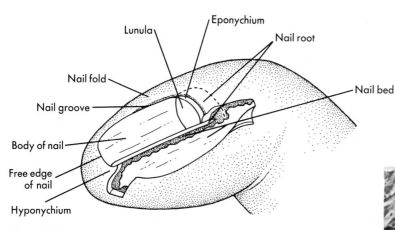

Fig. 4-14
Structure of fingernail and its associated tissues.

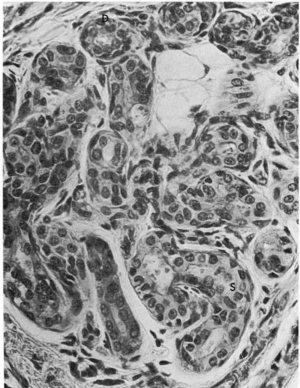

Fig. 4-15
Sweat gland in dermis of skin.

Sebaceous glands (Fig. 4-16) are commonly associated with the body hairs. They are baglike (alveolar or saccular) structures filled with cells. The peripheral darker staining cells are designated *indifferent cells* and transform into the large *sebaceous cells* filling the glands. Breakdown of these latter cells produces *sebum,* an oily material that lubricates the hair and that contributes to the surface film on the skin surface. The ducts of sebaceous glands usually empty into the space between a hair and its follicle.

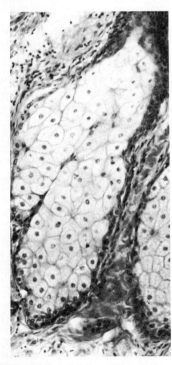

Fig. 4-16
Sebaceous gland in dermis of skin.

Summary

1. The skin develops from ectoderm (the epidermis) and mesoderm (the dermis and hypodermis).
   a. The epidermis is initially a layer of simple squamous or cuboidal cells. Proliferation produces a layer of flat surface cells (periderm) and a basal layer. Further proliferation produces the several strata characteristic of the skin at birth.
   b. Mesenchyme cells differentiate into fibroblasts that form the fibers of the dermis.
   c. Melanocytes develop from ectodermal cells that migrate into the epidermis and produce pigments that color the skin.
2. Hairs develop from downgrowths of the epidermal basal layer. Mesodermal tissue forms part of the sheath around the hair. Lanugo, vellus, and terminal hairs are recognized.
3. Nails develop from a nail field and the proximal nail fold.
4. Sweat and sebaceous glands are derived from epidermis and walls of the hair downgrowth respectively.
5. Gross features of the skin include:
   a. Flexion creases, as on the palm of the hand
   b. Flexion lines, as on the back of the hand
   c. Friction ridges, exemplified by the "fingerprints"
   d. Cleavage lines established by the elastic nature of the skin
   e. Color, as determined by pigments, amount of blood and its level of oxygenation, liver health, and exposure to sunshine.
6. Skin is classed as thick or thin, according to total thickness and structure. Thick skin has the following layers from the depths toward the surface:
   a. Epidermis
      (1) The stratum basale, the lowest layer has cylindrical cells and is capable of high rates of mitosis.
      (2) The stratum spinosum consists of cells that have "spines."
      (3) The stratum granulosum consists of cells losing internal structure and filled with dark-staining granules.
      (4) The stratum lucidum is a layer of clear cells showing no membranes or internal structure. (Thin skin has a poorly developed granulosum and usually lacks a lucidum.)
      (5) The stratum corneum consists of flat dead scalelike cells filled with keratin. Surface cells are shed.
   b. Dermis
      (1) The dermis is composed of dense irregular connective tissue, in an upper papillary and a lower reticular layer.
      (2) The dermis contains blood vessels, sensory corpuscles, their nerves, glands, and hair follicles.
   c. Hypodermis
      (1) Composed of loose connective tissue, the hypodermis contains much fat and may contain hair follicles and sensory corpuscles.
      (2) The hypodermis can accommodate injected volumes of fluids.
7. Blood and lymph vessels of the skin are organized into a definite pattern oriented mainly toward temperature regulation.

a. Both arteries and veins form a sinuous network in the hypodermis.
   b. Subpapillary plexuses occur at the papillary-reticular junction.
   c. Capillary loops penetrate into the papillae of the dermis.
   d. Arteriovenous shunts can cause blood to bypass the capillary loops.
   e. Lymph vessels originate in the dermal papillae, form networks at the dermis-hypodermis junction, and the larger vessels follow blood vessels.
8. The skin protects and is involved in temperature regulation, absorption, excretion, and reception of stimuli.
9. Burns are described by extent (a percentage of body surface) and depth (first, second, and third degree).
10. Hairs and their follicles have several layers.
    a. Hair consists of medulla, cortex, and cuticle.
    b. A follicle consists of an inner root sheath (cuticle, Huxley's layer, Henle's layer) an outer root sheath (corresponds to basale plus spinosum of epidermis), and a connective tissue sheath (vitreous, middle, and outer layers).
    c. Arrector pili are smooth muscle bundles that elevate the hair.
11. Nails have roots, bodies, and free edges; they rest on nail beds and are provided with eponychium, hyponychium, nail grooves, and nail folds.
12. Glands of the skin include the following:
    a. Eccrine sweat glands for cooling
    b. Apocrine sweat glands producing organic-rich secretions
    c. Sebaceous glands producing lubricating and waterproofing secretions

Questions

1. How do the three layers of tissue that form the integument develop?
2. What are the two major layers of the skin, and what is their relationship to the subcutaneous tissue?
3. What are the differences between thick and thin skin, and where is each found on the body?
4. What accounts for variations in skin color?
5. How do hairs develop? What are the parts of a hair and its follicle? What types of hair develop on the body?
6. What are the skin glands, how do they develop, and what do they produce?
7. What are:
   a. Flexion creases?
   b. Flexion lines?
   c. Friction ridges?
8. Describe the structure of thick skin; give the functions of the skin.
9. Describe the vasculature of skin and correlate this with temperature regulation by the organ.
10. Describe the structure of a nail and its surrounding tissues.
11. What sensory receptors are formed in the skin, and what sensation does each serve?

Readings

Elden, H.R., editor: Biophysical properties of the skin, New York, 1971, John Wiley & Sons, Inc.

Ferriman, D.: Human hair growth in health and disease, Springfield, Ill., 1965, Charles C. Thomas, publisher.

Marback, H.I., and Gavin, H.S.: Skin bacteria and their role in infection, New York, 1965, McGraw-Hill Book Co.

Marples, M.S.: The ecology of the human skin, Springfield, Ill., 1965, Charles C. Thomas, Publisher.

Moncrief, J.A.: Burns, N. Engl. J. Med. **228**:444, 1973.

Montagna, W.: The epidermis, New York, 1964, Academic Press, Inc.

Nicoll, P.A., and Cortese, T.A., Jr.: The physiology of skin, Annu. Rev. Physiol. **34**:177, 1972.

Tregear, R.T.: Physical functions of skin, New York, 1966, Oxford University Press, Inc.

5

The skeleton

Objectives

Tissues of the skeleton
   General remarks
   Development of bone (osteogenesis)
   Blood vessels of bone

Bones of the skeleton
   General remarks
   Skull
   Vertebral column
   Thorax
   Pectoral girdle
   Upper appendage
   Pelvic girdle
   Lower appendage

Ossification of the skeleton

Fractures

Questions

Readings

OBJECTIVES After studying this chapter, the reader should be able to:

List the functions of the skeleton.

Describe and compare the two processes by which bone is formed in the body, give examples of where each type of process occurs, and list some of the vitamins and hormones essential to bone formation.

Describe the blood supply to a bone.

List the types of bones according to shape that occur in the body and give examples of each type.

Demonstrate knowledge of the bones of the body, their features, and sexual differences as revealed by examination procedures determined by your instructor(s).

Describe some of the differences in times of maturation of the skeleton's parts; specifically, be able to point out which bones mature early and late and relate this to age.

Describe some of the types of fractures bones may suffer and some causes of fractures.

Tissues of the skeleton

General remarks

The skeleton of the human body may be defined as its bones and related cartilages. These tissues give **support** for the soft organs of the body that may attach to it via connective tissues. The skeleton serves as the basis for **movement** of the body through space and for the often complicated movements we carry out in our activities. Power to move the bones is, of course, supplied by the muscles of the body. The typical **form** or **shape** of the body is provided by the skeleton, which also **protects** many vital organs. The brain is encased in the cranium, the spinal cord is located in the vertebral column, and the heart and lungs are surrounded by the structures of the thorax. The skeleton provides a **storehouse** for the bulk of the calcium and phosphate of the body, and individual bones may contain marrow tissue. *Red bone marrow*, confined mainly to the ends of long bones and the interiors of other bones, is an active site for **blood cell production**. *Yellow bone marrow* is rich in fat, occupies the central portion of long bones, and serves as a **storehouse of nutrients**.

In toto, we come to appreciate the bones as areas of great activity and not the dry lifeless structures commonly studied in the anatomy laboratory.

The structure of bone and cartilage that constitutes the skeleton has been considered in Chapter 2. The next section will discuss the development of osseous tissue.

Development of bone (osteogenesis)

There are two methods by which bone is formed. In **intramembranous** formation, bony tissue is formed in "direct" fashion in tissue lying between two fibrous layers. In **intracartilaginous** *(endochondral)* formation, a hyaline cartilage model is first formed and is subsequently replaced by bony tissue. Because of this "two-step" process, this type of formation is often characterized as "indirect."

Intramembranous formation. Intramembranous formation (Fig. 5-1) is found in the flat bones of the cranium (frontal, parietal, occipital, temporal), the clavicle, part of the mandible, and on the outer surfaces of all bones. The term *membrane* or *dermal* bones is often used to reflect the membranous origin of these bones.

The process begins with an area of mesenchyme between two fibrous layers. The whole affair constitutes the "membrane" within which ("intra") bone formation occurs. The area of mesenchyme becomes richly vascularized. Next, the more centrally placed mesenchyme cells assume the *characteristics* of osteoblasts and lay down a layer of dense matrix material and collagenous fibrils that is termed **osteoid** (bonelike). It seems to be similar to the organic part of bony tissue. The next step sees the differentiation of mesenchyme cells into other cells recognizable as *osteoblasts,* their alignment into rows along the osteoid, and calcification (deposition of inorganic substance) of the osteoid to form bone. The pattern of osteoid formation is initially in the form of plates and bars that lie between the osteoblasts, and when this tissue is calcified, the resulting bony tissue is also in the form of plates and bars; that is, it becomes **spongy (cancellous) bone.** Cells trapped in the calcifying matrix become osteocytes. Sections of this process as viewed on a slide usually show the process as having advanced to the stage of bone formation midway between the fibrous membranes. As we proceed from center to either membrane, the steps are earlier in the process (see Fig. 5-1). Osteoid is usually visible as an area of clear material on the surface of the newly formed bone, with rows of osteoblasts outside that. Vascularization of the primitive tissue is usually reflected by the presence of many large thin-walled vessels, commonly containing shiny red erythrocytes (red blood cells). If any mesenchyme is still present, it is found next to the fibrous membranes. The process extends the network of spongy bone toward the membranes. Beneath the membranes, rows of osteoblasts differentiate and lay down osteoid material parallel to the membranes. Calcification of *this* osteoid produces layers of lamellar bone on either side of the spongy mass. Thus, in the cranium, **tables,** the outer and inner layers of lamellar bone, are produced, whereas the spongy central bone becomes known as the **diploë** (Fig. 5-2). In the other bones, lamellar bone surrounds a core of spongy bone. Lastly, as the bones form, the membranes covering them become a **periosteum.** The latter may be seen to be two layered: an inner **osteogenic layer** containing many cells capable of forming bone and an outer **fibrous *(vascular)* layer** rich in blood vessels and collagenous fibers. The whole periosteum is attached to the bones it covers by collagenous fibers that penetrate into

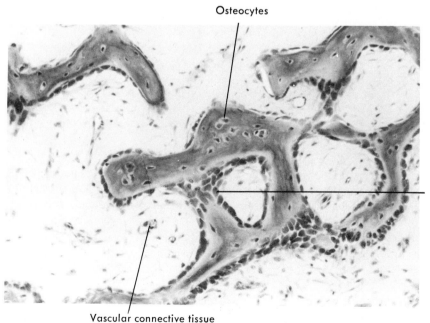

Fig. 5-1
Intramembranous bone formation.

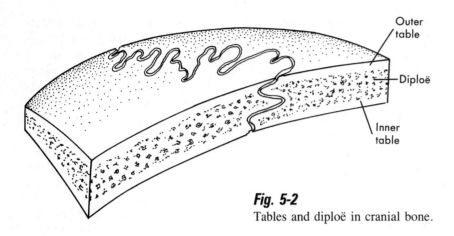

Fig. 5-2
Tables and diploë in cranial bone.

the bone itself. These fibers are called **Sharpey's fibers**.

Intracartilaginous formation (Fig. 5-3). All other bones of the body are formed from a model first laid down in hyaline cartilage. Surrounding this cartilage is a perichondrium. The events in this type of bone formation are best followed in a long bone of the body (for example, femur, tibia, humerus).

1. The initial changes appear in the center of the cartilage model (the *diaphysis* or *shaft*) as chondrocytes enlarge *(hypertrophy)* and increase in

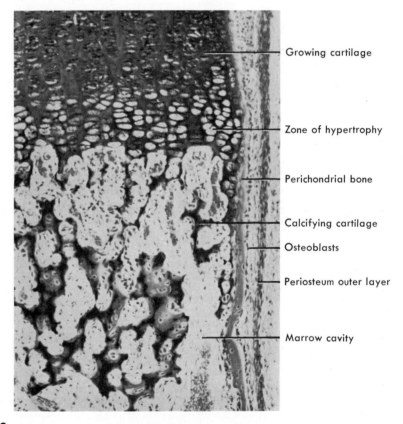

Fig. 5-3
Above, Photomicrograph of endochondral bone formation. **Right,** Diagrammatic representation of process in typical long bone. *A* to *J*, Longitudinal sections. A_1 to D_1, Cross-sections at indicated levels. *A*, Cartilage model of bone. *B*, Bone collar is formed. *C*, Cartilage calcification. *D*, Periosteal bud invades calcified cartilage. *E*, Progression of buds toward bone ends. *F* and *G*, Secondary ossification center appears in bone end and enlarges. *H*, Formation of other secondary center. *I* and *J*, Epiphyseal cartilage is ossified and bone length is fixed.

Modified from Bloom and Fawcett, and Moore.

number *(hyperplasia)*, causing a **compression of the cartilage matrix** into thin plates and bars between the enlarging cells.

2. The reduced matrix material becomes **calcifiable,** and inorganic calcium phosphate is deposited in the cartilage. Note that only calcified cartilage results, *not* bone. This calcification is termed a **provisional calcification.**

3. As these changes are occurring in the center of the model, the primitive perichondrial cells next to the model differentiate into osteoblasts and lay down, *intramembranously,* a **bone collar** completely around the center of the model. The membrane now overlying the collar is a *periosteum*.

4. From the blood vessels in the outer periosteal layer, **periosteal buds,** vascular loops, develop and penetrate through the bone collar, presumably by the differentiation of **osteoclasts,** or bone-destroying cells. The bud penetrates into the cartilage following the path of least resistance, which happens to be through the cartilage cells. Some calcified matrix is also destroyed.

5. Brought in with the bud are primitive cells *(mesenchyme)* that differentiate into osteoblasts and bone marrow cells. The osteoblasts form osteoid on the surface of any surviving calcified matrix or *de novo* in the spaces where there were cartilage cells. This osteoid is then calcified to form bone. This first bone formed is of the spongy variety. The bone marrow cells produce blood cells of various types.

6. Proliferation of the vascular buds and the formation of side branches that move toward the ends of the cartilage model occur next. Ahead of these buds the same changes occur as happened in the center of the model, that is, enlargement and multiplication of cartilage cells, squeezing of matrix, and calcification of matrix, so that the cartilage is "eaten away" and replaced with bone as the bud advances.

7. As the buds advance toward the model ends, the bone collar is extended along the bone, keeping just ahead of the buds inside the bone. Thus the whole structure is strengthened outside as supporting tissue disappears inside.

8. Cells within the buds differentiate into osteoblasts and come to line the cavities of the vessels. These form osteoid and calcify it *from the periphery toward the center* of the vascular lumen. Thus layers (lamellae) of bony tissue are formed around a central cavity containing the blood vessel. These cavities become the haversian canals, the layers of bone become the lamellae, and trapped cells become the osteocytes in their lacunae. Osteons are forming. The osteocytes are initially star shaped with long processes. As bone forms around these processes, they come to lie encased in bone. Withdrawal of these processes creates the small channels called canaliculi that allow nutrients to pass from the vessels in the haversian canals to the osteocytes.

9. At a later time than the changes occurring in the center of the model, similar areas of bone formation occur in the ends (epiphyses) of the developing bone. The bone-forming process here tends to move toward the process

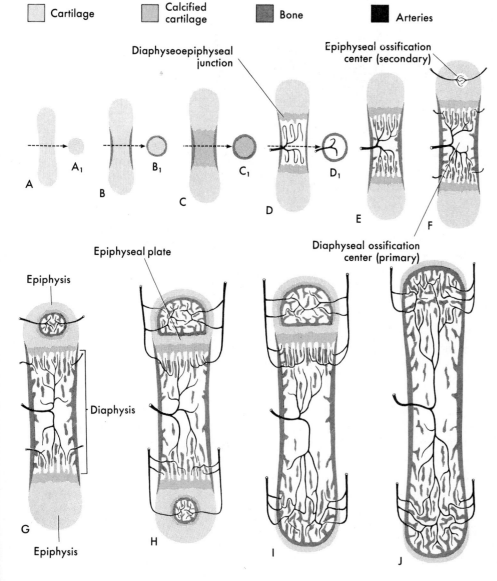

advancing from the center of the bone. As long as an area of cartilage remains between these areas, bone will be added on either side of the cartilage, in effect pushing the ends farther away from the center. The bone *grows in length*.

10. Not all the cartilage on the very end of the bone is replaced; some remains as the *articular cartilage* that ensures smooth joint function.

11. Growth in diameter of the bone occurs by intramembranous formation of new bone on the outer surface. Inside, osteoclasts will destroy bone, and thus the thickness of the wall of the bone will remain fairly constant. Otherwise the bones would become very heavy.

12. When the full length of the bone has been reached, cartilage formation slows, and the plates of cartilage between the center and bone ends are filled in by bone. Length growth ceases, but diameter can still increase.

We may note again that formation of long bones and indeed most bones of the body involves *both* types of formation, whereas flat skull bones and certain others involve only *one* type of formation.

The same *vitamins (A, C, D)* mentioned in connection with cartilage formation are utilized also in bone formation. Likewise, **hormones** (parathyroid and growth, calcitonin) are essential to normal bone formation and remodeling. Sex hormones play a vital role in determining when bone length growth ceases. Estrogens, female sex hormones, cause closure of epiphyseal cartilages by bone, whereas androgens prolong closure. This may account for the usually shorter length of female long bones. Estrogens also seem to play a role in maintaining normal content of inorganic substance. **Osteoporosis** occurs in (primarily) postmenopausal women and, apparently as a result of diminished estrogen production, the bones suffer a loss of osteoblast activity. Bone replacement is exceeded by osteoclast remodeling with loss of calcium and phosphate, and the bones become brittle and more liable to fracture. They may appear "moth-eaten" on x-ray films. Increasing calcium intake appears to slow the process. Paget's disease is characterized by areas of greater and lesser bone formation, particularly in the skull, extremities, and pelvis. No certain cause is advanced for the disorder. Infections of bone include osteomyelitis, often caused by the bacterium *Staphylococcus aureus* (staph) or by the tuberculosis bacterium. The administration of antibiotics is the usual mode of treating these disorders.

Bone cancers form another category of bone disorders. **Osteogenic sarcoma** (bone-forming–connective tissue cancer) results from osteoblast neoplasm. **Myeloma** refers to cancer of the bone marrow.

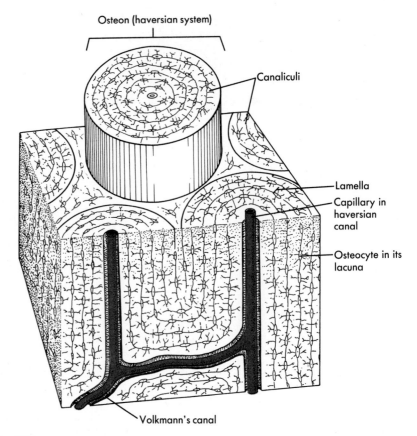

Fig. 5-4

Arrangement of blood vessels in long bone. Vessels will empty into capillary spaces in marrow cavity located to right of diagram.

Blood vessels of bone (Fig. 5-4)

The vessels *inside* bones are derived from the periosteal buds that penetrate through the bone collar. The original channels of bud penetration are at oblique angles to the bone surface, are relatively large, and remain in the mature bone as **Volkmann's canals**. The opening on the bone surface of a Volkmann's canal is often called a **nutrient foramen** (hole). As mentioned previously, the side branches of the bud remain as haversian canals. The bud also reaches what will become the marrow cavity of a long bone and the spaces in the spongy bone of the shaft and bone ends. Thus a route for nourishment and passage of blood cells from the marrow is ensured.

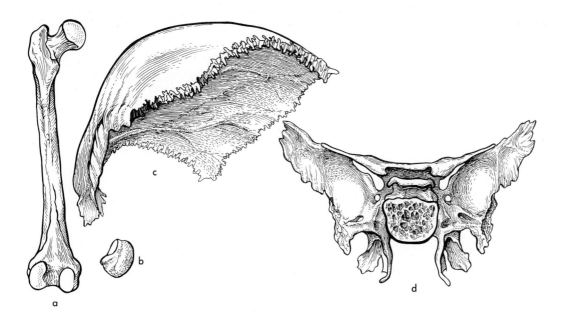

Fig. 5-5
Bone classification by shape. *a*, Long bone (femur); *b*, short bone (carpal); *c*, flat bone (parietal); *d*, irregular bone (sphenoid).

Bones of the skeleton
General remarks

Bones, as organs, are classified into four groups according to their shapes (Fig. 5-5).

long bones have greater length than width. Examples include the humerus, radius, ulna, femur, tibia, fibula, and phalanges. These bones only have a marrow cavity inside their shafts; the epiphyses contain spongy bone (Fig. 5-6).

short bones have a length and width that are about equal. They have lamellae of bone around a spongy center. Examples include the *carpal* (wrist) and *tarsal* (ankle) bones.

irregular bones have complex shapes. The vertebrae and several of the skull bones (ethmoid, sphenoid) are of this type.

flat bones are exemplified by the bones of the cranium (frontal, parietal, occipital) and by the "blade" of the scapula. Not all flat-shaped bones are formed intramembranously, only those of the skull and the others previously mentioned.

The skeleton is usually said to be composed of 206 separate named bones. In the skull there are small "extra" bones along the courses of the major cranial sutures. Such bones are called **wormian (sutural) bones** and are not usually counted in the total.

They arise from small separate areas of intramembranous formation. Also, in tendons, where there is a lot of pressure, small **sesamoid bones** may form for protective purposes. The "ball" of the foot is a common site for such bones. These also are not counted in the total.

The 206 bones are divided as follows:

Axial skeleton, 80
  Skull,* 29
    Cranium, 8
    Face, 14
    Hyoid, 1
    Ossicles (ear bones), 6
      Malleus, 2
      Incus, 2
      Stapes, 2
  Vertebral column,* 26
    Cervical (neck) vertebrae, 7
    Thoracic (chest) vertebrae, 12
    Lumbar (lower back) vertebra, 5
    Sacrum (5 fused bones), 1
    Coccyx (3 to 5 fused bones), 1
  Thorax,* 25
    Sternum, 1
    Ribs (12 pairs), 24

*These parts compose the axial skeleton. The other bones form the appendicular skeleton and its supporting girdles.

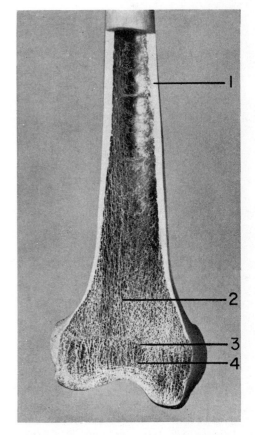

Fig. 5-6
Distal portion of frontally sectioned adult human femur. *1*, compact bone of diaphysis; *2*, spongy bone proximal to epiphyseal line; *3*, epiphyseal line where cartilage was; *4*, spongy bone of epiphysis.

Appendicular skeleton, 126
  Shoulder girdle, 4
    Clavicle, 2
    Scapula, 2
  Upper appendage, 60
    Humerus, 2
    Ulna, 2
    Radius, 2
    Carpals (wrist), 16
    Metacarpals (hand), 10
    Phalanges (digits), 28
  Pelvic girdle, 2
    Os coxae, 2
  Lower appendage, 60
    Femur, 2
    Fibula, 2
    Tibia, 2
    Patella (kneecap), 2
    Tarsals (ankle), 14
    Metatarsals (foot), 10
    Phalanges (digits), 28

An aid to understanding how parts of bones are named includes the fact that many structures are named according to their *position;* thus words derived from superior, inferior, medial, and lateral are used. For example, *supra*orbital would lead us to look above the orbit. Also there are descriptive terms for the various projections, elevations, and depressions on the bones. Sometimes positional and descriptive terms are combined, as in *medial condyle.* The following list of descriptive terms in alphabetical order is presented as an aid to what various terms imply:

condyle (Gk. *kondylos,* knuckle) is a rounded protuberance at the end of a bone *forming a joint.*

crest (L. *crista,* tuft) is a ridge or elongated prominence on a bone, often located on one of its edges. It may provide a point for muscular or tendon *attachment* or be a *border* of the bone.

epicondyle (Gk. *epi-,* above, + condyle) is a projection from a bone, above a joint condyle, affording a point for tendon attachment.

facet (Fr. *facette,* small face) is a small, smooth, flat or shallow surface on a bone, particularly a vertebra or rib, which usually *forms an articulation.*

fissure (L. *fissus,* a cleft) is a groove or narrow cleftlike opening, typically *passing blood vessels, nerves,* or both.

foramen (L. *foramen,* an opening) is a passageway through a bone; a hole that is used to *transmit vessels* and *nerves.*

fossa (L. *fossa,* a ditch) is a furrow or shallow depression, usually having more surface area than a facet, which commonly *contains* a muscle or some other organ.

fovea (L. *fovea,* a pit) is a small cuplike depression, typically affording *attachment* for a ligament.

head (A.S. *heafod,* head) the proximal (usually) end of a bone, typically having a rounded form, *forming a joint.*

line (L. *linea,* line) is a long narrow ridge-like structure, typically acting as a *tendon attachment.*

malleolus (L. *malleolus,* little hammer) is a projection from the distal end of tibia and fibula that is used as a *pulley* to change direction of muscle pull.

meatus (L. *meatus,* passage; pronounced mē-ā'-tus) is a passageway or short canal running through a bone. It may *transmit* another organ or be air filled.

notch (A.S. *nocke,* notch) is a deep indentation or gap in the edge or margin of a bone, typically *passing a vessel or nerve.*

process (L. *processus,* a going before) is a projection or outgrowth from a bone. It usually has considerable length and may be part of the *protection* for a joint or give *attachment* for tendons.

sinus (L. *sinus,* a curve) is a *cavity* within a bone. The term is also used in connection with the large venous blood vessels within the skull. The first definition is used here.

spine (L. *spina,* spine) is a sharp and usually fair-sized process from a bone, affording *attachment* for a tendon.

sulcus (L. *sulcus,* a groove; pronounced sul'kus) is a shallow furrow or groove. It usually *contains* a tendon, blood vessel, or nerve.

trochanter (Gk. *trochanter,* a runner; pronounced trō'kan-ter) refers to large processes occurring on the proximal end of the femur, used for *tendon attachment.*

tubercle (L. *tuberculum,* a little swelling) is a small rounded eminence on a bone, typically serving for *tendon attachment.*

tuberosity (L. *tuberositas,* a swelling) is an elevated round process from a bone that is usually larger than a tubercle (the word is often used as a synonym for tubercle). Also used for *tendon attachment* or sometimes *in a joint.*

Lastly, where there are several tuberosities (for example) on the same bone, terms such as greater or lesser may be used to distinguish the parts.

The *size* of those processes or features used for tendon attachments depends to a great degree on how much muscular force is applied to it and for how long. The processes of men are usually larger than those of women because their muscles are more powerful. Such differences form one part of determining the sex of a human skeleton.

Skull

It would be good at this time to review the surface features of the skull in the living human (Chapter 3) before proceeding with the study of the bones of the skull.

Examination of the exterior of the skull discloses the many **sutures** that form the joints or junctions between the skull bones. These are joints that are held by interlocking bony prominences and by fibrous membranes between the involved bones. They are thus *strong* and *immovable* joints, offering the greatest *protection* for the organs contained. The cranial vault sutures receive special or specific names, usually according to the plane in which they run or according to a particular area or part that is joined to another. The remaining sutures of the skull are usually named according to the bones involved (for example, nasomaxillary, sphenopalatine).

Cranial vault sutures. The cranial vault of the skull (Fig. 5-7) shows the *coronal suture* running in the coronal plane (from side to side) about one third of the way back from the forehead. In the midline following the median plane is the *sagittal suture*. It runs from the midpoint of the coronal suture to terminate at the *lambdoidal suture* (silent "b"). The latter suture arches across the posterior aspect of the cranium. Running between the coronal and lambdoidal sutures on the sides of the skull are the *squamosal sutures*. Wormian bones may occur along the courses of these sutures. The fetal skull (Fig. 5-8) shows the *frontal suture* (*metopic*, forehead) in the midline following the course of the sagittal suture. It fuses before birth. Skull sutures remain obvious until about 22 years of age. The sagittal suture closes first, followed by the coronal suture (twenty-fourth year) and the lambdoidal suture (twenty-sixth year). Thus a basis for judging the age of a skull is provided.

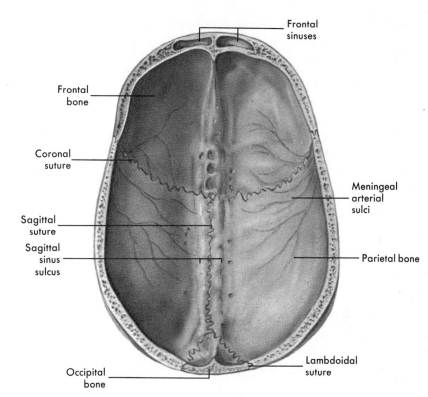

Fig. 5-7
Internal surface of "cap" of skull.

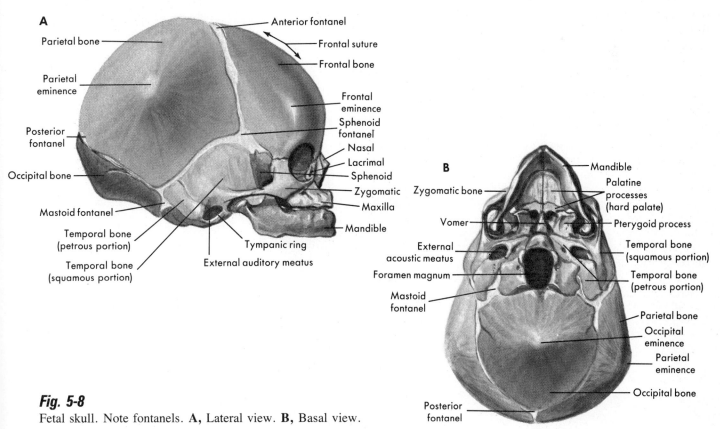

Fig. 5-8
Fetal skull. Note fontanels. **A,** Lateral view. **B,** Basal view.

Fontanels. In the fetal and newborn skulls (Fig. 5-8) there are large areas of membrane where bony tissue has not yet formed. Such areas are called fontanels, or "soft spots." Table 5-1 describes the names and locations of the several fontanels. Usually all but the frontal fontanel and occipital fontanel are closed by birth, at least as far as palpation is concerned. The frontal fontanel closes by 10 to 18 months after birth, the occipital fontanel by 2 months after birth. The presence of the fontanels and the fact that the cranial sutures do not fuse until adulthood allow growth of the brain to occur.

Bones of the skull and their features

Anterior view. An anterior view of the skull (Fig. 5-9) is dominated by the *orbits* or *orbital fossae* that house the eyes, the *nasal fossae (cavities)* that form part of the upper respiratory system, and the teeth contained within the jaws. The *squama of the frontal bone* forms the forehead and turns posteriorly to form the anterior part of the roof of the cranial cavity housing the brain. On both sides of the midline of the forehead are the *frontal eminences*, and below each eminence is a transversely oriented ridge, the *superciliary ridge* or *arch*. Between the arches the frontal bone shows a flattened area, the *glabella*. It extends downward to the junction with the nasal bones. Below the arches the frontal bone forms two *supraorbital margins*, commonly pen-

Table 5-1. Fontanels of the fetal skull

Names of fontanels	Number	Found at junctions of sutures
Frontal or anterior or bregma	1	Sagittal and coronal
Mastoid or posterolateral	2	Squamosal and lambdoidal
Occipital or posterior or lambda	1	Sagittal and lambdoidal
Sphenoid or anterolateral	2	Coronal and squamosal

Adapted from McClintic, J.R.: Basic anatomy and physiology of the human body, ed. 2, 1980. Copyright © 1980. Reprinted by permission of John Wiley & Sons, Inc.

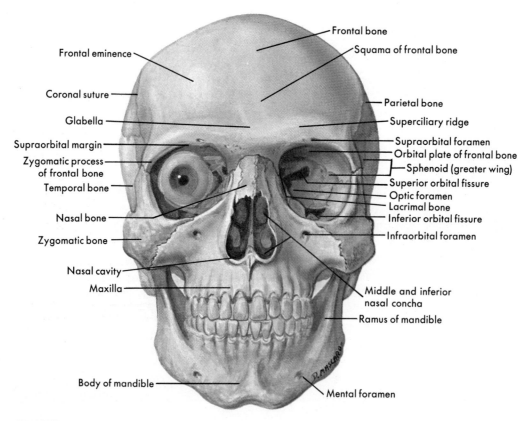

Fig. 5-9
Anterior view of skull.

etrated in the center by a **supraorbital notch** or **foramen**. The bone then curves posteriorly to form the **horizontal** or **orbital plate**. This plate is at the same time the roof of the orbit and a large part of the floor of the **anterior cranial fossa**, the latter containing the frontal lobes of the brain. The **zygomatic process of the frontal bone** projects laterally to articulate with the zygomatic (malar or "cheek") bone. The **frontal sinuses** (Fig. 5-10) lie within the frontal bone beneath the medial portions of the superciliary ridges.

The **orbit** *(orbital fossa)* has the shape of a pyramid with the apex directed inward. Its roof, as previously mentioned, is formed by the frontal orbital plate. The medial wall of the orbit is formed by the **frontal process of the maxilla,** the **lacrimal bone,** and the **orbital plate of the ethmoid bone.** The floor is formed by the **orbital plates of the maxilla** and the **zygomatic bone,** and the lateral wall by the zygomatic bone (anteriorly) and the **greater wing of the sphenoid bone** (posteriorly). Running along the floor of the orbit toward its apex and lying between the maxillary plate and the greater wing of the sphenoid is the **inferior orbital fissure**. It passes nerves and blood vessels to the structures lying in the **infratemporal fossa**. The **optic foramen,** transmitting the optic nerve and ophthalmic artery, rises from the apex of the orbit and joins the optic canal to the cranial cavity. The **superior orbital fissure** lies laterally to the optic foramen and carries branches of the oculomotor, trochlear, trigeminal, and abducent cranial nerves and ophthalmic arteries and veins to the orbit.

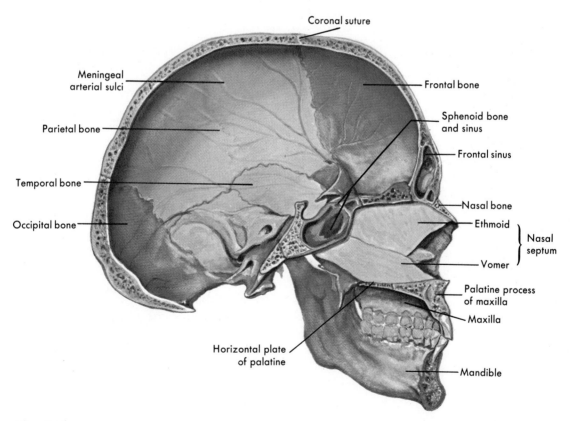

Fig. 5-10
View of median section of skull.

The *lacrimal bone* (Fig. 5-11), located in the anteromedial orbital wall, is the smallest skull bone. It is best identified by its *posterior lacrimal crest,* anterior to which lies the *lacrimal sulcus.* The latter houses the lacrimal sac and part of the nasolacrimal duct, portions of the system carrying the tears to the nasal cavity.

The *nasal fossa* is triangular in shape. Its margins are formed anteriorly and superiorly by the nasal bones and by the maxillas laterally and inferiorly. The fossa is internally divided in its center into two nasal cavities by the *perpendicular plate of the ethmoid bone* (Fig. 5-12) (superior two thirds or so) and the blade of the *vomer* (Fig. 5-13) (lower one third or so). Projecting into the nasal cavities from its medial aspect are the *middle* and *superior nasal conchae,* scroll-like portions of the ethmoid bone. The *inferior nasal conchae,* separate bones, project into the nasal cavities from their lateral walls. Each concha has a corresponding *meatus* below it. The roof of the nasal cavities is penetrated by a number of foramina passing branches of the olfactory nerve through the *cribriform plate of the ethmoid bone.* The floor of the nasal cavities is formed by the *horizontal processes of the maxillas* and *palatine bones* (hard palate) (Fig. 5-14).

The *maxillas* (Fig. 5-15) form the upper jaws and contain large *maxillary sinuses (antrums of Highmore).* Their *alveolar processes* contain sockets for the upper teeth. On the anterior surface below the inferior orbital margin is an *infraorbital foramen* passing nerves and blood vessels to the face and anterior teeth. A *frontal process* projects superiorly, and the aforementioned horizontal (palatine) process forms part of

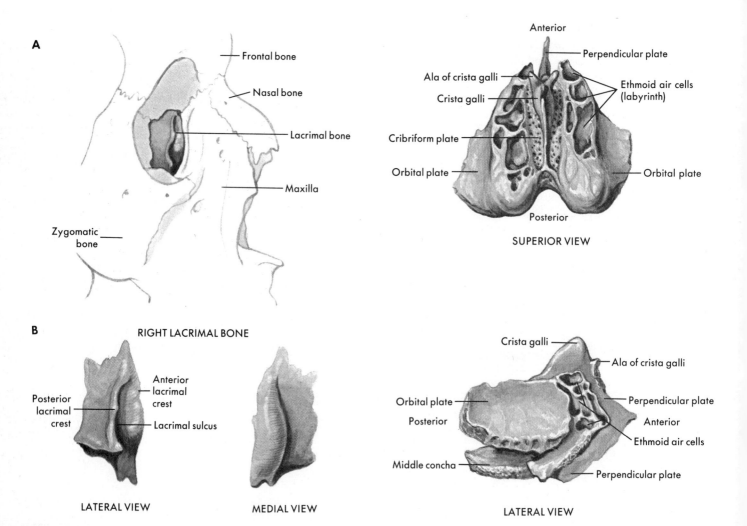

Fig. 5-11
A, Right lacrimal bone in situ and, B, in lateral and medial views.

Fig. 5-12
Superior and lateral views of ethmoid bone.

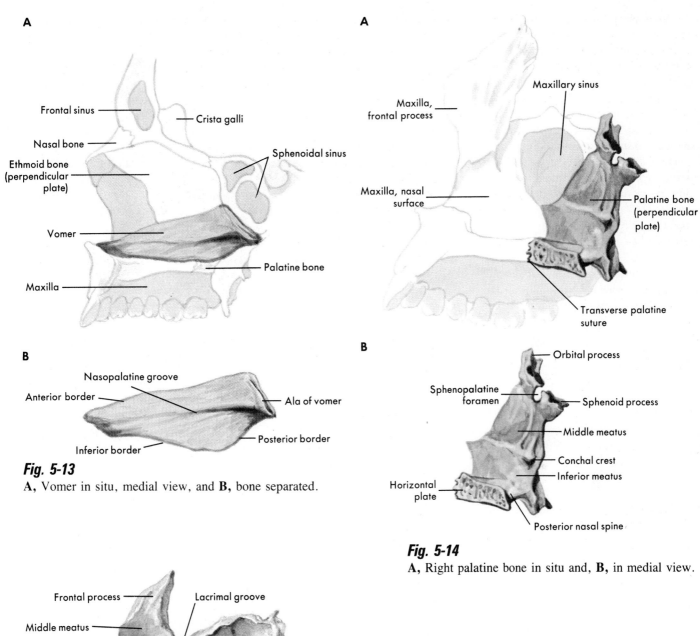

Fig. 5-13
A, Vomer in situ, medial view, and **B,** bone separated.

Fig. 5-14
A, Right palatine bone in situ and, **B,** in medial view.

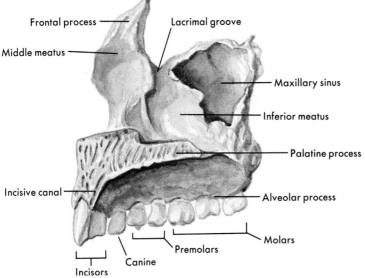

Fig. 5-15
Medial view of right maxilla.

the hard palate. A *zygomatic process* joins the zygomatic bone.

The *mandible* (Fig. 5-16) is the only skull bone possessing a freely movable joint. It has a horseshoe-shaped *body*, from which arise vertically directed *rami*. The rami join the body at the *angle*. Each ramus bears a posterior *condyloid process* forming the joint (temporomandibular) with the rest of the skull, an anterior *coronoid process*, and the *mandibular notch* between the processes. Its *alveolar margin* contains sockets for the lower teeth. The *oblique line* runs downward from the anterior border of the coronoid process on the exterior. Near the *mandibular symphysis* where the two halves of the bone join is an exteriorly placed *mental foramen* on each side. Interiorly, the bone shows the *mandibular foramen* about one third of the way down the ramus. Beginning anteriorly to the mandibular foramen and running onto the body is the *mylohyoid line,* serving to attach the mylohyoid muscle. *Mental spines* lie on the posterior aspect of the symphysis. Slight depressions (fossae) on the interior body mark the location of submandibular and sublingual salivary glands.

Lateral view. Some of the previously mentioned bones and their features are evident in a lateral view of the skull (Fig. 5-17) and may be reviewed.

At the coronal suture the frontal bones articulate with the *parietal bones.* Beginning above the orbit and running on the frontal *and* parietal bones are two lines, the *superior* and *inferior temporal lines.* These lines serve as attachments for the fan-shaped temporalis muscle. The area on the side of the skull enclosed by these lines is the *temporal fossa.* The fossa is sometimes divided into a supratemporal and an infratemporal part, the division occurring at about the midorbital level. The parietal bones articulate anteriorly with the frontal bone, inferiorly with the *greater wing of the sphenoid bone* (anterior one fourth or so), at the squamosal suture with the *squama of the temporal bone* (posterior three fourths), and posteriorly at the lambdoidal suture with the *occipital bone.*

The *temporal bone* (Fig. 5-18) has four portions to it. The *squama* forms the "temple" of the skull, most of the lower lateral cranial vault. At its inferior aspect a long slender *zygomatic process* arises that articulates anteriorly with the *temporal process of the zygomatic* bone. Together, the zygomatic process and zygomatic bone form the *zygomatic arch.* The *tympanic portion* contains the *external auditory meatus.* The *mastoid portion* exhibits the *mastoid process* that is easily felt behind and below the pinna. It contains mastoid air cells that communicate with the middle ear cavity. Anterior to this process is the *stylomastoid foramen;* anterior to that is the *styloid process,* commonly broken off on specimens. The last portion, the *petrous portion,* forms a prominent ridge *inside* the cranial cavity. It houses the structures of the inner ear. The *internal auditory*

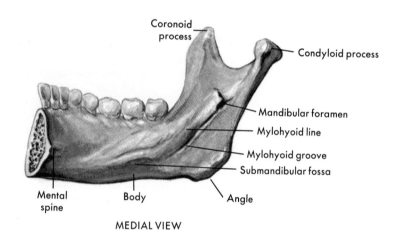

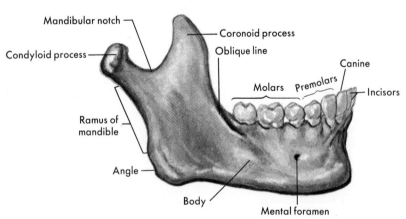

Fig. 5-16
Lateral and medial views of mandible.

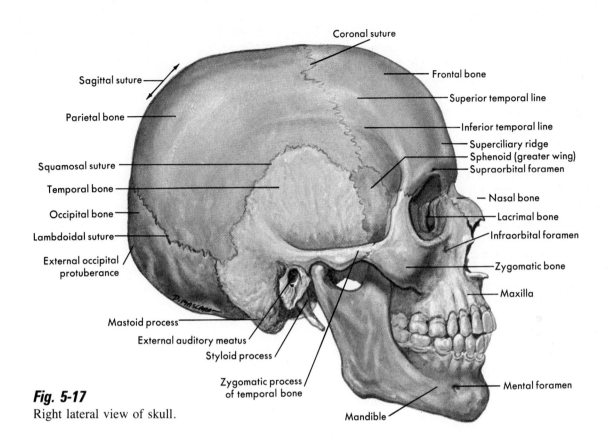

Fig. 5-17
Right lateral view of skull.

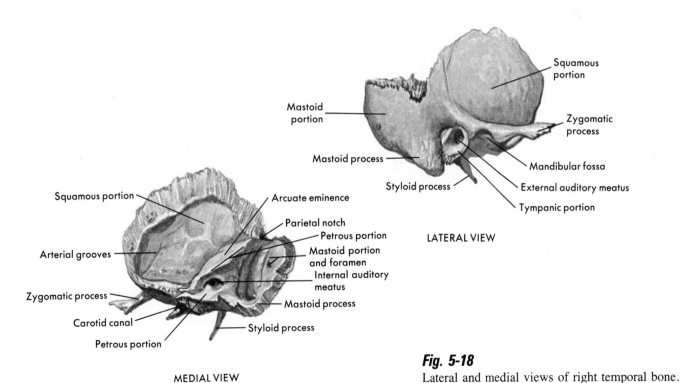

Fig. 5-18
Lateral and medial views of right temporal bone.

meatus lies on the posterior side of the petrous portion. The ear ossicles, *malleus, incus,* and *stapes,* lie in the *middle ear cavity* contained within the tympanic portion.

The *zygomatic (malar) bone* forms what is commonly called the "cheek." A roughly square *body* sends *frontal, maxillary,* and *temporal processes* to articulate with the respective bones. A prominent *zygomaticofacial foramen* lies on the body's external surface.

Base of the skull. Fig. 5-19 shows the base of the skull, dominated anteriorly by the maxillas bearing the teeth. At the anterior end of the hard palate, just behind the incisor teeth, is the *incisive fossa*. In some skulls the fossa may house two midline foramina, the *foramina of Scarpa,* and two laterally placed foramina, the *foramina of Stensen.* The horizontal plates of maxillas and palatine bones form the hard palate. The palatine portion shows two obvious foramina, the *greater* and *lesser palatine foramina.* In the midline lies the wedge-shaped body of the *vomer.* Reaching from side to side on the skull just behind the hard palate is the complicated-shaped *sphenoid bone.* The *greater wings* extend laterally up the sides of the skull. The paired *pterygoid processes* project toward the viewer from the area just posterior to the hard palate. Each process has a *medial* and a *lateral lamina,* between which lies the *pterygoid fossa.* Several prominent foramina are visible in the sphenoid bone to either side of the origins of the pterygoid processes. The *foramen ovale* (oval) lies a bit anterior, followed by the *foramen spinosum.* Posterior to the origin of the process is a jagged-edge foramen, the *foramen lacerum.* In the lower edge of the temporal bone's petrous portion, visible in this view posterior to the sphenoid bone, is the large smooth-edged *carotid canal* and posterior and a bit lateral is a large rather rough-edged *jugular foramen.* These two foramina pass the

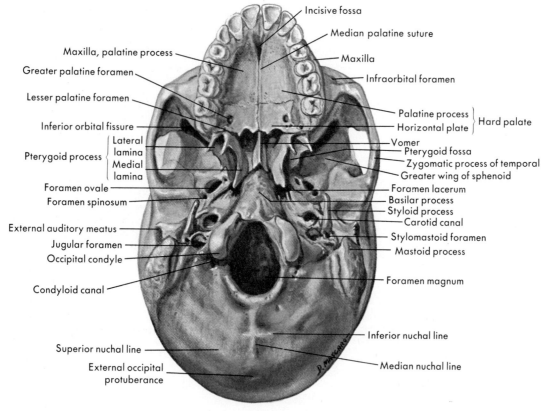

Fig. 5-19
Skull as viewed from below.

internal carotid artery to the brain and the *internal jugular vein* from the brain.

The posterior portion of the basal view of the skull is dominated by the **occipital bone.** The central rather narrow portion that articulates anteriorly with the sphenoid bone is the **basilar process.** Immediately posterior to the process is the **foramen magnum,** through which passes the spinal cord. On either side of the foramen are paired **occipital condyles** forming an articulation with the first cervical vertebra of the spinal column. A look *beneath* the condyles brings to view the **hypoglossal canals** that transmit a cranial nerve.

Condyloid canals lie at the posterior edge of the condyles. Extending posteriorly and upward in the midline from the foramen magnum is the **median nuchal line.** This line is intersected by laterally extending **inferior** and **superior nuchal lines.** Where superior and median lines intersect is found the **external occipital protuberance,** an easily palpable feature in the living person.

Internal view of the skull (Fig. 5-20). Remove the calvarium of the skull and look at its inner surface. Running along the sagittal suture between the parietal bones is a shallow sulcus, the **groove for the superior sagittal sinus,** a large venous-type blood vessel. Tree-like patterns of grooves on the parietal bones are for the middle meningeal arteries located in the dura mater, one of the membranes surrounding the brain. Viewing the floor of the cranial vault shows first the three cranial fossae. The **anterior cranial fossa** floor is formed mainly by the orbital plates of the frontal bone. Behind these plates is a narrow projection of bone that is the **lesser** (smaller) **wing of the sphenoid bone.** There is then an abrupt "step-down" into the **middle cranial fossa.** It is delimited posteriorly by the ridge of the petrous temporal bone; the **posterior cranial fossa,** encompassed mainly by the occipital bone, can then be seen.

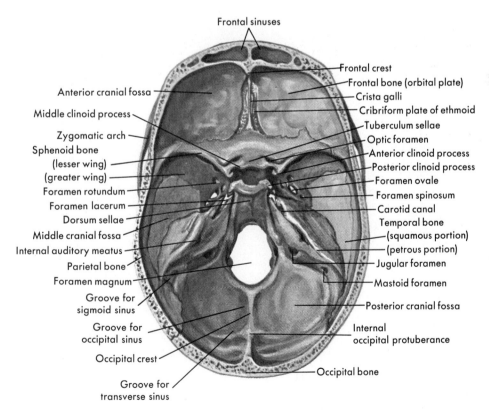

Fig. 5-20
Floor of cranial cavity as viewed from above.

In the midline of the anterior cranial fossa is the *frontal crest* that passes inferiorly to terminate at the *cribriform plates* of the ethmoid bone. The *foramen cecum* is located at the base of the crest. Between the two cribriform plates is an elevated crest of bone, the *crista galli*. Find again the lesser sphenoid wings (Fig. 5-21). Note the optic foramina again and the *optic groove* between them. Projecting posteriorly from the medial portion of each lesser wing is an *anterior clinoid process*, between which lies the *tuberculum sellae*. The clinoid processes encompass between them a depression housing the pituitary gland, which is known as the *sella turcica*. The sella is bounded anteriorly by two elevations, the *middle clinoid processes*. Posteriorly, the sella is bordered by a vertically projecting square-shaped piece of bone, the *dorsum sellae*. Laterally, the dorsum carries the *posterior clinoid processes*. The several foramina in the floor of the cranium as they appear in this view should be noted again. The occipital bone shows a medially located crest with a groove to one side. These are, respectively, the *internal occipital crest* and the *groove for the occipital sinus*. The superior sagittal sinus groove may be seen coming downward to the occipital sinus groove. At the junction is the *internal occipital protuberance*. From the protuberance the *grooves for the transverse sinuses* pass laterally along the occipital bone. They terminate at an S-shaped groove that leads to the jugular foramen. This groove is the *groove for the sigmoid sinus*. Blood collected from the brain is

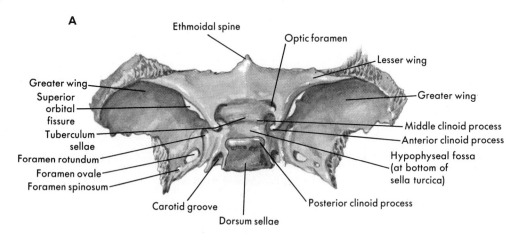

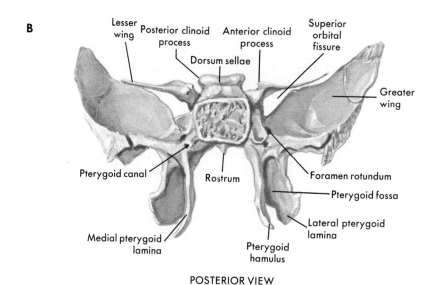

Fig. 5-21
Superior and posterior views of sphenoid bone.

carried out of the skull in these venous sinuses.

The *hyoid bone* (Fig. 5-22), which articulates with no other bone but is connected by ligaments to the styloid process of the skull, lies in the neck at about the level of the lower border of the mandible. Its **body** and *greater* and *lesser cornua* (horns) afford attachment for muscles of the tongue and those muscles used in swallowing.

Tables 5-2 and 5-3 present summaries of the skull bones and foramina.

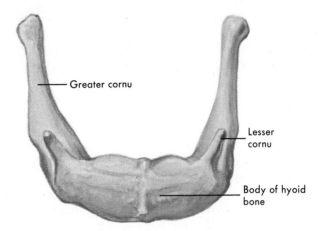

Fig. 5-22
Anterior view of hyoid bone.

Table 5-2. Summary of skull bones and their major features

Bone name and quantity	Major feature on bone	Description or location of bone or feature
Ethmoid (1)		Lies between orbital fossae
	Cribriform (horizontal) plate	Perforated plate in roof of nasal cavity
	Crista galli	Crest in center of cribriform plate
	Perpendicular plate	Forms upper two thirds of septum separating the two nasal cavitites
	Conchae (lateral masses)	Scroll-like projections into nasal cavities
Frontal (1)		Forms roof of orbital cavity, forehead, and nasal cavities
	Supraorbital margin	Superior edge of orbit
	Superciliary ridge	Blunt ridge above supraorbital margin
	Frontal sinuses	Cavities inside bone behind superciliary ridges
	Glabella	Central flat portion of forehead

Adopted from McClintic, J.R.: Basic anatomy and physiology of the human body, ed. 2. Copyright © 1980. Reprinted by permission of John Wiley & Sons, Inc.

Continued.

Table 5-2. Summary of skull bones and their major features—cont'd

Bone name and quantity	Major feature on bone	Description or location of bone or feature
Hyoid (1)		U-shaped bone lying above larynx
	Body	Anterior portion of bone
	Greater cornu (horn)	Posteriorly projected process
	Lesser cornu (horn)	Small and superiorly directed process
Inferior concha (2)		Scroll-like projections form lateral and inferior wall of anasal cavity
Lacrimal (2)		Small bones in anteromedial orbit wall
	Lacrimal canal or groove	Passage along bone
Mandible (1)		Lower jaw
	Body	Horseshoe-shaped portion carrying teeth
	Ramus (*pl.* rami)	Vertically directed process from each side of body
	Angle	Where body and ramus join
	Coronoid process	Anterior process of ramus
	Condyloid process	Articular process—posterior to coronoid process
	Mandibular foramen	Hole on medial aspect of ramus
	Mylohyoid line	Rough ridge anteriorly from mandibular foramen
	Mental foramen	Hole in anteroexterior body
	Alveolar margin	Bear sockets for lower teeth
Maxillary (2)		Carry upper teeth (upper jawbone)
	Orbital plate	Floor of orbital cavity
	Palatine process	Forms anterior two thirds of hard palate
	Infraorbital foramen	Hole below orbital cavity
	Maxillary sinus (antrum of Highmore)	Large cavity within the bone
	Alveolar process or margin	Border carrying sockets for upper teeth
Nasal (2)		Form root or bridge of nose

Table 5-2. Summary of skull bones and their major features—cont'd

Bone name and quantity	Major feature on bone	Description or location of bone or feature
Occipital (1)		Forms posterior aspect and part of base of skull
	Foramen magnum	Large hole in base of bone
	Occipital condyles	Articular surfaces lateral to foramen magnum
	Nuchal lines (superior, inferior, median)	Horizontal and vertical lines on posterior aspect of bone
	External occipital protuberance	Elevation on exterior of posterior part of bone
	Internal occipital protuberance	Opposite external protuberance on interior aspect of bone
Palatine (2)		L-shaped bone forming posterior part of hard palate and sides of nasal cavity
	Horizontal plate	Forms hard palate
	Perpendicular process	Forms part of lateral nasal cavity wall
Parietal (2)		Forms roof and side walls of cranial cavity
	Superior and inferior temporal lines	Arched lines on lateral aspect of each bone
Sphenoid (1)		Butterfly-shaped bone in center of skull, reaches across skull
	Body	Central portion of bone
	Sphenoid sinus(es)	One or two cavities in body
	Sella turcica	Saddle-shaped depression on superior surface of body
	Clinoid processes (anterior, middle, posterior)	Form anterior, center, and posterior limits of sella
	Lesser wings	Extend laterally from anterior clinoid processes
	Greater wings	Extend up lateral sides of skull from body
	Pterygoid processes	Extend inferiorly from body
	Optic foramen	Hole at the base of each lesser wing

Continued.

Table 5-2. Summary of skull bones and their major features—cont'd

Bone name and quantity	Major feature on bone	Description or location of bone or feature
Sphenoid (1)—cont'd	Superior orbital fissure	Cleft between greater and lesser wings
	Foramen ovale	Oval hole at base of greater wing
	Foramen spinosum	Rounded hole posterolateral to foramen ovale
	Foramen rotundum	In posterior aspect of greater wing base
Temporal (2)		Forms part of side wall and floor of cranial cavity
	Squama or squamous portion	Flattened lateral wall of skull
	Zygomatic process	Anterior projection from squama
	Mandibular fossa	Articular surface for mandible at base of zygomatic process
	Tympanic portion	Inferior and posterior to squama
	External auditory meatus	Canal of external ear
	Styloid process	Often broken, it is a spikelike projection inferior on skull
	Mastoid portion	Posterior to meatus
	Mastoid process	Blunt process projecting inferiorly behind meatus
	Mastoid air cells	Air-filled cavities inside process
	Sytlomastoid foramen	Hole between mastoid and styloid processes
	Petrous portion	Laterally running ridge inside cranial cavity
	Internal auditory meatus	Located on anterior and medial aspect of petrous portion
	Jugular foramen	Halfway along suture between occipital bone and petrous temporal bone
	Carotid canal	Anterior to jugular foramen
Vomer (1)		Plow-shaped bone forming lower third of nasal septum
Zygomatic (2) (malar)		Bone of the cheek
	Processes (frontal, zygomatic, maxillary)	Articulate with respective bones

Table 5-3. Foramina and fissures of the skull of humans

Foramina or fissure(s)	Location	Structures passing through
Cranial bones		
Carotid canal	Temporal, petrous portion	Internal carotid artery
Condyloid canal	Occipital, posterior to condyles	Vein from transverse sinus
Foramen magnum	Occipital	Medulla oblongata and its meninges; accessory nerves (XI); vertebral arteries
Hypoglossal canal	Occipital, beneath condyles	Hypoglossal nerve (XII); meningeal artery
Inferior orbital fissure	Sphenoid, zygomatic, maxilla, palatine form its limits	Trigeminal (V), maxillary nerve; infraorbital vessels
Internal auditory meatus	Temporal, petrous portion	Facial nerve (VII); auditory nerve (VIII); internal auditory artery; nervus intermedius
Jugular	Temporal, petrous portion; occipital	Glossopharyngeal (IX); vagus (X); accessory (XI); internal jugular vein
Lacerum	Sphenoid, greater wing and body; apex of petrous temporal; base of occipital	Meningeal branch of the ascending pharyngeal artery; internal carotid artery
Mastoid	Temporal, mastoid portion	An emissary vein
Olfactory	Ethmoid, cribriform plate	Olfactory nerves (I)
Optic	Sphenoid, superior surface	Optic nerves (II); ophthalmic arteries
Ovale	Sphenoid, greater wing	Mandibular nerve (V); accessory meningeal artery; lesser petrosal nerve
Rotundum	Sphenoid, greater wing	Maxillary nerve (V)
Spinosum	Sphenoid, greater wing	Mandibular nerve, recurrent branch (V); middle meningeal vessels
Stylomastoid	Temporal, between mastoid and styloid processes	Facial nerve (VII)
Superior orbital fissure	Sphenoid, between small and great wings	Oculomotor (III); trochlear (IV); trigeminal (V) ophthalmic branch; abducens (VI); orbital branches of middle meningeal artery; superior ophthalmic vein; branch of lacrimal artery

Adapted from McClintic, J.R.: Basic anatomy and physiology of the human body, ed. 2. Copyright © 1980. Reprinted by permission of John Wiley & Sons, Inc.

Continued.

Table 5-3. Foramina and fissures of the skull of humans—cont'd

Foramina or fissure(s)	Location	Structures passing through
Facial bones		
Greater palatine	Palatine bones at posterolateral angle of hard palate	Greater palatine nerve
Incisive (Stensen; Scarpa, sometimes)	Posterior to incisor teeth in hard palate	Stensen, anterior branches of descending palatine vessels; Scarpa, nasopalatine nerves
Infraorbital	Maxilla, body	Infraorbital nerve and vessels
Lesser palatine	Palatine bones at posterolateral angle of hard palate	Lesser palatine nerves
Mandibular	Mandible, about center on medial side of ramus	Inferior alveolar vessels and nerves
Mental	Mandible, lateral anterior surface, inferior to second premolar	Mental nerve and vessels
Supraorbital (or notch)	Frontal bone, above orbit	Supraorbital nerve and vessels
Zygomaticofacial	Zygomatic bone, malar surface (cheek surface)	Zygomaticofacial nerve and vessels
Zygomaticoorbital	Zygomatic bone, orbital surface	Zygomaticotemporal and zygomaticoorbital nerves
Zygomaticotemporal	Zygomatic bone, temporal surface	Zygomaticotemporal nerve

Vertebral column (Fig. 5-23)

The *vertebral column* ("backbone," spine, spinal column) houses the spinal cord and supports the head and thorax and the girdles that attach the appendages to the trunk. It is composed of 24 separate bones called *vertebrae,* the *sacrum* that develops as five distinct bones that fuse together, and the *coccyx,* three to five fused rudimentary "tail bones." In the column above the sacrum the vertebrae are separated by the *intervertebral discs* composed of fibrocartilage. Each disc has a central softer center, the *nucleus pulposus,* and an outer tough capsule of collagenous fibers called the **annulus fibrosus.** The discs are of different size in different areas of the spine, usually being thicker and larger in the lumbar area and thinner higher in the column. They are also thicker anteriorly than posteriorly.

The vertebrae occur in groups: there are seven *cervical* (neck); twelve *thoracic* (chest); and five *lumbar* (lower back) vertebrae. All have a typical structure (Fig. 5-24). The **body** serves as the weight-bearing portion, and it is between vertebral bodies that the discs are placed. The *vertebral* or *neural arch* lies posterior to the body and is composed of two anterior **pedicles** and two posterior **laminae.** Each pedicle contains a superior and inferior notch that encloses an *intervertebral foramen* when vertebrae are articulated. The arch encloses the spinal cord that lies in the **vertebral canal.** Two **transverse processes** and one **spinous process** arise from the arch. The processes afford attachments for muscular and ligamentous structures that support and move the spine. **Superior** and **inferior articular processes** above and below form articulations between vertebrae.

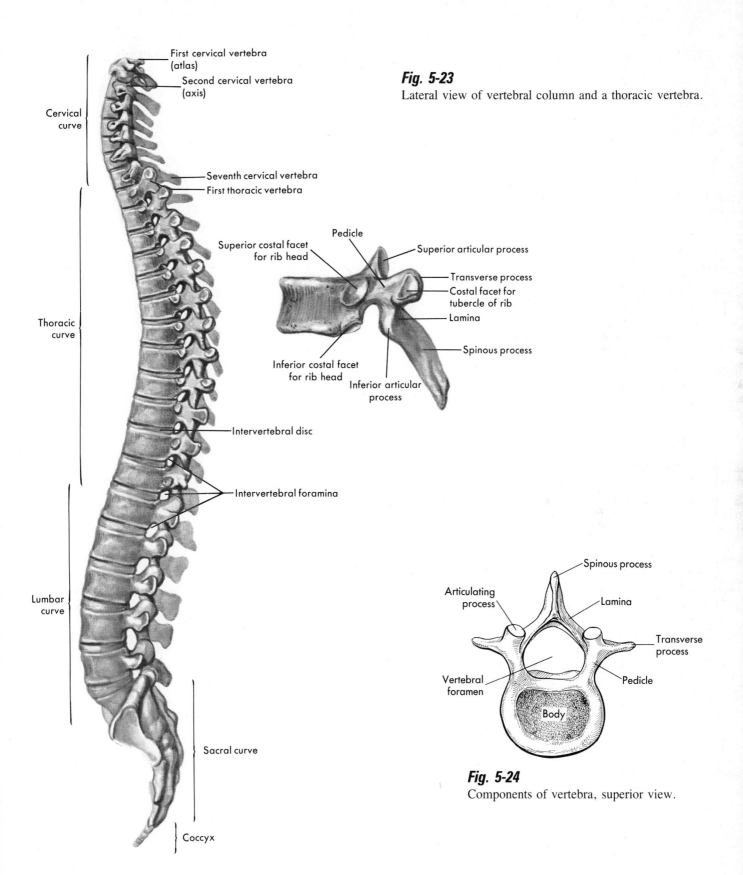

Fig. 5-23
Lateral view of vertebral column and a thoracic vertebra.

Fig. 5-24
Components of vertebra, superior view.

Curvatures of the spine. Viewed from the side, the adult spinal column shows four curvatures (see Fig. 5-23). From the top down they are:

Cervical curve: convex anteriorly
Thoracic curve: concave anteriorly
Lumbar curve: convex anteriorly
Sacral curve: concave anteriorly

The fetal spinal column has only a single curve that is concave anteriorly. Since the thoracic and sacral curves follow the same curvature as the fetal spine, they are called *primary curves*. The cervical and lumbar curves have a curvature opposite to that of the primary curve and are called *secondary curves*. The cervical curve is believed to develop as the infant learns to raise its head while in a prone position, whereas the lumbar curve develops as sitting and standing postures are assumed.

Viewed from in front or from behind, the column should be aligned vertically. Vertical misalignment is called *scoliosis*. Exaggerated thoracic and lumbar curves are called *kyphosis* and *lordosis* respectively (Fig. 5-25).

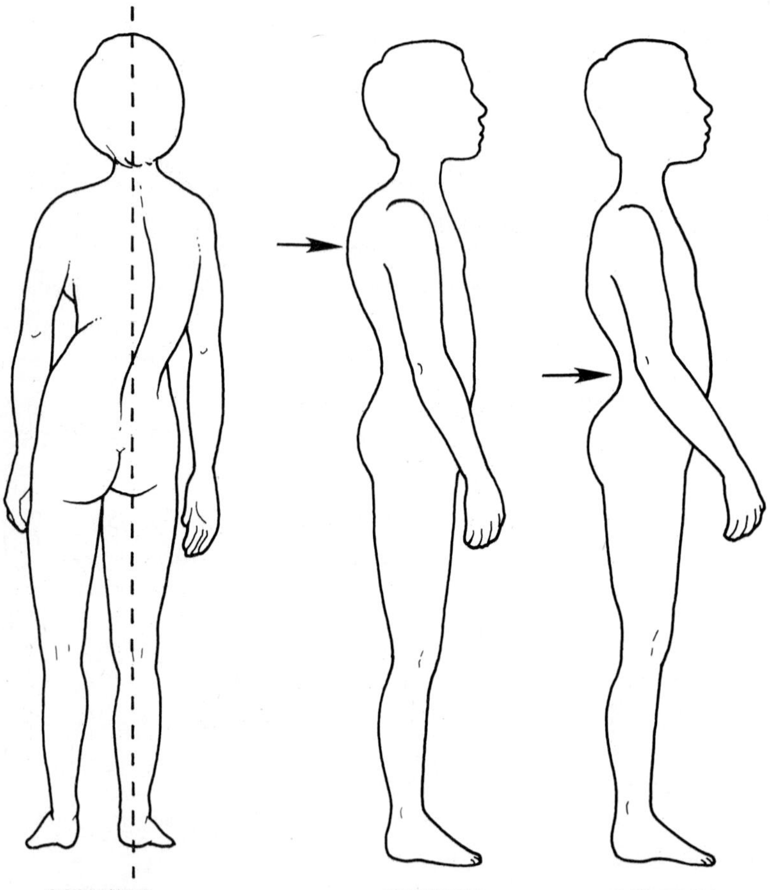

SCOLIOSIS   KYPHOSIS   LORDOSIS

Fig. 5-25
Abnormal curvatures of spine.

Ligaments of the spine. Helping to hold the vertebrae together and to maintain proper alignment of the column are a variety of ligaments (Fig. 5-26) that ultimately attach to the skull.

anterior longitudinal and *posterior longitudinal ligaments* run the length of the column along the anterior and posterior aspects of the vertebral bodies. These ligaments are rich in tough collagenous fibers and give a stability to the column as a whole.

ligamenta flava are oriented vertically between the laminae. They are rich in elastic fibers and aid in returning the column (extension) to the normal position from a flexed ("bent forward") position.

interspinous ligaments lie between spinous processes. They are also elastic and aid in extension of the spine.

supraspinous ligament connects the tips of the spinous processes and is elastic also.

ligamentum nuchae represents an enlarged supraspinous ligament in the neck. It extends to the skull, separates the muscle masses of the neck into right and left portions, and aids in holding the head "up."

intertransverse ligaments, also containing much elastic tissue, lie between the transverse processes. They aid in maintaining lateral stability of the column.

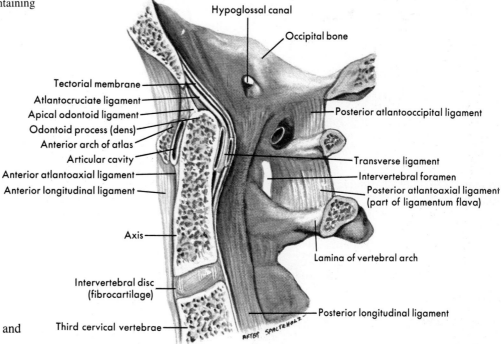

Fig. 5-26
Median section of occipital bone and first three cervical vertebrae.

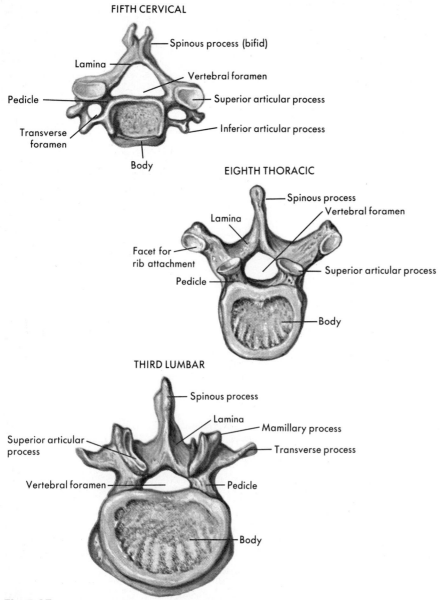

Fig 5-27
Regional variations of vertebrae to illustrate major differences (drawings to same scale).

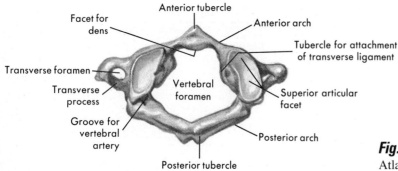

Regional variations in the vertebral column (Fig. 5-27). Bodies* that appear compressed with elevated margins characterize the *cervical vertebrae*. Vertebral foramina are large and triangular in outline. The bodies have jutting "rims" that aid in containing the disc. The spinous processes are commonly *bifid* (divided). The seventh cervical vertebra has an elongated spine that is easily felt at the posterior base of the neck *(vertebra prominens)*. The transverse process contains the *transverse foramen* that passes the vertebral arteries and veins to and from the brain.

The first cervical vertebra, or *atlas* (Fig. 5-28), has no body: its two *lateral masses* are joined by an *anterior* and *posterior arch* to form a ring. The superior surfaces of the lateral masses act as the superior articulating processes with the occipital condyles. The anterior arch is twice as long as the posterior arch and bears the midline *anterior tubercle* anteriorly and a midline *facet for the dens* (see following paragraph) posteriorly: its posterior and distal ends that connect with the lateral masses bear *tubercles for attachment of the transverse ligament*. The posterior arch bears a midline *posterior tubercle* that represents the spine of the vertebra. Transverse foramina are present in the blunt transverse processes.

The second cervical vertebra (Fig. 5-29) or *axis* has as its characteristic feature the *dens* (odontoid process) attached to the superior aspect of the body. It is regarded as what remains of the atlas body that has become attached to the axis. The transverse ligament of the atlas passes behind the dens to create a pivot joint permitting rotation of the skull. Otherwise, the axis does not differ greatly from the third to seventh cervical vertebrae.

---

*Except the atlas or first cervical vertebra, to be described in greater detail in the next paragraph.

Fig. 5-28
Atlas (first cervical vertebra), superior view.

Thoracic vertebrae have a rounded vertebral foramen, and the bodies carry **facets for rib attachment.** The bodies of the first, tenth, eleventh, and twelfth thoracic vertebrae carry full facets, the rib head articulating with only one vertebra. The other vertebrae share a rib, and strictly speaking the articular surfaces are *demifacets.* Pedicles are flattened from side to side. Spinous processes are three-sided and incline downward to overlap the spine below. The transverse processes are strong and carry a **costal facet** for attachment to the tubercle of a rib.

Lumbar vertebrae have bodies that are heavy and slightly kidney shaped. Vertebral foramina are triangular in shape. The spinous processes are almost square in shape, whereas the transverse processes are slender and inclined slightly backward. The posterior articulating processes are faced laterally and are clasped by the superior articulating processes of the vertebra below to give strength to the lumbar area. A small rounded **mamillary process** is seen on the posterior rim of the superior process.

The ***sacrum*** (Fig. 5-30) is a flattened triangular bone originating as five sacral vertebrae that have fused to form the single bone. These lines of fusion are easily seen on the anterior surface of the bone. The four pairs of **anterior sacral foramina** pass the anterior divisions of the sacral nerves. The posterior surface of the bone is convex, and its midline is marked by the **middle sacral crest,** the evidence of fusion of the sacral spinous processes. The **vertebral canal** is flattened, opening superiorly at the *sacral canal* and inferiorly at the *hiatus of the sacral canal.* Laterally to the hiatus are paired **sacral cornua.** The lateral processes, or **alae,** bear four pairs of **posterior sacral foramina:** on the upper side of the processes is the **auricular articulating surface** for the os coxae. The **sacral promontory** is the anterior landmark of the sacrum. It is the anterior and superior edge of the first sacral vertebra.

The *coccyx* (see Fig. 5-30) consists of three to five rudimentary and more or less fused vertebrae. The superior one carries ***coccygeal cornua,*** whereas the remainder consist of little more than reduced bodies.

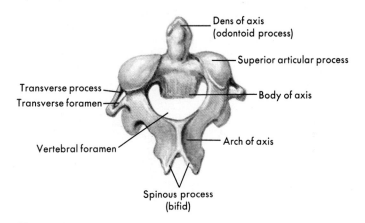

Fig. 5-29
Axis (second cervical vertebra).

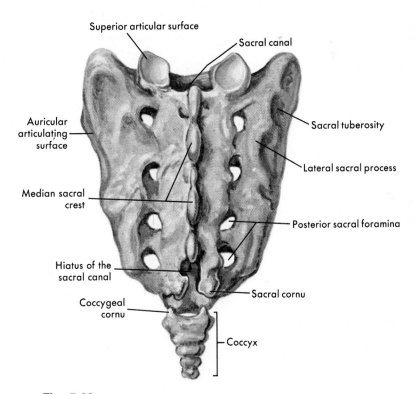

Fig. 5-30
Posterior view of sacrum and coccyx.

Thorax

The thorax or chest is composed of the 12 pairs of *ribs,* their *costal cartilages,* and the *sternum.* Viewed anteriorly or posteriorly, the thorax has a conical shape with a narrow superior *inlet* and the broad inferior *outlet.* In cross-section the thorax presents a kidney-shaped outline. (In persons habituated to high altitudes from infancy, the thorax has a "barrel" shape with wide inlet and a more circular cross-section. This reflects the deeper breathing and larger lungs.)

Ribs. There are 12 pairs of ribs (Fig. 5-31) in the normal adult ("extra" cervical ribs are sometimes present in the neck; presumably only Adam had 11). The first seven pairs connect, via their costal cartilages, *directly* to the sternum and are thus called "true ribs." The next three pairs have costal cartilages that connect to the costal cartilage above and not directly to the sternum. The last two pairs have no cartilages at all and "float" in the posterior abdominal musculature. The last five pairs are called "false ribs" and have no direct or no sternal attachment at all; the last two pairs are called "floating" as well as "false" ribs.

General morphology of a rib. The parts of a "typical" or central rib such as the sixth or seventh will be described first, followed by a discussion of the morphology of the first, second, eleventh, and twelfth ribs.

A typical rib articulates with the vertebral body (or bodies) by way of its *head.* The head contains two *demifacets* for articulation with two adjacent vertebrae separated by an *interarticular*

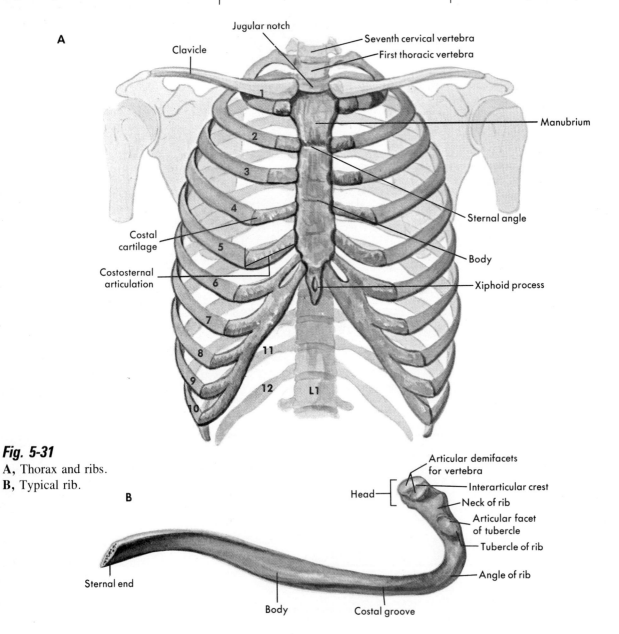

Fig. 5-31
A, Thorax and ribs.
B, Typical rib.

crest. (Ribs 1, 10, 11, and 12 articulate with only one vertebral body and thus have single facets and no crest.) A narrowed **neck** lies distal to the head, terminating at the **tubercle**. The tubercle has an *articulating portion* that is larger and that attaches to the transverse process of the vertebra and a smaller *nonarticulating portion* that serves as the attachment for the **costotransverse ligament** connecting the rib to the transverse process. Beyond the tubercle is the **body** or **shaft** of the rib. The body is composed of parts of different curvature; where the two parts meet is the **angle**. The body curves downward and forward, and the rib cannot be "flattened" on a surface. The inner side of the body bears a **costal groove** for passage of the *intercostal vessels* and *nerves*. A blunt or rounded **superior surface** of the body affords attachment for *intercostal* muscles (except the first rib), whereas the **inferior surface** is sharper and also attaches intercostal muscles (except the twelfth rib). Ribs 1 to 10 carry a depression on their distal ends for attachment to the costal cartilage.

The first rib is flat, not curving downward, and has no angle. Its body is horizontal and presents, on its superior aspect about halfway down the body, the **scalene tubercle** for attachment of a scalene muscle. The superior surface also is grooved for the *subclavian artery*. A single articular facet on the head articulates with the first vertebral body.

The second rib is larger, has its body placed at about a 45-degree angle to the vertical plane, and shows only a slight angle. From ribs 1 to 7 lengths increase, and the body is oriented vertically. The ribs then become shorter to the twelfth and have smaller angles.

The eleventh and twelfth ribs have no necks, no tubercles, and no distal depressions for a costal cartilage, since they have no cartilages. The twelfth rib may be shorter than the first rib.

Sternum. The sternum (Fig. 5-32) forms the anterior midline of the upper thorax. Three portions compose it: a superior **manubrium**, a central **body** *(gladiolus)*, and an inferior **xiphoid process.** The process is cartilaginous in younger individuals and may be absent on a sternum specimen.

The manubrium is roughly triangular in shape and bears on its superior surface the **suprasternal (jugular) notch.** On either side of the notch are articular surfaces for the medial ends of the clavicles *(sternoclavicular joint).* The lateral borders of the manubrium carry **facets** for the costal cartilages of the first ribs. Lower on the lateral sides are *demifacets* for the costal cartilage of the second rib. Anteriorly the manubrium joins the body at a ridge called the **sternal angle,** a feature readily palpable about 5 cm below the suprasternal notch on the anterior thorax. The upper end of the body bears laterally the other two demifacets for articulation of the second costal cartilages. The lateral sides of the body bear **facets** for attachment of the third through seventh costal cartilages. The xiphoid process is variable in shape in different individuals.

The bones of the vertebral column and thorax and their features are summarized in Table 5-4.

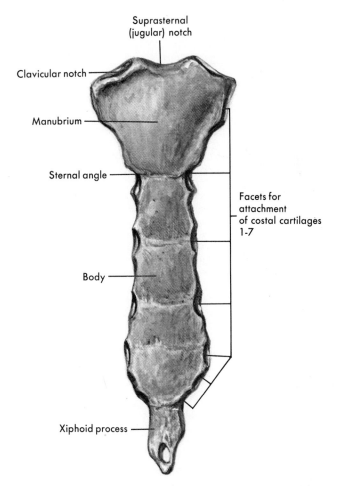

Fig. 5-32
Sternum.

Table 5-4. Summary of the features of the vertebral column and thorax

Name of bone or region	Distinguishing feature(s)	Description/location of bone/feature
Vertebra		
General structure	Body, with discs between adjacent bodies; neural arch, composed of pedicles and laminae; transverse, spinous, and articulating processes	Body, anterior; arch, posterior; transverse processes laterally, spinous process posteriorly, pedicles between body and transverse processes, laminae between transverse and spinous processes; articulating processes superiorly and inferiorly from upper pedicle
Cervical vertebrae (7)	Compressed bodies, transverse foramina, often bifid spinous processes	Transverse foramina in base of transverse processes
Atlas (first cervical)	No body, anterior and posterior arches; superior articulating processes large and concave	Ringlike bone
Axis (second cervical)	Dens	Dens projects superiorly from body
Thoracic vertebrae (12)	Facet or demifacet on body for rib attachment; facets on transverse processes for rib attachment	—
Lumbar vertebrae (5)	Lack of transverse foramina and facets for ribs; mamillary processes	Heavy bodies; articulating processes "clasp" one another
Sacrum	Triangular shape, sacral canal and foramina; articulating surface for ilium	Five fused and flattened vertebrae
Coccyx	Three to five fused bones consisting of bodies	No canal; "tail bone"
Ribs (24)		
General structure	Head with (demi-) facets for articulation with vertebral body (bodies); neck, tubercle for articulation with transverse process; angle, shaft; costal groove on shaft; depression distally for costal cartilage	—
First rib	Flattened, no angle	—
Second rib	Oriented 45 degrees to vertical plane; slight angle	—
Eleventh and twelfth ribs	No neck; no depression for cartilage	—
Sternum	Manubrium, body, xiphoid process; sternal angle	Manubrium carries notch superiorly; angle between manubrium and body anteriorly; xiphoid cartilaginous in youth

Pectoral girdle

The upper appendage attaches to the axial skeleton by the *pectoral girdle*, which consists of four bones: paired anterior *clavicles* (collarbones) and posterior *scapulae* (shoulder blades). The girdle attaches to the axial skeleton at only one point, the sternoclavicular joint. This arrangement permits the great mobility of the upper appendage.

Clavicles. Viewed from the superior aspect, the clavicle (Fig. 5-33) presents an S-shaped outline with the medial half convex anteriorly. The lateral half is concave anteriorly. The *medial (inner or sternal) end* is broad and articulates with the sternum. The *lateral (outer or acromial) end* is flattened and articulates with the acromion process of the scapula (acromioclavicular joint). On the inferior surface, about 2.5 cm from the lateral end, is the *conoid tubercle (coracoid tuberosity)*. It affords attachment for the conoid ligament, holding the clavicle to the coracoid process of the scapula. The clavicle is responsible for maintaining the normal height of the shoulder. If it is broken, the shoulder falls and is not maintained at the proper distance from the thorax. Falling on the outstretched arm is the most common cause of clavicular fracture, since the force of the fall is transmitted directly to the length of the clavicle.

Scapulae. The scapulae (Fig. 5-34) lie over the posterosuperior aspect of the thorax. A triangular flattened *body* gives rise posteriorly to a ridgelike *spine* that terminates laterally in the *acromion process*. The body presents three borders: the *medial (vertebral) border*, the *lateral (axillary) border*, and the *superior border*. The latter bears laterally the *scapular notch*. Where the medial and lateral borders come together is the *inferior angle*. Where medial and superior borders meet is the *superior angle*. The shallow *glenoid fossa* is located at the *lateral angle* where lateral and superior borders meet. It affords an articular surface for the head of the humerus.

Projecting anteriorly over the glenoid fossa is the *coracoid process*. A *supraglenoid tubercle* lies above the glenoid fossa and an *infraglenoid tubercle* below. The anterior or *costal surface* of the body contains the *subscapular fossa*. The posterior surface of the body has two fossae: the *infraspinous fossa* below the spine and the *supraspinous fossa* above the spine. These three fossae contain muscles operating the shoulder joint.

Fig. 5-33
Anterior view of right clavicle.

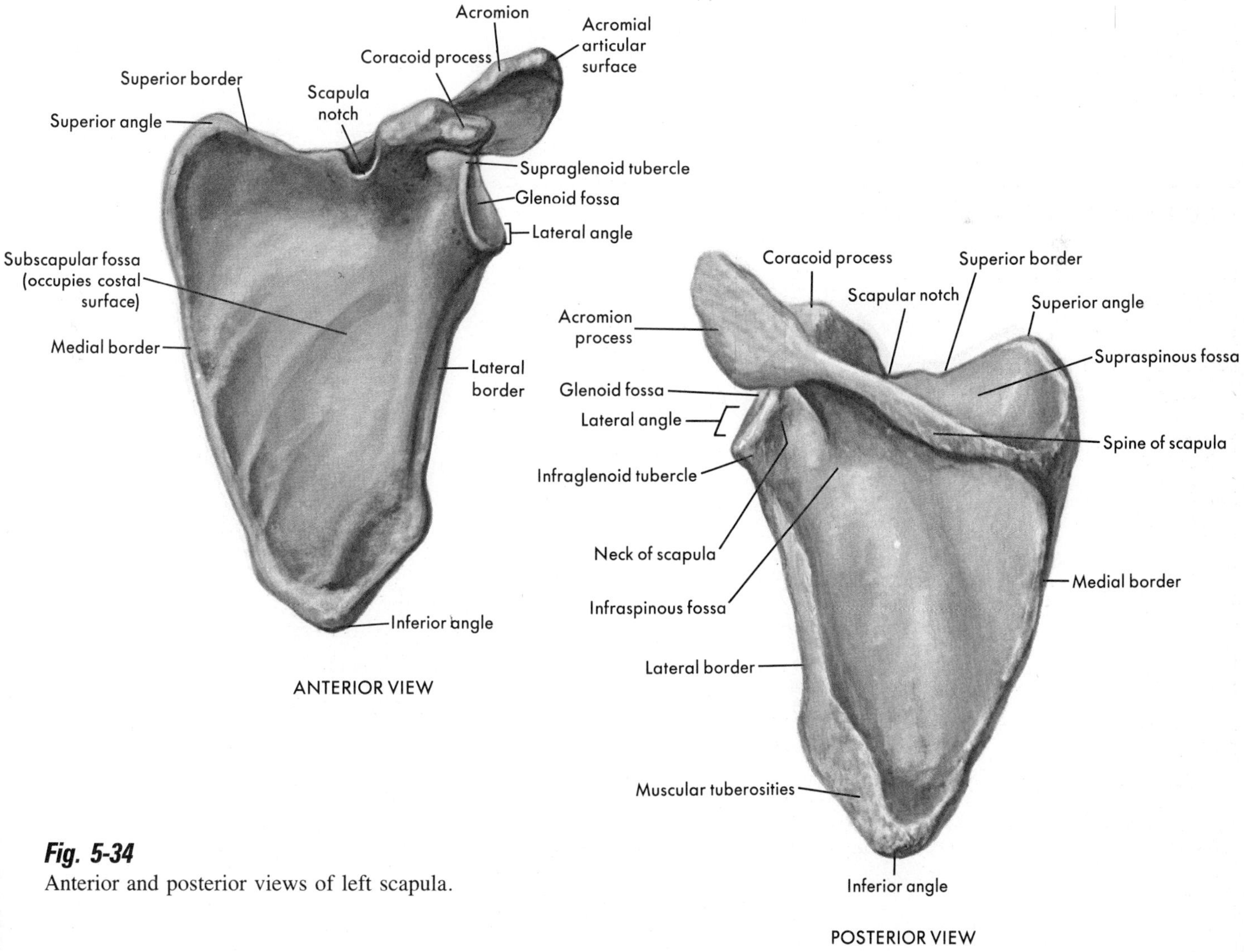

Fig. 5-34
Anterior and posterior views of left scapula.

Upper appendage

Humerus. The bone of the arm is the *humerus* (Fig. 5-35). It articulates proximally with the glenoid fossa of the scapula and distally with the bones of the forearm, the radius and ulna. A large rounded **head** forms the proximal end of the bone, under which is a thick **anatomical neck**. The **greater tubercle** *(tuberosity)* lies on the lateral aspect of the proximal end. Anterior is the **lesser tubercle** *(tuberosity)*, and between the two tubercles, on the anterior side of the proximal end, is the **intertubercular sulcus** (or *bicipital groove*). The latter contains the tendon of the long head of the biceps brachii muscle as it passes to attach to the supraglenoid tubercle. Below the bases of the greater and lesser tubercles lies the thin **surgical neck,** so named because of its liability to fracture. The **shaft** of the humerus is cylindrical above and prismatic below, where it presents a **rounded anterior border** and sharper **medial** and **lateral borders**. About halfway down the lateral side of the shaft is the **deltoid tuberosity,** affording attachment of the deltoid muscle of the shoulder. At the base of the deltoid tu-

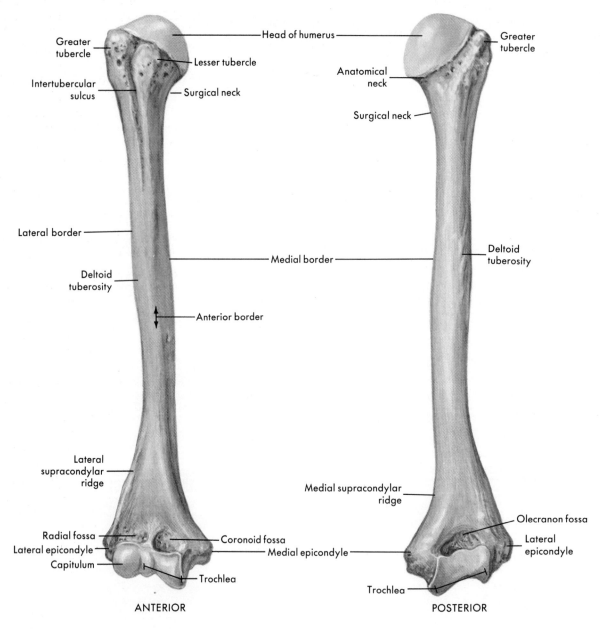

Fig. 5-35

Anterior and posterior views of right humerus.

berosity is a large **nutrient foramen** passing blood vessels into and out of the bone. The distal end bears two *condyles*. The medial one is spool shaped and is known as the **trochlea**. It articulates with the ulna. The lateral condyle is rounded, is named the **capitulum,** and articulates with the radius. A large, easily palpable, triangular-shaped process projects medially above the trochlea. It is the **medial epicondyle**. The medial border of the distal humerus forms the **medial supracondylar ridge** above the epicondyle. A small, more blunt **lateral epicondyle** is continued upward by the **lateral supracondylar ridge**. The epicondyles and ridges provide attachments for many of the forearm muscles. Anterior on the distal humerus, lying above the trochlea, is the **coronoid fossa**. Above the capitulum is the small **radial fossa**. When the elbow is flexed (bent), the ulna's coronoid process and radius' head contact the fossae to stop motion. Posterior on the distal end is the larger and deeper **olecranon fossa**. When the elbow is extended (straightened), the ulna's olecranon process contacts the fossa to limit extension.

Ulna. The medial bone of the forearm is the **ulna** (Fig. 5-36). Its proximal end bears the large **olecranon process** posteriorly and the more pointed **coronoid process** anteriorly. Between the two processes is the **semilunar (trochlear) notch** that articulates with the trochlea of the humerus. The joint so formed only permits flexion and extension of the elbow. On the lateral side of the coronoid process is the **radial notch** that articulates with the head of the radius. The **shaft** has three borders: the **interosseous border** faces laterally and affords attachment to the **interosseous membrane** that helps hold the ulna to the radius; the **anterior border** is smooth and round, whereas the **posterior border** is sharp above and more rounded below. The distal end of the ulna bears an expanded **head** with a fossalike condyle on it for articulation with the carpal (wrist) bones. A short **styloid process** projects from the memdial side of the head.

Radius. The lateral bone of the forearm is the radius (see Fig. 5-36). Its proximal end bears a **head**, a **neck** distal to the head, and a large **radial tuberosity** on the anterior surface below the neck. The **shaft** is bowed laterally, giving resilience to the bone, and is narrow above and thicker below. It also presents an **interosseous border,** directed medially, and **anterior** and **posterior borders**. The distal end is expanded, bearing a concave **carpal surface** and a lateral **styloid process**. The medial margin of the lower end bears the **ulnar notch** into which fits the head of the ulna. **Colles' fracture** refers to a transverse break of the distal radius about 2.5 cm above the wrist joint. It is the most frequent site of fracture in the entire skeleton.

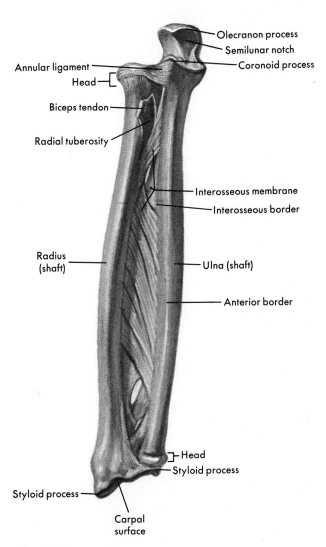

Fig. 5-36
Anterior view of proximal, intermediate, and distal articulations of right forearm and features of radius and ulna.

Carpal bones. The eight *carpal (wrist)* bones (Fig. 5-37) are arranged in a proximal and distal row of four bones each. Viewed from the posterior side of the hand, the bones, from lateral to medial, are the **scaphoid** *(navicular,* which resembles a boat; the **lunate** (a "half-moon"); the ***triquetrum*** *(triangular);* and the ***pisiform*** (pealike). The distal row, from lateral to medial, consists of the ***trapezium*** *(greater multangular);* ***trapezoid*** *(lesser multangular);* ***capitate***, with a rounded head; and ***hamate***, bearing a curved hooklike process. The proximal row presents a curved surface articulating with the distal and concave ends of the radius and ulna, whereas the distal row bears surfaces for articulation with the metacarpals.

Metacarpals. The five ***metacarpals*** (see Fig. 5-37) form the bones of the hand. They are numbered 1 to 5 beginning with the lateral or thumb side. The proximal ***base*** of each metacarpal is enlarged and carries a concave surface for articulation with the carpals. The ***shaft*** is narrowest in its middle third and then enlarges slightly to terminate distally in a rounded ***head.*** The rounded heads are felt as the knuckles when the digits are flexed. The first metacarpal is the shortest, the second the longest.

Phalanges. The hand bears five digits having a total of 14 ***phalanges*** (see Fig. 5-37). Four *fingers* each carry three phalanges, whereas the *thumb* has two phalanges. In the fingers the phalanges are named *proximal* (first), *middle* (second) and *distal* (third or terminal). In the thumb they are named *proximal* (first) and *distal* (second or terminal). Proximal and middle phalanges show a ***base*** proximally, a ***shaft,*** and a distal ***head.*** The terminal phalanges have a *base,* a *shaft,* and an expanded helmet-shaped *head* containing a *tuberosity* for the pad of the fingertip.

Sesamoid bones are common at the anterior aspects of the joints of the metacarpal and the first phalanx of the thumb, index finger, and little finger.

Table 5-5 summarizes the bones of the pectoral girdle and upper appendage.

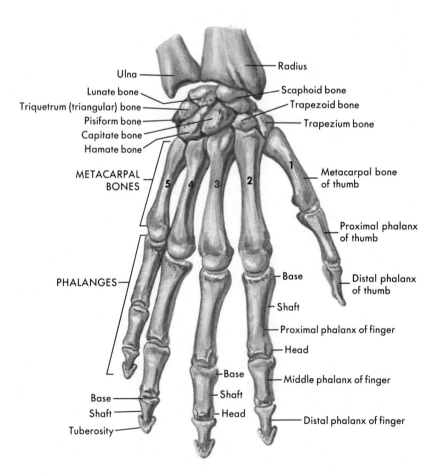

Fig. 5-37
Posterior view of bones of right hand and wrist.

Table 5-5. Summary of the bones of the upper appendage and girdle and their major features

Name and number of bones in *each* appendage or half of girdle	Major feature on bone	Description or location of bone or feature
Clavicle (1)		The "collarbone"
	Medial (sternal) end	Articulates with sternum
	Lateral (acromial) end	Flattened, articulates with scapula
Scapula (1)		"Shoulder blade"
	Body	Bladelike triangular portion of the bone
	Superior border	Upper border of the bone
	Axillary (lateral) border	Border facing the armpit
	Vertebral (medial) border	Border facing the spine
	Spine	Horizontal ridge on posterior surface
	Supraspinous fossa	Shallow depression above the spine
	Infraspinous fossa	Shallow depression below the spine
	Subscapular fossa	Shallow depression on anterior surface
	Glenoid fossa	Lateral depression for articulation with humerus
	Acromion process	Lateral projection from spine; overlies glenoid fossa
	Coracoid process	Beaklike projection anterior to glenoid fossa
	Inferior angle	Where vertebral and medial borders meet; inferior
	Superior angle	Where superior and vertebral borders meet
	Lateral angle	Where superior and lateral borders meet

Adapted from McClintic, J.R.: Basic anatomy and physiology of the human body, ed. 2. Copyright © 1980. Reprinted by permission of John Wiley & Sons, Inc.

Continued.

Table 5-5. Summary of the bones of the upper appendage and girdle and their major features—cont'd

Name and number of bones in *each* half of girdle	Major feature on bone	Description or location of bone or feature
Humerus (1)		The bone of the (upper) arm
	Head	Rounded, proximal articular surface
	Anatomical neck	Thick portion just distal to head
	Surgical neck	Proximal area of shaft
	Greater tuberosity	Lateral to anatomical neck
	Lesser tuberosity	Anterior and inferior to anatomical neck
	Bicipital groove (intertubercular sulcus)	Between greater and lesser tuberosities
	Deltoid tuberosity	On lateral aspect of humerus, about one third the way along shaft
	Medial epicondyle	Medial sharp process on distal end of humerus
	Lateral epicondyle	Lateral process on distal end of humerus
	Capitulum	Rounded articular surface on distal humerus
	Trochlea	Spool-shaped articular surface on distal humerus
	Coronoid fossa	Anterior depression above articular surfaces
	Olecranon fossa	Posterior deep depression above articular surfaces
Ulna (1)		Medial bone of forearm
	Semilunar notch	Half-moon–shaped surface on distal ulna; articulates with trochlea
	Olecranon process	Posterior process forming "point" of elbow
	Coronoid process	Anterior process forming inferior limit of semilunar notch
	Radial notch	Laterally placed smooth articular surface for radius head
	Styloid process	Pointed process on distal end of the ulna

Table 5-5. Summary of the bones of the upper appendage and girdle and their major features—cont'd

Name and number of bones in *each* appendage or half of girdle	Major feature on bone	Description or location of bone or feature
Radius (1)		Lateral bone of the forearm
	Head	Rounded proximal end
	Neck	Distal to the head
	Radial tuberosity	Anteriorly directed rough elevation distal to neck
	Styloid process	Triangular-shaped distal end of bone
Carpals (8)		Bones of wrist
	Navicular (scaphoid)	Shaped like a boat
	Lunate	Has half-moon–shaped surface
	Triquetrum (triangular)	Shaped like a right triangle
	Pisiform	Shaped like a pea
	Trapezium (greater multangular)	A "little table"
	Trapezoid (lesser multangular)	"Table shaped"
	Capitate	Has rounded head
	Hamate	Has hooklike process
Metacarpals (5)		Bones of hand, distal ends form "knuckles," numbered 1-5 from the thumb side
Phalanges (14)		Bones of the digits: 2 in thumb, 3 in each finger; named by position as proximal, medial, distal (except thumb—no medial bone)

Pelvic girdle

The sacrum attaches at the sacroiliac joints to the two *os coxae* ("hip bones") that form the *pelvic girdle* or pelvis. Fetally, each os coxae develops as three separate bones (Fig. 5-38): a large superior *ilium*, an inferior and anterior *pubis*, and an inferior and posterior *ischium*. The three bones meet at the *acetabulum*, the socket forming part of the hip joint. The acetabulum shows an inferior *acetabular notch*, a *lunate area* or *surface*, and an *acetabular fossa* (Fig. 5-39).

Ilium. The ilium (see Fig. 5-39) is bounded superiorly by the *iliac crest*, which ends in the *anterior superior iliac spine* and the *posterior superior iliac spine*. The anterior margin of the ilium runs downward and presents the *anterior inferior iliac spine* just above the acetabulum. The posterior border runs downward from the posterior superior spine and terminates at the *posterior inferior iliac spine*. Beneath the latter spine the posterior border turns abruptly forward and then down to form the *greater sciatic notch*. The lateral surface of the *wing* of the ilium is marked by three *gluteal lines: anterior* (middle), *posterior*, and *inferior*. These serve as attachments for the gluteal muscles. The medial surface of the iliac wing presents the large *iliac fossa* that contains the iliacus muscle. Posteriorly is found the roughly ear-shaped *articular surface* for forming a joint with the sacrum. Beginning at the anterior and inferior portion of the articular surface is a rounded ridge that continues to the pubic symphysis. It is the *arcuate (iliopectineal) line*.

Ischium. The posterior and inferior bone forming the os coxae is the *ischium* (see Fig. 5-39). It forms the posterior two fifths of the acetabulum and then projects downward and backward as the *body* of the bone. Its *medial border* completes the greater sciatic notch that is bounded inferiorly by the *ischial spine*. The *lesser sciatic notch* lies inferior to the spine, and then comes a larger roughened *ischial tuberosity*. Projecting anteriorly from the tuberosity is the *ramus* of the ischium.

Pubis. The anterior and inferior bone forming the os coxae is the *pubis* (see Fig. 5-39). Its *inferior ramus* joins with the ischial ramus posteriorly and with the pubis *body* anteriorly. The medial surface of the body presents a roughened *symphyseal surface* that forms, with the opposite bone, a joint held by fibrocartilage called the *pubic symphysis*. The *pubic crest* is directed anteriorly from the body. The *superior ramus* is directed upward and laterally and expands to form the anterior one fifth or so of the acetabulum. The ischium and pubis enclose the *obturator foramen*, the largest foramen in the body.

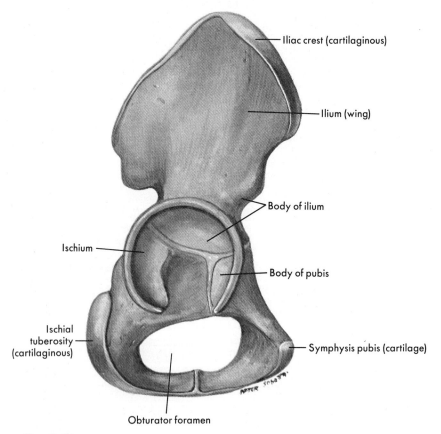

Fig. 5-38
Right os coxae of child showing limits of three bones composing it.

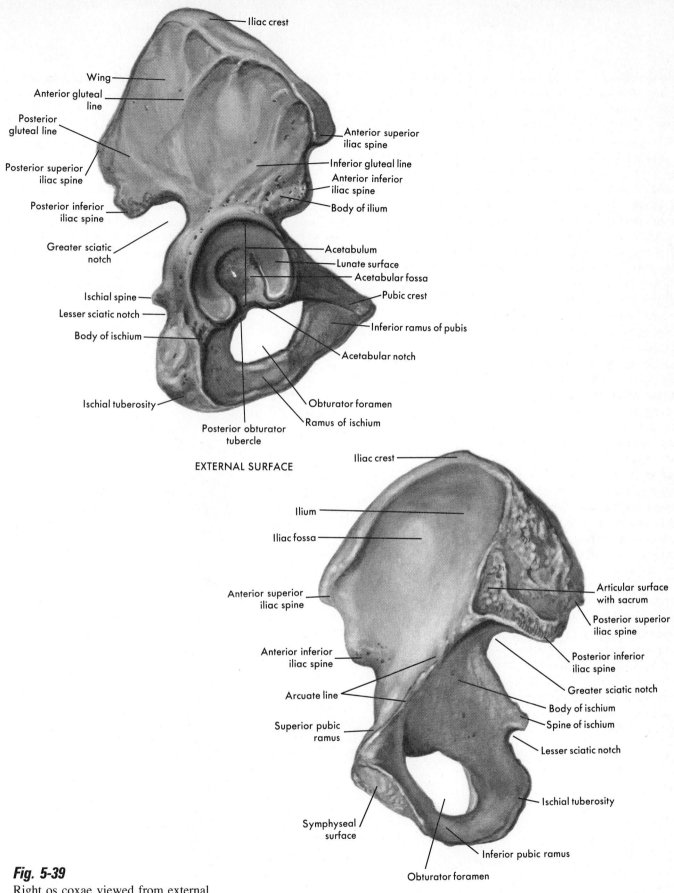

Fig. 5-39
Right os coxae viewed from external and internal surfaces.

Bony pelvis. The oval formed by the arcuate line of the ilium and extending to the pubic symphysis, and posteriorly across the sacrum as the pelvic brim, separates the *greater* or *false pelvis* superiorly between the iliac wings from the *lesser* or *true pelvis* enclosed mostly by the ischial and pubic bones. The inferior aperture or *outlet* of the true pelvis forms the birth canal. If we place the anterior superior iliac spine and the pubic symphysis in the same vertical plane, the normal position of the pelvis is established. It may then be seen that the plane of the inlet to the true pelvis inclines backward and downward at about 30 to 40 degrees from the vertical plane. The *transverse diameter of the inlet* is between the widest points of the arcuate lines and measures about 12.5 cm. A line drawn between the sacral promontory (Fig. 5-40) and the superior border of the symphysis defines the *anteroposterior* or *true conjugate diameter*. It measures about 11.5 cm. The angle seen to be formed by the superior rami of the pubic bones, having its apex at the symphysis, is the *subpubic angle*.

Characteristics of male and female pelves (Fig. 5-40). The male pelvis is generally more massive than that of the female, has somewhat more obvious processes and lines, and has a smaller diameter. The male pelvis is deeper in the vertical plane but narrower across the iliac crests than that of the female and is sometimes described as heart shaped (♡), as compared with the more kidney-shaped (⌒) female pelvis. In the male the subpubic angle is typically 90 degrees or less, whereas in the female the more widely flaring ilia and kidney-shaped inlet to the false pelvis result in an angle that is usually greater than 90 degrees (110 to about 120 degrees is common).

Other differences include a larger acetabular fossa in the male, a hooklike greater sciatic notch in the male and a wider notch in the female, and a more triangular obturator foramen in the female versus a rounder one in the male.

To reach a decision as to the sex of a skeleton, all the factors mentioned must be considered. Race, heredity, and nutritional and hormonal factors all combine to define the final form of the pelvis.

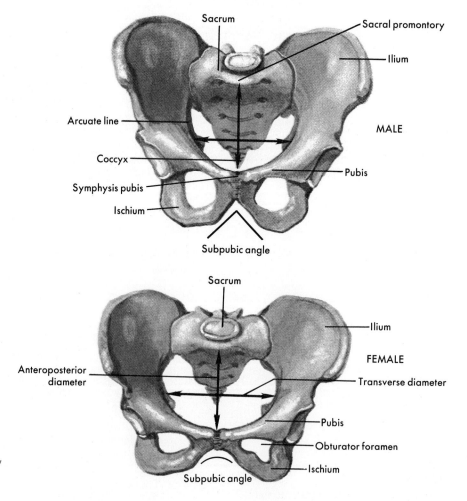

Fig. 5-40
Male and female pelves to show differences.

Lower appendage

Femur. The bone of the thigh is the *femur* (Fig. 5-41), which is the longest and strongest bone of the body. The **head** articulates with the acetabulum and bears a centrally placed ***fovea capitis*** for attachment of the *ligamentum teres* (round ligament) that carries blood vessels into the bone. Beneath the head lies the **neck**. Projecting upward and laterally at the base of the neck is the ***greater trochanter***. A ***lesser trochanter*** projects medially and slightly posteriorly below the neck. Anteriorly, the two trochanters are connected by a roughened ***intertrochanteric line***; posteriorly they are connected by a larger ***intertrochanteric crest***.

The ***shaft (body)*** of the femur is narrowest in the middle and widens toward both ends. The posterior aspect is marked by a prominent roughened ridge occupying approximately the middle third of the shaft. This is the ***linea aspera***. At the upper end of the linea the ***gluteal tuberosity*** extends toward the base of the greater trochanter. Extending toward the base of the lesser trochanter from the linea is the ***pectineal line***, and inferior to this is a line that passes from the linea around the medial aspect of the shaft toward the anterior surface. The latter is the ***spiral line***. Inferiorly the linea gives rise to two lines, the ***medial*** and ***lateral supracondylar lines*** (ridges) that terminate distally at the ***medial*** and ***lateral epicondyles*** of the femur. The ***adductor tubercle*** lies at the junction of medial epicondylic line and medial epicondyle.

The distal end of the femur displays a ***medial*** and ***lateral condyle*** and, anteriorly, the ***patellar surface***, into which fits the patella or kneecap. The ***intercondylar notch*** (or *fossa*) lies posteriorly between the condyles and bears the ***intercondylar line*** at its upper border. A triangular ***popliteal surface*** is enclosed between the intercondylar line and the two supracondylar lines.

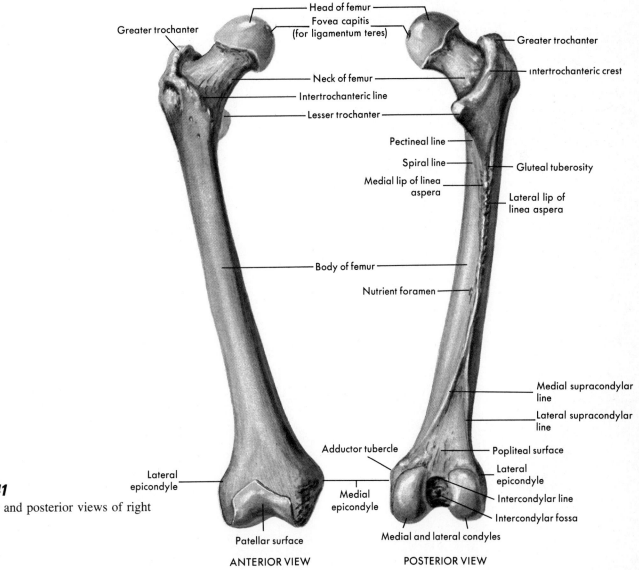

Fig. 5-41
Anterior and posterior views of right femur.

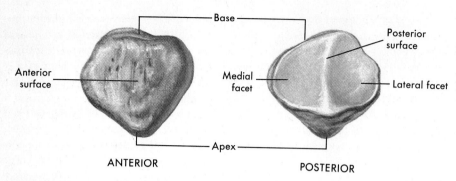

Fig. 5-42
Anterior and posterior views of right patella.

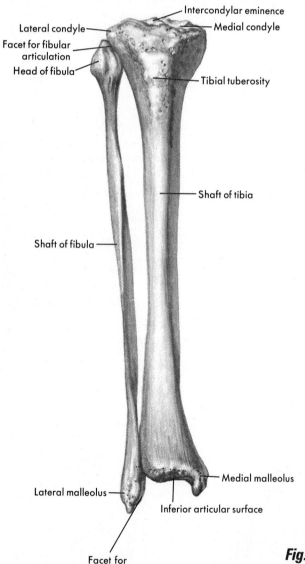

Fig. 5-43
Anterior view of right tibia and fibula.

Patella. The *patella* or kneecap (Fig. 5-42) is roughly triangular in shape with a superior *base* and an inferior *apex*. It is a sesamoid bone, developing in the tendon of the quadriceps muscle. The *anterior surface* is ridged vertically by the connective tissue fibers of the tendon. The *posterior surface* presents a smaller *medial facet* for articulation with the medial condyle of the femur and a larger *lateral facet* for articulation with the lateral condyle of the femur. The patella tends to protect the knee joint when it is in the flexed state and also changes the direction of muscle pull on the tibia to a more efficient angle. Its loss does not seriously impair limb function.

Tibia. The medial, larger, and weight-bearing bone of the leg is the *tibia* or shin bone (Fig. 5-43). The proximal end bears *medial* and *lateral condyles* that articulate with the femoral condyles. The lateral condyle carries a *facet* for articulation with the fibula. A central, bifid, *intercondylar eminence* lies between the condyles. Anteriorly, the *tibial tuberosity* lies about 6 cm below the condyles.

The *shaft* is triangular in cross-section with the base of the triangle posteriorly. Thus there are *posterior, medial,* and *lateral surfaces* and *anterior, medial,* and *interosseous borders*. The distal end of the bone bears an *inferior articular surface* with a medially oriented *medial malleolus* that bears a *facet* for the talus (an ankle bone).

Fibula. The slender lateral bone of the leg is the *fibula* or "splint bone" (see Fig. 5-43). Its fragility lends little to weight bearing, but, along with the interosseous membrane that connects it to the tibia, it extends the area for muscular attachment.

The proximal end is called the **head**, is more cuboidal in shape, and carries an articular facet for connection to the tibia. The **shaft** is triangular in cross-section and bears the same surfaces and borders as the tibia. The medial surface of the shaft is obviously grooved to accommodate muscular attachments.

The distal end of the bone carries the pointed **lateral malleolus** that bears a *facet* for the talus on its medial aspect.

The tibia and fibula together form a cuplike articulating surface for the ankle.

In *Pott's fracture* the break occurs at the distal end of the fibula. The lateral malleolus may also be chipped. Blows to the lateral side of the distal leg or sudden violent eversion of the foot may cause such breaks.

Tarsal bones. In each lower appendage there are seven *tarsal* or ankle bones (Fig. 5-44). Only one, the **talus** (astragalus), is involved in forming the ankle joint. It is recognized by its large superiorly and laterally placed **articular surfaces** that connect to the tibia and fibula. Anterior is a rounded **head**. The **calcaneus**, or heel bone, is the largest tarsal bone, with a roughened posterior surface for attachment of the calf muscles. The **navicular bone** lies anterior to the talus and has a definite boatlike shape. The **cuboid bone** lies anterior to the calcaneus, and its name indicates its general shape. Three **cuneiform** bones, numbered 1 to 3 from the medial side, are designated as *medial, intermediate,* and *lateral* and are wedge-shaped bones that contribute to the formation of the transverse arch of the foot. The medial cuneiform bone is the largest, the intermediate bone is smallest, and the lateral bone lies between the other two in size.

Metatarsals. There are five *metatarsals* (see Fig. 5-44), numbered 1 to 5 from the medial (big toe) side. Each has a proximal **base**, a **shaft**, and a distal rounded **head**.

The first metatarsal is the heaviest and carries much of the weight-bearing function of the metatarsals. Bones 2 to 5 are slender and provide stability to the foot like an "outrigger." The fifth metatarsal is distinguished from 2 to 4 by its backward-projecting **tuberosity**.

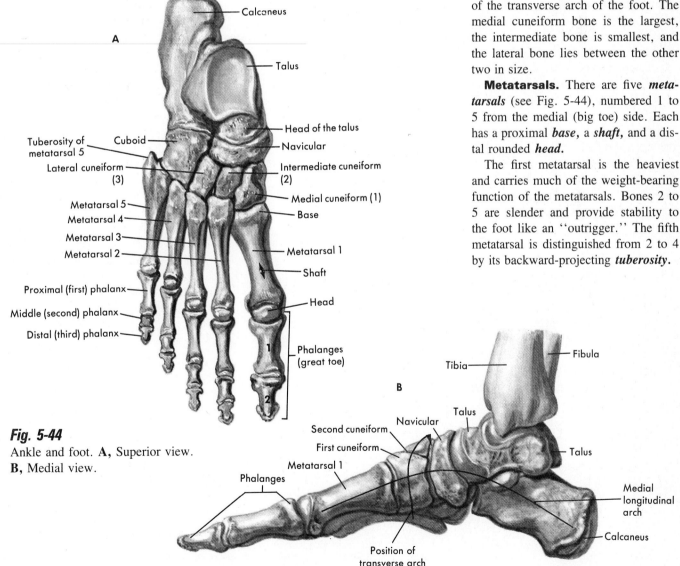

Fig. 5-44
Ankle and foot. **A,** Superior view. **B,** Medial view.

Phalanges. The *phalanges* of the toes (see Fig. 5-44) are arranged like those of the hand. Proximal phalanges are the longest, and middle and distal phalanges are very short. Sesamoid bones are commonly found at the head of the first metatarsal; its inferior surface usually shows two grooves for the bones.

Table 5-6 summarizes the bones of the pelvic girdle and lower appendage.

Arches of the foot. There are three arches recognized as occurring in the foot. The **medial longitudinal arch** is the highest and begins with the *calcaneus*. It rises to a high point at the head of the *talus*, then descends through the *navicular bone, cuneiform bones,* and *first metatarsal*. The calcaneus is said to form the posterior pillar of the arch, the talus the keystone, and the remaining bones the anterior pillar.

The **lateral longitudinal arch** is low and utilizes the *calcaneus* as the posterior pillar, the *cuboid* as the keystone, and the *fourth and fifth metatarsals* as the anterior pillar.

The **transverse arch** is best seen at the level of the cuboid and cuneiform bones and is most obvious when the foot is not bearing weight. The cuneiform bones fit so as to create the curve to the arch. High points of transverse and medial longitudinal arches coincide, and this point also is where the line of gravity of the body falls. Thus the maximal amount of "spring" is given to the step. A "flat foot" is one in which medial longitudinal and transverse arches have become depressed. Weight bearing is shifted toward the inside of the foot and may aggravate the condition.

Table 5-6. Summary of the bones of the lower appendage and girdle and their major features

Name and number of bones in *each* appendage or half of girdle	Major feature on bone	Description or location of bone or feature
Os coxae (1)	Formed from ilium, ischium, and pubis; all three form cuplike acetabulum	Bone of the pelvic girdle attaching to sacrum
Ilium (1)		Superior component of os coxae
	Crest	Superior border of bone
	Anterior superior spine	Anterior termination of crest
	Anterior inferior spine	Anterior projection just superior to acetabulum
	Posterior superior spine	Posterior termination of crest
	Posterior inferior spine	Inferior to posterior superior spine
	Gluteal lines (anterior, posterior, inferior)	Three small ridges on posterior surface of bone
	Greater sciatic notch	Lies inferior to posterior inferior spine
	Iliac fossa	Anterior surface of bone
	Articular surface	Ear-shaped surface for sacral articulation
Ischium (1)		Posterior and inferior bone of os coxae
	Ischial spine	Pointed, posteriorly directed process at lower end of greater sciatic notch
	Ischial tuberosity	Large roughened elevation "on which we sit"
	Lesser sciatic notch	Between spine and tuberosity
	Ramus	Anteriorly directed part of bone
Pubis (1)		Anterior bone of os coxae
	Inferior ramus	Joins ischial ramus
	Body	Gives rise to inferior ramus
	Symphyseal surface	Articular surface with other pubis
	Pubic crest	Anteriorly from body
	Superior ramus	Upward and laterally from body
Ischium and pubis enclose the obturator foramen, the largest foramen in the human body		

Adapted from McClintic, J.R.: Basic anatomy and physiology of the human body, ed. 2. Copyright © 1980. Reprinted by permission of John Wiley & Sons, Inc.

Table 5-6. Summary of the bones of the lower appendage and girdle and their major features—cont'd

Name and number of bones in *each* appendage or half of girdle	Major feature on bone	Description or location of bone or feature
Femur (1)		Bone of the thigh
	Head	Proximal rounded end articulating with acetabulum
	Neck	Distal to head
	Greater trochanter	Large laterally projecting process below neck
	Lesser trochanter	Medial and posteriorly projecting process below neck
	Body or shaft	Bone length between trochanters and distal end
	Linea aspera	Rough ridge on posterior and middle third of shaft
	Gluteal tuberosity	Rough ridge from upper end of linea to base of greater trochanter
	Pectineal line	From lesser trochanter to linea
	Spiral line	From upper end of linea medially around shaft
	Supracondylar lines or ridges (medial and lateral)	From lower end of linea toward condyles of femur
	Medial condyle	Medial articular surface on distal end of bone
	Lateral condyle	Lateral articular surface on distal end of bone
	Intercondylar notch or fossa	Depression on posterior surface between condyles
Patella (1)		"Kneecap"; presents medial and lateral facets on posterior surface
Tibia (1)		Medial bone of leg ("shin")
	Medial condyle	Medial articular surface on proximal end of bone
	Lateral condyle	Lateral articular surface on proximal end of bone
	Intercondylar eminence	Projection between the condyles on proximal end of bone
	Tibial tuberosity	Anterior elevation below condyles
	Shaft	Triangular length of bone between its ends
	Medial malleolus	Prominence on medial side of distal end

Continued.

Table 5-6. Summary of the bones of the lower appendage and girdle and their major features—cont'd

Name and number of bones in *each* appendage or half of girdle	Major feature on bone	Description or location of bone or feature
Fibula (1)		Lateral bone of leg ("splint bone")
	Head	Proximal end of bone; more blunt
	Lateral malleolus	More pointed distal end of bone
	Shaft	Between proximal and distal ends
Tarsals (7)		Bones of the ankle
	Calcaneus	Largest; "heel bone"
	Talus	Articulates with tibia and fibula to form ankle joint
	Cuneiform	Wedge-shaped; numbered 1-3 starting from medial side
	Cuboid	Anterior to calcaneus
	Navicular	Anterior to talus; boat shaped
Metatarsals (5)		Bones of foot; numbered 1-5 beginning with great toe
Phalanges (14)		Two in great toe, 3 in each of the other toes

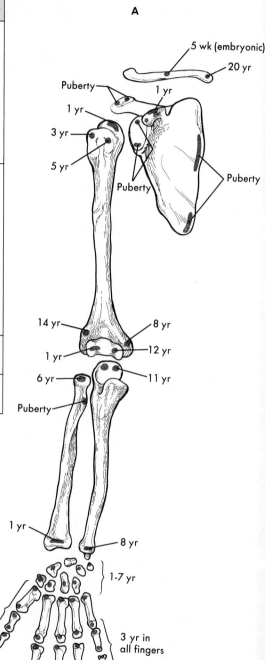

Ossification of the skeleton

For medical or legal purposes it is often required that the age of a skeleton be determined. On the living or fleshed person, x-ray examination can determine the extent to which various bones have been ossified. Different bones develop their primary and secondary ossification centers and complete their ossification (fusion) at different ages. It is more difficult to determine the age of fossilized bones, because cartilaginous parts may have disappeared. Determining the age of a bone in terms of the time epochs requires dating by sophisticated techniques.

Fig. 5-45 provides information as to when bones develop their centers and when the bones mature.

Fig. 5-45
Times of ossification of skeleton.
A, *Appearance* of ossification centers of upper limb and girdle. **B,** *Fusion* of bones of upper appendage and girdle. **C,** *Appearance* of ossification centers of os coxae and lower appendage. **D,** *Fusion* of bones of os coxae and lower appendage.

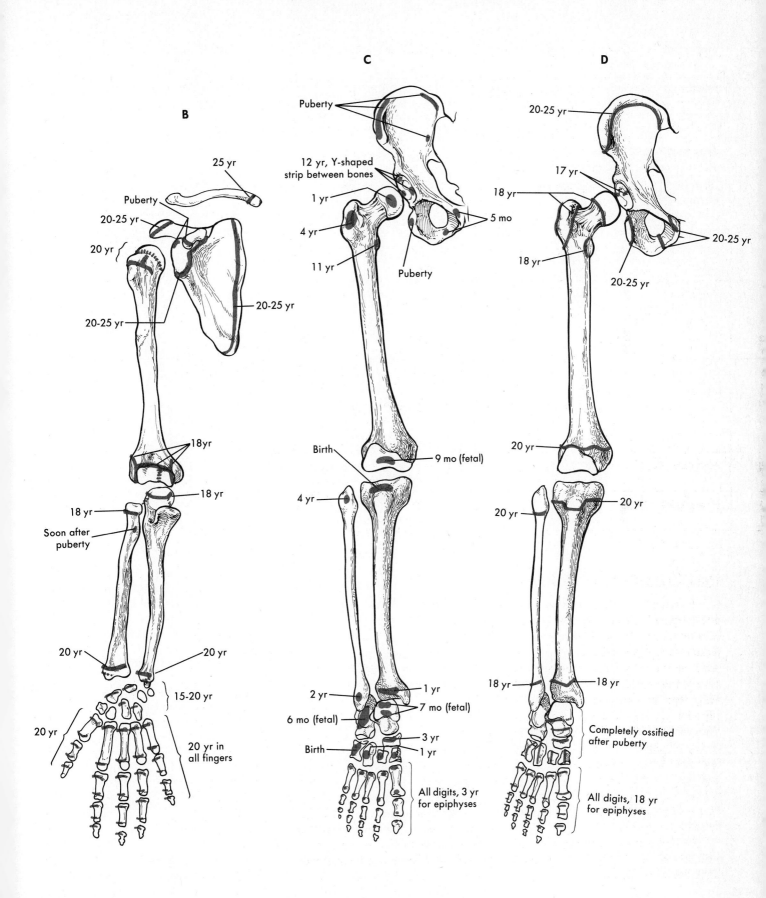

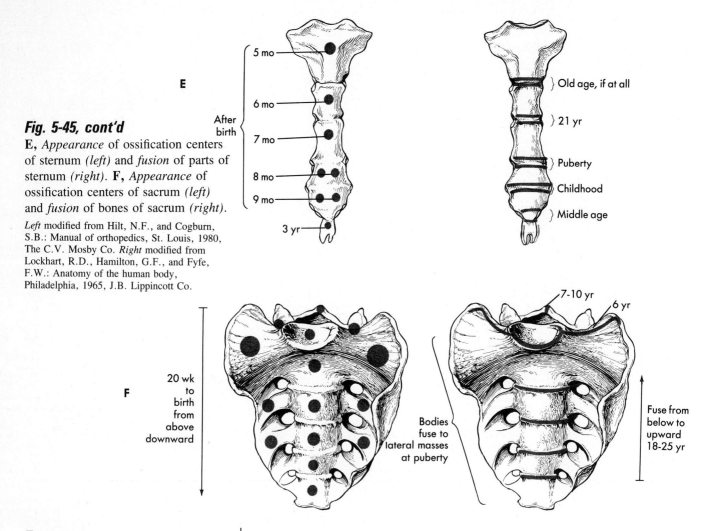

Fig. 5-45, cont'd
E, *Appearance* of ossification centers of sternum *(left)* and *fusion* of parts of sternum *(right)*. **F**, *Appearance* of ossification centers of sacrum *(left)* and *fusion* of bones of sacrum *(right)*.

Left modified from Hilt, N.F., and Cogburn, S.B.: Manual of orthopedics, St. Louis, 1980, The C.V. Mosby Co. *Right* modified from Lockhart, R.D., Hamilton, G.F., and Fyfe, F.W.: Anatomy of the human body, Philadelphia, 1965, J.B. Lippincott Co.

Fractures

The term *fracture* implies disruption of the continuity of a given bone. Breakage of a bone may be caused by *trauma* such as blows, by *pathological processes* that weaken the structure of a bone and render it more liable to spontaneous breaks, or by sudden *muscular contractions* that impart a violent force to the bone. Twisting (torque) or tension seems more liable to cause fractures than do compressional forces if bone structure is normal.

The terms used to describe fractures generally reflect the nature of the break. Since jagged bone ends may do damage to internal body structures, treatment should be left to physicians. Uninformed but eager hands may do more harm than good.

The following list of terms and Fig. 5-46 are presented to inform the reader about fractures:

simple fracture is one in which there is a break but *no external wound;* that is, the bone ends do not pierce the skin. Both this and the next type may cause internal organ injury and are called *complicated fractures* if there is proven internal organ injury, as in a fractured rib piercing a lung.

compound fracture is one in which there *is* an *external wound*, as in bone ends piercing the skin.

transverse fracture is one in which the bone is broken straight across.

oblique fracture is one in which the bone is broken at an angle to its long dimension.

spiral fracture is one in which the bone is broken along a twisting line.

comminuted fracture is one in which the bone is fragmented into several pieces.

impacted fracture is one in which the bone ends are driven into one another.

depressed fracture is one in which a piece of bone is driven inward; usually occurs in the skull.

incomplete fracture is one in which the whole bone is not involved. A *greenstick fracture*, such as occurs in the not yet completely ossified bones of children, is a good example of an incomplete fracture.

epiphyseal separation may occur when epiphyseal cartilages are still present, and the end of a long bone is separated from its shaft.

Healing of a fracture involves the formation of a *blood clot* between the bone ends. *Granulation tissue*, consist-

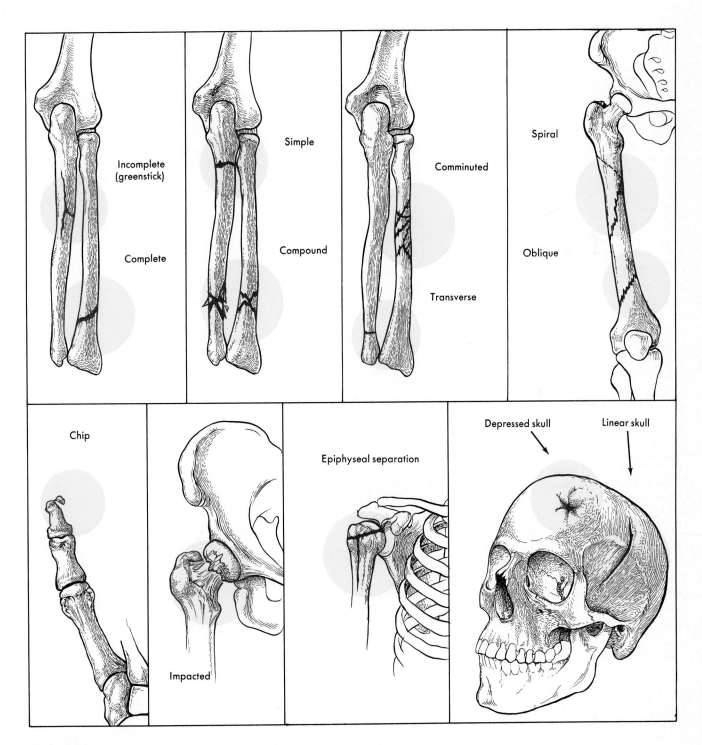

Fig. 5-46
Some types of fractures.

ing of fibroblasts and vascular elements, is next formed. *Transformation of cells* into osteoblasts, formation of osteoid, and calcification of the osteoid forms a *bony callus* that knits the ends together. Formation is by the intramembranous method.

• • •

No attempt is made to summarize this chapter, since it consists mainly of anatomical terms.

Questions

1. What functions does the skeleton serve in humans?
2. It is possible that some books list the human skeleton as being composed of more than the usually quoted 206 bones. What accounts for such differences?
3. Describe the structure of a flat skull bone, a long bone.
4. What functions do the various foramina of the skull serve? Name several foramina and what they pass.
5. Describe the manner in which blood vessels supply the interior of a long bone.
6. Compare the major steps in intramembranous and endochondral bone formation.
7. What are the contributions of the organic and inorganic parts of a bone to its total strength?
8. How do bones grow in diameter? In length?
9. What are the tasks of osteoblast? Osteocyte? Osteoclast?
10. What bones:
    a. Form the orbit?
    b. Form the walls of the cranial cavity?
    c. Form the pectoral girdle?
    d. Protect the heart and lungs?
    e. Form the wrist?
    f. Form the ankle?
    g. Form the ear ossicles?
11. What is a suture? A fontanel?
12. What are the differences between infant and adult vertebral columns?
13. In what ways do the vertebrae in the different regions of the spine differ?
14. What are true ribs? False ribs? Floating ribs?
15. Compare male and female pelves.

Readings

Bourne, G.H.: The biochemistry and physiology of bone, 4 vols., New York, 1971-1976, Academic Press, Inc.

Brown, B.J.: Complete guide to prevention and treatment of athletic injuries, Englewood Cliffs, N.J., 1972, Parker Publishing Co.

Committee on Trauma of American College of Surgeons: An outline of the treatment of fractures, ed. 8, Philadelphia, 1966, W.B. Saunders Co.

Menczel, J., and Harell, A., editors: Calcified tissue: structural, functional and metabolic aspects, New York, 1971, Academic Press, Inc.

Sobotta, J., and Figge, F.: Atlas of descriptive human anatomy, ed. 8, vol. 1, New York, 1963, Hafner Press.

Zorab, P.A., editor: Scoliosis, New York, 1973, Academic Press, Inc.

6
Articulations (joints)

Objectives

Classification of articulations (joints)
    Fibrous joints
    Cartilaginous joints
    Synovial joints

Specific joints of the body
    Temporomandibular joint
    Joints of the vertebral column
    Joints of the thorax
    Joints of the shoulder girdle
    Shoulder joint
    Elbow joints
    Joints of the wrist and hand
    Sacroiliac joint
    Hip joint
    Knee joint
    Tibiofibular articulations
    Joints of the ankle and foot

Some selected disorders of joints and their associated structures
    Traumatic damage to a joint
    Inflammation of joints
    Inflammation of associated structures

Questions

Readings

OBJECTIVES After studying this chapter, the reader should be able to:

Define an articulation.

Present a classification of the types of joints that are found in the body and give examples of each type.

Diagram the general structure of a synovial joint, giving the characteristics of synovial fluid.

Describe the structure of the major joints of the body, including the most important ligaments involved in maintaining their security.

Describe some common disorders of joints.

Classification of articulations (joints)

An *articulation (joint)* may be defined as an area of junction between two bones in the body or an area of junction between a bone and a cartilage in the body. According to the nature of the junction, whether or not movement is permitted, and if permitted, the type of movement occurring, the joints of the body fall into three types or categories: ***fibrous, cartilaginous,*** and ***synovial.***

Fibrous joints

In *fibrous joints* (synarthroses) the adjacent bones or cartilages are held together by fibrous tissue (usually of the collagenous type) that directly unites the two structures involved. Thus there is *no joint cavity,* the two structures are in close contact, and the joint *usually permits no movement.* There are two varieties of fibrous joints: syndesmoses and sutures.

In a *syndesmosis* the bones rest on one another or are a short distance apart and are held firmly by an *interosseous ligament* that runs directly between or partially surrounds the two bones (see Fig. 5-43). The reader may recall the interosseous ligaments that connect radius and ulna or tibia and fibula.

In a *suture* the bones are about as close together as they can get, and a thin membrane, the *sutural ligament,* holds the bones firmly together. This type of joint occurs only in the skull and may take several forms (Fig. 6-1) as follows:

dentate sutures consist of two bones that may possess rather long interlocking processes resembling teeth. The sagittal suture is an example.
serrated sutures also have processes that interlock, but they are smaller and shorter like the teeth of a fine saw. The fetal frontal suture is an example.
limbose sutures have interlocking processes plus beveling between the two bones. An example is the coronal suture.

Fig. 6-1
Various forms of skull sutures.

squamous sutures have beveled but smooth opposing surfaces. The squamous suture of the skull is an example of this type.

harmonious sutures have nearly flat surfaces that are opposed, as in the sutures of the hard palate. These are obviously the weakest of the sutures.

schindyleses have a cleftlike structure on one bone that receives a bladelike projection from another bone. An example is the suture between the perpendicular plate of the ethmoid bone and the vomer.

gomphoses consist of a conical part of one structure fitting into a pocket in another bone, as a nail would fit into a hole in a board. The teeth in their sockets in the jawbones are examples of gomphoses.

Obviously sutures are usually very strong and are used where the prime requisites for the joint are **strength** and **protection**.

Cartilaginous joints

Cartilaginous joints (amphiarthroses) (Fig. 6-2) use some variety of cartilage as the connecting substance between the two bones. The joints may *permit slight movement* because of the compressible nature of the connecting cartilage but *have no cavity*. Two types of cartilaginous joints are recognized:

synchondroses are cartilaginous joints in which the type of tissue effecting the connection is hyaline cartilage. *Temporary* synchondroses are seen in long bones between the shaft and the ends of the bone. It may be recalled that these epiphyseal discs permit growth in length of the bone as long as they are cartilaginous but may suffer separation. The cartilage discs are usually replaced by bone by 25 years of age in the whole skeleton. *Permanent* synchondroses occur between the first pair of ribs and the sternum.

symphyses are cartilaginous joints in which the connecting material is fibrocartilage. This type of joint permits more movement, typically of a compressional nature, because fibrocartilage is more compressible than hyaline cartilage. The joints between the vertebral bodies and the symphysis pubis are examples of symphyses.

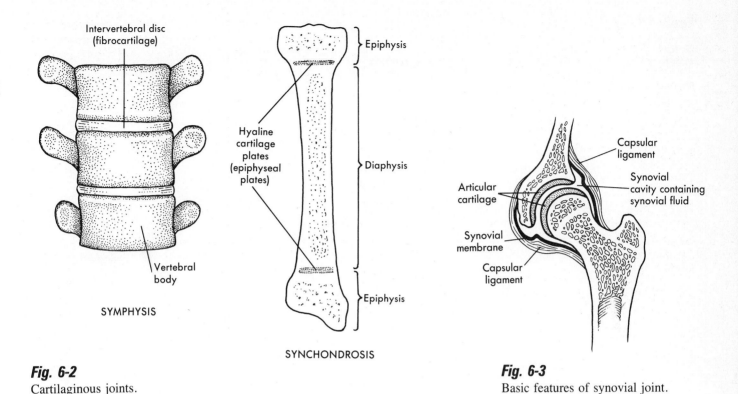

Fig. 6-2
Cartilaginous joints.

Fig. 6-3
Basic features of synovial joint.

Synovial joints

In *synovial joints* (*diarthroses*) the two bones involved are not directly connected by cartilage or fibrous tissue, although fibrous *discs* or *ligaments* may run between or around the articular surfaces. There is a *joint* (synovial) *cavity*, the bone ends are covered by *articular cartilage* (hyaline in type), the articular surfaces are lubricated by a viscous *synovial fluid* secreted by a *synovial membrane* lining the cavity except on the articular surfaces, and *capsular ligaments* (often receiving specific names) surround the joint like a sleeve. Such joints permit varying degrees of angular or rotational movements and are sometimes called freely movable joints. The generalized structure of a synovial joint is depicted in Fig. 6-3. Synovial joints are subdivided into six types (Fig. 6-4)

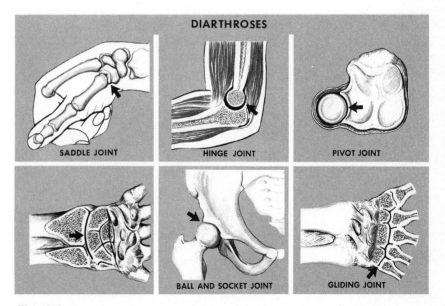

Fig. 6-4
The six basic types of synovial joints, illustrated by joints of body.

Table 6-1. Summary of articulations

Major category	Type	Characteristics	Axes of motion permitted	Examples
Fibrous joints		No cavity immovable	None	Sutures of skull syndesmoses
	Suture	A nearly bone-bone joint; minimal ct between	None	Skull; commonly becomes synostosis
	Syndesmosis	Bones held together by collagenous tissue	None	Tibia-fibula articulation at distal end
	Synostosis	Bone-bone junction	None	Arise by aging of fibrous joints, with loss of fibrous tissue
Cartilaginous joints		No cavity; slightly movable; a bone-cartilage-bone joint	Compression only, if connecting tissue is flexible	Synchondroses, symphyses
	Symphysis	Bone-fibrocartilage-bone joint	Slight compression	Symphysis pubis, vertebral column (discs)
	Synchondrosis	Bone-hyaline cartilage-bone joint	None	Shaft and ends of long bones, temporary; between ribs and sternum (costal cartilages), permanent
Synovial joints		Cavity, freely movable	1-3	Six types
	Ball and socket	Ball in cup	Triaxial; all movements possible	Hip, shoulder
	Gliding	Nearly flat surfaces opposed	Biaxial; abduction-adduction and flexion-extension	Intercarpal joints, intertarsal joints
	Hinge	Spool in half-moon describes most hinge joints	Uniaxial; flexion-extension	Knee, elbow, ankle, fingers
	Ovoid	Egg in depression	Biaxial; abduction-adduction and flexion-extension	Wrist
	Pivot	Cone in depression	Uniaxial; rotation medial and lateral	Radiohumeral, atlas-axis
	Saddle	"Saddle on horse"	Biaxial; abduction-adduction and flexion-extension	Carpometacarpal joint of thumb

Adapted from McClintic, J.R.: Basic anatomy and physiology of the human body, ed. 2. Copyright © 1980. Reprinted by permission of John Wiley & Sons, Inc.

according to the type of movement permitted:

- *gliding joints,* exemplified by the joints between the wrist and ankle bones (intercarpal and intertarsal joints respectively), are formed of surfaces that may move in a front-to-back or side-to-side motion but cannot rotate because of ligaments that hold the bones together. Such joints are thus *biaxial,* permitting movement in two planes.
- *pivot joints* involve a conical, pointed, or rounded surface articulating with a shallow depression on the other bone. Only a rotational or turning movement is allowed, and the joints are thus *uniaxial.* The joint between the dens of the axis and the atlas that permits "no" movements of the head and the radioulnar joint that permits rotation of the forearm are examples of pivot joints.
- *hinge joints* usually involve a spool-like surface on one bone fitting a half-moon or shallow depression on the other bone. Because of the nature of the bony surfaces, only a single plane of front-to-back movement is permitted, and the joints are thus *uniaxial.*
- *ovoid joints* involve egg-shaped surfaces provided by one or a series of bones that fits into a shallow cuplike surface on the other bone. These joints permit a side-to-side and front-to-back motion but no rotation. They are thus *biaxial.* The wrist joint, between the proximal row of carpals and the ends of the radius and ulna, is an ovoid joint.
- *saddle joints* (or reciprocal reception joints) resemble a saddle on the back of a horse with the saddle representing one bone and the horse's back the other. The surfaces are reciprocally shaped in a concave-convex match. Such joints will permit to-and-fro and side-to-side movement but no rotation and are thus *biaxial.* The carpometacarpal joint of the thumb is of this type.
- *ball-and-socket joints* involve a rounded head on one bone fitting into a cuplike surface on the other bone. These joints permit to-and-fro, side-to-side, and rotational movement and are thus *triaxial.* The hip and shoulder joints are of this type and are the most maneuverable of all body joints. Table 6-1 presents a summary of the types of articulations.

More specific descriptions or names are given to the movements synovial joints permit than was indicated previously.

The term **rotation** may be defined as the turning of a long bone around a longitudinal axis. Since the motion may occur in two directions, terms such as *medial* (inward) or *lateral* (outward) rotation may be included in the name of the motion.

Pronation is a term used in connection with medial rotation of the forearm only, and *supination* refers to lateral rotation of the forearm. The term *inversion* is used to refer to rotation of the foot to face the soles of the feet toward one another. The opposite movement is called *eversion.*

Angular movements change the angle between two bones or between a bone and another bone part:

- *flexion* decreases the angle between two bones, as in bending the elbow or knee. Since the foot is at right angles to the leg, the terms *dorsiflexion* and *plantarflexion* are used to refer to movements toward the upper surface or sole of the foot, respectively.
- *extension* increases the angle between bones, as in straightening the elbow or knee.
- *hyperflexion* is a term that is applied to the upper arm only. If the humerus is flexed beyond a vertical position, it is considered to be hyperflexed.
- *hyperextension* is extension beyond anatomical position as when the elbow is extended beyond a straight line.
- *abduction* removes a bone from the midline of the body or body part, as in raising the upper appendage directly sideward from the trunk or "fanning" the fingers.
- *adduction* is the opposite of abduction, returning the part(s) to a midline.
- *circumduction* occurs when a conical movement is made about a pivoting point. Describing a circle with the hand while the shoulder joint serves as the pivot point is an example of this type of motion.

Synovial fluid. Synovial fluid was mentioned as the lubricating fluid secreted by the synovial membranes of the synovial joints. It forms a thin fluid film over the articulating surfaces and has a total volume in all the body's synovial joints of about 100 ml. It is formed from the tissue fluid outside the body cells but differs from that fluid in several ways: it contains fewer dissolved salts; it contains mucus and more protein that give it a viscosity about 200 times greater than that of tissue fluid; it contains a substance called hyaluronic acid that is involved, with its enzyme hyaluronidase, in aiding in controlling viscosity of the fluid. More enzyme means a lower viscosity. Excessive production of synovial fluid, as after injury to a joint, may severely limit joint movement and lead to swelling.

Specific joints of the body

Temporomandibular joint

The *temporomandibular joint* (Fig. 6-5) is the only synovial joint of the skull and is a combined hinge and gliding joint. The mandibular fossa of the temporal bone and the condyle of the mandible are the bony parts involved in formation of the joint. The **capsular ligament** *(articular capsule)* originates from the rim of the mandibular fossa and attaches to the neck of the condyle to surround the joint. It is thin and loose and serves mainly to contain the other structures of the joint. The **lateral ligament** extends from the zygomatic arch to the mandibular neck. The **sphenomandibular ligament** extends from the base of the great wing of the sphenoid bone (sphenoidal spine) to the inner surface of the ramus just posterior to the mandibular foramen. The **stylomandibular ligament** extends from the styloid process to the angle and posterior border of the ramus. An *articular disc* composed of fibrocartilage lies between the mandibular fossa and condyle, separating the synovial cavity of the joint into two parts, one above and one below the disc.

Movements of the mandible include *opening and closing* the jaws, basically a hinge motion; *protrusion* (moving the jaw forward); *retraction* (moving the jaw backward); and *lateral displacement;* the latter three movements involve a gliding motion. Opening and closing the jaw causes the condyle to move on the disc like a hinge. Protrusion involves the anterior movement of both discs in the mandibular fossa. Lateral displacement, as occurs in grinding food between the teeth, causes one disc to glide forward while the other stays in place; which disc glides is determined by which side of the mouth is being used to chew with.

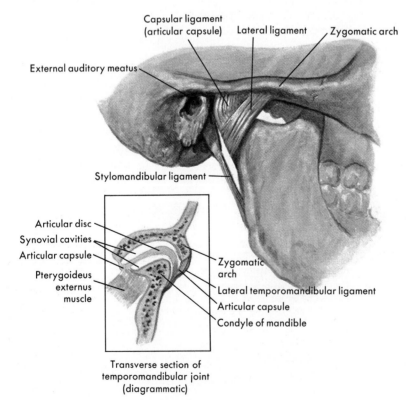

Fig. 6-5
Temporomandibular joint (sphenomandibular ligament lies medial to ramus of mandible and is not depicted).

Joints of the vertebral column

Articulation of the atlas with the skull. The *atlantooccipital joint* (Fig. 6-6) consists of two *capsular ligaments* that connect the occipital condyles to the superior articular processes of the atlas, *anterior* and *posterior atlantooccipital membranes* that connect the margin of the foramen magnum to the anterior and posterior arches of the atlas, and two *lateral ligaments* that run between the occipital bone and the transverse processes of the atlas. Two synovial cavities, one for each joint, are present.

Movements permitted by the joint include *flexion* ("nodding" the head or pulling the chin downward), *extension* (raising the chin), and *slight motion to the side* (bending the head to the side).

Atlantoaxial joint (Fig. 6-7). Three joints are involved in the connection of the atlas and axis. The **median atlantoaxial joint** occurs between the dens of the axis and a ring formed by the anterior arch and transverse ligament of the atlas. There are two synovial cavities associated with this joint, one anterior to and one posterior to the dens. This articulation is a pivot joint, permitting the shaking of the head ("no" movement). The lateral *atlantoaxial joints* (two) occur between the articulating process of the two bones. Two *capsular ligaments* connect the lateral masses of the atlas with the articular processes of the axis; an *anterior* and a *posterior atlantoaxial ligament* run between the anterior arch and axis body and between the posterior arch and axis laminae. The *transverse ligament*, mentioned previously, arches across the posterior surface of the dens. All the last ligaments are very strong and constitute the major forces holding

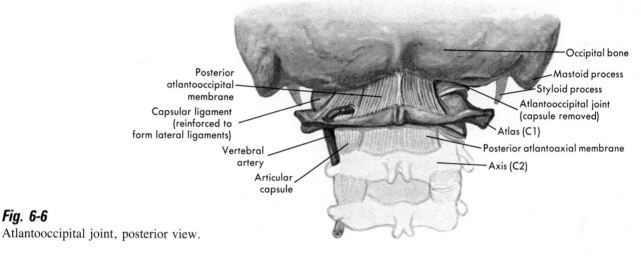

Fig. 6-6
Atlantooccipital joint, posterior view.

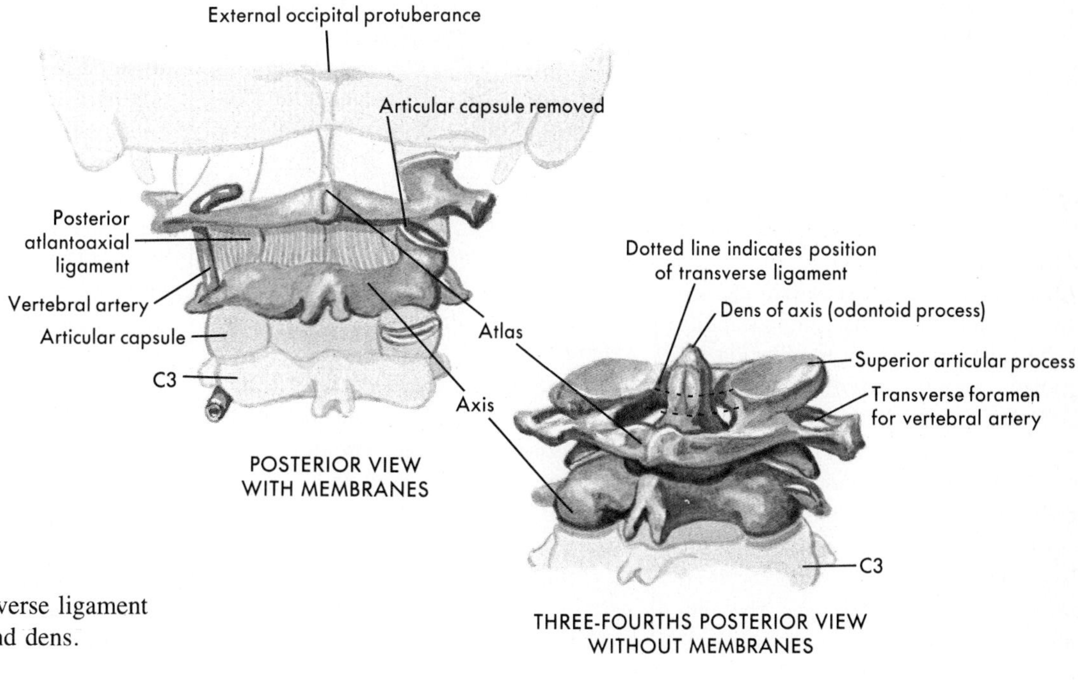

Fig. 6-7
Atlantoaxial joint. Transverse ligament (not shown) circles behind dens.

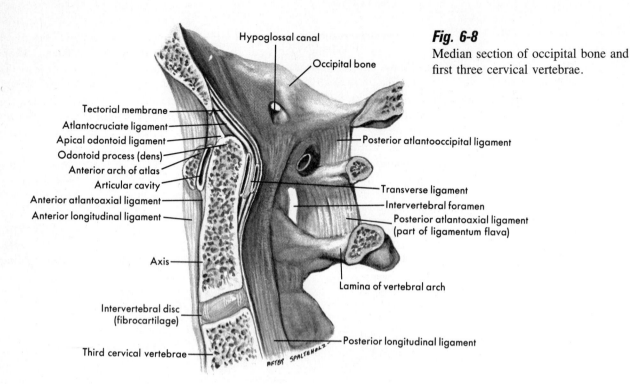

Fig. 6-8
Median section of occipital bone and first three cervical vertebrae.

the bones together. Synovial membranes line the joints between the articulating processes.

The joints in the vertebral column below the axis involve a series of symphyses between the vertebral bodies and joints between the vertebral arches. The ligaments and discs involved in the spine were described in Chapter 5. The joints between the articular processes of the column will be considered next.

The joints between the articular processes of the column are a series of synovial gliding joints (Fig. 6-8). The *capsular ligaments* are thin and loose and attach to the articular processes of adjacent vertebrae.

Movements of the column are diverse, permitted by the compressibility of the intervertebral discs. *Flexion,* or forward bending; *extension,* or backward bending; *abduction,* or sideward movement; *adduction,* or return to erect posture; *circumduction;* and *rotation* all occur in the column. Rotation is permitted by slight twisting of each disc; when they are added together, they produce great motion. The thoracic region has the least motion, presumably to prevent interference with breathing. In general the column can extend to a greater degree than it can flex.

Joints of the thorax

Costovertebral articulations (Fig. 6-9). There are two sets of joints to be considered here: the *costocentral joints,* between the head of the rib and the vertebral body, and the *costotransverse joints,* between the neck and tubercle of the rib and the transverse process of the vertebra.

The joints between the rib head and vertebral body are synovial gliding joints. A *capsular ligament* connects the rib head to the rim of the vertebral facet. Between the two surfaces is the synovial cavity. A *radiate ligament* runs between the head of the rib and the bodies of two adjacent vertebrae. The *intraarticular ligament* lies within the joint, joins the head to the intervertebral disc, and divides the synovial cavities of all but the first, tenth, eleventh, and twelfth ribs into two parts.

The costotransverse joints involve first, gliding joints between the rib tubercle and the facet on the transverse process on all but the eleventh and twelfth ribs, and second, nonsynovial joints between the rib neck and transverse process. A *capsular ligament* encloses the synovial joint. A *superior costotransverse ligament* runs between the superior border of the rib neck and passes to the lower border of the transverse process above. (The first rib lacks this ligament, and the twelfth rib has a *lumbocostal ligament* to the lumbar vertebrae.) A *posterior costotransverse ligament* runs between the rib neck and the base of the transverse process above. The *ligament of the neck* connects the rib neck to the anterior surface of the adjacent transverse process. It is very strong. The *ligament of the tubercle* is short but thick and connects the apex of the transverse process to the nonarticular portion of the rib tubercle. Movements are limited between ribs and bodies of the vertebrae and are of slight gliding motion. The transverse articulation permits the ribs to be *elevated* during inspiration and *depressed* during expiration.

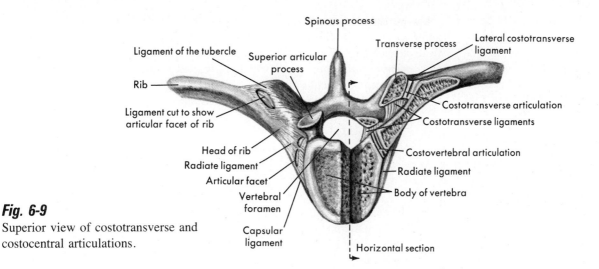

Fig. 6-9
Superior view of costotransverse and costocentral articulations.

Sternocostal articulations (Fig. 6-10). The cartilage of the first rib forms a synchondrosis with the sternum. The cartilages of ribs 2 to 7 form synovial gliding joints with the sternum. *Capsular ligaments* surround the joints of ribs 1 to 7 and the sternum. The *radiate sternocostal ligaments* begin on the costal cartilages and fan out to the manubrium and body of the sternum. An *intraarticular sternocostal ligament* occurs only between the second costal cartilage and the sternum. *Costoxiphoid ligaments* connect the anterior and posterior surfaces of the sixth and seventh costal cartilages to the sternum, and *interchondral ligaments* lie between the cartilages of ribs 5 through 10.

Only slight gliding movements are permitted at sternocostal joints.

Joints of the shoulder girdle

Sternoclavicular joint. Between the medial end of the clavicle and the manubrium of the sternum lies the sternoclavicular joint (Fig. 6-10). It is a double gliding joint. As in the temporomandibular joint, an *articular disc* of

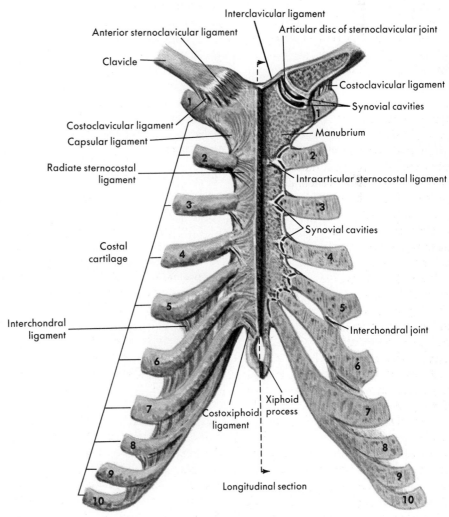

Fig. 6-10
Anterior view of sternocostal and sternoclavicular articulations.

fibrocartilage divides what would be a single synovial cavity into two parts, each with its own synovial membrane. A *capsular ligament* surrounds the joint and more closely resembles loose connective than collagenous tissue. The *anterior sternoclavicular ligament* is broad and connects the anterior aspect of the clavicle's medial end to the sternum. A *posterior sternoclavicular ligament* connects the posterior aspects of the same bones. A stout *interclavicular ligament* extends between the medial ends of the clavicles across the top of the manubrium. Finally a *costoclavicular ligament* runs between the inner end of the clavicle and the cartilage of the first rib.

Limited movement in any direction may occur at this joint. The major function of the joint is to attach the upper appendage to the axial skeleton; indeed, it is *only* at this joint that the upper appendage attaches. Thus firmness of attachment is more important than mobility. Also, any movement of the clavicle carries the scapula with it, and with the scapula the humerus. Thus this joint forms the center for all movements of the pectoral girdle.

Acromioclavicular joint. The acromioclavicular joint (Fig. 6-11) is constructed much like the sternoclavicular joint, with an **articular disc** (which may sometimes be absent). When present, the disc does not completely separate the synovial cavities of this gliding joint. The **capsular ligament** completely surrounds the joint. The **acromioclavicular ligament** binds the lateral end of the clavicle to the acromion process. The **coracoclavicular ligament** extends between the coracoid process of the scapula and the lateral portion of the inferior border of the clavicle. It maintains the spatial relationship between scapula and clavicle. The **trapezoid ligament** forms the lateral part of the preceding ligament, the **conoid ligament** the medial portion.

Movements permitted at this joint include a gliding motion of the clavicle on the acromion and rotation of the scapula on the clavicle. The joint allows the scapula to adjust its position to remain in close contact with the posterior thorax during movements of the upper appendage.

Shoulder joint

The *shoulder joint* (Fig. 6-11) is a triaxial ball-and-socket synovial joint formed by the head of the humerus and the glenoid fossa of the scapula. This arrangement allows much freedom of movement, and stability is provided by numerous ligaments and several scapular muscles that send their tendons to attach to the tuberosities.

The *glenoid labrum* is a fibrocartilaginous rim around the margin of the glenoid fossa. The labrum deepens the socket of the cavity without interfering with joint movement. The *capsular ligament* encircles the joint between the rim of the glenoid fossa and the anatomical neck of the humerus. It is very loose and permits the humerus to move widely without putting tension on the capsule. The *coracohumeral ligament* runs between the coracoid process and the greater tubercle. *Glenohumeral ligaments* thicken the capsular ligament on its ventral portion. A *transverse humeral ligament* runs between the greater and lesser tubercles across the bicipital groove and serves mainly to hold the tendon of the long head of the biceps in the groove.

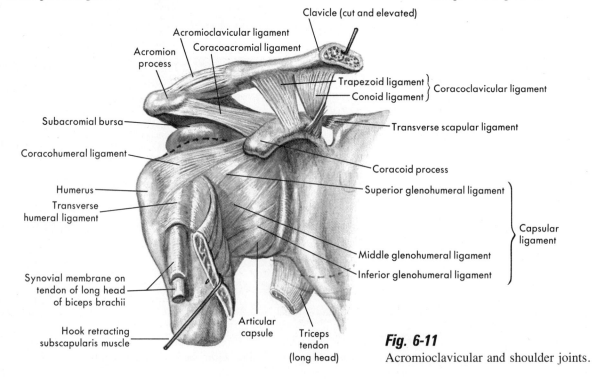

Fig. 6-11
Acromioclavicular and shoulder joints.

The synovial membranes of the joint are quite extensive; they not only surround the joint itself but also are prolonged around the biceps tendon as it lies in the groove. Several **bursae**, synovial sacs filled with fluid, cushion tendons and muscles as they pass over or around the joint. Posterior to the capsule lies the *subscapular bursa;* the *subdeltoid bursa* lies laterally over the joint, cushioning the large deltoid (shoulder) muscle. A *subacromial bursa* lies beneath the acromion process. Other small bursae may be present but are not of constant occurrence.

As suggested earlier, all possible ranges of movement are possible at the shoulder joint. Flexion, extension, abduction, adduction, medial and lateral rotation, and circumduction occur.

A "shoulder separation" results when the humerus is driven upward as when using the hand to break a fall. Labrum damage is common. Yanking small children upward by the arm can also dislocate the joint with the humerus head often overriding the labrum.

Elbow joints

The elbow includes three joints: the ulnohumeral joint, commonly called the elbow joint; the radiohumeral joint; and the radioulnar joint (Fig. 6-12).

The **ulnohumeral joint** occurs between the semilunar notch of the ulna and the trochlea of the humerus. It is a uniaxial hinge joint. Acting with the ulna, if the forearm is flexed without permitting rotation to occur, the **radiohumeral joint** between the radial head and the capitulum of the humerus also *acts* as a hinge. Rotation of the forearm causes the radial head to rotate on the capitulum as a pivot joint. Simultaneously the radial head will turn in the radial notch of the ulna, the annular ligament (see following paragraph) holding the head in place.

The **capsular ligaments** run anteriorly and posteriorly around the joints between the medial epicondyle and coronoid process and the olecranon fossa margin to the olecranon process. Medically the **radial collateral ligament** is continuous with and superficial to the capsular ligament; laterally the **ulnar collateral ligament** is present. The **annular ligament** is a strong band of fibers encircling the radial head and attaches to the anterior and posterior margins of the radial notch.

The synovial membrane of the joint is very extensive. It encloses the joints between the humerus, radius, and ulna and forms a pouch around the radial head beneath the annular ligament. Movements of the joint have already been mentioned.

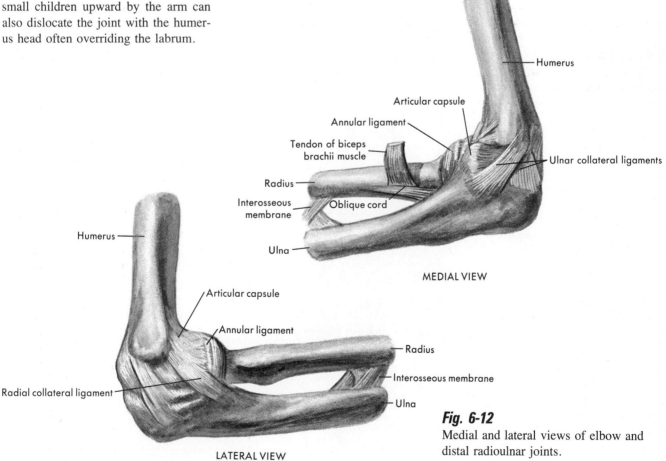

Fig. 6-12
Medial and lateral views of elbow and distal radioulnar joints.

Joints of the wrist and hand

Wrist joint. Between the proximal row of carpal bones (except the pisiform) and the ends of the radius and ulna is the *wrist joint* (Fig. 6-13), a biaxial ovoid joint. *Capsular ligaments* surround the joint, and four ligaments strengthen the joint: the *palmar radiocarpal ligament* anteriorly extends between the distal radius and ulna and the scaphoid, lunate, and triangular bones and also holds the distal ends of the radius and ulna together; the *dorsal radiocarpal ligament* lies between the posterior distal end of the radius and the three previously mentioned carpals; the *ulnar collateral ligament* lies on the medial side of the wrist between the ulnar styloid process and the triangular and pisiform bones; the *radial collateral ligament* laterally extends from the radial styloid process to the trapezium.

The synovial membrane of the wrist joint lines a single cavity between the triquetrum, lunate, scaphoid, and the distal ends of the radius and ulna.

Intercarpal joints (see Fig. 6-13). Three sets of articulations are included in the intercarpal joints. All are gliding joints as far as movement *permitted* is concerned. Articulations between the bones of the proximal row are made by *dorsal, palmar,* and *interosseous ligaments.* The names of the ligaments indicate their positions or locations. Articulations of the distal row are maintained by ligaments receiving the same names as in the proximal row. The articulations *between* the two rows of bones constitute the **midcarpal joint.** Here *palmar, dorsal, radial collateral,* and *ulnar collateral* ligaments are present. The latter two ligaments aid in connecting the distal ends of the two bones to the carpus as a group.

Synovial cavities are found between each bone and between the distal row and the metacarpals. The cavities appear to be continuous with one another.

The midcarpal joint permits flexion, extension, and a slight rotation.

Carpometacarpal joints (see Fig. 6-13). The carpometacarpal joint of the *thumb* is a saddle joint. The thumb itself is placed with its palmar surface at about a 90-degree angle to the fingers. It is this position that allows it to be opposed to the fingers in grasping or manipulative movements. Flexion, extension, abduction, adduction, circumduction ("twirling the thumbs"), and apposition are permitted.

The other four carpometacarpal joints are gliding joints. *Dorsal, palmar,* and *interosseous* ligaments unite the bones. The synovial cavities, as mentioned previously, are continuations of those of the carpals.

Metacarpophalangeal joints. The metacarpophalangeal joints are ball-and-socket in type, with rotation prevented by the ligaments involved. Each joint has a *palmar* and two *collateral ligaments.* Synovial cavities lie between the bone ends.

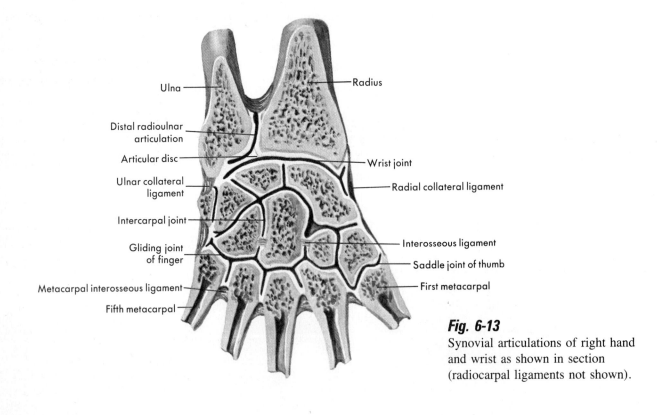

Fig. 6-13
Synovial articulations of right hand and wrist as shown in section (radiocarpal ligaments not shown).

Interphalangeal joints (see Fig. 6-13). Joints between phalanges are hinge joints, each possessing a *palmar* and two *collateral ligaments*.

Both these joints and the preceding ones appear to lack a supporting structure on their dorsal surfaces. No ligament is present, support being conferred by the tendons of the extensor muscles that have expansions covering the tops of the joints.

Sacroiliac joint

The *sacroiliac joint* (Fig. 6-14) occurs between the auricular surfaces of the sacrum and ilium. The joint is discussed here rather than with the vertebral column because of its functional relationship to movements of the lower limb. The joint provides a firm base from which the lower limb operates.

The joint is classed as a synchondrosis in which there is a space containing synovial fluid. The joint is held firmly by ligaments that cross its anterior and posterior surfaces and that extend between points on the sacrum and os coxae.

The *anterior sacroiliac ligament* extends between the anterior margins of the auricular surfaces. The *posterior sacroiliac ligament* is the most important ligament binding the bones together. It lies in the depression between the posterior surfaces of the sacrum and ilium. A *posterior sacrosciatic ligament* connects the posterior inferior iliac spine and the sides of the sacral crest to the ischial tuberosity. Other ligaments are present, but the ones presented are the more important ones.

A ''sacroiliac slip'' results when the ligaments stretch slightly as the result of lifting or other activities that apply weight to the sacrum in relation to the os coxae. Any slight movement may compress or put tension on the sacral nerves, with pain radiating down the thigh or leg. Rest and avoiding weight lifting aid recovery.

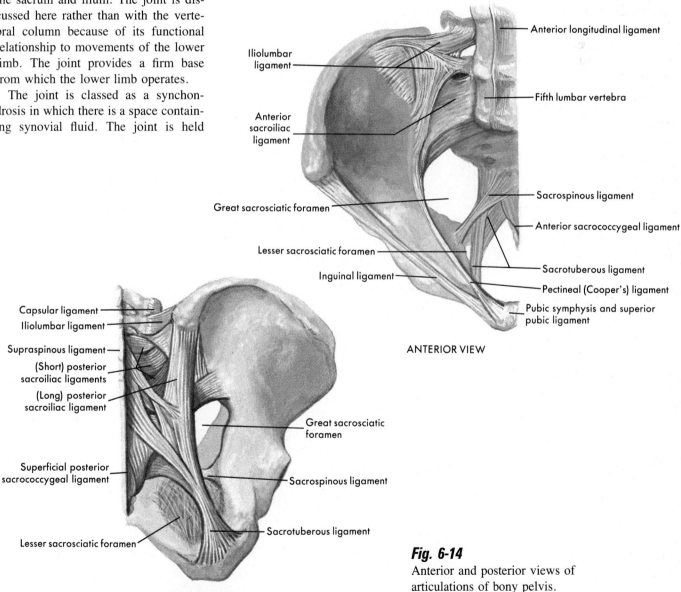

Fig. 6-14
Anterior and posterior views of articulations of bony pelvis.

Hip joint

The *hip joint* (Fig. 6-15) is a ball-and-socket joint between the acetabulum and the femoral head. The *capsular ligament* is very strong and is attached to the rim of the acetabulum and to the neck of the femur. Distinct reinforcing bands of fibrous tissue are applied to the capsular ligament and are named separately from it. The most important of these ligaments is the *iliofemoral ligament.* Between the anterior inferior iliac spine and the intertrochanteric line, it imparts great strength to the joint. The ischiofemoral ligament strengthens the posterior aspect of the capsule.

The synovial cavity is large in this joint but is confined to the area around the femoral head.

Since the joint is a ball-and-socket type, all ranges of motion are permitted. Because of the muscles associated with the area, the joint does not have the freedom of movement enjoyed by the shoulder joint. It is, however, much more secure than the shoulder joint.

Congenital malformation of the hip joint usually results in a low acetabular rim, and the femoral head easily slips out of it. The condition is easily diagnosed if there is limited sideward movement of the thigh when both hip and knee are flexed, and a "click" may be felt on the pubic side of the joint as the thigh is moved sideward (abducted). The click represents the actual dislocation of the joint (Fig. 6-16).

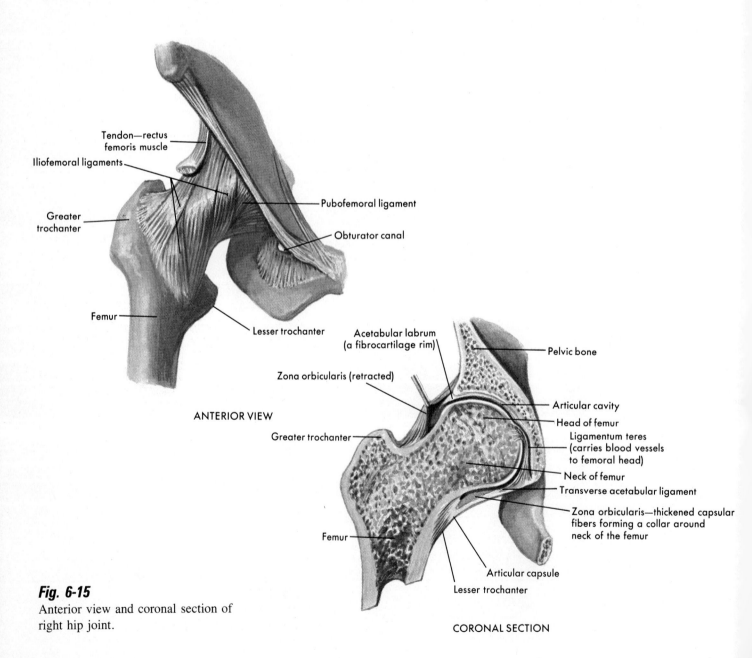

Fig. 6-15
Anterior view and coronal section of right hip joint.

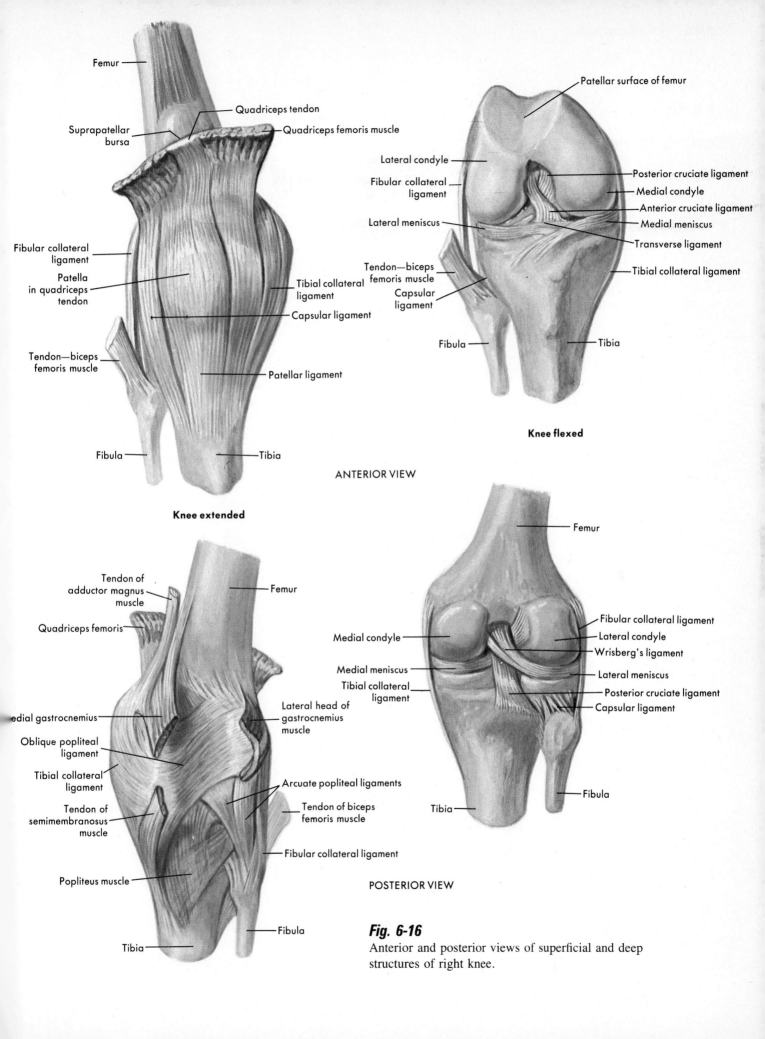

Fig. 6-16

Anterior and posterior views of superficial and deep structures of right knee.

162 Human anatomy

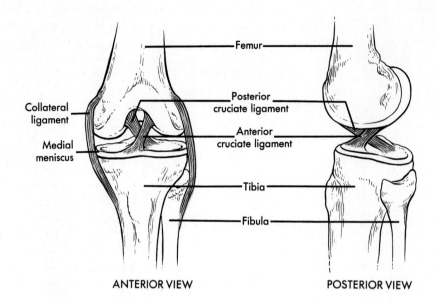

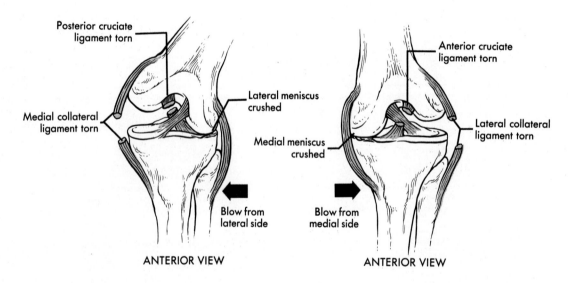

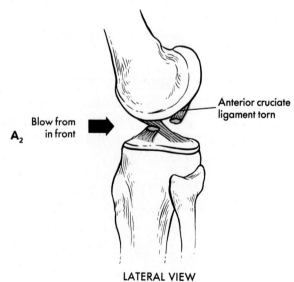

Fig. 6-17
Common disorders of knee and hip joints. **A₁** and **A₂**, Diagrams of knee structures injured by trauma.

Knee joint

The *knee joint* (Fig. 6-17) has classically been regarded as a hinge joint. However, when it is flexed and extended, these motions are combined with slight axial rotation and lateral movement. The axial rotation is most pronounced when the foot is swung from side to side at the ankle; the lateral rotation is inherent in the shape of the joint surfaces, mainly because the lateral condyle is larger than the medial condyle, and occurs when the joint is bent and straightened.

Because of the shapes of the bony parts, the knee is a catastrophe in terms of inherent stability. Thus various structures intended to strengthen, connect, and deepen the joint are used.

The *capsular ligament* is a thin but strong membrane surrounding the joint. A *tibial collateral ligament* forms a broad medially placed band of tissue between the superior margin of the medial femoral condyle and the medial tibial head. A *fibular collateral ligament* occupies a position on the lateral side of the joint between the lateral condyle margin and fibular head. There are no strengthening ligaments anteriorly and posteriorly. The quadriceps and patellar tendons and the hamstring tendons reinforce the joint in these directions.

Internally, two wedge-shaped *menisci*, *medial* and *lateral*, are attached to the tibial condyles. The thick portion of each meniscus is outward to deepen the surfaces of the tibia. Also, two *cruciate ligaments*, *anterior* and *posterior*, form an X within the joint. The anterior ligament runs between the upper part of the tibial tuberosity and the posterior part of the medial surface of the lateral femoral condyle. The posterior ligament attaches to the posterior surface of the tibial head between the condyles and runs to the anterior aspect of the medial femoral condyle. These ligaments tend to limit front-back movement of the tibia on the femur. A *transverse ligament* connects the menisci anteriorly. Several smaller ligaments, actually portions of the capsule, complete the ligaments of the joint. They are not discussed here.

The synovial membrane of the knee is the largest and most extensive of all in the body. In addition to surrounding the articular surfaces, the membrane extends anteriorly and superiorly beneath the quadriceps tendon, posteriorly toward the fibular head, and around the outer rims of the menisci.

Four bursae lie anterior. One is between the patella and the skin, and the other three are placed between the femur, tibia and fibula ends, and the quadriceps tendon or skin. There are four lateral bursae between the tendons of muscles and bone; five bursae lie medial, again between the bones and muscles or their tendons. Thus it should not be surprising to note the speed with which the joint or its bursae swell when injured.

The knee joint may be damaged by strong rotation, as when the foot is planted and cannot move and the lower appendage is twisted. Blows from any direction can tear the cruciate or capsular ligaments and crush menisci. Such types of injuries are inherent in football. Fig. 6-17 also shows some of the combinations of joint damage that may occur with blows to the joint.

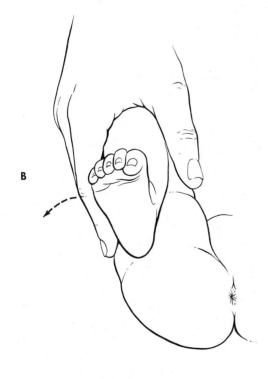

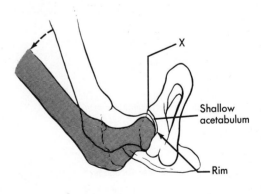

Fig. 6-17, cont'd
B, Testing infant for congenital hip dislocation. A "click" may be felt at X as thigh is abducted.

Tibiofibular articulations
(Fig. 6-18)

The tibia is connected to the head of the fibula by a gliding joint. **Capsular**, **anterior**, and **posterior ligaments** hold the bones together. An **interosseous membrane** holds the shafts of the bones together, and a *syndesmosis* holds the distal ends to one another. Only the superior joint has a synovial membrane.

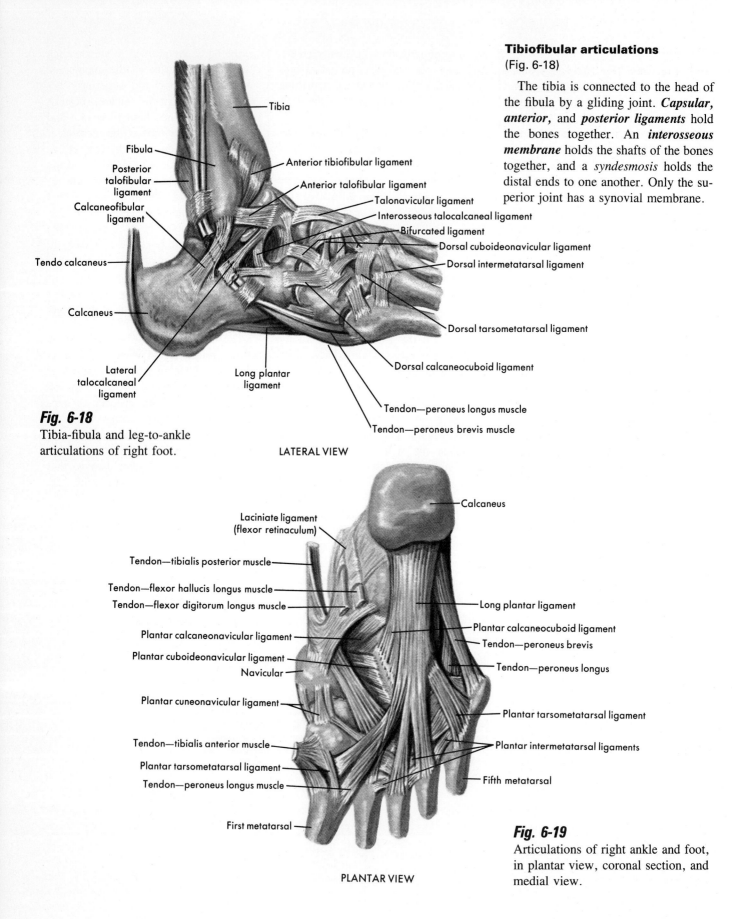

Fig. 6-18
Tibia-fibula and leg-to-ankle articulations of right foot.

Fig. 6-19
Articulations of right ankle and foot, in plantar view, coronal section, and medial view.

Joints of the ankle and foot

Ankle joint. Lying between the talus and the distal ends of tibia and fibula is a hinge joint, the ankle joint (Fig. 6-19). The reader is referred to the illustration for the names and locations of the several supporting ligaments. The usual capsular ligaments are present, and a series of strengthening ligaments stitch the leg bones to the tarsals. The synovial membrane invests the tibiofibular-talar surfaces.

Intertarsal joints. The intertarsal joints (see Fig. 6-19) are gliding in nature. Again, a host of ligaments are present and may be best studied in Fig. 6-18. Synovial membranes line a continuous series of cavities termed collectively the *great tarsal cavity*.

Tarsal-metatarsal joints. The tarsal-metatarsal joints (see Fig. 6-19) are gliding in nature and are joined by *dorsal*, *plantar*, and *interosseous ligaments*. The synovial membranes and cavities are continuous with the intertarsal structures.

Metatarsophalangeal and interphalangeal joints. The metatarsophalangeal and interphalangeal joints (see Fig. 6-19) are constructed like those of the hand. The joints are ball-and-socket anatomically but ovoid functionally.

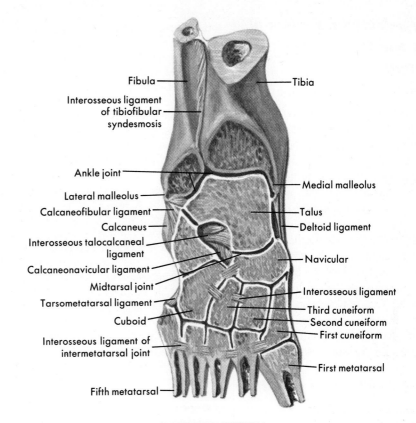

CORONAL SECTION

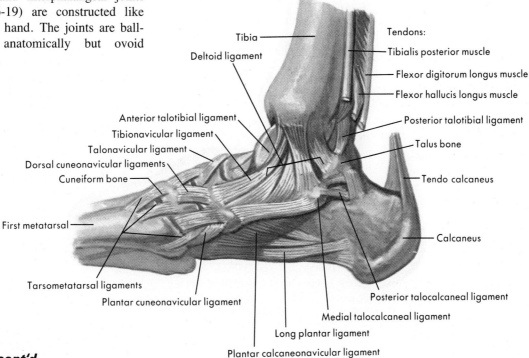

MEDIAL VIEW

Fig. 6-19, cont'd
For legend see opposite page.

Some selected disorders of joints and their associated structures

Traumatic damage to a joint

A *dislocation* is defined as loss of continuity of the joint structures. It may occur as the result of a blow or because of malformation. A *sprain* occurs when ligaments and muscles are torn and tendons are stretched as the result of trauma. Pain is common and often intense, and rapid development of edema (swelling) is usual. Hemorrhages into the skin (ecchymoses) or into the joint itself are initially black or bluish and change to greenish brown or yellow with time. The bruise changes color with time because of the breakdown of blood pigments to bilirubin (yellow) and biliverdin (green). A *strain* is caused by overstretching or pulling muscles and tendons without tearing.

Inflammation of joints

Arthritis is a term used to designate joint inflammation. There are many types of arthritis. A few of the more common types are as follows:

rheumatoid arthritis is a disease for which no specific cause is known. It may occur at any age. The synovial membranes become inflamed and infiltrated with lymphocytes from the bloodstream and antibody-producing plasma cells. These cells produce a *rheumatoid factor,* an immune globulin that represents an antibody to the chemicals produced by the inflamed membrane; the factor further increases the inflammation of the joint. As cells die, their lysosomes release their enzymes into the joint tissues, and ligaments, bone, and articular cartilage are eroded. The condition may develop suddenly or slowly and may involve one joint or several in a progressive fashion. Symmetrical involvement of joints of the hands, feet, wrists, and elbows is common. Hip and knee joint involvement may occur later, making locomotion extremely difficult. The joints are very tender, and the patient often complains of depression, fatigue, and morning stiffness of the joints. Deformities such as enlargement of the affected joint are common and develop rapidly. Treatment of the disorder may involve resting the affected joints, the use of salicylates (aspirin) for relief of pain, and the use of steroids (cortisone) to reduce inflammation. Surgery to remove the affected synovial membranes sometimes offers relief, and severe cases, in which joint mobility is lost, may be treated by replacement of the entire joint with a plastic or stainless steel prosthesis (artificial structure).

osteoarthritis is a chronic degenerative disease, especially of weight-bearing joints. It is characterized by fissuring and destruction of articular cartilage, without great inflammation, and overgrowth of bone at the margins of the joint, often with twisting and enlargement of the affected appendage (such as fingers). Pain is minimal but may increase with hard joint usage. This type of arthritis rarely occurs before 40 years of age, is slow to develop, and usually does not cripple the subject unless it occurs in the hip joints. If advanced, the disorder may be treated with antiinflammatory substances or surgical intervention to remove excess bone or replace the joint.

gouty arthritis is the result of a metabolic error involving the degradation of nitrogenous bases, such as the purines adenine and guanine that are constituents of nucleic acids. Uric acid is a product of purine metabolism, and when blood levels reach high values, the crystals are deposited in certain joints (such as ankle, knee, and hands) with development of intense pain and swelling. Relief may be afforded by decreasing purine intake in the diet and by the use of drugs (such as colchicine) that reduce inflammation and deposition of crystals in the joints.

septic arthritis arises as a result of the presence of bacteria or their products in the bloodstream, with involvement of the joints. Gonococcal (gonorrhea), nonhemolytic streptococcal (strep), and staphylococcal (staph) infections are most commonly involved as causative agents.

Inflammation of associated structures

Bursae are simple synovial sacs filled with synovial fluid placed between two frictional surfaces. They serve to cushion structures or to increase motility between two moving parts. They may or may not communicate with a joint and may be placed beneath the skin (such as the bursa in front of the patella), beneath muscles (as in the shoulder region where the deltoideus muscle overlies the shoulder joint), or beneath tendons (as beneath the patellar tendon). The large bursae found associated with the major body joints (hip, knee, and shoulder) are usually named by their positions. Small bursae without specific names are associated with most other synovial joints.

Bursitis refers to acute or chronic inflammation of a bursa. Bursitis may result from injury to a joint that has bursae around it, rheumatoid inflammation, gout, or bacterial infection (syphilis, tuberculosis). The most common forms include *olecranon bursitis* (tennis elbow), *prepatellar bursitis* (housemaid's knee), *ischial bursitis* (tailor's bottom), and *subdeltoid bursitis* (found in football quarterbacks and baseball pitchers). Chronic bursitis may result in eventual deposition of calcium in the sac with extreme pain resulting from movement of the area.

• • •

As with the skeletal chapter, no attempt will be made to summarize this chapter, since it is mainly anatomical terms.

Questions

1. What are the categories of joints and the types within each category?
2. What features do all synovial joints have in common?
3. What are bursae, and what are their functions?
4. For each of the following joints, describe the particular features that are involved in giving the joint security or permitting its mobility.
    a. Temporomandibular
    b. Atlantoaxial
    c. Shoulder
    d. Elbow
    e. Sacroiliac
    f. Hip
    g. Knee
5. What features distinguish rheumatoid arthritis, gouty arthritis, and osteoarthritis?

Readings

Broer, M.R.: Efficiency of human movement, Philadelphia, 1966, W.B. Saunders Co.

Harris, W.H.: Current concepts: total joint replacement, N. Engl. J. Med. **297**:650, 1977.

Hartung, E.F., et al., editors: Arthritis: manual for nurses, physical therapists, and medical social workers, New York, n.d., Arthritis and Rheumatism Foundation.

Heppenstall, R.B.: Fracture: treatment and healing, Philadelphia, 1980, W.B. Saunders Co.

Sokoloff, L.: The joints and synovial fluid, 2 vols., New York, 1978, Academic Press, Inc.

Sonstegard, D.A., Mathews, L.S., and Kaufer, H.: The surgical replacement of the human knee joint, Sci. Am. **238**:44, Jan. 1978.

7

Muscular tissue

Objectives

Development of muscular tissue

Types of muscular tissue
  Smooth muscle
  Striated muscle

Skeletal muscle
  Structure of the fibers
  Arrangement of fibers within a muscle
  Types of fibers
  Contractile mechanisms in skeletal muscle
  Sources of energy for contraction
  Properties of skeletal muscle

Connective tissue components of skeletal muscle
  Fasciae
  Organization of fibrous components of a muscle
  Tendons and tendon sheaths

Blood and nerve supply to skeletal muscle

Clinical conditions associated with skeletal muscle

Summary

Questions

Readings

OBJECTIVES After studying this chapter, the reader should be able to:

Outline the development of muscular tissue.

Describe the microscopic and ultramicroscopic structure of smooth and skeletal muscle.

Discuss the manner in which fibers may be arranged in skeletal muscles, so as to match the muscle to its location and tasks.

Discuss the differences implied in the structure of red, white, and intermediate skeletal muscle fibers.

Describe how muscle contracts, list its sources of energy, and give properties of skeletal and smooth muscle.

List and give the locations and characteristics of the three components of the body's fascial system.

Describe the location of the connective tissue components within and around a body muscle.

Show how a muscle connects to its tendon and how the connection to a bone is made.

Diagram the organization of tendon sheaths and state their function.

Describe in general terms the distribution of blood vessels and nerves to a skeletal muscle.

Define some of the terms used to describe disorders of skeletal muscle.

Development of muscular tissue

All types of muscle originate from mesoderm. The cells that specialize in forming all muscular tissue are called *myoblasts*. In the case of cardiac (heart) and smooth (visceral) muscle, one myoblast gives rise to one muscle cell, each with a single nucleus. In skeletal muscle many uninucleate myoblasts are believed to fuse together to form the multinucleate fibers of this type of muscle.

Myofibrils are among the earliest intracellular structures to develop in muscle tissue. They are the contractile units within muscle. In skeletal and cardiac muscle they develop cross-bands or *striations*, leading to early identification of myoblasts forming these types of muscle. The myofibrils remain unstriated in smooth muscle, and the lack of striations is what gives smooth muscle its name.

Lastly, skeletal muscle fiber *numbers* are fixed by about the fifth month of development, and any growth of a whole muscle beyond this time is the result of increase in fiber size, not number.

Types of muscular tissue

Muscular tissue through its ability to *contract* or shorten provides for the ability of the body to move through space (locomotion), for the propulsion of material through the body (digestive system, circulation of blood), and for changes in caliber of organs (blood vessels). In all types of muscular tissue the cells or *fibers* are elongated structures that can shorten to varying degrees of their original length.

Two general categories of muscular tissue are recognized, and they are called *smooth* (visceral) and *striated*. Striated muscle may be subdivided into two types called *skeletal* and *cardiac*. Thus three varieties or types of muscular tissue are found in the human body.

Smooth muscle

Smooth or visceral muscle forms the contractile layers of the digestive system from the midesophagus to the anus, forms part of the walls of the respiratory system below the larynx, and is found in the walls of urinary and reproductive organs and around many blood vessels. Small amounts are found in the skin (arrector pili) and in the iris of the eye.

Smooth muscle cells (Fig. 7-1) are long and spindle shaped. They range in length from 20 μm in small blood vessels to 500 μm in the pregnant uterus. In the intestine, a typical location, the length of the fibers is about 200 μm. Diameters range between 3 and 8 μm. A single nucleus occupies the widest portion of the cell and is an elongated structure containing several nucleoli and dispersed chromatin granules. Since they are held together in sheets by reticular fibers, the thin portions of one cell overlap the thicker portions of adjacent cells. The cytoplasm, called *sarcoplasm,* is contained by a membrane called the *sarcolemma.* In the sarcoplasm are sparse longitudinally oriented nonstriated *myofibrils* that represent the contractile elements of the fibers (Fig. 7-2). They appear to attach to dense portions of the sarcoplasm that act as "fixation points" to enable the fiber to shorten.

In longitudinal section (Fig. 7-3) the tissue shows many oval nuclei not arranged in rows; in cross-section the tis-

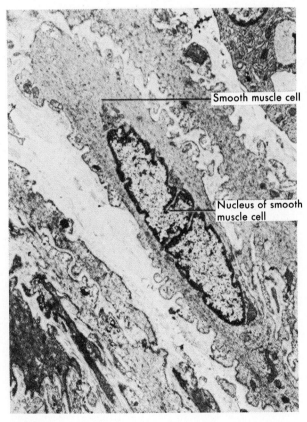

Fig. 7-1
Electron micrograph of smooth muscle. (×8000.)

From McClintic, J.R.: Basic anatomy and physiology of the human body, ed. 2, New York, 1980, John Wiley & Sons, Inc.

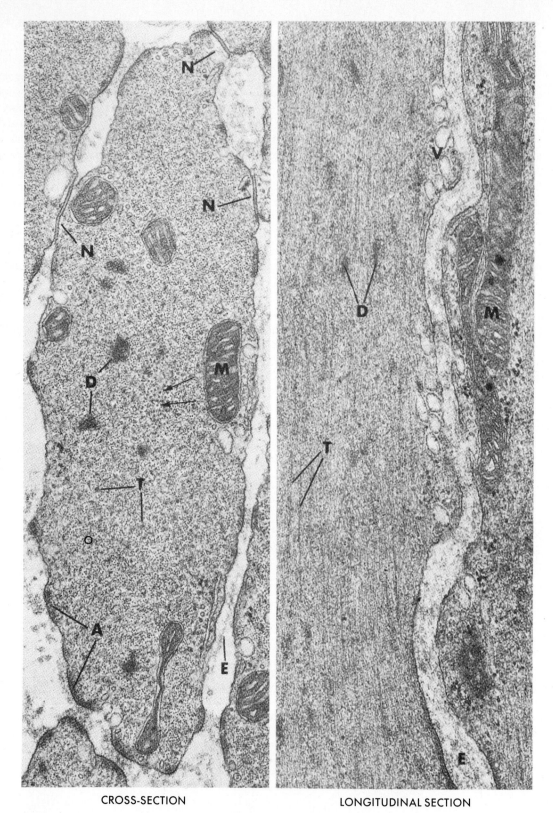

CROSS-SECTION   LONGITUDINAL SECTION

Fig. 7-2

Cross- *(left)* and longitudinal *(right)* sections of smooth muscle viewed under the electron microscope at ×67,200 magnification. *T*, Thick filaments interspersed among many thin filaments that appear as small dots or fine lines. Arrows show intermediate-sized filaments. *N*, Tight junctions that allow depolarization to travel from one cell to another. *A*, Plaques affording peripheral attachment for filaments. *D*, Cytoplasmic dense bodies for filament attachment. *V*, Pinocytic vesicles. *E*, External lamina separating one cell from another. *M*, mitochondrion. Thin filaments circled.

From Copenhaver, W.M., et al.: Bailey's textbook of histology, ed. 17, Baltimore, 1976, The Williams & Wilkins Co.

sue shows sections that are more or less regular in outline but varying in diameter, with some cells showing nuclei and others not. The latter phenomenon is due to the fact that some cells are sectioned through their thicker portions that contain nuclei, whereas others are cut through their tapering ends.

Smooth muscle found in the gut, uterus, ureters, and small blood vessels is of the *unitary* variety. It may show "spontaneous" activity in response to chemical stimuli or stretching. An area of depolarization and an electrical "current" are developed that spread through the entire mass. Contraction is slow, and this is good because fluids are not moved so rapidly as to preclude absorption (gut), or volume changes do not occur so rapidly as to cause sudden changes in pressure (blood vessels). This type of smooth muscle can also maintain a given degree of contraction for long periods of time—this is called "tone."

Smooth muscle of the eye (sphincter and dilator pupillae) is served by nerves and depends on them to cause its contraction. This type of smooth muscle is called *multiunit* smooth muscle. When a light is shined into the eye, the pupil *rapidly* narrows as the result of a nervous reflex. This rapid response protects the delicate retinal cells from damage.

In some areas, such as the urinary bladder, the muscle exhibits characteristics of both types of muscle described previously. As the bladder fills, the muscle behaves like a unitary muscle; to empty the organ requires nerve stimulation, and micturition results.

Striated muscle

The remainder of this chapter will be directed to a discussion of skeletal muscle. Cardiac muscle will be discussed in Chapter 11.

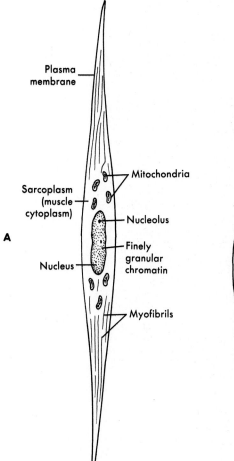
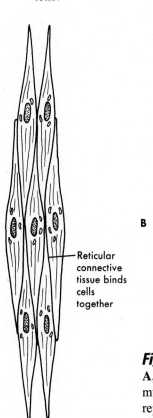

Fig. 7-3
A, Morphology of single smooth muscle cell. **B,** Diagrammatic representation of organization of smooth muscle cells into sheets of tissue.

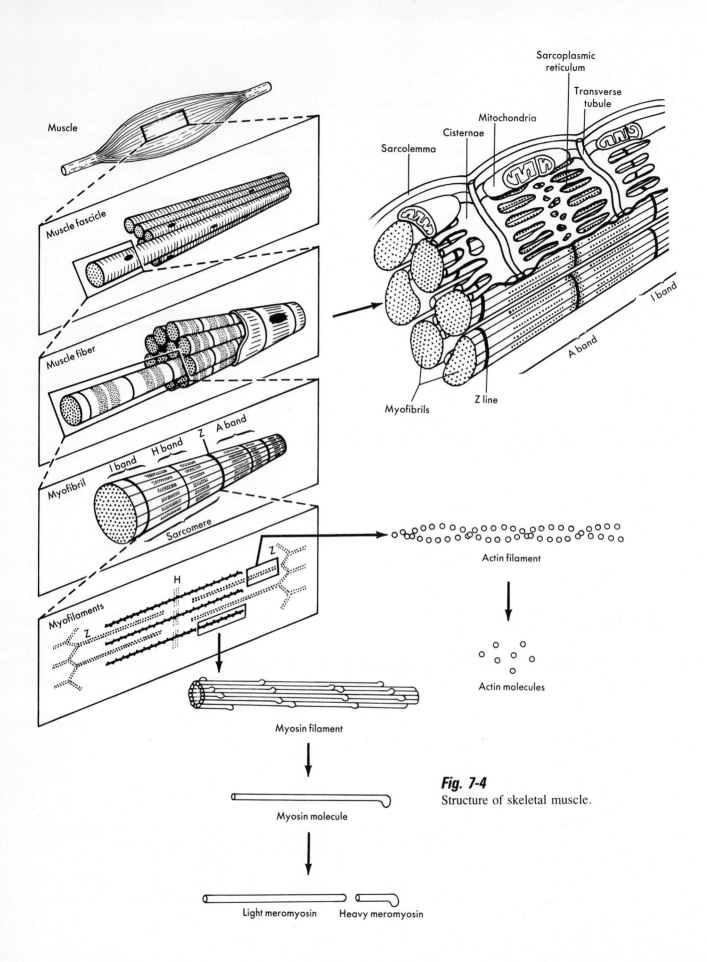

Fig. 7-4
Structure of skeletal muscle.

Skeletal muscle

Structure of the fibers (Fig. 7-4)

The unit of the skeletal muscles is the *fiber*. Each fiber is a cylindrical unit with a diameter between 10 and 100 μm and lengths between 2 mm and 7.5 cm. In general, the smaller the muscle in which the fibers are found, the smaller will those fibers be. The fibers are surrounded by a *sarcolemma*, and this membrane encloses the *sarcoplasm*, or muscle cytoplasm. Peripherally located in the sarcoplasm are the muscle nuclei. Each fiber is *multinucleated*, containing from several to several hundred nuclei. The most prominent feature in the sarcoplasm are many closely packed longitudinally oriented fibrillar units known as *myofibrils*. The myofibrils are *striated*, exhibiting alternating light and dark bands that in the muscle fiber as a whole lie above one another. This gives the pronounced striation to the entire fiber. At the ultramicroscopic level, the myofibrils may be seen to alternate with rows of mitochondria. Additionally, the myofibrils may be seen to be composed of *myofilaments*, proteins that represent the contractile mechanism of the muscle. Proteins designated as *actin, troponin,* and *tropomyosin* (Fig. 7-5) form the thin filaments of the myofibril; one called *myosin* forms the thick filaments of the muscle. These proteins are organized in an orderly arrangement in skeletal muscle (see Fig. 7-4), not randomly as in smooth muscle (see Fig. 7-2). The arrangement of the filaments accounts for the striations the myofibrils exhibit. For example (see Fig. 7-4) the dark band or *A line* contains both myosin and actin complexes; the *H line* that divides the A line contains only myosin; the light *I line* contains only actin complexes; the *Z line* that divides the I line contains only actin that appears to *attach* to the Z line. Additionally, Z lines appear to extend all the way across a given myofibril, and

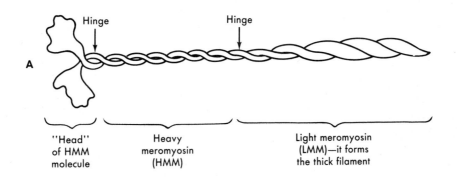

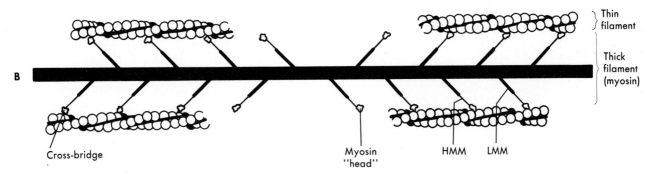

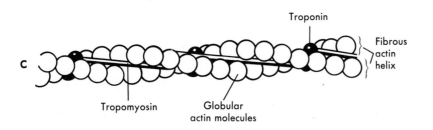

Fig. 7-5
Arrangement of proteins in skeletal muscle. **A,** Myosin molcule, showing "hinges" or areas of greater flexibility where bending may occur; **B,** formation of cross-bridges between myosin and thin filaments; **C,** structure of thin filaments.

perhaps across the entire fiber. When the muscle shortens, Z lines are drawn closer together. For this reason, the Z-to-Z distance is sometimes designated a *sarcomere,* the "contractile unit" of the myofibril.

Additional features of skeletal muscle at the ultramicroscopic level include the *transverse tubules (T-tubules)* that begin at the sarcolemma and penetrate the fiber to form *annuli* or rings around each myofibril and the *sarcoplasmic reticulum* that runs longitudinally around the myofibrils and provides a storehouse for the calcium essential for muscle contraction. The T-tubule membranes provide a route for depolarization of the sarcolemma to penetrate to the interior and a route for delivering fluids and solutes to the fiber interior. Depolarization of the fiber membranes is a prerequisite for the contractile process.

Arrangement of fibers within a muscle

According to where the muscles are found on the body, the form of the muscle will be determined by the fiber arrangement (Fig. 7-6). On the appendages the muscle fibers all run parallel to one another and terminate in tendons at either end that connect the muscle to the skeleton. Such a muscle is called a *fusiform* muscle and presents in cross-section a rounded or oval outline. On the thorax and abdomen such muscles would be too bulky to be accommodated on the broad surfaces of these body areas, and here the muscle fibers converge on tendons like the plumes of a feather. Such a muscle is thus termed a *pennate (penniform)* muscle. These present rectangular or flattened outlines in cross-section. If only one band of fibers sweeps, on one side, to the tendon that runs the length of the muscle, it is said to be *demipennate,* or "half a feather." *Bipennate* muscles have two bands of fibers that converge on the tendon from opposite sides. *Multipennate* muscles have three or more groups of fibers converging on the tendon. *Circumpennate* muscles have their fibers circularly oriented (as around the eye or the mouth) around a body orifice.

Types of fibers

When muscles are examined in the fresh condition, they differ somewhat in their color. Three types of fibers are recognized on the basis of their color: *red, intermediate,* and *white.* Red fibers are rich in mitochondria, are small, and contain significant amounts of a pigment known as *myoglobin.* The myoglobin can, like the blood hemoglobin, "store" oxygen within the fibers. Such red fibers are thus less subject to fatigue and are found in many extensor muscles of the body that are responsible for maintaining the erect posture. White fibers contain fewer mitochondria and little myoglobin. They are found chiefly in the upper appendage, where speed is important, but where fatigue may occur sooner. Intermediate fibers lie between the other two types in characteristics. All three types of fibers may occur in a given body muscle, and prolonged exercise or training can increase the proportion of red fibers.

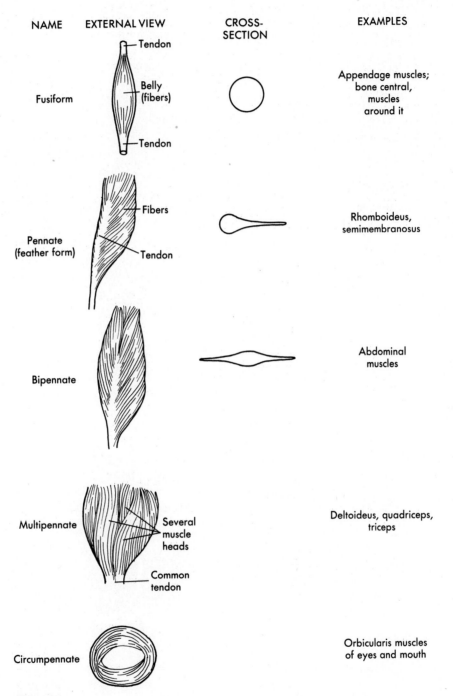

Fig. 7-6
Various arrangements of fibers in muscles of body.

Contractile mechanisms in skeletal muscle

All varieties of muscle have a common mechanism for contraction and share common energy sources. My comments will be directed to the contraction of skeletal muscle with the understanding that what is said about contraction applies to all types. Skeletal muscle has a special structure that delivers nervous stimuli to the muscle; it is called a neuromuscular junction.

The **neuromuscular junction** (Fig. 7-7) consists of an expanded end of a nerve fiber that contains small membrane-surrounded chemical-containing **vesicles**. The chemical is acetylcholine, synthesized within the nerve cell and packaged by its Golgi body. On the sarcolemma opposite where one of the end knobs is located is a depression, the **synaptic trough**. The trough walls are folded, forming **subneural clefts,** and bear receptor sites for acetylcholine. When a nerve impulse arrives at the nerve fiber's ending, it causes a change in the nerve membrane that allows tissue fluid calcium ion (Ca^{2+}) to flow into the fiber. The Ca^{2+} causes the acetylcholine to be released from the vesicles or causes release of whole vesicles to the outside of the nerve fiber where they then release their chemical. The acetylcholine crosses the space between the nerve fiber and the subneural clefts and binds to the receptor sites. This causes the sarcolemma to depolarize, that is, to set up an electrical field on the muscle fiber membrane. An enzyme, acetylcholinesterase (cholinesterase), then cleaves the acetylcholine, and it is no longer effective as a stimulus. Meanwhile, the depolarization spreads along the sarcolemma, follows the T-tubule membranes into the fiber, and eventually causes depolarization of the sarcoplasmic reticulum (SR) membranes that are in close association with the T-tubules (see Fig. 7-4).

Ca^{2+}, stored within the SR cavities when the muscle is at rest, is released and is bound to troponin molecules (see Fig. 7-5). The troponin molecules are then believed to change shape, pulling on tropomyosin molecules to which they are attached. This causes the tropomyosin molecules to "slide" on actin molecules, uncovering a binding site to which the "heads" of the myosin molecules (see Fig. 7-5) can attach. This forms a cross-bridge that moves to draw the actin molecules closer together in the A band. The cross-bridges are then broken and the myosin head returned to its original position; this requires energy provided by adenosine triphosphate (ATP). The myosin then attaches to new actin-binding sites further along the actin molecules. Again, by movement of the head, the actin filaments are drawn still closer together. The movement is analogous to pulling in a rope using one hand and then the other. When all binding sites are occupied, the muscle can shorten no further. Relaxation is accomplished by returning the Ca^{2+} to the SR by active processes, and the initial events (sliding and troponin shape change) cover the binding sites.

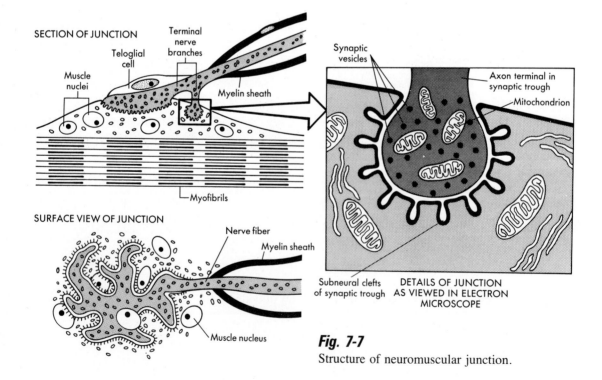

Fig. 7-7
Structure of neuromuscular junction.

Sources of energy for contraction

The essential chemical in contraction after Ca^{2+} exerts its effect is ATP. It is used to ensure continued cross-bridge formation and movement. Therefore it is accurate to say that maintenance of ATP stores is essential to continued muscular activity.

Energy for synthesis of ATP comes from:
- Degradation of glucose, the energy released being used to synthesize ATP from adenosine diphosphate (ADP) and phosphate.
- Degradation of fatty acids with the energy used as in the previous source.
- Breakdown of creatine phosphate, a compound unique to muscle. The energy released is used as stated previously. Creatine phosphate breakdown requires only one step, whereas the other schemes have many steps and take a short time to "get going." The muscle need not suffer a shortage of energy for its initial activity.

Properties of skeletal muscle

Skeletal muscles are composed of groups of muscle cells served by a single nerve cell process (axon). Such an assemblage constitutes a **motor unit**, and many thousands of such units form the whole muscle. *Each* motor unit, when stimulated, *contracts maximally or not at all* (according to the strength required to depolarize it). *Strength of contraction can therefore be varied* in the whole muscle according to how many motor units are activated, matching strength to task. The whole muscle can be *tetanized*, or thrown into a sustained contraction, an essential in postural muscles that hold our bodies erect against the force of gravity. Contraction of skeletal muscles is very rapid, one series of contraction-relaxation being accomplished in less than one tenth of a second. Rapid contractions are essential for the muscles of the body, particularly those of the appendages so that we can carry out the many activities we pursue.

Connective tissue components of skeletal muscle

Connective tissue components are necessary in a muscle to maintain its organization and to allow for skeletal attachment of the muscle. Planes of connective tissue surround muscles, binding them to one another or to other body structures such as skin.

Fasciae

The dissectible fibrous connective tissues, other than those that are specifically organized into ligaments, tendons, and aponeuroses, are called *fasciae* (sing., fascia).

The entire fascial system consists of three components:

subcutaneous *(superficial) fascia,* or *hypodermis,* is continuous over the entire body between the skin and the deeper lying structures. It is two layered: the outer or superficial layer contains much fat and forms the *panniculus adiposus;* the inner or deeper layer usually contains no fat and is rich in elastic fibers. Superficial arteries, veins, and nerves are carried between the two layers.

deep fascia invests the muscles of the body, divides them from one another, and forms compartments within which muscles lie. It holds muscles in proper positions.

subserous fascia lies between the deep fascia and the serous membranes lining the true body cavities. It may contain much fat, as around the kidneys, or be thin and devoid of fat, as around the chest (thoracic) cavity.

All fasciae are continuous with one another to maintain body structure.

Organization of fibrous components of a muscle
(Fig. 7-8)

The deep fascia around muscles may be a denser layer surrounding a finer connective tissue coat immediately around the muscle, or it may form the outer limits of the whole muscle. In either case the most external connective tissue layer on the muscle is termed the *epimysium.* From the epimysium, partitions extend inward as the *perimysium* to surround groups of fibers called *fascicles.* These two components are composed of collagenous and some elastic fibers. From the perimysium, more delicate reticular fibers and collagenous fibrils tie individual fibers together as the *endomysium.* Again, all components connect with one another.

Tendons and tendon sheaths

Tendons are dense masses of collagenous fibers that are continuous with the previously described connective tissue components of a muscle and that allow attachment to skeletal or fascial structures. An *aponeurosis* is a broad flat tendon. Where a muscle joins a tendon, the fibers may abruptly end, taper, or be split into digitations. The reticular fibers of the endomysium are increased in amount at the junction and often form a "cuplike" mass over the end of the muscle fiber (Fig. 7-9). The reticular fibers are then seen to be continuous with the collagenous fibers of the tendon. Fibers of the tendon then pass to fuse with the periosteum of the bone, and the periosteum then attaches *into* the bone by means of Sharpey's fibers.

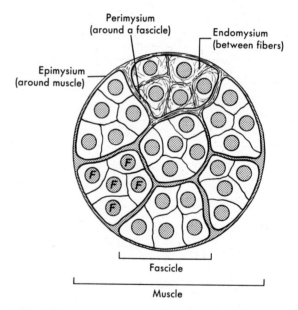

Fig. 7-8

Diagrammatic representation of connective tissue components of skeletal muscle. *F* indicates muscle fibers.

Thus the whole system is firmly affixed in its various parts.

Tendons are found in the body where strength of attachment is required, where there is not room for fleshy portions of a muscle, where there is a long distance to span, or where joint operation would destroy the fleshy part of a muscle.

At the wrist and ankle particularly, where tendons pass over rather mobile joints or where the tendons themselves move considerably as the joint works, the tendons are enclosed in *tendon sheaths* (Fig. 7-10). These are actually special synovial sacs separate from those of the joints. The sheaths are doubled, and as the tendon moves, its passage is lubricated by synovial fluid between the two layers.

Blood and nerve supply to skeletal muscle

The blood vessels to skeletal muscles (Fig. 7-11) course in the connective tissue septae that separate fascicles and fibers from one another. Capillary beds are formed around and between the individual fibers. The capillaries are tortuous to permit accommodation to changes in fiber length, and even in untrained muscles, there is a ratio of 1:1.5 between capillaries and muscle fibers. In the trained muscle the ratio becomes 1:1.

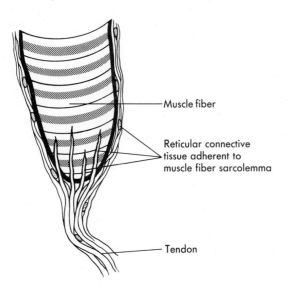

Fig. 7-9
Attachment of muscle to tendon.

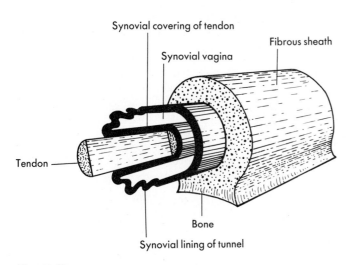

Fig. 7-10
Organization of tendon sheaths.

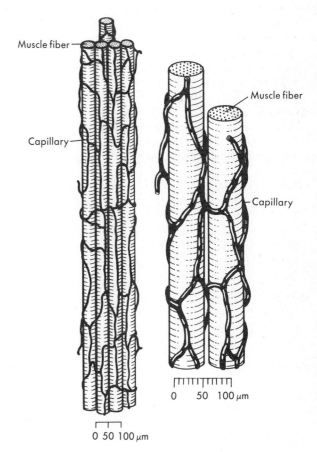

Fig. 7-11
Blood vessel arrangement in skeletal muscle.

Efferent or motor nerves to skeletal muscle have their cell bodies in the ventral gray column of the spinal cord. One axon passes to a group of fibers (a motor unit) within the muscle, and if cut, that group of fibers will no longer contract voluntarily. A nerve to a muscle is, of course, composed of many axons that ultimately supply all fibers within the muscle. If a nerve is cut, the whole muscle is paralyzed.

Sensory nerves from the muscle are derived from complex sensory organs within the muscle called *muscle spindles* (Fig. 7-12). The nerve fibers are wrapped around the tiny muscle fibers of the spindle and are "fired" by changes in tension on the muscle fibers. These impulses are integrated into coordinated and purposeful motion.

Clinical conditions associated with skeletal muscle

It is obviously impossible to discuss all, or even very many, of the pathological conditions associated with skeletal muscle. This section is intended to briefly explain or clarify some of the terms that reach us through the mass communications media. It must also be realized that nervous and muscular function are intimately related and that it is often difficult to determine wherein the fault lies when a muscle loses strength.

The term *atrophy* implies wastage or loss of bulk of a muscle, regardless of cause. If used in the name of a disease, the implication is that the atrophy is related to a neural or genetic disorder. A *myopathy* is a disorder not caused by neural or emotional factors. *Dystrophies* are a type of myopathy that have two special characteristics: they are genetically determined and muscular weakness is progressive. In the muscular dystrophies there is no demonstrable enzyme or structural protein defect, as might be suspected from a genetic disease. Muscle membrane abnormalities are indicated in some forms of dystrophy, but no direct proof for this thesis is available.

In human muscular dystrophies the pathological picture is one of random death of muscle fibers and their replacement by connective tissue and fat. Weakness progresses as more and more fibers are involved, and eventually the patient may be confined to bed or a wheelchair. There is no specific treatment for any form of dystrophy. Genetic counseling is the only technique available to control the incidence of dystrophy at present.

Myasthenia gravis is usually defined as a state of abnormal fatigability in which there is no sign of a neural lesion. The disease has a predilection for ocular and facial muscles, causing them to "sag," which creates an inability to keep the eyelids open or creates excessive jowls or drooping mouths. No specific cause for the disorder can be advanced at this time. Some theories that have been advanced over the years as to cause include the following:

Failure of synthesis of the neuromuscular transmitter acetylcholine
Lack of receptor sites for acetylcholine on the trough membranes or excessive cholinesterase activity that destroys the chemical before much can bind
Development of antibodies to muscular tissue that results in fiber destruction

Paralysis is generally used to describe the inability to voluntarily contract a muscle and may be traced to a nerve lesion. If the neuron from the brain to the spinal cord is damaged, the muscle will be paralyzed in a contracted state (spastic paralysis). If the

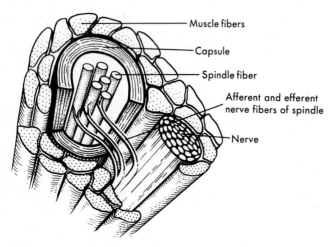

Fig. 7-12
Muscle spindle in section.

neuron from cord to a muscle is damaged, the muscle is limp (flaccid paralysis).

Fasciculations and *fibrillations* refer to involuntary rapid contractions of groups of skeletal muscle fibers. The usual cause is nerve damage.

These days, with preoccupation of large segments of the population with jogging or other activities demanding muscular contraction, a host of painful muscular disorders have surfaced. *Muscle spasms* most commonly occur as the result of chemical imbalances within the muscle, usually traceable to abnormal blood levels of calcium, sodium, or potassium, or to chemical toxins (for example, tetanus). Massaging the muscle may increase blood supply and speed recovery. *Muscle cramps,* although similar to spasms, may be caused by intense sensory nerve stimulation from the exercise that results in reflex contraction. A *pulled muscle* usually refers to tearing of muscle fibers as a result of severe activity. *Shin splints* involve the muscles on the anterolateral aspect of the leg. Severe activity, as in running, may result in production of large quantities of osmotically active metabolites that cause water to enter the muscle fibers. Swelling compresses the blood vessels against the walls of the fascial compartment, the latter being nonexpansive on the shin. *Ischemia* (decreased blood flow) results. Thus accumulation of products of metabolism may stimulate sensory nerves, and ischemic muscles are painful on their own. Graduated exercise and rest when the muscles hurt will usually overcome the condition.

Torticollis or *wryneck* occurs when there is a spasmodic contraction of the neck muscles that draws the head to one side with the chin pointing to the other side. Some causes advanced for the condition include vertebral disease, tonsillitis, "rheumatism," and trauma to the neck area that results in extensive scar tissue formation.

The discussion of cardiac muscle structure will be deferred to Chapter 11.

Summary

1. Muscular tissue is of mesodermal origin.
   a. The cell of origin is the myoblast.
   b. Cardiac and skeletal muscle acquire striated myofibrils; smooth muscle does not.
   c. Growth of skeletal muscles after the fifth month is the result of increase in size, not number, of fibers.
2. Muscular tissue is contractile and does work in the body.
   a. All types of muscle possess fibers or cells.
   b. There are three varieties of muscle in the body: smooth (visceral), skeletal, and cardiac.
3. Smooth muscle is distinguished by the following characteristics:
   a. It occurs in the digestive system, respiratory system, urinary and reproductive systems, and in the walls of blood vessels.
   b. It has cells that are uninucleate and spindle shaped, that are surrounded by a sarcolemma, and that contain sarcoplasm with a few scattered nonstriated myofibrils.
   c. It may be stimulated via nerves (iris of eye) or exhibit spontaneous contraction (gut).
4. Skeletal muscle is characterized by the following:
   a. It attaches to the skeleton.
   b. It is organized into multinucleated, cylindrical, striated fibers. Myofibrils are present, as are sarcolemma and sarcoplasm, myofilaments, sarcoplasmic reticulum, and transverse tubules.
5. Muscle fibers are organized into muscles in several patterns.
   a. Fusiform muscles, exemplified by those of the appendages, are cylindrical in shape with tendons at either end.
   b. Fan- or feather-shaped muscles are called pennate. Demipennate, bipennate, multipennate, and circumpennate arrangements are seen.
6. There are three types of skeletal muscle fibers distinguished by color and chemical content.
   a. Red fibers are rich in mitochondria and myoglobin and are more resistant to fatigue. Postural muscles are mainly red in type.
   b. White fibers are poor in myoglobin, are more easily fatigued, and occur chiefly in the upper limb.
   c. Intermediate fibers are between the other two in myoglobin content.
   d. A given muscle contains mixtures of the three types of fibers.
7. Contraction of skeletal muscle depends on stimulation by a nerve fiber.
   a. The neuromuscular junction uses acetylcholine as the chemical that is released by the nerve fiber and that causes depolarization of the muscle membrane, the T-tubule membranes, and the SR membranes in turn.
   b. Calcium ion is released from the SR cavities. It binds to troponin that changes shape, pulls on tropomyosin, and uncovers binding sites on the actin molecules. Myosin heads attach to the actin-binding sites and pull the actin molecules closer together to cause contraction.
   c. Relaxation is accomplished by return of the calcium ion to the SR, covering of actin-binding sites, and loss of cross-bridges.
8. Maintenance of ATP stores in the muscle is accomplished by the metabolism of glucose, fatty acids, and creatine phosphate.
   a. The energy released by this metabolism is used to synthesize ATP.
   b. Creatine phosphate is used first, because its metabolism requires only one step, whereas metabolism of the other sources requires a series of steps and provides energy for the continued activity of the muscle.
9. Skeletal muscle is composed of motor units that consist of a single nerve fiber and the muscle fibers it supplies.
   a. The motor unit follows the all-or-none law, and strength of contraction depends on how many motor units are active.
   b. Skeletal muscle contracts most rapidly of all types.
   c. Skeletal muscle can be tetanized, or caused to sustain a contraction.
10. There are several connective tissue components associated with the body generally and muscles in particular.
    a. The fasciae are those connective tissues not organized into tendons

and ligaments. There are three components to the fasciae.
  (1) The subcutaneous fascia lies beneath the skin.
  (2) The deep fascia invests the body muscles.
  (3) The subserous fascia lies beneath the linings of the body cavities.
  b. Connective tissue of muscles is organized in three ways.
   (1) Epimysium surrounds the whole muscle. It may be continuous with or separate from the deep fascia.
   (2) Perimysium surrounds groups of fibers (fascicles).
   (3) Endomysium lies between individual fibers.
  c. Tendons, composed of collagenous tissue, afford muscle attachments to bones or other body structures.
   (1) A flat tendon is an aponeurosis.
   (2) Tendons are continuous with the inner connective tissues of the muscle and with the periosteum of the bones.
   (3) Tendon sheaths, synovial-lined sleeves, are found around tendons where tendon movement occurs as the muscles contract.
11. Blood vessels and nerves are abundant in skeletal muscles.
12. Junction of a nerve on a skeletal muscle is made by a neuromuscular junction. Sensory structures in the muscle convey information about the contraction.
13. There are several well-known disorders associated with skeletal muscle.
  a. Atrophy refers to wastage.
  b. A myopathy is a disorder resulting from nonnervous or nonemotional causes.
  c. Dystrophies are genetically determined.
  d. Myasthenia gravis may be due to neuromuscular, immunological, or genetic causes.
  e. Paralysis involves motor nerve damage.
  f. Spasms, cramps, and associated disorders may have chemical bases.

Questions

1. In what ways are the three types of muscle similar? Different?
2. What are the ultimate contractile elements of muscle? How are they arranged in skeletal muscle?
3. How is arrangement of fibers in a muscle correlated with location of the muscle on the body?
4. What is the functional significance of red and white muscle fibers?
5. List the three subdivisions of the body fasciae and tell where each is found.
6. What are the names and locations of the connective tissues that bind fibers into a muscle?
7. Describe the musculotendinous junction and the reasons tendons are present in the body.
8. Describe the structure and use of tendon sheaths.
9. Describe the blood supply to a skeletal muscle.
10. What is the structure of the neuromuscular junction, and how do impulses cross it to the muscle?
11. Describe several disorders associated with the skeletal muscles.

Readings

Felig, P., and Wahren, J.: Fuel homeostasis in exercise, N. Engl. J. Med. **298:**1078, 1975.

Homsher, E., and Kean, C.J.: Skeletal muscle energetics and metabolism, Annu. Rev. Physiol. **40:**93, 1978.

Hoyle, G.: How is muscle turned on and off? Sci. Am. **227:**84, April 1970.

Huddart, H., and Hunt, S.: Visceral muscle: its structure and function, New York. 1975, Halsted Press.

Prosser, C.L.: Smooth muscle, Annu. Rev. Physiol. **36:**503, 1974.

8

The skeletal muscles

Objectives

Bones and muscles as body lever systems

How muscles are named

Muscles of the head
    Facial muscles
    Cranial muscles

Muscles of mastication

Muscles of the tongue

Muscles associated with the hyoid bone and larynx

Muscles of the neck

Muscles of the back

Muscles of the anterior and lateral abdominal walls

Muscles of the posterior abdominal wall

Muscles of respiration

Muscles operating the scapula and clavicle

Muscles moving the shoulder joint or humerus
    Flexors of the humerus
    Extensors of the humerus
    Abductors of the humerus
    Rotators of the humerus

Muscles moving the forearm
    Flexors of the forearm
    Extensors of the forearm
    Rotators of the forearm

Muscles moving the wrist and fingers
    Flexors of the wrist and fingers
    Extensors of the wrist and fingers
    Abduction and adduction of the wrist

Muscles of the thumb

Muscles within the hand

Muscles moving the hip and knee joints

Muscles moving the ankle and foot

Muscles within the foot

Readings

OBJECTIVES After studying this chapter, the reader should be able to:

Describe the several classes of levers occurring in the body and understand the use of the bones of the body as levers by the muscular system.

Relate muscle shape, fiber arrangement, and origin, insertion, and action to the naming of muscles.

Define the various terms used to describe muscular actions.

Gain an understanding of where muscles are located on the body and how they are organized in groups having different and often opposite actions on body joints.

Achieve an understanding of the manner of innervation of the skeletal muscles.

Recognize that muscles cooperate to achieve the complicated movements of which our bodies are capable and that a stated muscle action is not always the one it exhibits as we move.

Bones and muscles as body lever systems

Movement of the body occurs when muscles contract and pull at a particular point on a bone. The bones always move with reference to a joint, whose construction determines the type of movement permitted. The bones and joints thus act as levers to achieve body movement.

To understand the several types of levers, let us first examine the parts of a lever (Fig. 8-1) as follows:

fulcrum (F) is usually considered to be a point or line about which the lever moves and in the body is provided by a joint.

effort (E) represents the force required to move the lever and is the point at which the muscle inserts on the bone to exert its contractile force.

resistance (R) is the weight that muscular contraction must overcome and is usually considered to be concentrated in a small area on the lever.

effort arm (EA) is a distance or length along the lever and lies between F and E. Short EAs usually mean that a greater force is required to move the lever, but a larger movement of the resistance occurs for a given distance of EA movement.

resistance arm (RA) is also a distance along the lever, in this case from F to R. Short RAs mean that less muscular force is required to move the lever.

The placement of *F*, *E*, and *R* in relationship to one another determines the **class** of the lever. Three classes are recognized (see Fig. 8-1):

first class lever has *the fulcrum lying between the effort and the resistance*. Many everyday tools are first class levers, including prybars, scissors, and manually operated posthole diggers. In the body, few first class levers are found, because such a lever would require projections on a bone on either side of the joint. The olecranon process of the ulna has the triceps tendon attached to it, and when the forearm is extended (straightened relative to the arm), the ulnohumeral joint is seen to lie between the process and the resistance formed by the forearm and hand.

second class lever has the fulcrum at or toward one end with *resistance between fulcrum and effort*. Lifting the handles of a loaded wheelbarrow is an everyday example of a second class lever; the wheel is the fulcrum with the load between it and the operator. Some consider a body example of a second class lever to be rising on the toes. Effort is applied at the heel, the "ball" of the foot forms the fulcrum, and the body weight concentrated at the high point of the transverse arch constitutes the resistance.

third class lever is the most common type of lever in the body and has *effort between resistance and fulcrum*. Placing the base of a ladder next to a building and raising it by pushing on the rungs between both ends illustrates an everyday application of a third class lever. The reason for the abundance of this type of lever in the body should be obvious—a joint is at one end and nearly the whole bone length is available for muscle attachment. The biceps, whose tendon attaches to the radial tuberosity, in flexing the forearm is an obvious example of this class of lever in the body.

To illustrate the relationship of force required versus resistance to overcome, the formula

$$E \times EA = R \times RA$$

may be used. The formula suggests that force is applied over a distance and that resistance is also exerted over a distance. If force and resistance are balanced, no movement will occur. Changing *E* or *R* will permit motion. Let us make some calculations of the force required to move a given resistance using the three classes of levers as we find them in the body and see what we can discover.

First class (triceps acting to extend forearm)

E = ?
EA = 2.5 cm (the distance from the center of the ulnohumeral joint to the tip of the olecranon process)
R = A 5-kg weight held in the hand
RA = 25 cm (the distance from the center of the elbow joint to the center of the palm of the hand)

$$E \times EA = R \times RA$$
$$E = \frac{R \times RA}{EA} = \frac{5 \times 25}{2.5} = 50 \text{ kg}$$

Note a 10:1 ratio of power to resistance. Make two more calculations: in one, lengthen *EA* to 5 cm and leave everything else the same; in the second, shorten *RA* to 20 cm; what happens to *E*?

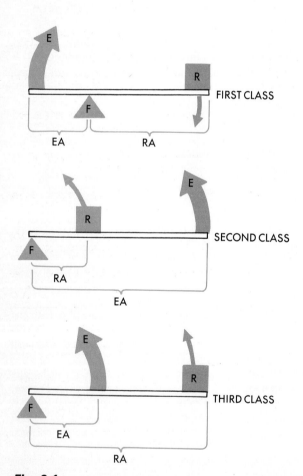

Fig. 8-1
Basic parts of lever for three classes of levers. Note that *EA* and *RA* may overlap, and observe direction resistance moves when force is applied. See text for explanation of symbols.

Second class (rising on the toes)
E = ?
EA = 25 cm (the distance from the ball of the foot to the tip of the heel)
R = A 5-kg body weight on the center of the foot
RA = 20 cm (the distance from the center of the ankle joint to the ball of the foot)

$$E = \frac{R \times RA}{EA} = \frac{5 \times 20}{25} = 4 \text{ kg}$$

Note that the effort is *less* than the resistance.

Third class (the biceps in flexion of the forearm)
E = ?
EA = 5 cm (the distance from the center of the elbow joint to the radial tuberosity)
R = 5 kg
RA = 30 cm (the distance from the center of the elbow joint to the center of the hand)

$$E = \frac{R \times RA}{EA} = \frac{5 \times 30}{5} = 30 \text{ kg}$$

Better than a first class lever, but still 6:1.

How muscles are named

It often aids in remembering facts about muscles to have some notion of how they are named. The following illustrates some of the criteria used to name muscles:

Naming by **shape**. Using their resemblance to geometric figures, we note the *trapezius* and the *rhomboideus*.

Naming by **origin** and **insertion**. The *origin* of a muscle is the less movable bone to which the muscle is attached; the *insertion* is the more movable bone. The *sternocleidomastoideus* (origin—sternum and clavicle; insertion—mastoid process of temporal bone) illustrates this well.

Naming by **location**. The rectus *abdominis* (abdomen) and latissimus *dorsi* (back) illustrate the use of the location criterion.

Naming by **fiber arrangement**. Fibers oriented circularly around a body orifice are said to have an orbicular (circular) arrangement. Thus *orbicularis oculi* (around the eyes) and *orbicularis oris* (around the mouth) are named for fiber arrangement.

Naming by **action**. Action is probably the most useful criterion, for often we can learn the muscle name, the action, and the body part moved all at once. The muscles of the forearm and leg illustrate well this criterion:

Flexor carpi radialis (flexes wrist; on radial side of forearm)
Flexor digitorum profundus (flexes the fingers, a deep muscle)

We have explained the terms *origin* and *insertion* as applied to the skeletal muscles; the term *action* is, of course, the movement a body part or joint undergoes when a muscle contracts. The box to the right presents definitions of actions, and these should be thoroughly learned to understand how muscles achieve movement of body parts. Remember that these actions assume an initial starting position that is *anatomical position*. Discussion of the muscles is carried out in a regional fashion and is general in nature. Specific origins, insertions, actions, and innervations are given in a table correlated with the illustration showing each group of muscles.

As an introduction to the muscles, look at Fig. 8-2, depicting the entire body musculature as it would appear with the skin and subcutaneous fat layer removed. Superficial muscles and muscle groups are indicated. Review of surface anatomy (Chapter 3) may be helpful at this time.

Muscle actions

Action	Definition
Flexion	Decrease of angle between two bones
Extension	Increase of angle between two bones
Abduction	Movement away from midline (of body or part)
Adduction	Movement toward midline (of body or part)
Elevation	Upward or superior movement
Depression	Downward or inferior movement
Rotation	Turning about the longitudinal axis of the bone
Medial	Toward midline of body ("inward")
Lateral	Away from midline of body ("outward")
Supination	To turn the palm up or anterior
Pronation	To turn the palm down or posterior
Inversion	To face the soles of the feet toward each other
Eversion	To face the soles of the feet away from each other
Dorsiflexion (flexion)	At the ankle, to move the top of the foot toward the shin
Plantar flexion (extension)	At the ankle, to move the sole of the foot downward, as in standing on the toes

From McClintic, J.R.: Basic anatomy and physiology of the human body, ed. 2. Copyright © 1980. Reprinted by permission of John Wiley & Sons, Inc.

186 Human anatomy

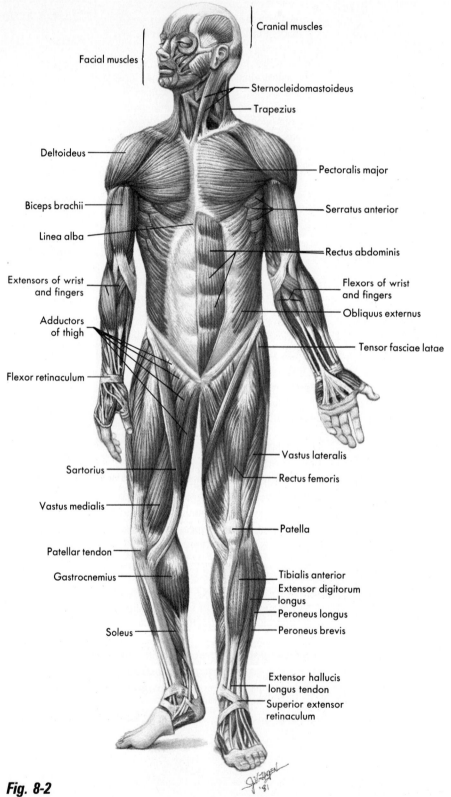

Fig. 8-2
Anterior view of human muscular system, with major groups or muscles labeled.

8   *The skeletal muscles*   187

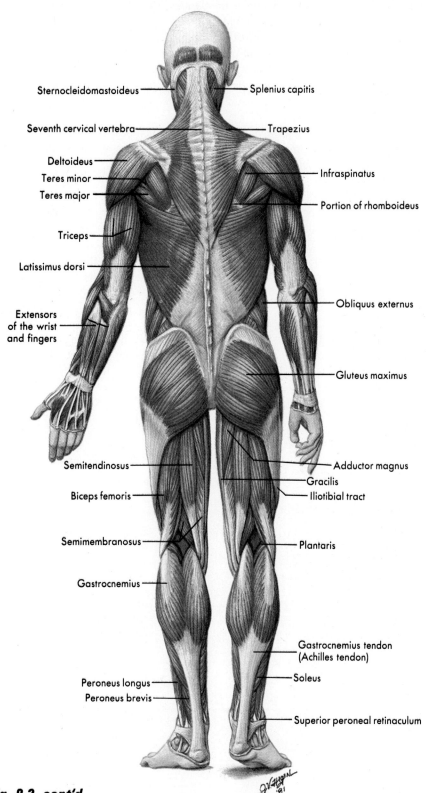

Fig. 8-2, cont'd
Posterior view of human muscular system, with major groups or muscles labeled.

Muscles of the head
Facial muscles

Facial muscles (Fig. 8-3) are the muscles by which our various emotions are expressed; they are involved in activities such as blowing, blinking, and kissing.

There are three orbicular muscles on the face. After considering these, we will continue the discussion of the facial muscles by starting with the nose and proceeding around one corner of the mouth to the chin.

Paired *orbicularis oculi* muscles encircle the orbits. The greater part of each muscle lies in this circular orientation *without* passing fibers through the eyelids. These fibers are known as the *orbital portion* and close or wink the eyelids in response to bright light, romantic interest, or voluntary activity. Those fibers that pass through the eyelids are known as the *palpebral portion* and are mainly under reflex control that results in a blinking action. Drying or irritation of the cornea is the primary stimulus that triggers the blink reflex that then spreads tears across the eyeball for cleansing or moistening the eye.

The **orbicularis oris** muscle encircles the mouth and lips and closes the mouth as its main action. Puckering of the lips as in whistling and whispering and aiding in keeping food within the mouth are other activities in which the muscle is involved.

The **procerus** muscle runs between the nasal bone and the skin between the eyebrows and causes transverse wrinkles to appear across the root of the nose.

The **nasalis** muscle has two portions. The **transverse portion** (also called the compressor naris) lies across the sides and top of the cartilage-supported part of the nose and slightly "pinches" the nostrils together while drawing the cartilaginous part of the nose downward. The **alar portion** (also called the dilator naris) attaches to the margin of the cartilages around the nostrils (alar cartilages) and "flares" the nostrils, as in anger or strong breathing. It also prevents collapse of the nostrils during inhalation.

The name **quadratus labii superioris** is applied to a group of four muscles lying above the upper lip. Each of the four muscles may be given a separate name or may be considered a head of the quadratus (which will be abbreviated q.l.s.). Both terminologies will be given.

The **levator labii superioris alaeque nasi** muscle (q.l.s., angular head) lies along the sides of the nose next to the point where it joins the skin below the

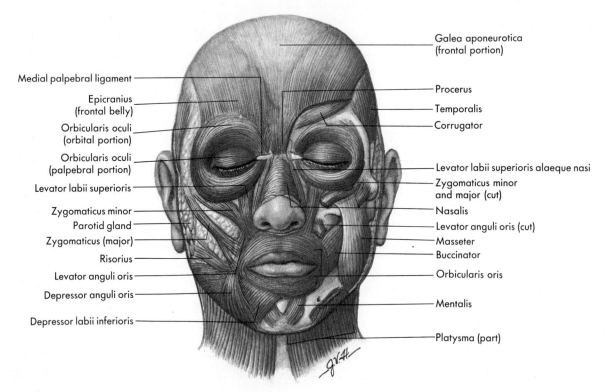

Fig. 8-3
Anterior view of muscles of face and anterior cranium.

orbits. It also dilates the nostrils and deepens the nasolabial groove, as when a person is sad.

The **levator labii superioris** (q.l.s., infraorbital head) is the muscle that runs between the infraorbital foramen and the upper lip. It is the strongest elevator of the upper lip.

The **zygomaticus minor** muscle (q.l.s., zygomatic head) runs between the zygomatic bone and the angle of the mouth and aids in elevating the upper lip and deepening the nasolabial groove.

The **levator anguli oris** (caninus) is the muscle that lies beneath the two previously listed muscles and, as its name suggests, elevates the angle of the mouth.

The **zygomaticus major** muscle runs at an angle from the zygomatic bone to the corners of the mouth. It is our "smiling muscle," drawing the corners of the mouth upward and backward when we are smiling or laughing.

The **risorius** is a muscle that is not always present. If present, it runs nearly horizontally between the corners of the mouth to the front part of the masseter. It retracts the corners of the mouth, as in a grimace.

The **depressor anguli oris** (or triangularis) muscle runs between the oblique line of the mandible and the corner of the mouth. Its contraction pulls the corner of the mouth downward, as in sadness.

Fibers of the **depressor labii inferioris** muscle run almost vertically between the mandible and the lower lip and draw the lip downward (depression).

The **mentalis** is the muscle that runs between the lateral side of the "point" of the chin to the lower lip. It raises and protrudes the lower lip and wrinkles the chin, as in doubt or contempt.

The **buccinator** is the major muscle of the cheek and is beneath the other facial muscles. Insertion is around the lips, and the muscle basically compresses the cheek. It thus aids in holding food between the teeth and acts strongly in blowing. It is sometimes called the "trumpeter's muscle," and buccinator in Latin means "a trumpet player" (alluding to the blowing action).

The **corrugator** is a small muscle that lies on the forehead at the medial end of the eyebrow. Perhaps better considered with the cranial muscles because of its location, this muscle produces "frown lines" in the middle forehead and may thus be included as a muscle of facial expression.

The facial muscles are presented in Table 8-1.

Table 8-1. Facial muscles

Muscle	Origin	Insertion	Action	Innervation
Buccinator	Alveolar processes of maxillae and mandible	Fibers of orbicularis oris in upper and lower lips	Compresses cheek, as in blowing ("trumpeter's muscle")	Seventh cranial nerve,* buccal branches
Corrugator	Inner end of superciliary ridge	Medial half of eyebrow	Produces vertical lines on forehead, as in frowning	Seventh cranial nerve, temporal and zygomatic branches
Depressor labii inferioris	Oblique line of mandible	Skin of lower lip	Depresses lower lip	Seventh cranial nerve, buccal and mandibular branches
Mentalis	Mental symphysis and incisive fossa of mandible	Skin of chin and lower lip	Depresses lower lip, wrinkles chin	Seventh cranial nerve, buccal and mandibular branches
Nasalis (compressor naris)	Maxilla, above incisor teeth and greater alar cartilage	Skin along side of nose and tip of nose	Narrows nostrils	Seventh cranial nerve, buccal branches
Orbicularis oculi	Medial surface of orbit	Skin of eyelids, circularly around orbit	Closes and winks eye(s)	Seventh cranial nerve, temporal and zygomatic branches

Adapted from McClintic, J.R.: Basic anatomy and physiology of the human body, ed. 2. Copyright 1980. Reprinted by permission of John Wiley & Sons, Inc.
*Facial nerve.

Continued.

Table 8-1. Facial muscles—cont'd

Muscle	Origin	Insertion	Action	Innervation
Orbicularis oris	Skin of lips, other facial muscles	Corners of mouth and median line below nose and lower lip	Closes and puckers lips	Seventh cranial nerve, buccal branches
Procerus	Nasal bones	Skin between eyebrows	Produces transverse wrinkles at root of nose	Seventh cranial nerve, buccal branches
Quadratus labii superioris (a muscle with four parts or heads)	As a whole, from nose to below orbit to zygomatic bone	As a whole, skin of upper lip	As a whole, elevates upper lip	Seventh cranial nerve, buccal branches
Levator labii superioris alaeque nasi	Nasal process of maxilla	Skin of upper lip (all four parts)	Elevates upper lip as in sneering (all four parts)	
Levator labii superioris	Below orbit			Seventh cranial nerve, buccal branches
Zygomaticus minor	Anterior aspect of zygomatic bone			
Levator anguli oris	Below infraorbital foramen			
Risorius	Fascia of masseter muscle	Skin of angle (corner) of mouth	Pulls mouth laterally, as in a grimace	Seventh cranial nerve, buccal and mandibular branches
Triangularis (depressor anguli oris)	Oblique line of mandible	Skin of lower lip at angle (corner) of mouth	Draws corner of mouth downward, as in sadness	Seventh cranial nerve, buccal and mandibular branches
Zygomaticus (zygomaticus major)	Lateral aspect of zygomatic bone	Skin of angle (corner) of mouth	Draws mouth up and back, as in smiling	Seventh cranial nerve, buccal branches

Cranial muscles

The cranial muscles (Fig. 8-4; see also Fig. 8-3) are the muscles of the scalp, and they are positioned on the forehead, the back of the skull, and around the ears. No muscles lie over the top of the skull; the skin lies on a broad tendon, the *galea aponeurotica*. The muscles of the scalp use the galea as origin or insertion, and it may move as the muscles contract.

The term *epicranius* is applied to the muscles and galea together. Three subdivisions are commonly made in the epicranius: an *occipitofrontal group,* a *temporoparietal group,* and an *auricular group* associated with the pinna of the ear.

The occipitofrontal group consists of the posterior *occipitalis* and the anterior *frontalis* muscles. The occipitalis muscle lies on the lateral aspect of the occipital bone between the superior nuchal line and the posterior galea. Its contraction draws the scalp backward. The frontalis muscle raises the eyebrows, as in surprise or fright, and produces transverse wrinkles ("thought lines") in the forehead. It may also draw the scalp forward.

The *temporoparietalis* is a very thin sheet of muscle formerly included with the anterior and superior auricularis muscles. Its contraction pulls the skin of the temple area backward.

The *auricularis* muscle, consisting of an *anterior, superior,* and *posterior* muscle on each side, draws the pinna forward, upward, or backward respectively.

The cranial muscles are presented in Table 8-2.

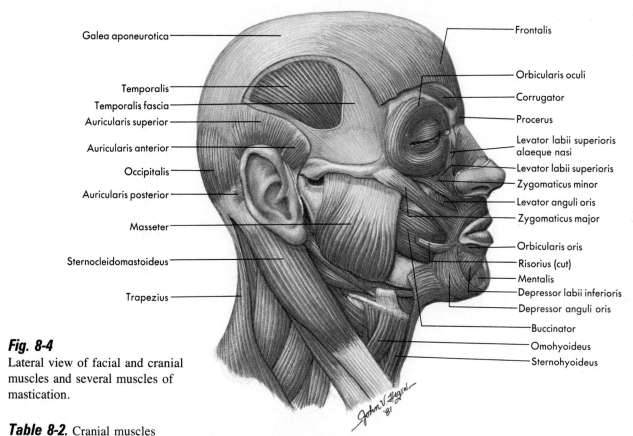

Fig. 8-4
Lateral view of facial and cranial muscles and several muscles of mastication.

Table 8-2. Cranial muscles

Muscle	Origin	Insertion	Action	Innervation
Epicranius				
Frontalis (frontal belly)	Tissue above the supraorbital margin, (no bony attachments)	Galea aponeurotica	Raises eyebrows, (surprise or fright), produces transverse furrows in skin of forehead, and pulls scalp forward	Seventh cranial nerve, temporal branches
Occipitalis (occipital belly)	Superior nuchal line and mastoid portion of temporal bone	Galea aponeurotica	Pulls scalp backward	Seventh cranial nerve, posterior auricular branches
Temporoparietalis	Temporal fascia around ears	Lateral border of galea aponeurotica	Tightens scalp, draws pinna upward	Seventh cranial nerve, temporal branches
Auricularis				
Anterior	Fascia of scalp (all three)	Pinna of ear (all three)	Draws pinna forward	Seventh cranial nerve, temporal branches
Posterior			Draws pinna backward	Seventh cranial nerve, posterior auricular branches
Superior			Draws pinna upward	Seventh cranial nerve temporal branches

Adapted from McClintic, J.R.: Basic anatomy and physiology of the human body, ed. 2. Copyright © 1980. Reprinted by permission of John Wiley & Sons, Inc.

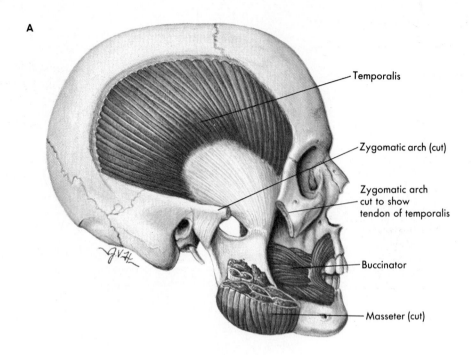
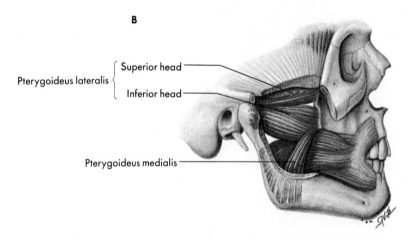
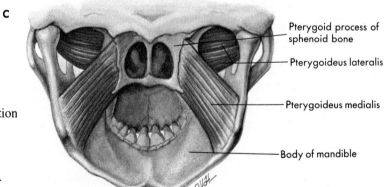

Fig. 8-5
A, Temporalis and masseter muscles of mastication (buccinator shown to demonstrate its extent and heads). **B,** Lateral view of skull to show pterygoideus muscles, muscles of mastication (ramus of mandible has been partially removed). **C,** Pterygoideus muscles viewed from below.

Muscles of mastication

The muscles of mastication (Fig. 8-5) are four in number on each side, three of which elevate the mandible (closing the jaw) and one that depresses the mandible (opening the jaw) and brings about the other movements the mandible exhibits.

The *temporalis* muscle occupies the temporal fossa and inserts primarily on the mandible's coronoid process. It elevates the mandible.

The *masseter* is a thick muscle that runs from the zygomatic arch to the angle and inferior ramus of the mandible. It elevates the mandible.

The *pterygoideus medialis (internal)* muscle runs between the medial pterygoid lamina and the ramus and elevates the mandible.

These three muscles, in closing the jaw, have been estimated to be able to exert a maximal "squeeze" of 245 kg.

The *pterygoideus lateralis (external)* muscle has two heads that run more or less horizontally between the lateral pterygoid lamina and the front of the mandible's neck. This muscle protrudes the mandible and moves it from side to side.

The *mandibular sling* is formed by the masseter and pterygoideus medialis muscles to "cradle" the mandibular angle in their tendons of insertion. The sling forms a center of rotation when the jaw is opened.

The muscles of mastication are presented in Table 8-3.

Table 8-3. Muscles of mastication

Muscle	Origin	Insertion	Action	Innervation
Masseter	Zygomatic process of maxilla, and zygomatic arch	Coronoid process, ramus, and angle of mandible	Elevates mandible (and closes mouth)	Mandibular nerve (a branch of fifth cranial nerve [trigeminal])
Pterygoideus lateralis	Lateral pterygoid lamina and greater wing of sphenoid bone	Anterior surface of neck of condyloid process of mandible	Protrudes and opens mouth; moves mandible side to side	Mandibular nerve (a branch of fifth cranial nerve [trigeminal])
Pterygoideus medialis	Medial pterygoid lamina and pterygoid fossa	Medial aspect of ramus and posterior body of mandible	Elevates mandible (and closes mouth)	Mandibular nerve (a branch of fifth cranial nerve [trigeminal])
Temporalis	Temporal fossa and temporal fascia	Coronoid process and ramus of mandible	Elevates mandible (and closes mouth)	Mandibular nerve (a branch of fifth cranial nerve [trigeminal])

Adapted from McClintic, J.R.: Basic anatomy and physiology of the human body, ed. 2. Copyright © 1980. Reprinted by permission of John Wiley & Sons, Inc.

Muscles of the tongue

The tongue is primarily a muscular organ used in speech, swallowing, and guiding food between the teeth during mastication. The lingual (tongue) muscles (Fig. 8-6) are divided in half by a median fibrous septum that extends the entire length of the tongue and is attached to the hyoid bone. Each half contains both intrinsic (origin and insertion inside the tongue) and extrinsic (origin outside, insertion within the tongue) muscles. Both sets create a distinctive vertical, lateral, and longitudinal pattern of muscle fibers.

The intrinsic muscles are named by position as *longitudinal, transverse,* and *vertical lingual* muscles. Extrinsic muscles carry the suffix *-glossus* in their name.

The **genioglossus** muscle runs between the mental spines and hyoid bone and passes to the inferior aspect of the entire length of the tongue. The muscle's posterior fibers protrude the tongue from the mouth; the anterior fibers draw it back into the mouth. Both acting together depress the center of the tongue to form a channel, as in sucking, through which fluids pass.

The **hyoglossus** muscle extends from the greater horns of the hyoid bone to the sides of the tongue. The tongue is depressed and drawn backward in the mouth, and its sides are drawn downward by the hyoglossus.

The **styloglossus** muscle runs from the skull's styloid process longitudinally into the tongue. Its contraction pulls the tongue upward and backward and raises its sides.

The **palatoglossus** muscle raises the base of the tongue, narrowing the opening into the pharynx (throat) by ''closing the curtain'' of the soft palate.

These muscles are presented in Table 8-4.

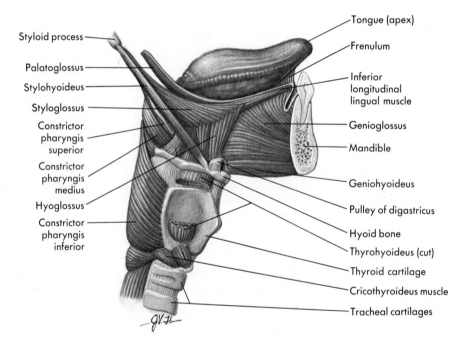

Fig. 8-6
Extrinsic muscles of tongue and some associated structures.

Table 8-4. Extrinsic muscles of the tongue

Muscle	Origin	Insertion	Action	Innervation
Genioglossus	Genial tubercles on either side of posterior mandibular symphysis	Tongue	Raises tongue and pulls it forward (acting together), protrudes tongue to opposite side (acting singly)	Twelfth cranial nerve*
Hyoglossus	Body and greater horn of hyoid bone	Tongue	Pulls tongue backward and arches it	Twelfth cranial nerve
Palatoglossus	Palatine aponeurosis (on posterior aspect of the hard palate)	Tongue	Narrows opening of mouth into pharnyx	Twelfth cranial nerve
Styloglossus	Styloid process	Tongue	Pulls tongue upward and backward and raises sides of tongue	Twelfth cranial nerve

Adapted from McClintic, J.R.: Basic anatomy and physiology of the human body, ed. 2. Copyright © 1980. Reprinted by permission of John Wiley & Sons, Inc.
*Hypoglossal nerve.

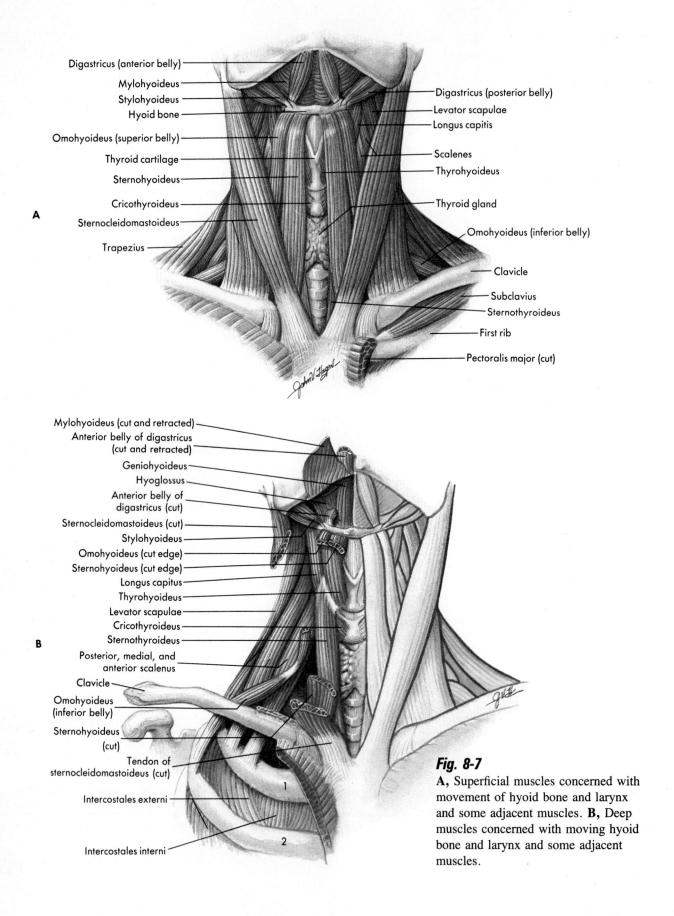

Fig. 8-7

A, Superficial muscles concerned with movement of hyoid bone and larynx and some adjacent muscles. **B,** Deep muscles concerned with moving hyoid bone and larynx and some adjacent muscles.

Muscles associated with the hyoid bone and larynx

Two groups of muscles are associated with the hyoid bone. The *suprahyoidei muscles* generally lie above the hyoid bone, the *infrahyoidei muscles* below (Fig. 8-7). The suprahyoidei group includes the mylohyoideus, digastricus, stylohyoideus, and geniohyoideus muscles.

The **mylohyoideus** muscle forms the muscular floor of the mouth cavity. It runs from the mylohyoid lines of the mandible to the hyoid body posteriorly and to a median fibrous septum anteriorly. It raises the hyoid body and is active in sucking, blowing, chewing, and swallowing movements.

The **digastricus** muscle has two bellies: an *anterior belly* runs between the anterior mandible and the hyoid body; a *posterior belly* runs between the mastoid process and the hyoid body. The anterior belly draws the hyoid body forward and aids in opening the jaw. The posterior belly draws the hyoid body backward.

The **stylohyoideus** muscle rises from the styloid process of the skull, lies alongside the posterior digastric belly, and draws the hyoid body backward and slightly upward.

The **geniohyoideus** muscle runs between the mental spines and the hyoid body and draws the hyoid body forward.

The suprahyoidei muscles are presented in Table 8-5.

The infrahyoidei group includes the sternohyoideus, sternothyroideus, thyrohyoideus, and omohyoideus muscles.

The **sternohyoideus** muscle runs between the medial clavicle and the manubrium of the sternum to the hyoid

Table 8-5. Muscles associated with the hyoid bone and larynx

Muscle	Origin	Insertion	Action	Innervation
Infrahyoidei group				
Omohyoideus	Superior border of scapula	Body of hyoid bone	Depresses hyoid	Cervical spinal nerves 1-3
Sternohyoideus	Manubrium of sternum and medial end of clavicle	Body of hyoid bone	Depresses hyoid	Cervical spinal nerves 1-3
Sternothyroideus	Manubrium and first costal cartilage	Thyroid cartilage of larynx	Depresses larynx	Cervical spinal nerves 1-3
Thyrohyoideus	Thyroid cartilage of larynx	Body and greater horn of hyoid bone	Depresses hyoid	Cervical spinal nerves 1-3
Suprahyoidei group Digastricus				
Anterior belly	Digastric fossa of mandible (below posterior mandible symphysis)	Body of hyoid bone	Draws hyoid forward	Mylohyoid branch of mandibular nerve (branch of fifth cranial nerve [trigeminal])
Posterior belly	Mastoid notch of temporalis bone	Body of hyoid bone	Elevates hyoid and aids mouth opening	Facial nerve
Geniohyoideus	Genial tubercles on each side of posterior mandibular symphysis	Body of hyoid bone	Elevates hyoid	C1* and C2* by way of hypoglossal nerve (twelfth cranial nerve)
Mylohyoideus	Mylohyoid lines of mandible	Median raphe in center of muscle	Elevates hyoid bone	Mylohyoid branch of mandibular nerve (branch of fifth cranial nerve [trigeminal])
Stylohyoideus	Styloid process of temporal bone	Greater horn of hyoid bone	Elevates and retracts hyoid	Facial nerve

Adapted from McClintic, J.R.: Basic anatomy and physiology of the human body, ed. 2. Copyright © 1980. Reprinted by permission of John Wiley & Sons, Inc.
*First and second cervical spinal nerves.

body inferior surface. It draws the hyoid bone downward (depression) and steadies the bone during swallowing.

The **sternothyroideus** muscle is deep to the preceding muscle, rises from the manubrium and first costal cartilage, and passes to the side of the thyroid cartilage of the larynx. It draws the thyroid cartilage (and therefore the larynx) downward after swallowing and when singing low notes.

The **thyrohyoideus** muscle seems to continue the course of the previous muscle, running from the thyroid cartilage to the greater horn of the hyoid body. It either raises the larynx, or if that is fixed, depresses and steadies the hyoid.

The **omohyoideus** muscle has two parts or bellies (superior and inferior) separated by the *intermediate tendon*. It acts as a unit and draws the hyoid body downward (depression).

The infrahyoidei muscles are also presented in Table 8-5, and Fig. 8-8 may aid the reader in understanding the naming, position, and actions of the hyoidei muscles.

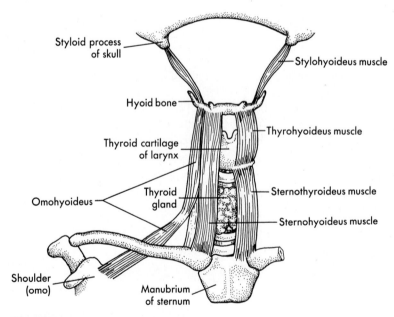

Fig. 8-8
Diagram illustrates how muscles associated with hyoid bone are named.

Muscles of the neck

Included in this discussion of muscles of the neck are those muscles whose fleshy bellies lie on the neck and that would be appreciated in a cross-section of the neck (Figs. 8-9 to 8-11; Table 8-6). Some of these muscles reach the face or cranium; thus their actions may complement those of muscles described previously. Other muscles may originate below the neck proper (which is actually a part of the spinal column) and may be met again in later sections concerned with the vertebral column. If a muscle *inserts* on the neck, regardless of its origin, the name *cervicis* will usually appear as part of its name; it it reaches the skull, *capitis* (head) will usually be included in naming it. It is suggested that this section and the next one be considered together for muscles that operate the vertebral column so that a picture of the actions of the whole structure may be achieved.

The *platysma* is a broad and thin sheet of muscle that begins on the anterior chest and sweeps up each side of the neck to terminate along the mandibular body and the corners of the mouth. The action draws the corners of the mouth downward, as in sadness, screaming, or an expression of horror. It is a muscle that is variable as to thickness and ultimate insertion and may be absent in some individuals.

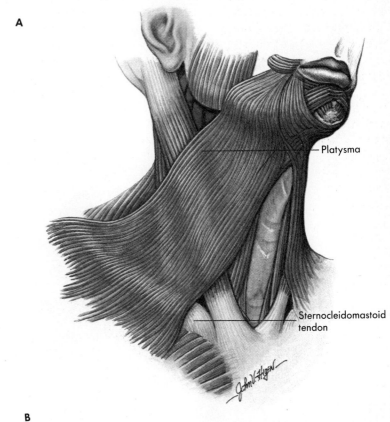

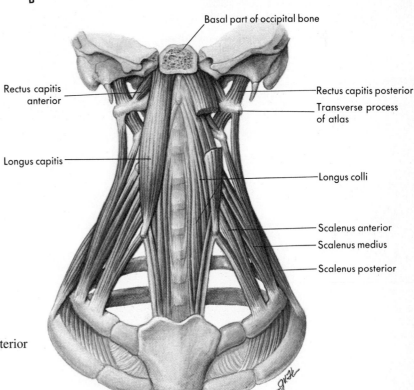

Fig. 8-9

A, Platysma muscle. **B,** Some anterior and lateral neck muscles.

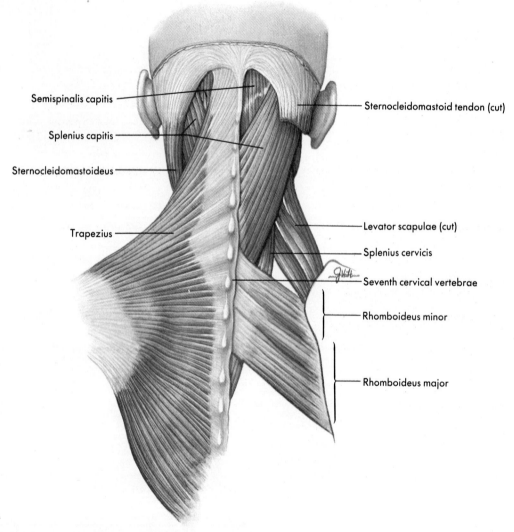

Fig. 8-10
Superficial muscles of posterior neck and upper back.

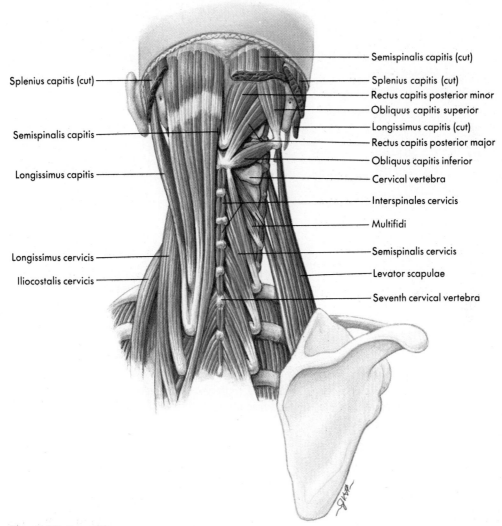

Fig. 8-10, cont'd
Deep muscles of posterior neck and upper back.

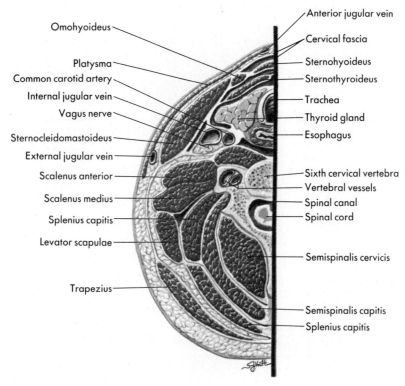

Fig. 8-11
Cross-section of neck at level of sixth cervical vertebra to show arrangement of muscles and fascial planes.
Adapted from Gray.

Table 8-6. Muscles of the neck and back

Muscle	Origin	Insertion	Action	Innervation
Superficial				
Platysma	Fascia of chest wall below clavicle and of deltoideus muscle	Muscles of angle of mouth; lower lip; margin of mandible	Pulls corners of mouth down; "screaming"	Facial nerve, cervical branches
Sternocleidomastoideus	Manubrium of sternum, medial one third of clavicle	Mastoid process of temporal bone	Flexes neck (acting together); rotates head to opposite side, or tilts to same side (acting singly)	Accessory nerve (eleventh cranial nerve) and C2
Deep				
Interspinales, intertransversarii	Lie between spinous and transverse processes in virtually only the cervical and lumbar areas		Steady column and prevent its "folding"	Spinal nerves C3-L4*
Longus colli (cervicis)	Transverse processes and bodies of vertebrae C3-T7	Atlas transverse processes and bodies of cervical vertebrae	Flexes and rotates neck	Cervical spinal nerves 2-8
Longus capitis	Transverse processes C3-6	Occipital bone	Flexes neck	Cervical spinal nerves 1-4
Rectus capitis	Atlas	Base of occipital bone	Flexes, extends, abducts, or rotates the head	Cervical spinal nerves 1, 2
Scalenus (anterior, middle, posterior)	Transverse processes, cervical vertebrae	Ribs 1 and 2	Flex neck and bend it to the side; elevate ribs 1 and 2	Cervical spinal nerves 3-8
Semispinalis (capitis, cervicis, thoracis)	Transverse processes of lower cervical and all thoracic vertebrae	Spinous processes of cervical vertebrae and occipital bone	Extend and rotate spine	Spinal nerves C1-T6† (capitis, cervicis) T7-T12 (thoracis)
Splenius (capitis, cervicis)	Spinous processes C8-T5	Transverse processes C1-4, superior nuchal line and mastoid process	Extend and rotate neck and head	Cervical spinal nerves C2-T6
Transversospinalis	Articular processes of vertebrae C5-T12, lumbar vertebrae, and sacrum	Spines of vertebrae above and occipital bone	Extends spine and rotates it	Spinal nerves C2-L4

Adapted from McClintic, J.R.: Basic anatomy and physiology of the human body, ed. 2. Copyright © 1980. Reprinted by permission of John Wiley & Sons, Inc.
*Lumbar spinal nerves.
†Thoracic spinal nerves.

The remaining muscles of the neck are often divided into anterior, lateral, and posterior groups according to their position relative to the spinal column. Such a division enables us to predict what the action of a muscle should or might be —anterior muscles flex, lateral muscles abduct or adduct, and posterior muscles extend the neck or head.

The anterior group includes the sternocleidomastoideus, longus colli, longus capitis, and rectus capitis muscles.

The **sternocleidomastoideus** muscle runs from the sternum and inner ends of the clavicle to the mastoid process of the skull. On contraction, one muscle will rotate the head to the opposite side and tilt the chin upward to the opposite side. Acting together, the muscles flex the neck on the chest.

The *longus colli* muscle (also called *l. cervicis*) originates along the last six cervical vertebral bodies and the upper seven thoracic vertebral bodies. Insertions are on the transverse processes and bodies of the upper cervical vertebrae. Their contraction flexes (together) and rotates (singly) the neck to the same side.

The *longus capitis* muscle runs from the transverse processes of cervical vertebrae 3 to 6 and reaches the occipital bone. It flexes the head on the chest.

The *rectus capitis* (r.c.) muscle includes several parts, named *r.c. anterior, r.c. lateralis, r.c. posterior minor,* and *r.c. posterior major*. All run between the atlas and the occipital bone. The anterior muscles flex the head on the chest, the lateral muscles tilt it to the side, and the posterior muscles (which will be mentioned again with the posterior group of muscles) extend the head.

The lateral neck muscles are the three scalenus muscles on each side. They are named **scalenus anterior, medius,** and **posterior muscles** from front to back.

The scalenus anterior muscle lies deep to the sternocleidomastoideus muscle between the third and sixth cervical transverse processes and the first rib.

The scalenus medius muscle rises from the transverse processes of cervical vertebrae 2 to 7 and attaches to the first rib. Both muscles raise the first rib and bend the neck laterally.

The scalenus posterior muscle rises from the last two or three cervical transverse processes and attaches to the second rib. It raises the second rib and slightly bends the neck to the side.

It is obvious that these muscles aid inspiration by helping to elevate the ribs ("acting from above"). If the ribs are fixed, their action is then exerted on the neck ("acting from below").

The posterior muscles of the neck are the *semispinalis capitis* and *cervicis,* the *longissimus capitis* and *cervicis,* and the *splenius*. These extend the neck and will be discussed in the next section so that their extent can be better appreciated.

Muscles of the back

The spinal column is capable of undergoing flexion or forward bending, extension or backward bending, abduction and adduction, and rotation to both sides (Fig. 8-12).

The muscles of the back (Fig. 8-13) may be best understood if viewed as a series of slips that begin low and attach to various parts of vertebrae higher in the column. Some of these slips may reach the skull and cause movements that were described in the previous section.

The *transversospinalis* muscle lies in the groove formed between the transverse and spinous processes of the vertebrae. Deeper lying fibers connect the transverse processes of one vertebra with the spinous process of the vertebra above. The more superficial fibers span several vertebrae. The name *multifidus* is given to the deep fibers and all those arising in the lumbosacral area, whereas the more superficial fibers are designated *semispinalis thoracis* if they rise below the sixth thoracic vertebra or *semispinalis cervicis* if they originate above the sixth thoracic vertebra. The *semispinalis capitis* begins above *T6* and inserts on the back of the skull.

The term *erector spinae* or *sacrospinalis* applies to the muscle that generally covers the transversospinalis below and lies lateral to it above. It is again a general name for a group of muscles that have longer bundles that run more vertically than the transversospinalis. The bundles of the erector begin at the iliac and lower lumbar area, and at least one of the bundles reaches the skull.

The most lateral column of muscle is the *iliocostalis* (*i.*). It is divided into the *i. lumborum* that attaches to the last six ribs, the *i. thoracis* from the last six ribs to the upper six ribs, and the *i. cervicis* from the first six ribs to the transverse processes of *C4* to *6*.

The middle column of muscle is the *longissimus* (*l.*), taking origin from the sacrum, lumbar vertebrae, and the processes of thoracic and cervical vertebrae. The insertion determines the three parts of this muscle. The *l. thoracis* inserts on the last 10 ribs and thoracic transverse processes; the *l. cervicis* attaches to the transverse processes of C2 to 6; the *l. capitis* reaches the mastoid portion of the temporal bone.

The medial column of muscle is the *spinalis* (*s.*), which arises from lower thoracic and upper lumbar transverse processes. It also has three parts, defined by insertion: the *s. thoracis* reaches the upper thoracic spinous processes; the *s. cervicis* attaches to cervical processes; the *s. capitis* reaches the occipital bone.

The *splenius* is a broad and straplike muscle that runs between the last cervical and upper thoracic spinous processes to the upper four cervical transverse processes (*splenius cervicis*). The *splenius capitis* reaches the superior nuchal line and the mastoid portion of the temporal bone.

In terms of action, all the muscles described act in opposing the force of gravity when the column is flexed and aid in regaining erect posture. Lateral bending is accomplished mainly by the iliocostalis muscle of the same side, whereas regaining anatomical position uses the opposite muscle. The transversospinalis muscle is instrumental in rotation of the column, as is the semispinalis. The splenius muscle is important in extension and rotation of the head and neck.

Very small muscles called the *interspinales* and *intertransversarii* lie between adjacent spinous and transverse processes and are regarded as preventing the column from buckling and steadying it for the larger muscles to cause movement.

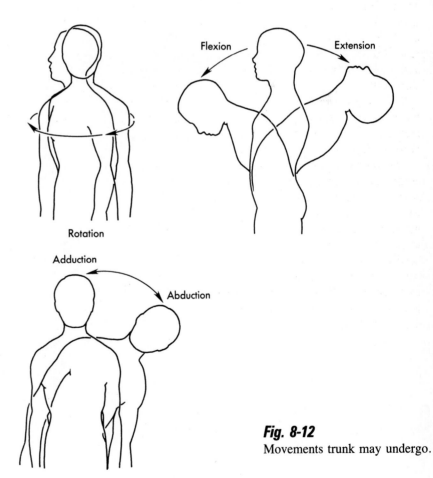

Fig. 8-12
Movements trunk may undergo.

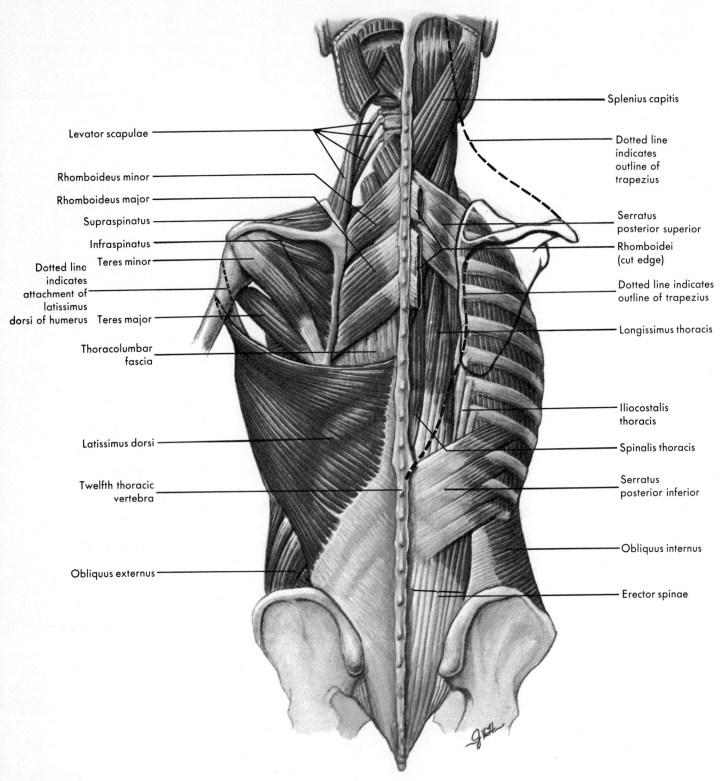

Fig. 8-13
Some superficial muscles of back.

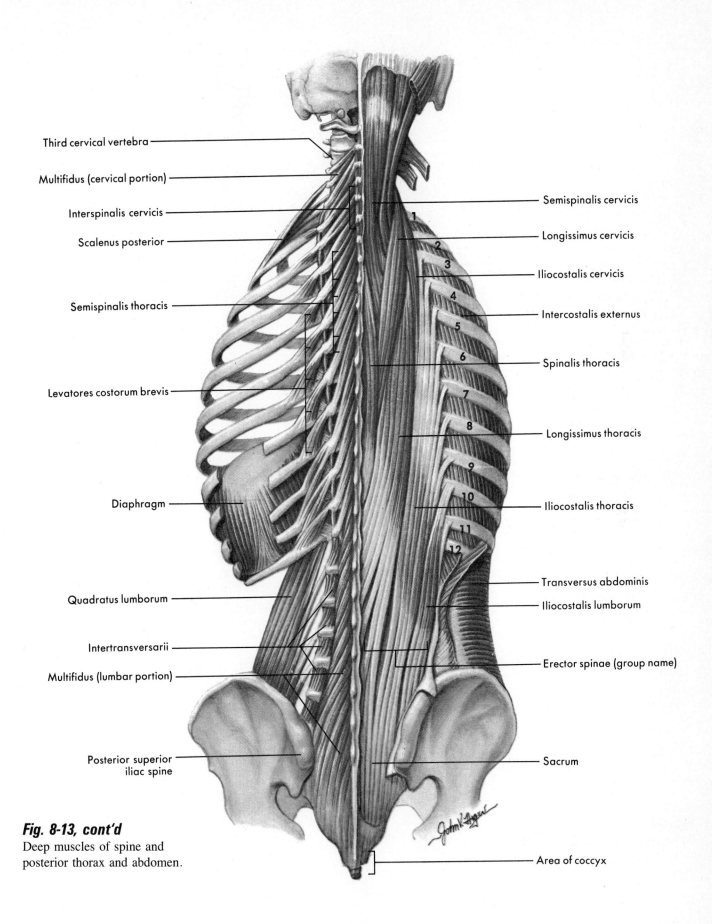

Fig. 8-13, cont'd
Deep muscles of spine and posterior thorax and abdomen.

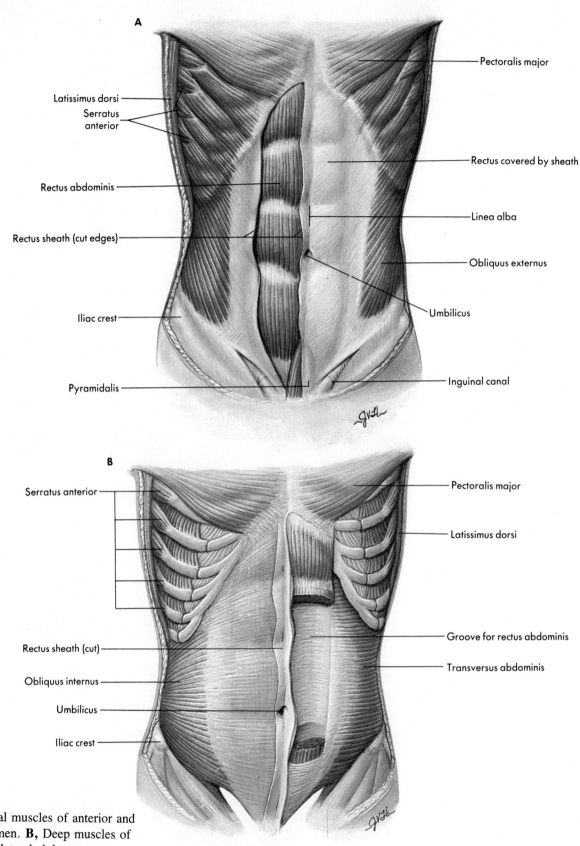

Fig. 8-14
A, Superficial muscles of anterior and lateral abdomen. **B,** Deep muscles of anterior and lateral abdomen.

Muscles of the anterior and lateral abdominal walls

(Fig. 8-14)

The abdomen is generally regarded as that portion of the trunk that is not bounded by bony structures. In general, the more anteriorly placed muscles flex the spine, whereas the more lateral muscles rotate the spine and compress the abdomen. Fig. 8-15 shows the manner in which these muscles interdigitate to form the abdominal wall; the legend draws attention to the important facts of wall construction.

The *rectus abdominis* muscle alone forms the anterior wall of the abdomen. It originates as a rather narrow band from the pubic crest and broadens as it ascends to attach to the xiphoid process and the fifth to seventh rib cartilages. Anteriorly the muscle is crossed by three tendinous "intersections," and one muscle is separated from the other by the *linea alba*. The latter represents the ligamentous connection between the tendons of the oblique and transverse abdominal muscles from both sides. The rectus is a powerful flexor of the spine, as in doing a "sit-up," and contracts powerfully during childbirth, or when a person coughs, sneezes, vomits, or defecates. The small *pyramidalis* muscle may lie on the rectus above the pubic symphysis.

The superficially placed *obliquus externus abdominis* muscle fibers originate from the lower eight ribs and pass downward and forward to the linea alba, pubis, and iliac crest. The insertion may be seen to lie medial to the origin, and thus if *one* muscle contracts, it will rotate the body to the *opposite side* from the muscle contracting, bringing the *same* shoulder forward. Together the muscles compress the abdomen as described for the rectus.

The *obliquus internus abdominis* muscle forms the middle layer of muscle of the lateral wall of the abdomen. *Its* origin is from the ilium and lumbar area, and the fibers sweep upward at about a 90-degree angle to the fibers of the obliquus externus. Insertion is on the linea alba, xiphoid, and last three ribs. This muscle's action in rotation is opposed to that of the obliquus externus; that is, one muscle rotates the spine to the *same side*, bringing the *opposite* shoulder forward. Compressing action is as for the rectus.

The *transversus abdominis* muscle runs nearly horizontally from the iliac crest, last six ribs, and lumbar area to the xiphoid, linea alba, and pubis. Its contraction mostly compresses the abdomen.

The *inguinal canal* (see Fig. 8-14, A) is an obliquely placed fissure about 4 cm long in the groin. It is an opening for passage of the spermatic cord or round ligament of the uterus. It is bordered anteriorly by the tendon of the obliquus externus, the inferior part of which is thickened to form the *inguinal ligament*. Its roof is formed by the fibers of the obliquus internus and the transversus abdominis muscles. The posterior wall is formed by the tendons of the two muscles just mentioned and the floor by the *transversalis fascia* that lines the inner surface of the transversus muscle. Weakness in the walls of the canal may give rise to inguinal hernia, in which the canal is enlarged, sometimes to the point at which intestines may enter the scrotum.

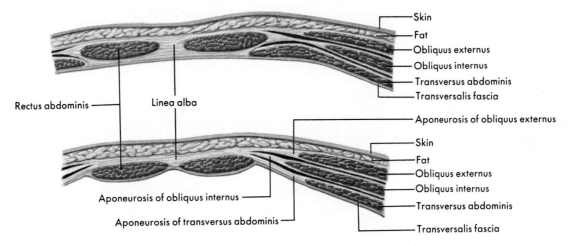

Fig. 8-15

Cross-sections of anterior abdomen to show arrangements of muscles and fascial planes. *Above*, section above umbilicus. *Below*, Section at level of iliac crest. Note that sheath of rectus abdominis is formed by aponeuroses of lateral abdominal muscles and that below iliac crest, aponeuroses of all three lateral abdominal muscles pass anterior to rectus abdominus.

Modified from Braune.

Muscles of the posterior abdominal wall

The muscles to be listed in this section occupy positions on the posterior or posterolateral aspects of the lower abdomen and set a limit of sorts to those parts of the abdomen. The spinal erectors lie external to these muscles.

The **quadratus lumborum** muscle passes from the iliac crest to the last rib. Its contraction fixes or holds the last rib against the pull of a part of the diaphragm, and it is considered by some anatomists to be a respiratory muscle. Additionally, its contraction may aid in lateral bending of the spine by pulling the twelfth rib closer to the ilium.

The **psoas major** and **iliacus** muscles have two origins but a common insertion. For this reason the muscle is often called the *iliopsoas*. The two muscles have a given action as a hip flexor, but if the thigh is fixed, the muscles help to flex the spine.

All the abdominal muscles are summarized in Table 8-7.

The muscles of the pelvic floor in the female and male are included in Chapter 15.

Table 8-7. Muscles of the trunk and abdomen

Muscle	Origin	Insertion	Action	Innervation
Abductors/adductors				
Quadratus lumborum	Iliac crest and lower lumbar vertebrae	Twelfth rib and upper lumbar transverse processes	Bends spine to sides	Spinal nerves T12-L4
Extensors				
Iliocostalis (lumborum, thoracis, cervicis)	Iliac crest and ribs	Ribs and transverse processes of C4-6	Extend and bend spine laterally	Spinal nerves T7-L2 (lumborum), T1-T12 (thoracis), C4-T6 (cervicis)
Longissimus (thoracis, cervicis, capitis)	Lumbar vertebrae and sacrum, processes of thoracic and cervical vertebrae	Lower ten ribs, thoracic and cervical transverse processes, mastoid process of temporal bone	Extend spine, bend it to the sides	C2-T6 (capitis, cervicis) T7-S3 (thoracis)
Spinalis (thoracis, cervicis, capitis)	Lower thoracic and upper lumbar transverse processes	Spinous processes of thoracic and cervical vertebrae; occipital bone	Extend spine	Spinal nerves C8-T12
Flexors				
Psoas ⎫ Iliopsoas Iliacus ⎭	Transverse processes of lumbar vertebrae (psoas) iliac fossa (iliacus)	Medial part of lesser trochanter	Flex thigh and (if thigh is fixed) flex spine	Spinal nerve L1 Spinal nerve L1
Rectus abdominis	Pubic crest and symphysis	Xiphoid process, cartilages of ribs 5-7	Flexes spine (sit-up) and compresses abdomen	Spinal nerves T7-T12 (intercostal nerves)
Rotators				
Obliquus externus	Anteroinferior surfaces of lower 8 ribs	Linea alba, pubis, iliac crest	Singly, rotates spine to bring same shoulder forward; acting together, compress abdomen	Intercostal nerves (spinal nerves T2-12)
Obliquus internus	Iliac crest, lumbodorsal fascia	Lower 3 ribs, linea alba, xiphoid process	Singly, rotates spine to bring opposite shoulder forward; acting together, compress abdomen	Intercostal nerves (spinal nerves T2-12)
Transversus abdominis	Iliac crest, lumbar fascia, last 6 ribs	Pubis, linea alba, xiphoid process	Compress abdomen	Intercostal nerves (spinal nerves T2-12)

Adapted from McClintic, J.R.: Basic anatomy and physiology of the human body, ed. 2. Copyright © 1980. Reprinted by permission of John Wiley, & Sons, Inc.

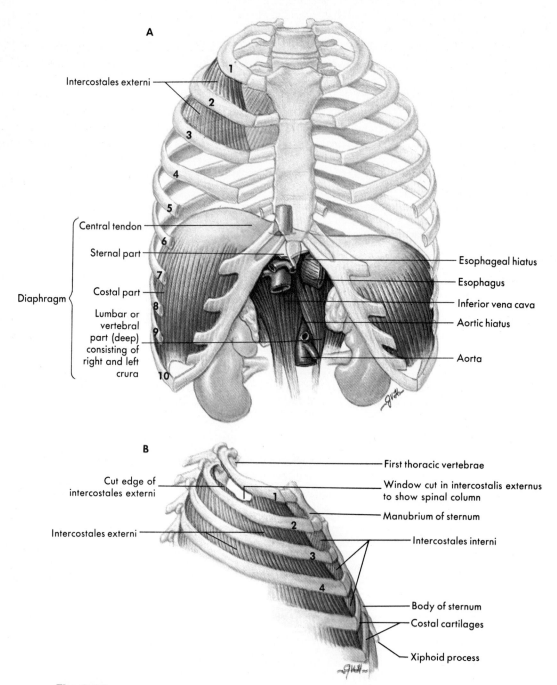

Fig. 8-16
A, Diaphragm and some associated muscles and organs. Hiatuses (openings) through diaphragm are provided for esophagus, aorta, and inferior vena cava. **B,** Intercostales muscles and associated skeletal structures.

Modified from Braune.

Muscles of respiration

(Fig. 8-16)

Breathing is subdivided into an inspiratory phase in which the anteroposterior, lateral, and vertical dimensions of the thorax are increased and an expiratory phase when these dimensions are returned to original size. The inspiratory phase in particular requires muscular activity. Fig. 8-17 depicts the changes in thoracic dimensions that occur as the muscles of respiration contract.

The **diaphragm** is a dome-shaped muscle whose arch reaches the level of the sternal attachment of the fourth costal cartilages. The fibers insert into a *central tendon* that is pulled downward during contraction to increase the vertical dimension of the thoracic cavity. The *sternal part* of the muscle takes its origin from the back of the xiphoid process; the *costal part* derives from the inner surfaces of the lower six costal cartilages and distal ribs; the *lumbar (vertebral) part* is formed by *right* and *left crura* (L., legs) that extend along the lateral bodies of the upper lumbar vertebrae and their discs. The muscle is penetrated by three large openings or *hiatuses*. The aorta, inferior vena cava, and esophagus are the organs passing through the openings. (A hiatus hernia involves the protrusion of the stomach through an enlarged esophageal hiatus).

The **intercostales externi** muscles are a series of 11 muscles extending from the outer inferior border of one rib to the superior border of the rib below. They extend from the rib tubercle as far as the costal cartilage (on false ribs they go to the rib end). Their contraction elevates the ribs that because of their curvature enlarge the thorax front to back and side to side.

The **intercostales interni** muscles lie internal to the aforementioned muscles, extending from the rib below to the one above. Their fibers begin at the sternum and continue to the rib angle, with their fibers at about 90 degrees to those of the externi muscles. Their contraction depresses the ribs.

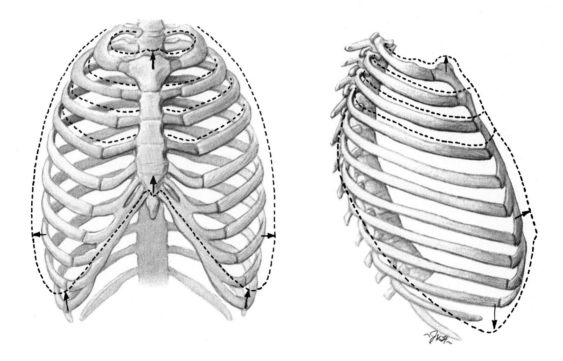

Fig. 8-17

Movements of thorax. As ribs are elevated, anterior-posterior and transverse dimensions are increased as shown by dashed lines.

The *levatores costarum breves* muscles are 12 pairs of muscles that run between the transverse processes of *C7* to *T11* and the ribs. They aid in rib elevation.

The *serratus posterior superior* muscle runs between the spinous processes of *C7 to T3* and the first four or five ribs. It aids rib elevation.

A *serratus posterior inferior* muscle connects the lumbar vertebrae and lower ribs, aiding in their depression.

The muscles of respiration are presented in Table 8-8.

Muscles operating the scapula and clavicle
(Fig. 8-18)

The bones attaching the upper appendage to the axial skeleton make up the *pectoral girdle* and consist of the paired scapulae and clavicles. The clavicle acts as a platform suspending the upper appendage at the proper distance from the sternum, whereas the scapula is really the movable bone of the girdle. The clavicle can be elevated, depressed, and moved forward and back.

The scapula may undergo a wider variety of movements, as shown in Fig. 8-19. For convenience of description, the muscles to be considered may be divided into an anterior group lying on or taking origin from the chest and a posterior group on the upper back and neck.

The anterior group includes the pectoralis minor, serratus anterior, and subclavius muscles.

The **pectoralis minor** muscle rises from the first and second ribs and attaches to the coracoid process of the

Table 8-8. Muscles of respiration

Muscle	Origin	Insertion	Action	Innervation
Diaphragm	Xiphoid process (sternal part), last 6 ribs (costal part), lumbar vertebrae (crura)	Central tendon	Increases vertical dimension of thorax by drawing central tendon downward	Phrenic nerve (from cervical plexus, C4)
Intercostales externi	Inferior border of upper 11 ribs	Superior border of rib below	Elevate ribs (if first rib is fixed), draw adjacent ribs together	Intercostal nerves (T2-12)
Intercostales interni	Inner surface of lower 11 ribs and corresponding costal cartilage (except 11, 12)	Superior surface of rib above	Depress ribs (if twelfth rib is fixed), draw adjacent ribs together	Intercostal nerves (T2-12)
Serratus posterior (two muscles designated s.p. superioris, and s.p. inferioris)	Superioris, between lower ligamentum nuchae and third thoracic spine	Ribs 2-5	Elevates ribs	Spinal nerves C8-T6
	Inferioris, spinous processes of T11 to L2 vertebrae	Ribs 9-12	Depresses ribs	Spinal nerves T10-L4

Adapted from McClintic, J.R.: Basic anatomy and physiology of the human body, ed. 2. Copyright © 1980. Reprinted by permission of John Wiley & Sons, Inc.

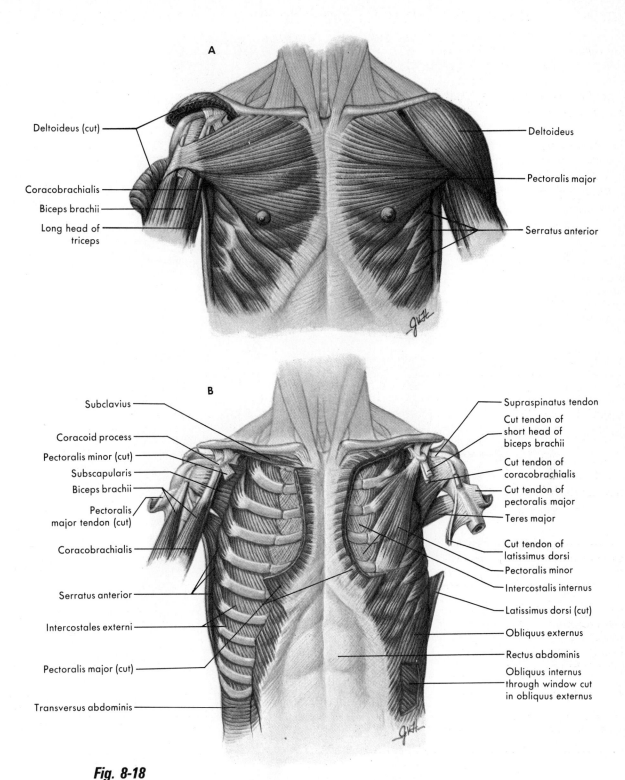

Fig. 8-18
A, Superficial muscles of upper chest and shoulder, with part of muscles of arm.
B, Deep muscles of upper chest and some associated muscles of scapula and arm.

Continued.

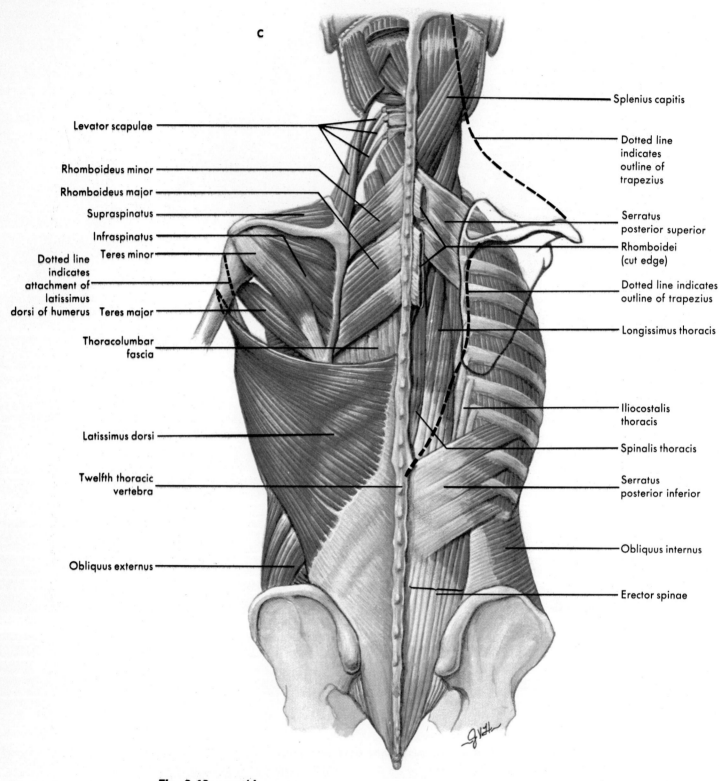

Fig. 8-18, cont'd
C, Muscles of posterior thorax.

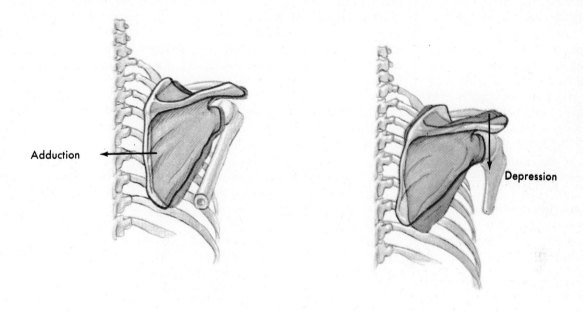
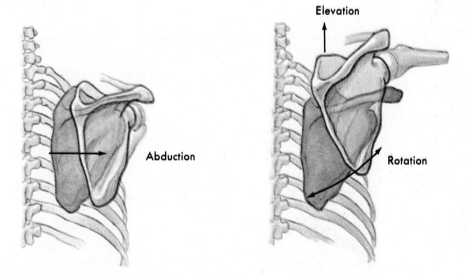

Fig. 8-19
Movements scapula may undergo. Shaded areas indicate starting positions; unshaded areas and arrows indicate movement.

scapula. It helps keep the scapula against the thorax and aids in raising the ribs during forced inspiration.

The **serratus anterior** muscle rises from the outer surfaces of the upper eight or nine ribs, passes *between* the scapula and the thorax, and attaches to the *medial border* of the scapula. Contraction thus draws the scapula laterally from the spine (abduction). The muscle is used strongly in pushing or punching motions. It also helps to keep the inferior angle of the scapula against the thorax to prevent "winging" of the scapula.

The **subclavius** muscle is a small muscle between the first rib and the inferior border of the clavicle. It acts to hold the clavicle steady as a platform for movements of the arm.

The posterior group includes the trapezius, rhomboideus, and levator scapulae muscles.

The **trapezius** muscle is placed like a cowl on the upper back and neck. Its origin extends from the occipital bone down the neck and back to the twelfth thoracic vertebra. Insertion is in a semicircle involving the lateral third of the clavicle and the acromion and spine of the scapula. Its fibers are disposed in three directions: nearly vertically, horizontally, and downward. The vertical fibers, sometimes designated as part 1 of the muscle, aid in elevation of the scapula, as in "shrugging" the shoulders. The horizontal fibers, part 2, are the heaviest and powerfully adduct the scapula. The downward directed fibers, part 3, depress the scapula. If the arm is abducted past a horizontal position, the upper two sets of fibers aid scapular rotation.

The **rhomboideus** (r.) muscle is usually considered to be composed of a small superiorly placed *r. minor* and a larger inferior *r. major*. The fibers run from the lower cervical and upper thoracic vertebral spines to the medial scapular border. The fibers incline slightly downward, and thus contraction adducts and slightly elevates the scapula.

The **levator scapulae** muscle is a straplike muscle running between the C1 to C4 transverse processes and the superior scapular angle. The fiber direction is nearly vertical and its action is inherent in the name—elevation of the scapula.

These muscles are presented in Table 8-9.

Table 8-9. Muscles of the shoulder (pectoral) girdle

Muscle	Origin	Insertion	Action	Innervation
Anterior				
Pectoralis minor	Outer surface ribs 3-5	Coracoid process	Pulls scapula forward (if arm is fixed, elevates ribs)	Spinal nerves C7, 8; T1
Serratus anterior	Outer surface ribs 1-9	Vertebral (medial) border of scapula	Abducts scapula and holds it against thorax	Spinal nerves C5-7
Subclavius	Junction of first rib and its cartilage	Middle third of clavicle	Depresses clavicle and draws it anteriorly	Spinal nerves C5, 6
Posterior				
Levator scapulae	Transverse processes C1-4	Between superior angle and spine of scapula on medial border	Elevates scapula	Spinal nerves C3-5
Rhomboideus (2 parts: major [upper], minor [lower]; separated at T2 spine)	Lower part of ligamentum nuchae, spines of C7-T5	Vertebral (medial) border of scapula between spine and inferior angle	Adducts and elevates scapula	Spinal nerve C5
Trapezius	External occipital protuberance, superior nuchal line, ligamentum nuchae, spines of vertebrae C7-T12	Lateral third of clavicle, acromion and spine of scapula	"Shrugs" shoulders, adducts and depresses scapula, rotates scapula in elevation of upper arm	Eleventh cranial nerve* and spinal nerves C3,4

Adapted from McClintic, J.R.: Basic anatomy and physiology of the human body, ed. 2. Copyright © 1980. Reprinted by permission of John Wiley & Sons, Inc.
*Accessory nerve.

Muscles moving the shoulder joint or humerus

(Fig. 8-20; see also Fig. 8-18)

It may be recalled that the shoulder joint is a ball-and-socket joint that permits all possible planes of motion. These movements are pictured in Figure 8-21. The muscles may be grouped according to their actions as flexors, extensors, abductors, adductors, and medial and lateral rotators of the humerus. The actions to be given are the *major* actions the muscles have, and additional movements a muscle may have will be described with each.

Flexors of the humerus

The *pectoralis major* muscle is the superficial muscle of the upper chest. It is fan shaped and has clavicular, sternal, and costal origins. The fibers sweep together in front of the shoulder joint, forming the anterior axillary fold, and insert into the lateral lip of the intertubercular sulcus (bicipital groove). The muscle draws the humerus forward (flexion), draws it across the chest (adduction), and medially rotates the bone. Throwing with an "inward spin" employs all three actions.

The *coracobrachialis* muscle is a small and weak flexor of the humerus that runs between the coracoid process and the medial portion of the humerus.

Extensors of the humerus

The *latissimus dorsi* muscle (broad muscle of the back) rises from the *T6* vertebral spine on a line to the sacrum. Its fibers converge to a broad ribbonlike tendon that passes across the front of the humerus to insert in the floor of the intertubercular sulcus. Its action extends the humerus, draws it across the back, and medially rotates it. It is active in climbing (for example, a rope) that requires a downward movement of the humerus and in swimming a backstroke.

The *teres major* muscle runs between the inferior angle of the scapula

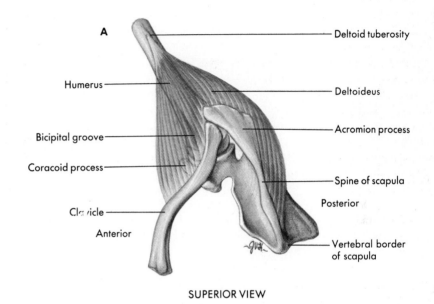

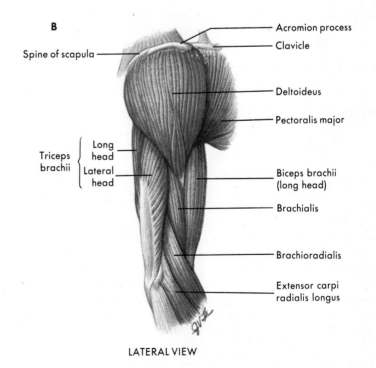

Fig. 8-20
A, Superior view of right shoulder to show relationships of deltoideus to scapula, clavicle, and humerus. **B,** Lateral view of right shoulder and arm.

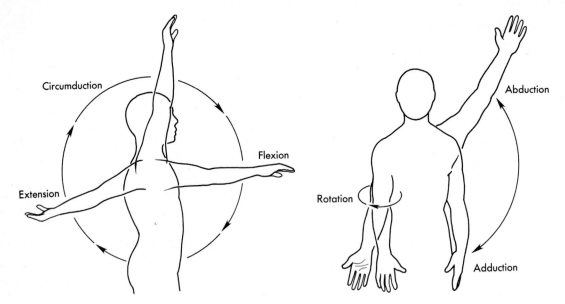

Fig. 8-21
Movements humerus undergoes.

and inserts essentially with the latissimus on the humerus. Its action is the same as that of the latissimus.

Abductors of the humerus

The *deltoideus* muscle covers the shoulder joint like a cupped hand. Its origin is opposite the insertion of the trapezius, that is, on the clavicle, acromion, and scapular spine. Accordingly the muscle shows three parts: anterior (clavicular), medial (acromial), and posterior (spinous). It inserts on the deltoid tuberosity of the humerus. The middle fibers are powerful abductors of the humerus, as are the other two if acting together with equal force. The anterior part alone flexes, whereas the posterior part alone extends the humerus.

The *supraspinatus* muscle fills the supraspinous fossa of the scapula and sends its tendon over the top of the shoulder joint to insert on the top of the greater tubercle. Its assigned action is abduction of the humerus—it is also important as an articular muscle of the shoulder (see following section).

There are no individual muscles to draw the humerus directly inward (adduct it). The pectoralis major and the latissimus must cooperate to achieve this motion.

Rotators of the humerus

In an earlier section the pectoralis major, latissimus dorsi, and teres major were described as causing medial rotation of the humerus. Three additional muscles are involved in rotating the arm. They are called the "rotator cuff" in sports medicine.

The *subscapularis* muscle fills the subscapular fossa, and its tendon passes across the front of the shoulder joint to the lesser tubercle. It causes medial rotation.

The *infraspinatus* muscle occupies the infraspinous fossa, and its tendon crosses behind the shoulder joint to the greater tubercle. This muscle *laterally* rotates the arm.

The *teres minor* muscle is almost a separate head of the infraspinatus. It courses from the inferior scapular angle to insert with the infraspinatus on the tubercle. Its action is to laterally rotate the humerus along with the infraspinatus.

The last three muscles mentioned insert so close to the fulcrum of the lever (that is, the center of the shoulder joint) that a very short effort arm is present, whereas the resistance arm makes up nearly the entire length of the upper appendage. Tremendous effort would thus be required to move the appendage in any action except rotation. The muscles appear to be most important in helping to hold the shoulder joint together. Contraction of all three muscles pulls the humerus into the glenoid fossa to resist dislocation. Because of this joint-stabilizing action, the muscles are often designated *articular muscles of the shoulder*.

The muscles concerned with the shoulder joint are presented in Table 8-10.

Table 8-10. Muscles moving the shoulder joint (humerus)

Muscle	Origin	Insertion	Action	Innervation
Abductors				
Deltoideus	Lateral one third of clavicle, acromion, spine of scapula	Deltoid tuberosity	Abducts humerus (acting as a whole)	Axillary nerve of brachial plexus
Supraspinatus*	Supraspinous fossa	Greater tubercle of humerus	Abducts humerus	Spinal nerves C5-6
Extensors				
Latissimus dorsi	Spinous processes T6-L5, iliac crest, last 3 ribs	Floor of bicipital groove (intertubercular sulcus) of humerus	Extension, adduction, medial rotation of humerus	Spinal nerves C6-8 (thoracodorsal nerve, branch of posterior cord of brachial plexus)
Teres major	Inferior angle of scapula	Medial lip of bicipital groove	Extension, adduction, medial rotation of humerus	Spinal nerves C6-8
Flexors				
Coracobrachialis	Coracoid process	Midhumerus	Flexes, adducts, and medially rotates humerus	Musculocutaneous nerve (C7)
Pectoralis major	Medial one third of clavicle, sternum, cartilages of ribs 1-7	Lateral lip of bicipital groove	Flexes, adducts, and medially rotates humerus	Spinal nerves C5-T1 (pectoral nerve)
Rotators				
Infraspinatus*	Infraspinous fossa	Greater tubercle of humerus	Laterally rotates humerus	Spinal nerves C5,6
Subscapularis*	Subscapular fossa	Lesser tubercle of humerus	Medially rotates humerus	Spinal nerves C7,8 (subscapular nerve)
Teres minor	Inferior angle of scapula	Greater tubercle of humerus	Laterally rotates humerus	Axillary nerve (C5)

Adapted from McClintic, J.R.: Basic anatomy and physiology of the human body, ed. 2. Copyright © 1980. Reprinted by permission of John Wiley & Sons, Inc.
*"Articular muscles" that aid in stabilizing and holding shoulder joint together.

Muscles moving the forearm

The muscles operating the forearm (Fig. 8-22) extend from the scapula or humerus to the radius or ulna. If it is recalled that the elbow region contains three joints, the forearm may be seen to be capable of flexion, extension, and medial and lateral rotation. These movements are depicted in Fig. 8-22, C.

Flexors of the forearm

The prominent *biceps brachii* muscle on the anterior arm rises by two heads: the laterally placed *long head* by a tendon attached to the supraglenoid tubercle; the *short head*, medially placed, by a tendon from the coracoid process. The tendon of the long head lies in the intertubercular sulcus. The two heads insert into a common tendon that passes to the radial tuberosity. If the forearm is in anatomical position, it is already in a full lateral rotated or supinated position, and the muscle flexes the forearm. If the forearm is partially or completely medially rotated (pronated), the muscle becomes a powerful supinator of the forearm. This is because the radial tuberosity is directed posteriorly in pronation, and the biceps contraction rotates the bone laterally.

The *brachialis* muscle occupies the lower half of the anterior humerus beneath the biceps and inserts on the coronoid process of the ulna. The ulna cannot rotate, and so this muscle only flexes the forearm.

The *brachioradialis* muscle forms the lateral curve of the forearm, running between the lateral supracondylar line and a point just above the radial styloid process. It is mainly a forearm flexor but can also return the forearm to a point midway between full supination and full pronation.

Extensors of the forearm

The *triceps brachii* muscle is a three-headed muscle that is the most powerful extensor of the forearm. The *lateral head* rises from the lateral side of the humerus, the *medial head* from

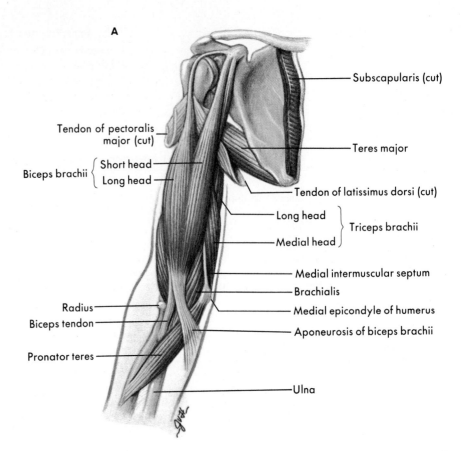

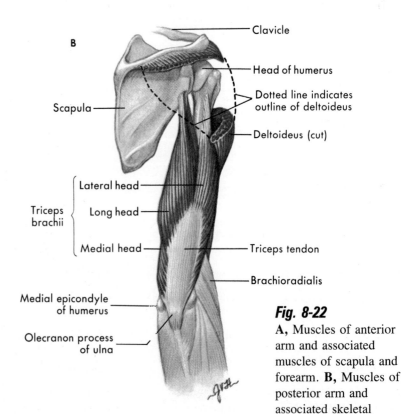

Fig. 8-22
A, Muscles of anterior arm and associated muscles of scapula and forearm. **B,** Muscles of posterior arm and associated skeletal structures.

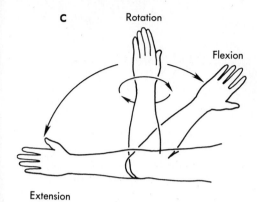

Fig. 8-22, cont'd
C, Movements of forearm.

the medial side of the humerus; the long head is placed between the other two and rises by a tendon from the infraglenoid tubercle. The heads join a common tendon that attaches to the olecranon process of the ulna. All three heads act together to extend the forearm.

The *anconeus* muscle appears to be a part of the medial triceps head but rises from the lateral epicondyle. It also attaches to the olecranon process and aids the triceps in forearm extension.

Rotators of the forearm

The biceps brachii and brachioradialis muscles have previously been mentioned as rotators of the forearm. Three other muscles, the supinator, pronator teres, and pronator quadratus muscles, rotate the forearm.

The *supinator* muscle crosses the forearm obliquely beneath the brachioradialis, extending from the *lateral* epicondyle to the middle area of the radius. Its name indicates its action.

The *pronator teres* muscle runs between the *medial* epicondyle and mid-radius.

The *pronator quadratus* muscle lies between the ulna and the radius in the distal one fourth of the forearm. The action of both the latter muscles is pronation.

The muscles moving the forearm are presented in Table 8-11.

Table 8-11. Muscles moving the forearm

Muscle	Origin	Insertion	Action	Innervation
Extensors				
Anconeus	Posterior aspect of lateral epicondyle of humerus	Lateral side of olecranon process	Extends forearm	Radial nerve (spinal nerves C7, 8)
Triceps brachii				
Lateral head	Posterior shaft of humerus			
Long head	Infraglenoid tubercle	Olecranon process (all three heads)	Extends forearm	Radial nerve (spinal nerves C7, 8)
Medial head	Posterior shaft of humerus			
Flexors				
Biceps brachii				
Long head	Supraglenoid tubercle	Radial tuberosity (both heads)	Flexes and supinates forearm	Musculocutaneous nerve (spinal nerves C5, 6)
Short head	Coracoid process			
Brachialis	Lower half of anterior surface of humerus	Coronoid process of ulna and ulnar tuberosity	Flexes forearm	Musculocutaneous nerve (spinal nerves C5, 6)
Brachioradialis	Upper two thirds of lateral surface of humherus	Lateral side of radius above styloid process	Flexes, semisupinates, semipronates forearm	Radial nerve (spinal nerves C5, 6)
Rotators				
Pronator quadratus	Anterior surface of ulna	Anterior surface of radius	Pronates forearm	Median nerve (spinal nerves C8, T1)
Pronator teres	Medial epicondyle humerus	Halfway down lateral surface of radius	Pronates forearm	Median nerve (spinal nerves C6, 7)
Supinator	Lateral epicondyle of humerus	Lateral aspect of radius	Supinates forearm	Radial nerve (spinal nerves C5, 6)

Adapted from McClintic, J.R.: Basic anatomy and physiology of the human body, ed. 2. Copyright © 1980. Reprinted by permission of John Wiley & Sons, Inc.

Muscles moving the wrist and fingers

The muscles that move the wrist and fingers (Fig. 8-23) take their origins from the distal end of the humerus or the proximal ends of the radius and ulna. Functionally they may be divided into two groups: those lying on the anterior aspect of the forearm are primarily *flexors* of the wrist and fingers; those lying on the posterior forearm are primarily *extensors* of the wrist and fingers. Abduction and adduction of the wrist are accomplished by cooperation of certain of the anterior and posterior muscles. The movements of the wrist and fingers are shown in Fig. 8-24. Most of the muscles to be described are named according to their position relative to the radius and ulna and carry the action in their name—thus the task of learning them is simplified.

Flexors of the wrist and fingers

The anterior group is disposed in three layers on the forearm.

The *flexor carpi radialis* is the most lateral of the superficial anterior group of muscles, rising from the medial epicondyle of the humerus and inserting on the second and third metacarpals. It flexes and slightly abducts the wrist. When the wrist is forcibly flexed, the tendon of this muscle is visible on the distal forearm.

The *palmaris longus* originates beside the flexor carpi radialis, is the medial muscle of the superficial layer, and inserts into the *palmar aponeurosis*, a broad flat tendon in the palm of the hand. It flexes the wrist.

The *flexor carpi ulnaris* muscle is two headed, with a humeral head rising from the medial epicondyle and an ulnar head rising from the base and medial aspect of the olecranon process. It inserts on the fifth metacarpal and pisiform bone and flexes and slightly adducts the wrist.

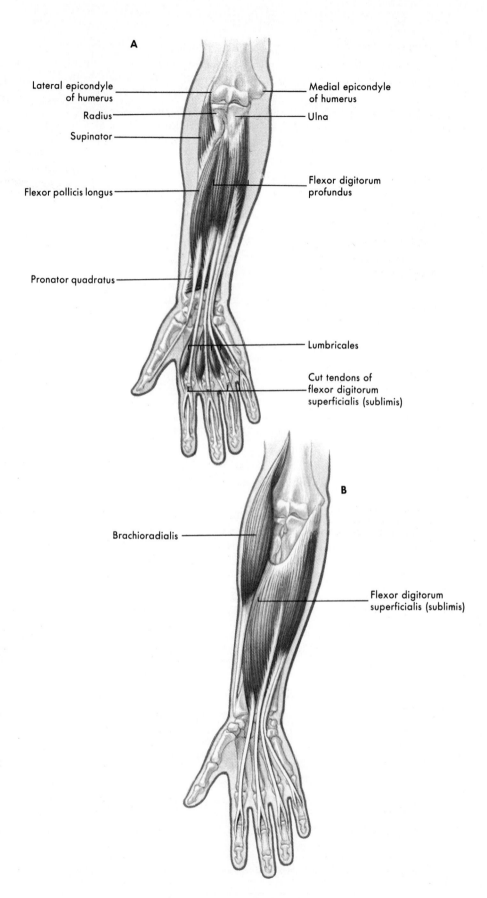

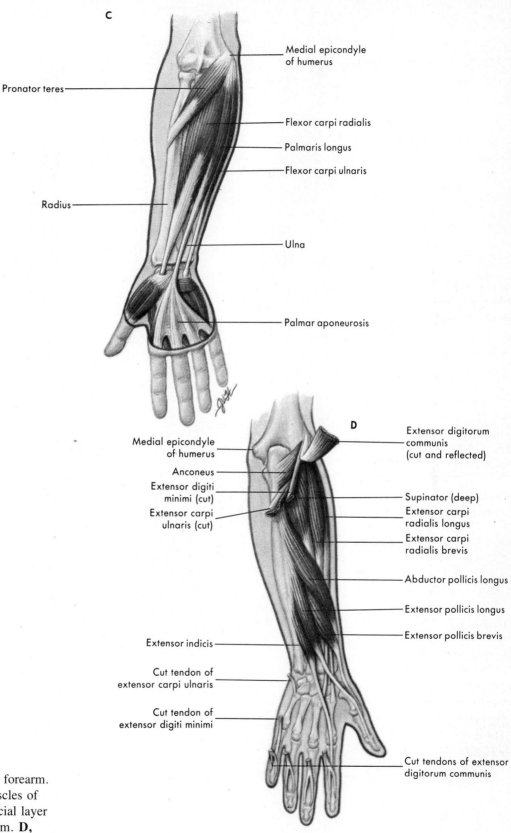

Fig. 8-23
A, Deep muscles of anterior forearm.
B, Intermediate layer of muscles of anterior forearm. **C,** Superficial layer of muscles of anterior forearm. **D,** Deep muscles of posterior forearm.

Continued.

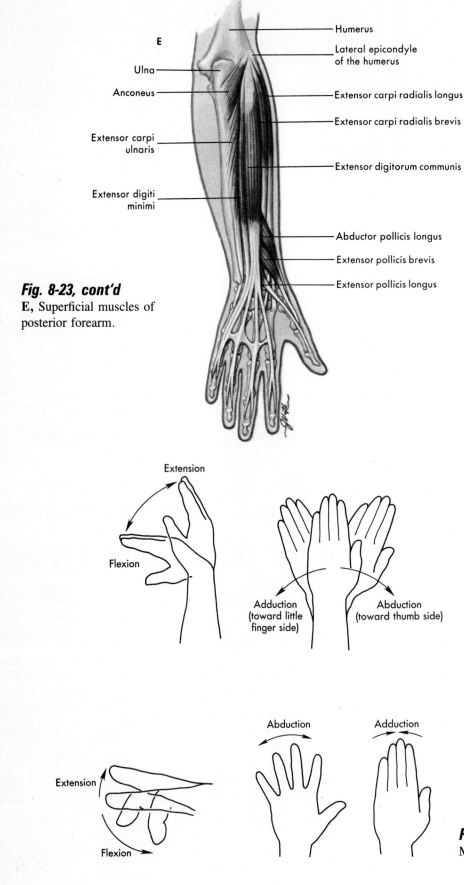

Fig. 8-23, cont'd
E, Superficial muscles of posterior forearm.

The intermediate layer of anterior forearm muscles is the *flexor digitorum superficialis* (or *f.d. sublimis*). Its origin is from the humerus, radius, and ulna, and it forms four heads, each sending a tendon to the second phalanx of each finger. Thus each finger may be operated more or less independently of the others, being flexed at the first interphalangeal joint.

The deep layer of anterior muscles is the *flexor digitorum profundus*. It has an extensive origin from the ulna and the forearm interosseous membrane and sends its tendons to the third phalanx of each finger. The second interphalangeal joints are thus flexed. Full flexion of the fingers requires the action of both muscles.

Extensors of the wrist and fingers

There are six muscles in a single layer across the posterior aspect of the forearm that extend the wrist and fingers. These muscles are presented in order from lateral to medial.

The *extensor carpi radialis longus* muscle rises from the lateral supracondylar ridge of the humerus and sends its tendon to the second metacarpal. It extends and slightly abducts the wrist.

The *extensor carpi radialis brevis* muscle inserts on the third metacarpal and has the same action as the previous muscle.

The *extensor digitorum communis* muscle rises from the lateral epicondyle of the humerus and divides to form three heads just above the wrist. From each head tendons pass to all three phalanges of the index, middle, and ring fingers. Its action extends both the first and second interphalangeal joints. A small muscle rising from the posterior ulna is called the *extensor indicis* and

Fig. 8-24
Movements of wrist and fingers.

provides an additional tendon to the index finger.

The *extensor digiti minimi* muscle also rises from the lateral epicondyle and inserts on the phalanges of the little finger, extending it.

The *extensor carpi ulnaris* muscle has humeral and ulnar origins and inserts on the fifth metacarpal. It extends and slightly adducts the wrist.

Abduction and adduction of the wrist

In the sections describing the flexors and extensors of the wrist, certain muscles were mentioned as having additional actions of abduction and adduction of the wrist. It may be recalled that the flexor carpi radialis muscle and the two extensor carpi radialis muscles would both abduct the wrist slightly but in a palmar or posterior direction respectively. *Direct* abduction at the wrist requires the simultaneous contraction of *both* muscles. The same principle holds for the flexor and extensor carpi ulnaris muscles to cooperate to achieve direct adduction of the wrist.

The muscles moving the wrist and fingers are presented in Table 8-12.

Table 8-12. Muscles moving the wrist and fingers

Muscle	Origin	Insertion	Action	Innervation
Anterior group				
Flex fingers				
Flexor digitorum superficialis (sublimis)	Medial epicondyle of humerus, coronoid process of ulna, shaft of radius	Bases of second phalanges of fingers	Flexes fingers	Median nerve (spinal nerves C7,8,T1)
Flexor digitorum profundus	Proximal three fourths ulna and interosseous membrane	Bases of distal (third) phalanges of fingers	Flexes fingers	Median (index and middle fingers) and ulnar nerve (ring and little fingers), spinal nerves C7-T1
Flex wrist				
Flexor carpi radialis	Medial epicondyle of humerus	Bases of metacarpals 2,3	Flexes wrist and abducts it	Median nerve (spinal nerves C6,7)
Palmaris longus	Medial epicondyle of humerus	Palmar aponeurosis	Flexes wrist	Median nerve (spinal nerve C6,7)
Flexor carpi ulnaris	Medial epicondyle of humerus and upper two thirds of posterior ulna	Base of fifth metacarpal and pisiform bone	Flexes wrist and adducts it	Ulnar nerve (spinal nerves C8,T1
Posterior group				
Extend fingers				
Extensor digitorum communis	Lateral epicondyle of humerus	Phalanges 2 and 3, first 3 fingers	Extends first 3 fingers	Radial nerve (spinal nerves C7,8)
Extensor digiti minimi	Lateral epicondyle of humerus	All phalanges of little finger	Extends little finger	Radial nerve (spinal nerves C7,8)
Extend wrist				
Extensor carpi radialis longus	Lateral supracondylar ridge of humerus	Base of second metacarpal	Extends wrist and abducts it	Radial nerve (spinal nerves C6,7)
Externsor carpi radialis brevis	Lateral epicondyle of humerus	Base of third metacarpal	Extends wrist	Radial nerve (spinal nerves C6,7)
Extensor carpi ulnaris	Lateral epicondyle of humerus	Base of fifth metacarpal	Extends wrist and adducts it	Radial nerve (spinal nerves C7,8)

Adapted from McClintic, J.R.: Basic anatomy and physiology of the human body, ed. 2. Copyright © 1980. Reprinted by permission of John Wiley & Sons, Inc.

Muscles of the thumb

Recall that the bones of the thumb are placed at a 90-degree angle to those of the fingers. This arrangement allows the thumb to be placed in opposition to the fingers for grasping, manipulating, and writing with objects. Eight muscles serve the thumb (Fig. 8-25; see also Fig. 8-23).

The *flexor pollicis longus* muscle rises from the anterior radius and inserts on the palmar surface of the base of the second phalanx. It flexes the interphalangeal joint, and continued action also flexes the metacarpophalangeal joint, drawing the thumb across the palm of the hand (as in touching the thumb to the little finger).

The *flexor pollicis brevis* rises from the trapezium and inserts on the lateral side of the base of the first phalanx of the thumb. It flexes the thumb.

The *abductor pollicis longus* muscle rises from the posterior and middle thirds of the ulna and radius and the interosseous membrane and inserts on the thumb's metacarpal. It abducts the thumb and extends the metacarpal.

The *abductor pollicis brevis* muscle rises from the scaphoid and trapezium and attaches to the base of the first phalanx. It abducts the thumb.

The *extensor pollicis brevis* muscle rises from the radius and the interosseous membrane and inserts on the first phalanx of the thumb. It extends the thumb.

The *extensor pollicis longus* muscle rises from the posterior ulna and inter-

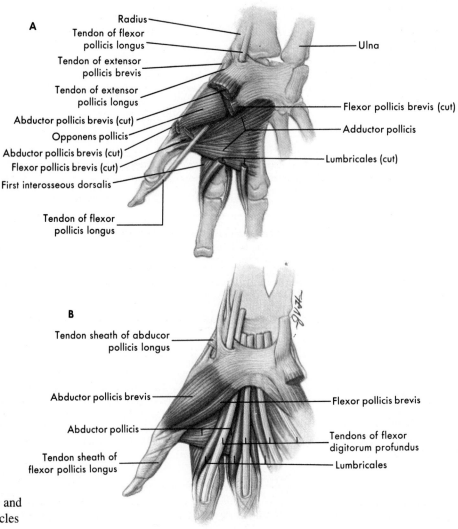

Fig. 8-25
A, Some deep muscles of thumb and fingers. **B,** Some superficial muscles of thumb and fingers.

osseous membrane and inserts on the second phalanx, extending the thumb.

The **adductor pollicis** muscle rises from the capitate, trapezoid, and second and third metacarpals and inserts on the first phalanx. It adducts the thumb.

Last, the **opponens pollicis** muscle originates from the trapezium and inserts on the first metacarpal. It draws the thumb across the palm, enabling it to be touched to the tip of any finger.

The muscles of the thumb are presented in Table 8-13.

Table 8-13. Muscles of the thumb

Muscle	Origin	Insertion	Action	Innervation
Abductor pollicis brevis	Trapezium and scaphoid	Lateral side of base of first phalanx of thumb	Abducts thumb	Median nerve (spinal nerves C8, T1)
Abductor pollicis longus	Posterior ulna, interosseous membrane, and middle one third of posterior radius	Lateral side of base of first metacarpal	Abducts thumb	Radial nerve (spinal nerve C7,8)
Adductor pollicis	Trapezoid, capitate, second and third metacarpals	Medial side of base of first phalanx of thumb	Adducts thumb	Ulnar nerve (spinal nerves C8,T1)
Extensor pollicis brevis	Posterior one third of radius and interosseous membrane	Base of first phalanx of thumb	Extends thumb	Radial nerve (spinal nerves C7, 8)
Extensor pollicis longus	Posterior ulna and interosseous membrane	Base of second phalanx of thumb	Extends thumb	Radial nerve (spinal nerves C7,8)
Flexor pollicis brevis	Trapezium	Lateral side of base of first phalanx of thumb	Flexes thumb	Median nerve (spinal nerves C8, T1)
Flexor pollicis longus	Anterior midradius and interosseous membrane	Palmar surface of base of distal phalanx of thumb	Flexes thumb and metacarpophalangeal joint	Radial nerve (spinal nerve C8,T1)
Opponens pollicis	Trapezium	Lateral half of anterior surface of first metacarpal	Flexes and adducts thumb (so that thumb may touch any flexed fingertip)	Median nerve (spinal nerves C8, T1)

Adapted from McClintic, J.R.: Basic anatomy and physiology of the human body, ed. 2. Copyright © 1980. Reprinted by permission of John Wiley & Sons, Inc.

Muscles within the hand

The intrinsic muscles of the hand are the *interossei* and the *lumbricales* (Fig. 8-26). They abduct and adduct the fingers and aid in flexing them.

The interossei muscles are seven in number, divided into four dorsal muscles and three palmar muscles according to their location. The dorsal muscles fill the spaces between metacarpals and attach to the first three fingers. The palmar muscles originate from metacarpals 1, 2, 4, and 5 and insert on the index, ring, and little fingers. The middle finger receives two dorsal muscles, the ring and index fingers have a palmar and a dorsal muscle, and the thumb and little finger have a palmar muscle only. Dorsal interossei muscles abduct and palmar interossei muscles adduct the digit.

The lumbricales muscles are four in number, rising from the palmar surfaces of metacarpals 2 to 5, and insert on the first phalanx of the fingers. They extend the interphalangeal joints and flex the metacarpophalangeal joints.

Intrinsic hand muscles are presented in Table 8-14.

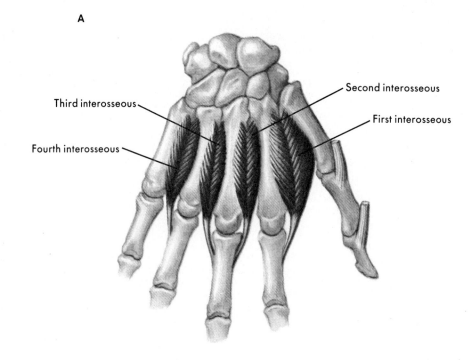

Fig. 8-26
Muscles of hand. **A,** Right dorsal interossei.

Table 8-14. Muscles within the hand

Muscle	Origin	Insertion	Action	Innervation
Interossei (4 dorsal and 4 palmar muscles; numbered from thumb side)	Dorsal: fill spaces between metacarpals; from adjacent surfaces of bones	Sides of first phalanges of first 3 fingers	Abduct fingers	Ulnar nerve (spinal nerves C8, T1)
	Palmar: metacarpals 2-5	Sides of first phalanges of first, third and fifth fingers	Adduct fingers	Ulnar nerve (spinal nerves C8, T1)
Lumbricales (4 muscles)	Tendons of flexor digitorum profundus over palmar surface of metacarpals 2-5	Dorsal surfaces of first phalanges of fingers	Flex metacarpophalangeal and extend interphalangeal joints	Two medial by ulnar nerve; two lateral by median nerve (spinal nerves C8, T1)

Adapted from McClintic, J.R. Basic anatomy and physiology of the human body, ed. 2. Copyright © 1980. Reprinted by permission of John Wiley & Sons, Inc.

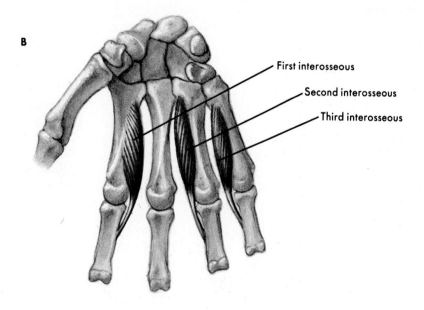

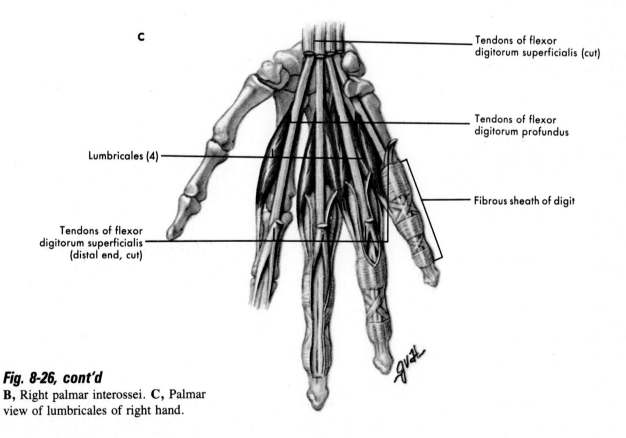

Fig. 8-26, cont'd
B, Right palmar interossei. **C,** Palmar view of lumbricales of right hand.

Muscles moving the hip and knee joints

The hip joint, as a ball-and-socket joint, permits three planes of motion: flexion-extension, abduction-adduction, and medial and lateral rotation. The knee is usually described as basically a hinge joint; its main action is flexion-extension. Because of the shape of the bones involved in the joint, the knee rotates slightly when flexed or extended and may be slightly abducted and adducted. The basic movements of both joints are depicted in Fig. 8-27.

The many muscles involved in operating these two joints may be conveniently divided into five groups according to the actions they have on the joints.

The first group of muscles (Fig. 8-28) lie on the anterior aspects of the iliac bones or the lumbar vertebral transverse processes; they flex the hip or extend the knee, or both.

The *iliacus* muscle fills the iliac fossa. The *psoas major* muscle rises from the lumbar processes. Both attach to a common tendon that passes to the area of the lesser trochanter. Because of a common tendon of insertion, some anatomists name the muscle *iliopsoas*, since it has an iliac and a lumbar head. Its action is flexion and medial rotation of the thigh (femur).

The name *quadriceps femoris* is given to the next four muscles. All have separate origins but a common insertion on the patella, from which the patellar tendon passes to the tibial tuberosity.

The *rectus femoris* muscle lies superficially in the anterior midline of the thigh. It rises from the anterior inferior iliac spine and inserts on the patella. It flexes the thigh *and* extends the knee, as in kicking.

The *vastus lateralis* muscle originates from the lateral aspect of the linea aspera and sweeps around the lateral side of the femur to insert on the patella. It extends the knee. The upper lateral part of the muscle is commonly used as a site for intramuscular injections.

The *vastus intermedius* muscle rises from the anterior femur, lies beneath the rectus femoris, and inserts on the patella. It extends the knee.

The *vastus medialis* muscle originates from the medial lip of the linea aspera and inserts on the patella. It also extends the knee.

The *sartorius* muscle originates from the anterior superior iliac spine and crosses the anterior thigh to insert on the *medial* side of the tibial head. It flexes the hip and rotates it laterally and also flexes the knee. Placing the ankle on the opposite knee well illustrates the combined actions of sartorius.

The anterior group of muscles is presented in Table 8-15.

The second and medial group of thigh muscles (Fig. 8-29) run between

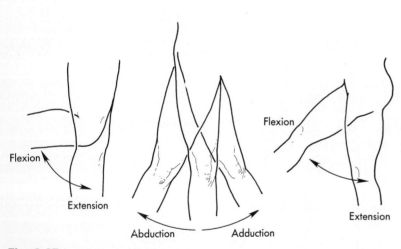

Fig. 8-27
Movements of hip and knee.

Fig. 8-28
Muscles of thigh. **A,** Vastus intermedius and neighboring muscles. **B,** Some muscles of anterior hip and thigh. **C,** Rectus femoris and neighboring muscles. **D,** Sartorius, tensor fasciae latae, and neighboring muscles.

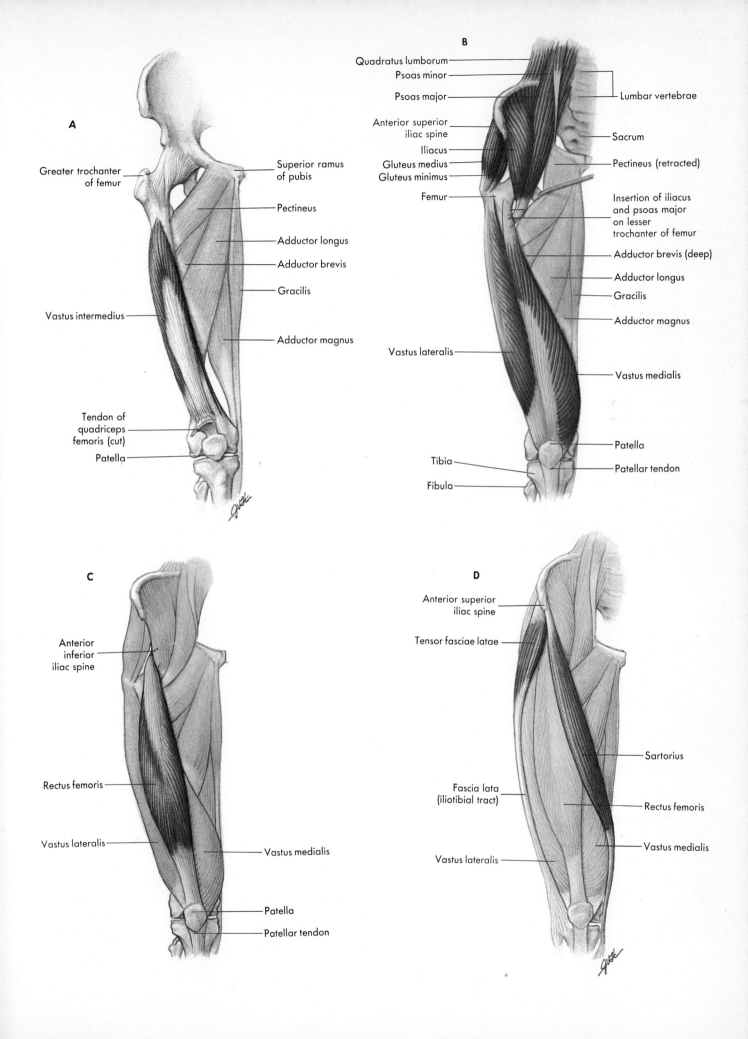

Table 8-15. Muscles moving the hip and knee joints

Muscle	Origin	Insertion	Action	Innervation
Anterior group Flex hip and/or extend knee				
Iliacus } Iliopsoas Psoas }	Iliac fossa	Below lesser trochanter	Flexes thigh and rotates hip medially	Femoral nerve (spinal nerves L2, 3)
	Transverse processes of lumbar vertebrae	Lesser trochanter	Flexes thigh and rotates hip medially	Spinal nerves L2, 3 from lumbar plexus
Rectus femoris	Anterior inferior iliac spine and rim of acetabulum	Patella	Flexes hip and extends knee	Femoral nerve (spinal nerves L2-4)
Sartorius	Anterior superior iliac spine	Upper part of medial tibia head	Flexes hip and knee, rotates thigh laterally	Femoral nerve (spinal nerves L2, 3)
Vastus intermedius	Anterior upper three fourths of femur	Patella	Extends knee	Femoral nerve (spinal nerves L2-4)
Vastus lateralis	Lateral border of linea aspera	Patella	Extends knee	Femoral nerve (spinal nerves L2-4)
Vastus medialis	Medial border of linea aspera	Patella	Extends knee	Femoral nerve (spinal nerves L2-4)
Medial group Adductors				
Adductor brevis	Body and inferior ramus of pubis	Upper half of linea aspera	Adducts and aids in flexion and medial rotation of femur	Sciatic nerve and obturator nerve (spinal nerves L2-4)
Adductor longus	Pubis, between symphysis and crest	Linea aspera	Adducts and aids in flexion and medial rotation of femur	Sciatic nerve and obturator nerve (spinal nerves L2-4)
Adductor magnus	Rami of ischium and pubis	Lower one third of linea aspera and adduction tubercle	Adducts and aids in flexion and lateral rotation of femur	Sciatic nerve and obturator nerve (spinal nerves L2-4)
Gracilis	Rami of ischium and pubis	Medial side of tibial head	Adducts, flexes, and medially rotates thigh	Obturator nerve (spinal nerves L2, 3)
Pectineus	Iliopectineal line and pubis	Base of lesser trochanter	Adducts and aids in flexing thigh	Femoral nerve (spinal nerves L2-4)

Adapted from McClintic, J.R.: Basic anatomy and physiology of the human body, ed. 2. Copyright © 1980. Reprinted by permission of John Wiley & Sons, Inc.

Table 8-15. Muscles moving the hip and knee joints—cont'd

Muscle	Origin	Insertion	Action	Innervation
Posterior group Extend hip and/or flex knee				
Gluteus maximus	Between posterior gluteal line and crest of ilium, sacrum and coccyx	Gluteal tuberosity of femur	Extends thigh and laterally rotates it	Sciatic nerve (spinal nerves L5, S1, 2)
Biceps femoris	Ischial tuberosity (long head) and lateral lip of linea aspera (short head)	Lateral side of fibular head	Extends thigh and flexes knee	Sciatic nerve (spinal nerves L5, S1, 2)
Semitendinosus	Ischial tuberosity	Medial surface of upper tibia	Extends thigh and flexes knee	Sciatic nerve (spinal nerves L5, S1, 2)
Semimembranosus	Ischial tuberosity	Posterior side of medial tibial condyle	Extends thigh and flexes knee	Sciatic nerve (spinal nerves L5, S1, 2)
Abductors				
Gluteus medius	Lateral superior surface of ilium	Greater trochanter	Abducts and medially rotates thigh	Sciatic nerve (spinal nerves L4, 5, S1, 2)
Gluteus minimus	Lateral inferior surface of ilium	Greater trochanter	Abducts and medially rotates thigh	Sciatic nerve (spinal nerves L4, 5, S1, 2)
Tensor fasciae latae	Lateral superior surface of ilium above anterior superior spine	Iliotibial tract that inserts into lateral tibial head	Abducts, flexes, and medially rotates thigh	Sciatic nerve (spinal nerves L4, 5, S1, 2)
Lateral rotators				
Piriformis	Anterior sacrum	Greater trochanter	Lateral rotation of thigh	Sacral spinal nerves 1, 2
Gemellus Superior Inferior	Ischial spine Ischial tuberosity	Greater trochanter (both heads)	Lateral rotation of thigh	Obturator nerve (spinal nerves L3, 4)
Obturator Internus Externus	Rim of obturator foramen, pubis, and ischium Rim of obturator foramen	Greater trochanter (both heads)	Lateral rotation of thigh	Obturator nerve (spinal nerves L3, 4)
Quadratus femoris	Ischial tuberosity	Below greater trochanter	Lateral rotation of thigh	Obturator nerve (spinal nerves L3, 4)

the pubis or ischium and the femur and are so positioned as to adduct the thigh as their major action.

The **adductor magnus** muscle, largest of the group, runs from the ischium to the medial side of the lower three fourths of the femur. It adducts and medially rotates the thigh and is a very important muscle in maintaining the lateral balance of the standing body.

The **adductor brevis** muscle is partially covered by the adductor magnus and runs from the pubis to the upper part of the linea aspera. It also adducts and medially rotates the thigh.

The **adductor longus** muscle lies above and superficial to both the adductor magnus and brevis and runs from the pubis to the upper linea aspera. Its actions are the same as the preceding

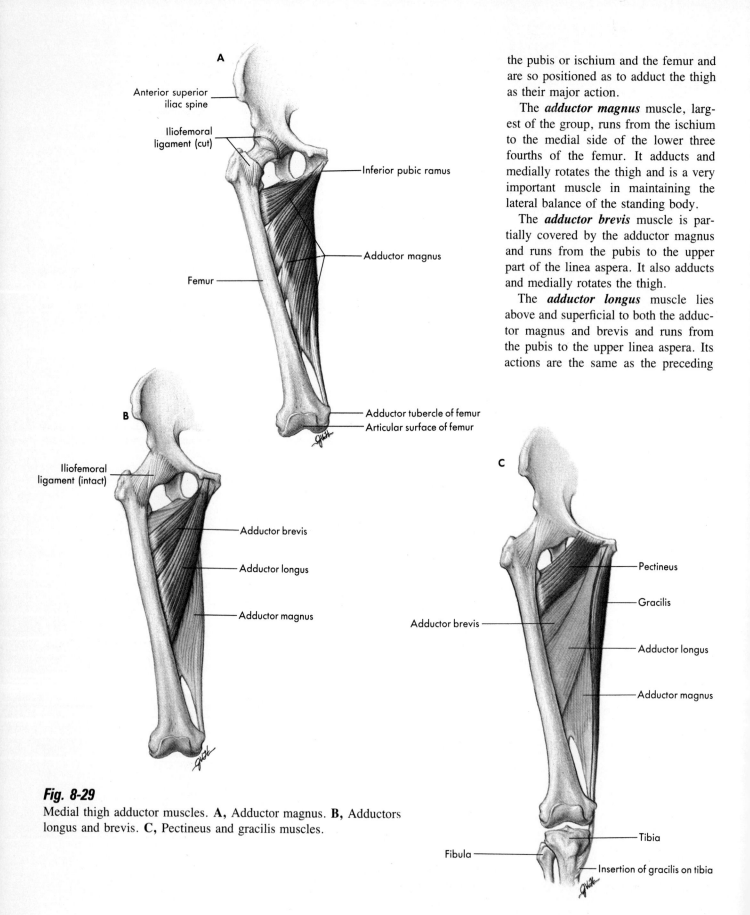

Fig. 8-29
Medial thigh adductor muscles. **A,** Adductor magnus. **B,** Adductors longus and brevis. **C,** Pectineus and gracilis muscles.

muscles, adduction and medial rotation of the thigh.

The *pectineus* muscle is located between the pubis and the pectineal line of the femur. It is also an adductor of the thigh.

The *gracilis* muscle lies on the medial aspect of the thigh and adducts it. After adduction, the gracilis aids knee flexion and medial rotation of the thigh.

These medial muscles are also presented in Table 8-15.

The third group of muscles lies on the posterior aspect of the pelvis and thigh (Fig. 8-30). These muscles are primarily extensors of the thigh or flexors of the knee, or both. They are antagonists to the rectus femoris and vastus muscles.

The *gluteus maximus* is familiar to

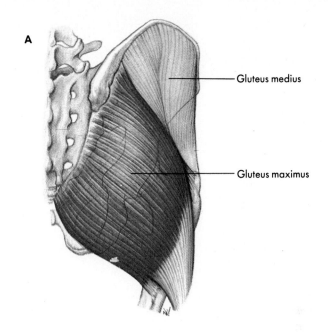

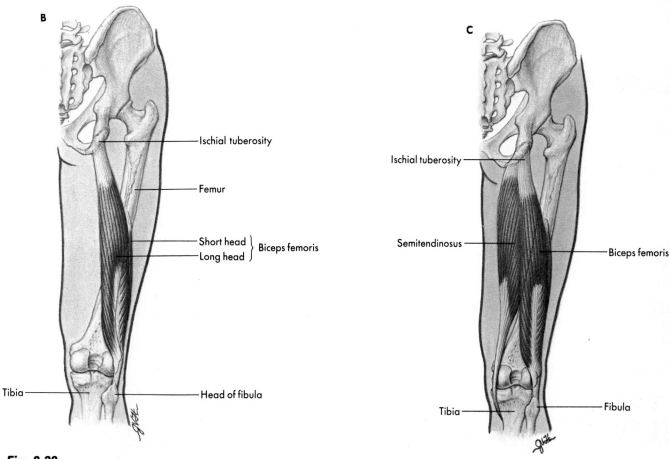

Fig. 8-30
Muscles of posterior pelvis and posterior thigh regions. **A,** Gluteus maximus. **B,** Biceps femoris. **C,** Biceps femoris and semitendinosus.   *Continued.*

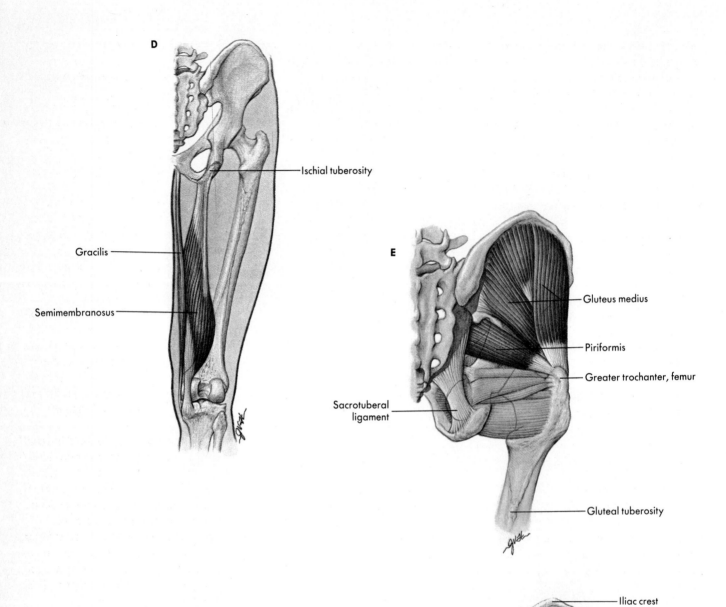

Fig. 8-30, cont'd
D, Semimembranosus, with gracilis as a reference point. **E,** Gluteus medius and piriformis. **F,** Gluteus minimus and remaining lateral rotators of thigh.

all as the major buttock muscle. It rises from the ilium between the posterior gluteal line and the iliac crest and passes to the gluteal tuberosity of the femur. The muscle is a powerful extensor of the thigh in such activities as stair climbing, rising from a squatting position, and bicycling. It also tends to laterally rotate the thigh and, when both muscles contract together, "pinches" the buttocks.

The "hamstrings" form a group of three muscles across the posterior thigh.

The *biceps femoris* muscle is the most lateral of the three hamstrings, rising by two heads from the ischial tuberosity *(long head)* and the shaft of the femur *(short head)*. A common tendon passes to the head of the fibula. It extends the thigh *and* flexes the knee and aids in lateral rotation of the lower appendage. It is used to maintain thigh extension against the force of gravity and in running or walking to lift the leg and extend the thigh. If a person finds it difficult to touch the toes without bending the knees, "tight hamstrings" is a common diagnosis.

The *semitendinosus* muscle is the middle of the three hamstrings, extending from the ischial tuberosity to the medial part of the tibial head. It is a powerful thigh extensor and knee flexor.

The *semimembranosus* muscle, the most medial of the three muscles, also rises from the ischial tuberosity and inserts on the medial tibial head. It is also a thigh extensor and knee flexor and aids in medial rotation of the lower appendage.

When the knee is flexed, the tendons of the hamstrings form the lateral and medial boundaries of the *popliteal fossa*, a "cavity" on the posterior aspect of the knee joint.

These posterior muscles are also presented in Table 8-15.

Abductors of the thigh (see Fig. 8-30) originate primarily on the posterior iliac surface and pass to the greater trochanter or to the lateral tibial head.

The *gluteus medius* muscle rises from the upper and posterior surface of the lateral ilium and inserts on the lateral side of the greater trochanter. The muscle abducts the thigh and medially rotates it and is important in maintaining balance when walking or running.

The *gluteus minimus* muscle lies beneath the medius and rises from the lower posterior aspect of the lateral surface of the ilium. It passes to the anterior surface of the greater trochanter, and its action is the same as that of the medius.

The *tensor fasciae latae* rises from the lateral iliac surface and inserts on the *iliotibial tract* just below the greater trochanter. The latter is a heavy band of connective tissue that runs along the lateral aspect of the femur (see Fig. 8-28) to attach to the head of the fibula. Contraction of the muscle pulls on the tract and abducts the lower appendage. It is important in balancing the thigh on the leg and stabilizes the knee during walking.

These abductor muscles are also presented in Table 8-15.

The fifth group of muscles run nearly horizontally from the sacrum to the greater trochanter (see Fig. 8-30). As a group they laterally rotate the thigh and act as articular muscles to help stabilize the hip joint.

The *piriformis* muscle rises from the anterior aspect of the sacrum, passes through the greater sciatic foramen (above the ischial spine), and inserts on the top of the greater trochanter. The most superior muscle of the group, it laterally rotates the thigh.

Below the piriformis lies the *gemellus* muscle, divided into *superior* and *inferior* parts. The superior gemellus originates from the ischial spine, the inferior from the ischial tuberosity. Both heads insert on the medial surface of the greater trochanter and laterally rotate the thigh.

The *obturator* muscle has *internal* and *external* parts. The obturator internus rises from the rim of the obturator foramen, the pubis, and the ischium and inserts near the trochanteric fossa. They too laterally rotate the thigh.

The *quadratus femoris* is the inferior muscle of the group, rising from the ischial tuberosity and passing to a point just below the greater trochanter. It is also a lateral rotator of the thigh.

These lateral rotator muscles are also presented in Table 8-15.

Muscles moving the ankle and foot

The *crural* muscles, which lie on the leg, move the ankle and foot and are conveniently divided by location into anterior, lateral, and posterior crural muscles. Anterior muscles generally dorsiflex and invert the foot and extend the toes. The lateral group plantarflexes and everts the foot, whereas the posterior group plantarflexes the foot and flexes the toes (Fig. 8-31).

The anterior crural muscles are shown in Fig. 8-32 and include the tibialis anterior, extensor hallucis longus, extensor digitorum longus, and peroneus tertius muscles.

The *tibialis anterior* muscle lies just lateral to the shin, rising from the upper two thirds of the tibia, and inserts on the *medial* cuneiform and *first* metatarsal. To arrive at its insertion, the tendon must cross the anterior aspect of the ankle from lateral to medial. Contraction dorsiflexes the foot *and* rotates the foot or inverts it. The muscle is extremely important in maintaining the medial longitudinal arch of the foot, "pulling from above."

The *extensor hallucis longus* muscle lies beneath and lateral to the tibialis anterior. It rises from the medial fibula, and its tendon crosses to the medial side of the foot to insert on the base of the second phalanx of the great toe. It extends the great toe, and continued action aids dorsiflexion of the foot to prevent the foot from dragging as the leg is swung forward during walking.

The *extensor digitorum longus* muscle lies lateral to the tibialis anterior. It originates from the lateral tibial condyle and the fibular head and gives rise to a single tendon that passes to the bases of the four outer metatarsals, where it divides to give four tendons that pass to the phalanges of the four outer toes. When the muscle contracts, all four outer toes are extended. This muscle is important in keeping the foot dorsiflexed when walking and is also important in creating a firm platform for maintaining the standing posture of the body.

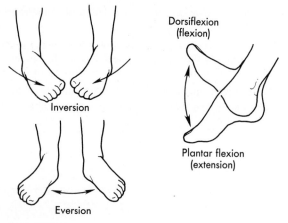

Fig. 8-31
Movements of ankle and foot.

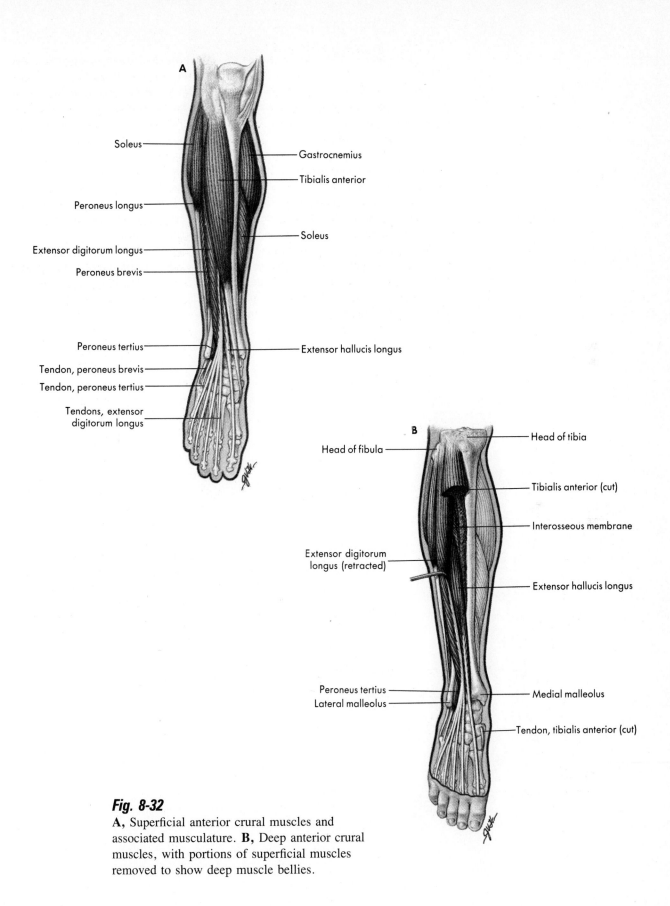

Fig. 8-32
A, Superficial anterior crural muscles and associated musculature. **B,** Deep anterior crural muscles, with portions of superficial muscles removed to show deep muscle bellies.

The **peroneus tertius** muscle may or may not be present. If present, it forms a clearly defined inferior band of the preceding muscle, sending its tendon to the fifth metatarsal. It dorsiflexes and slightly everts the foot.

The lateral crural muscles include the peroneus longus and peroneus brevis muscles (Fig. 8-33).

The **peroneus longus** muscle lies on the lateral border of the fibula. It sends its tendon *under* the lateral malleolus (a device to convert the direction of pull of the muscle) and across the sole of the foot to insert on the *first* cuneiform and metatarsal. Plantar flexion and eversion of the foot occur. The tendon also assists in maintaining the transverse arch of the foot.

The **peroneus brevis** muscle originates below the peroneus longus on the fibula. Its tendon also passes beneath the lateral malleolus to insert on metatarsal number five. It also plantarflexes and everts the foot.

The posterior crural muscles (Fig. 8-34) lie behind the tibia and fibula and include the gastrocnemius, soleus, tibialis posterior, flexor digitorum longus, and flexor hallucis longus muscles.

The **gastrocnemius,** the "calf" muscle, is two headed; lateral and medial heads rise from above the respective femoral condyles. A common tendon,

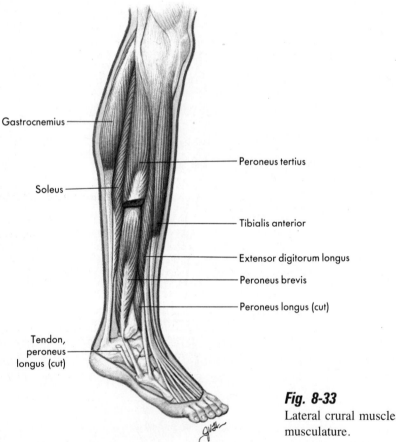

Fig. 8-33
Lateral crural muscles and associated musculature.

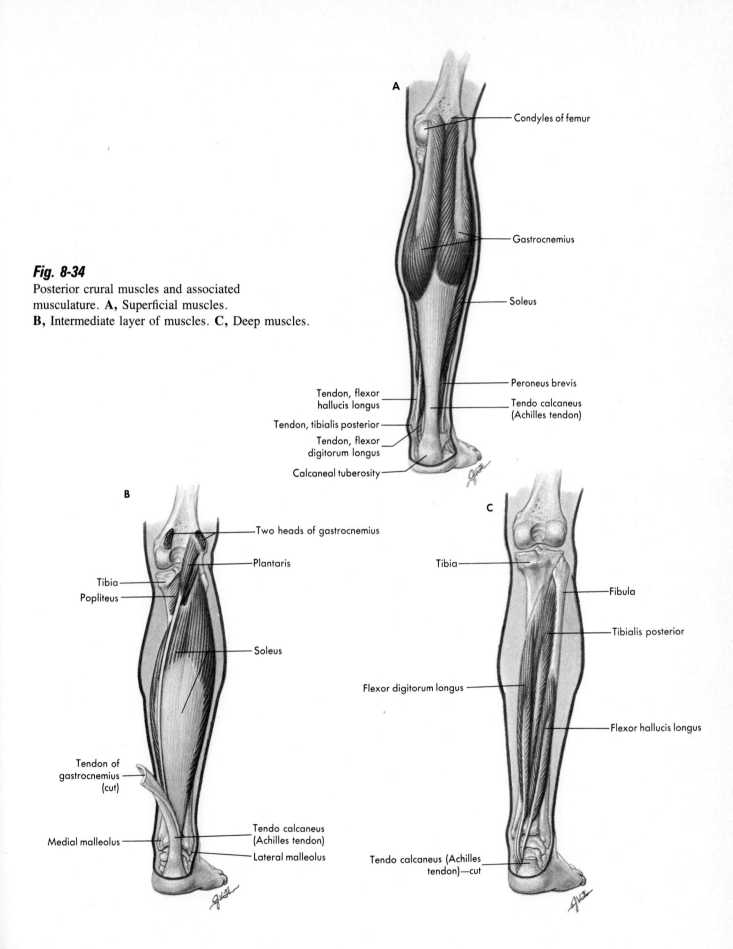

Fig. 8-34
Posterior crural muscles and associated musculature. **A,** Superficial muscles. **B,** Intermediate layer of muscles. **C,** Deep muscles.

the *tendo calcaneus* (Achilles tendon) inserts on the heel. It can flex the knee, but its most important action is in plantar flexion, as in giving "push off" for walking, running, or rising on the toes.

The *soleus* muscle is deep to the gastrocnemius and upper part of the Achilles tendon. It originates from the posterior upper parts of the tibia and fibula and inserts on the Achilles tendon. It acts with the gastrocnemius. The *plantaris* muscle is not present in all bodies. When it is, it rises from the lateral supracondylar ridge of the femur. Its long slender tendon joins the Achilles tendon at the back of the ankle to aid plantar flexion.

The *tibialis posterior* muscle occupies most of the posterior surface of the tibia and fibula. The tendon passes beneath the medial malleolus to insert on the medial cuneiform and navicular. It plantarflexes and inverts the foot and supports the medial longitudinal arch of the foot.

The *flexor digitorum longus* muscle rises from the posterior tibia. Its long tendon passes beneath the medial malleolus and divides at the midfoot into four tendons that insert on the plantar surfaces of the last phalanges of the four outer toes. The toes are flexed as a unit by the muscle, and it aids in balancing the body while standing and in giving "push" to walking.

The *flexor hallucis longus* muscle originates mainly from the posterior fibula and inserts on the plantar surfaces of the phalanges of the great toe. Aside from its obvious flexion of the great toe, the muscle aids in maintaining standing posture and helps maintain the medial longitudinal arch of the foot.

The crural muscles are presented in Table 8-16.

Table 8-16. Muscles moving the ankle and foot

Muscle	Origin	Insertion	Action	Innervation
Anterior crural				
Extensor digitorum longus	Lateral condyle of tibia and distal fibula	Second and third phalanges of 4 outer toes	Extends 4 outer toes	Anterior tibial nerve (spinal nerves L5, S1)
Extensor hallucis longus	Middle one third of anterior fibula	Top of second phalanx of great toe	Extends great toe	Anterior tibial nerve (spinal nerves L5, S1)
Tibialis anterior	Upper two thirds of lateral tibia	Medial cuneiform and base of first metatarsal	Dorsiflexion and inversion of foot	Anterior tibial and common peroneal nerves (spinal nerves L4, 5)
Peroneus tertius	Anterior fibula	Dorsal surface of fifth metatarsal	Dorsiflexes and everts foot	Anterior tibial nerve (spinal nerves L5, S1)
Lateral crural				
Peroneus brevis	Lower two thirds of fibula	Base of fifth metatarsal	Plantar flexion and eversion of foot	Common peroneal nerve (spinal nerves L5, S1)
Peroneus longus	Upper two thirds of fibula	Medial cuneiform and base of first metatarsal	Plantar flexion and eversion of foot	Common peroneal nerve (spinal nerves L5, S1)
Posterior crural				
Flexor digitorum longus	Posterior surface of tibial shaft	Distal phalanges of 4 outer toes	Flexes 4 outer toes	Posterior tibial nerve (spinal nerves S1, 2)
Flexor hallucis longus	Lower two thirds of fibula	Plantar surfaces of phalanges of great toe	Flexes great toe	Posterior tibial nerve (spinal nerves S1, 2)
Gastrocnemius	Posterior surfaces of femoral condyles	Calcaneus via Achilles tendon	Plantar flexion	Tibial nerve (spinal nerves L5, S1, 2)
Plantaris	Lateral supracondylar ridge	Calcaneus via Achilles tendon	Plantar flexion	Tibial nerve (spinal nerves L5, S1, 2)
Soleus	Upper portions of posterior tibia and fibula	Calcaneus via Achilles tendon	Plantar flexion	Tibial nerve (spinal nerves L5, S1, 2)
Tibialis posterior	Posterior shafts of tibia and fibula	Navicular and medial cuneiform	Plantar flexion and inversion of foot	Posterior tibial nerve (spinal nerves L5, S1)

Adapted from McClintic, J.R.: Basic anatomy and physiology of the human body, ed. 2. Copyright © 1980. Reprinted by permission of John Wiley & Sons, Inc.

Muscles within the foot

(Fig. 8-35)

The foot contains *interossei* and *lumbricales* muscles in an arrangement reminiscent of that in the hand. There are four *dorsal* and three *plantar* interossei. The dorsal interossei lie between the metatarsals—the first and second insert on the second toe; the third and fourth insert on the third and fourth toes. The plantar interossei rise from third to fifth metatarsals and insert on the third, fourth, and fifth toes. The second toe receives two dorsal interossei, the third and fourth a plantar interosseous *and* a dorsal interosseous, the fifth a plantar interosseous only, and the great toe receives *no* interossei.

The lumbricales muscles rise from the tendons of the flexor digitorum longus and are four in number. They insert on the medial sides of phalanx number one of the four outer toes. Both sets of muscles flex metatarsophalangeal joints and extend interphalangeal joints to help create a stable platform for the standing body.

The **quadratus plantae** (or *flexor digitorum accessorius*) muscle complements the flexor digitorum longus. It

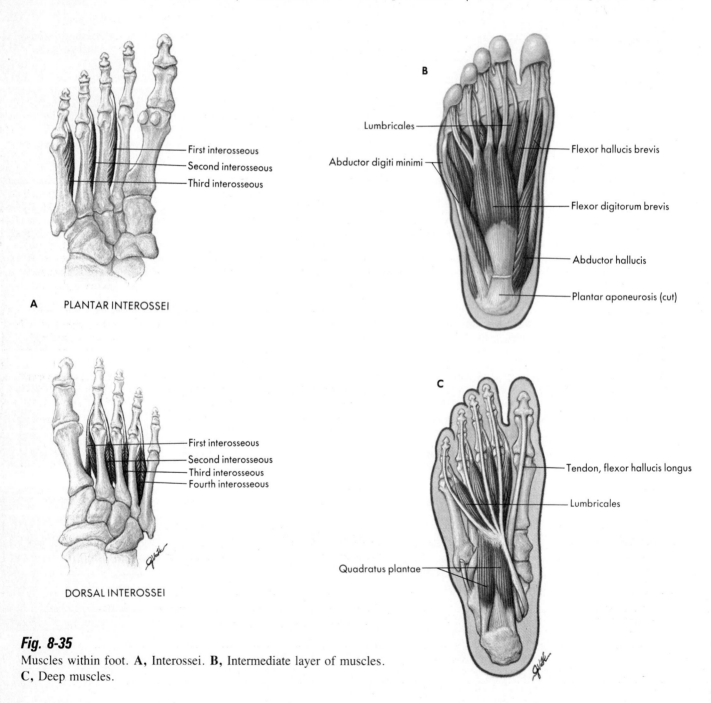

Fig. 8-35
Muscles within foot. **A,** Interossei. **B,** Intermediate layer of muscles. **C,** Deep muscles.

originates from the calcaneus and inserts by four tendons on the third phalanges of the four outer toes to flex them and give push off and support to the body.

The *abductor hallucis* muscle originates from the calcaneus and inserts on the base of the first phalanx of the great toe. The *adductor hallucis* muscle rises from the cuboid and second to fourth metatarsals and inserts with the abductor. The *flexor hallucis brevis* muscle rises from the cuboid and cuneiforms and inserts on the plantar surface of the first phalanx of the great toe. Basic actions are inherent in the names of the muscles; more importantly, the muscles work together to balance the body by stabilizing the great toe and assist in maintaining the arches of the foot.

The *flexor digitorum brevis* muscle aids flexion of the four outer toes, and the *abductor digiti minimi* muscle abducts the little toe.

These muscles are presented in Table 8-17.

No summary or questions are included with this chapter. Your knowledge of the muscles will be demonstrated by your ability to correctly name them and to give their origins, insertions, and actions.

Readings

Clemente, C.D.: Anatomy: a regional atlas of the human body, Philadelphia, 1975, Lea & Febiger.

McMinn, R.M.H., and Hutchings, R.T.: Color atlas of human anatomy, Chicago, 1977, Year Book Medical Publishers.

Table 8-17. Muscles within the foot

Muscle	Origin	Insertion	Action	Innervation
Abductor digiti minimi	Calcaneus	Lateral side, base of first phalanx of little toe	Abducts little toe	Lateral plantar nerve (spinal nerves S1, 2)
Abductor hallucis	Calcaneus	Base of first phalanx of great toe	Abducts great toe	Sciatic nerve (spinal nerves L5, S1, 2)
Adductor hallucis	Cuboid, metatarsals 2-4	Base of first phalanx of great toe	Adducts great toe	Sciatic nerve (spinal nerves L5, S1, 2)
Flexor digitorum brevis	Medial tubercle of calcaneus	Plantar surfaces of middle phalanges of four outer toes	Flexes four outer toes	Medial plantar nerve (spinal nerves L5, S1)
Flexor hallucis brevis	Cuneiforms and cuboid	Base of first phalanx of great toe	Flexes great toe	Sciatic nerve (spinal nerves L5, S1, 2)
Interossei Four dorsal	Dorsals: between metatarsals arising from facing surfaces faces of adjacent bones	First and second to sides of second toe, third and fourth to sides of third and fourth toes	Flex metatarsophalangeal joints and extend interphalangeal joints; Dorsal abduct and plantar adduct four outer toes	Common peroneal nerve (spinal nerves L5, S1)
Three plantar	Plantar: medial sides of third to fifth metatarsals	Medial side of corresponding toe		
Lumbricales (four muscles)	From tendons of flexor digitorum longus	Medial sides of first phalanges of the four outer toes	Flex metatarsophalangeal joints and extend interphalangeal joints	Common peroneal nerve (spinal nerves L5, S1)
Quadratus plantae (flexor digitorum accessorius)	Calcaneus	Distal phalanges of counter toes	Flex four outer toes	Posterior tibial nerve (spinal nerves S1, 2)

Adapted from McClintic, J.R.: Basic anatomy and physiology of the human body, ed. 2. Copyright © 1980. Reprinted by permission of John Wiley & Sons, Inc.

9

The nervous system

Objectives
Development of the nervous system
Organization of the nervous system
   Central nervous system
   Peripheral nervous system
Cells and tissues of the nervous system
   Neurons
   Glial cells
Synapse
Reflex arc
Nerves
Central nervous system
   Brain
   Spinal cord
   Meninges
   Ventricles and cerebrospinal fluid
Peripheral nervous system
   Cranial nerves
   Spinal nerves and plexuses
   Autonomic nervous system
Summary
Questions
Readings

OBJECTIVES After studying this chapter, the reader should be able to:

Outline the basic steps in the development of the nervous system.
Name the portions of the nervous system.
Describe the neuron and glial cell types that compose the nervous system.
Describe the structure of a synapse and a reflex arc.
Describe the construction of a large peripheral nerve.
Describe the gross and microscopic structure of the cerebrum, cerebellum, brainstem, and spinal cord.
Discuss the structure, location, and functions of the meninges around the central nervous system.
Locate and describe the form of the brain ventricles and the production and circulation of cerebrospinal fluid.
Name and give the peripheral distributions and functions of the cranial nerves.
Describe the organization of the autonomic nervous system.

Development of the nervous system

Conception, or fertilization of an ovum (egg) by a sperm, initiates the development of a new individual. The zygote, or fertilized ovum, next undergoes a series of mitotic divisions that forms first a solid morula and next a structure containing a mass of cells in one area and a cavity in the other area. It is known as a blastocyst. Within the mass of cells (the inner cell mass) in the blastocyst, two cavities appear, an upper gut cavity closest to the surface of the blastocyst and a deeper lying amniotic cavity. Between them a two-layered plate of cells forms, known as the *embryonic disc.* The layer of cells closest to the amniotic cavity is called the *ectoderm,* and the layer closest to the gut cavity is the *endoderm.* From the ectoderm will develop the epidermis of the skin and the entire nervous system. The length of time for this two-layered embryo to form is about 2 weeks.

At about 18 days of embryonic development, there appears on the ectodermal side of the embryonic disc an area of thickened ectoderm known as the *neural plate.* Formation of the nervous system begins at and from this plate. A groove, the *neural groove,* appears in the plate, its lips enlarge to form two *neural folds,* one on each side of the groove, and further growth of the folds then causes them to meet and fuse in the midline. A hollow tube,

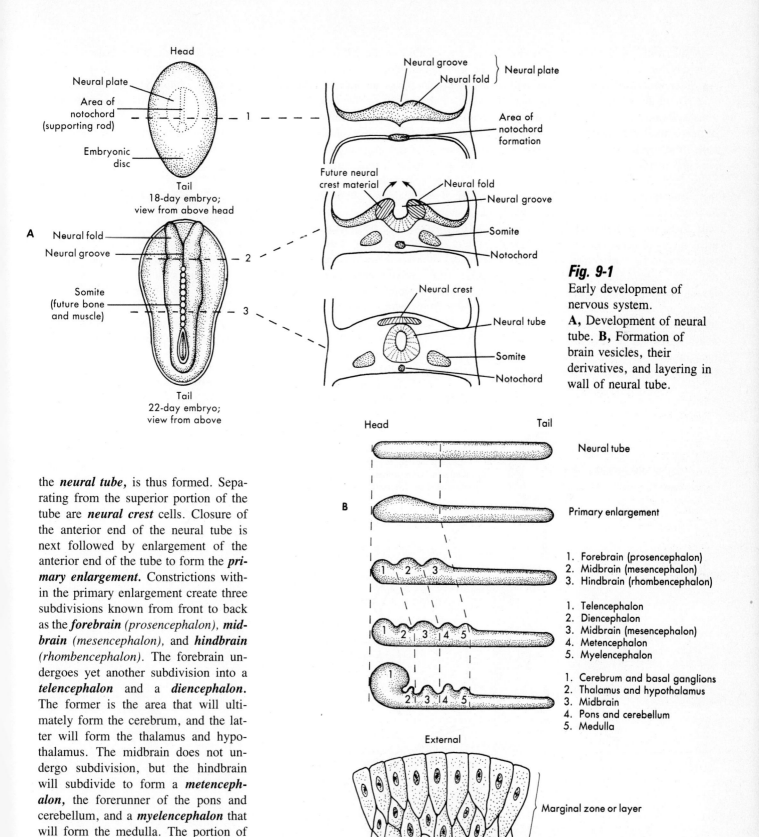

Fig. 9-1
Early development of nervous system. **A,** Development of neural tube. **B,** Formation of brain vesicles, their derivatives, and layering in wall of neural tube.

the *neural tube,* is thus formed. Separating from the superior portion of the tube are *neural crest* cells. Closure of the anterior end of the neural tube is next followed by enlargement of the anterior end of the tube to form the *primary enlargement.* Constrictions within the primary enlargement create three subdivisions known from front to back as the *forebrain (prosencephalon), midbrain (mesencephalon),* and *hindbrain (rhombencephalon).* The forebrain undergoes yet another subdivision into a *telencephalon* and a *diencephalon.* The former is the area that will ultimately form the cerebrum, and the latter will form the thalamus and hypothalamus. The midbrain does not undergo subdivision, but the hindbrain will subdivide to form a *metencephalon,* the forerunner of the pons and cerebellum, and a *myelencephalon* that will form the medulla. The portion of the tube posterior to the myelencephalon will become the spinal cord. These basic steps are depicted in Fig. 9-1.

In the walls of the neural tube the cells develop a three-layered structure. An outer **marginal zone** or *layer* will give rise to the white matter of the nervous system. This will consist mostly of nerve fibers carrying a fatty covering called the myelin sheath. In the spinal cord the marginal zone remains essentially free of cellular elements, but in the telencephalon and cerebellum it will be invaded by neuroblasts, cells that will form nerve cells. Thus in these areas layers of nerve cells (gray matter) will come to lie *outside* the white matter. An **intermediate zone** (*mantle layer*) develops neuroblasts that will become nerve cells. This zone remains internal to the white matter (marginal zone) in the spinal cord and serves as the source of cells that migrate to form the outer gray matter of the cerebrum and cerebellum. An inner layer of cells, the **ventricular zone** (*ependymal layer*), will remain as an inner epithelial lining of the cavity of the neural tube and its derivatives, the ventricles of the brain. In several regions of the tube, neuroblasts do not migrate and remain as the cells that will form nuclei of cranial and some peripheral nerves and the nuclei of the thalamus, hypothalamus, and cerebrum.

The fibers of the peripheral nervous system (Fig. 9-2) develop from several sources. Neurons that carry sensory impulses from outlying areas to (*afferent*) the central nervous system (brain/cord) develop from neural crest neuroblasts. From the crest one process grows peripherally, whereas another process grows centrally. Peripheral ganglia also develop from neural crest material. Outgoing (efferent) fibers that will supply muscle and glands originate from neuroblasts *within* the neural tube.

Although brief, this section should acquaint the reader with basic developmental processes and some essential terminology.

Organization of the nervous system

In the previous section, terms such as central nervous system and peripheral nervous system were used. For purposes of description and also for anatomical and functional reasons, the following divisions of the nervous system are described and depicted in Fig. 9-3.

Central nervous system

The central nervous system (CNS) consists of those organs that are surrounded by the bony structures of the skull and vertebral column, that is, the **brain** and **spinal cord**. These areas form the regions for integration, interpretation, thought, and transmission of messages to and from the periphery.

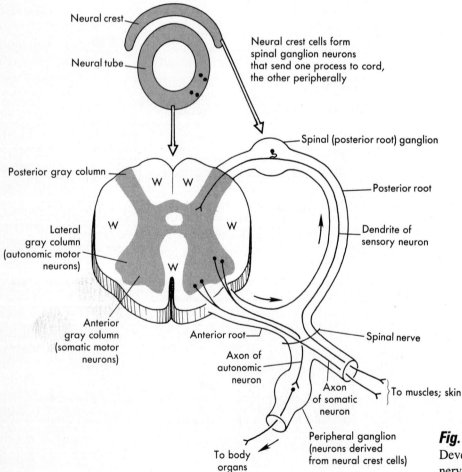

Fig. 9-2

Development of spinal portion of peripheral nervous system (*W*, white matter).

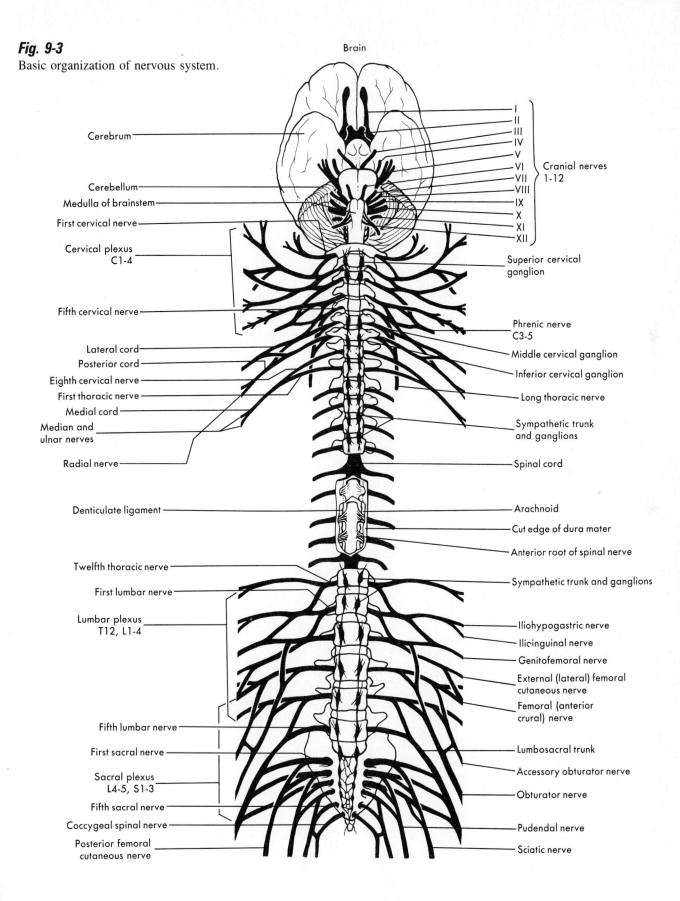

Fig. 9-3
Basic organization of nervous system.

Peripheral nervous system

Consisting of all nervous tissue lying outside the bony structures of skull and spinal column, the peripheral nervous system (PNS) provides a means of detection of changes in internal and external environments of the body, transmission of that information to the CNS for action, and then delivery of messages to muscles and glands for response. This portion includes not only the nerve fibers that carry impulses but also groupings of fibers (plexuses) and nerve cell bodies (ganglions) that are found in the periphery.

More specifically, the 12 pairs of **cranial nerves** that attach to the brain and their associated ganglions, the 31 pairs of **spinal nerves** and their ganglions, and the specialized **receptors** and **endings on muscle** make up the PNS.

Functionally speaking, the PNS is composed of **somatic fibers** *(somatic nervous system)* that innervate skeletal muscle and the skin's special receptors (for touch, pressure, heat, cold) and **autonomic fibers** *(autonomic nervous system)* that carry impulses *to* smooth and cardiac muscle and glands, and *from* visceral receptors. These fibers are involved in a series of reflexes that control breathing, heart rate, blood pressure, and other body functions in an automatic and continuous manner. Most body organs receive fibers from two functional subdivisions of the autonomic system. Formed from certain cranial nerves and several of the sacral spinal nerves, the *parasympathetic (craniosacral) division* provides nervous influences designed to conserve body resources and maintain normal levels of function. Formed from the thoracic and lumbar spinal nerves, the *sympathetic (thoracolumbar) division* provides impulses that result in elevation of body activity designed to tolerate or resist stressful or dangerous situations. Thus an organ may have its activity stepped up or slowed down according to what is required for survival or tolerance.

Cells and tissues of the nervous system

Primitive ectodermal cells of the embryo, called *neuroblasts*, give rise to the highly excitable and conductile cells of the nervous system designated as *neurons*. Originating primarily from ectoderm, cells known as *spongioblasts* give rise to a variety of cells called *glia* that are regarded as connective, supportive, and nutritive cells of the nervous system.

Neurons

Neurons occur in many diverse forms within the body (Fig. 9-4). One way to describe them is on the basis of the number of processes that extend from the cell body. According to this method, these are described:

unipolar neurons (see Fig. 9-4, **f**) have only a single process extending from the cell body. Found mainly in the PNS, these cells are involved in transmitting *sensory* information *to* the CNS. Some may be derived from neuroblasts that originally have two processes (hence bipolar, see next paragraph), and in which the two processes eventually fuse to form a single process. Neurons acquiring a single process in this manner are sometimes called *pseudounipolar*, since they did not begin life with a single process.

bipolar neurons (see Fig. 9-4, **d**) have two processes, typically from either end of an elongated cell body. Found mainly in organs of *special sense* (eye, ear, taste buds, olfactory epithelium), bipolar cells often have one of their processes highly modified to respond to particular forms of energy.

multipolar neurons have more than two processes from the cell body. Such neurons are most varied in terms of size and shape (see Fig. 9-4, **a-c, e, g, h**) and are mainly concerned with processing or transmitting *motor impulses* that lead to muscular contraction or gland secretion.

Neuron cell bodies vary from 5 to 135 μm in diameter. Their processes may be short or extend for several feet. Small short-process neurons are sometimes called *Golgi II* neurons; the larger ones with at least one long process are often called *Golgi I* neurons.

In terms of the more detailed structure of a neuron, I shall next describe a multipolar neuron of the type that extends from the spinal cord to skeletal muscle and a peripheral sensory neuron (Fig. 9-5).

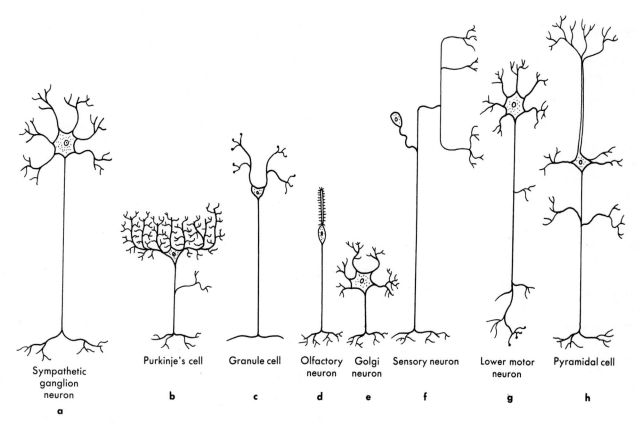

Fig. 9-4
Some different forms of neurons from human nervous system (not to same scale).

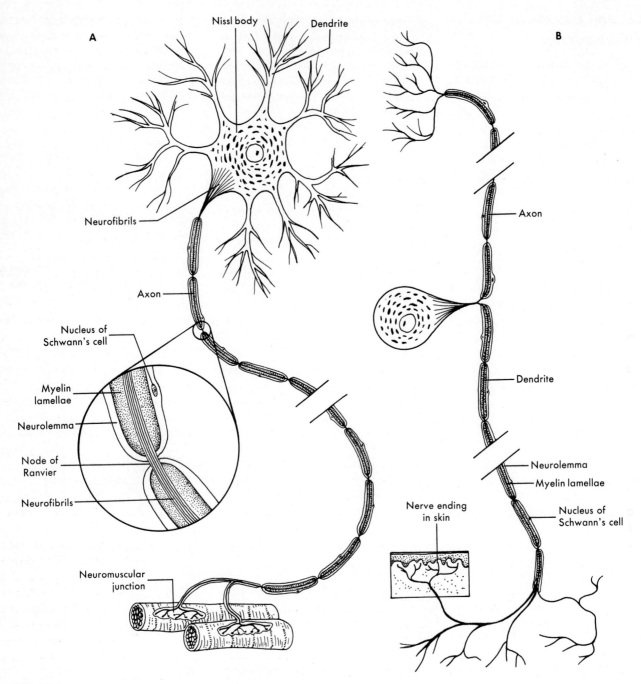

Fig. 9-5
Two major types of neurons in body using peripheral sensory neuron and lower motor neuron as examples. **A**, Motor neuron. **B**, Sensory neuron.

The cytoplasm of the neuron, its *perikaryon,* contains the usual cellular organelles, such as endoplasmic reticulum (ER), Golgi body, mitochondria, and ribosomes. There are conspicuous large clumped or granular structures in the cytoplasm known as **Nissl bodies.** They appear to be fragments of ER studded with ribosomes and are arranged in rows or in orderly fashion within the cytoplasm. They are unique to neurons and are probably concerned with protein synthesis by the cell. Also unique to neurons are **neurofibrils,** microtubules that may give structural strength to the cell body or that may carry materials throughout the neuron. Lacking or nonfunctional in mature neurons is the cell center; a missing or nonfunctional center implies inability of the mature neuron to reproduce itself.

The nucleus of the neuron is large, has a thin membrane, and usually has a large, round, obvious nucleolus. Chromatin is fine and usually uniformly distributed within the nucleus.

The processes of the multipolar neuron are of two types. **Dendrites** are multiple, highly branched, irregular processes that end close to the cell to which they are attached. They are *afferent* processes, conducting nerve impulses *toward* the cell body. A single long, sparsely branched, uniform diameter process, the **axon,** exits from the *axon hillock* of the perikaryon. The hillock is an area of the perikaryon that generally lacks cellular organelles and appears more clear than other regions of the cell cytoplasm. Neurofibrils pass into and through the length of the axon but are lacking in the dendrites. The axon is *efferent,* carrying impulses *away from* the cell body.

On the unipolar neuron there is a *peripheral process* extending from outlying areas to the cell body and a *central process* from the cell body to the CNS. *Both* processes of this neuron are structured like axons, that is, sparsely branched and of uniform diameter. The peripheral process *functions* as a dendrite (afferent), the central process as an axon (efferent).

Axons of multipolar neurons and both processes of unipolar neurons may be supplied with one or more **sheaths.** One such sheath lies next to the process and consists of layers or *lamellae* of membranes alternating with a lipid called *myelin* (Fig. 9-6). This covering is called the **myelin sheath,** and a process having such a sheath is said to be *myelinated* or *medullated.* Processes without these layers but that commonly have a thin layer or two of lipid molecules are said to be *nonmyelinated* or *nonmedullated* (see Fig. 9-6). On myelinated fibers the sheath is not continuous but is interrupted at intervals of 2 to 3 mm by **nodes** *(of Ranvier).* Where a side branch of the process occurs (a *collateral*), it always arises at a node. The portion of myelin between two nodes is termed an **internode.** The source of the myelin lamellae is one of the glial cells called **Schwann's cell** that is believed to wrap around the neuron process, laying down the lamellae as it goes (Fig. 9-7). The outermost layer of Schwann's cell membrane and cytoplasm forms the unbroken tubelike **neurilemma** *(neurolemma)* or **Schwann's sheath.** Nuclei of Schwann's cells may be seen at intervals along the course of the neurilemma.

The myelin sheath permits a nerve impulse to travel more rapidly along the process than if the sheath is lacking. The neurilemma tends to persist if the process is damaged and may serve as a guiding tube for regeneration of the process (axons can regenerate; if the cell body is destroyed, the whole unit dies).

Both sheaths are present on peripheral somatic neurons; fibers in the CNS have only the myelin sheath. Peripheral autonomic fibers usually possess a neurilemma only, and many fibers *within* organs lack sheaths, although the fibers from which they are derived have sheaths. In other words, the coverings are lost as the process enters the organ.

It was stated previously that neurons are the highly excitable and conductile cells of the nervous system. Excitability is a fundamental property of all cells, with neurons being more excitable than other cells. The *degree* of excitability is usually a function of the height of the transmembrane potential, an electrical difference that can be measured across a cell membrane by placing tiny recording electrodes inside and outside the cell. This difference or potential is about 70 millivolts (mv) on nerve cells, about half again larger than potentials on other cells.

Physiology of neurons
Basis of excitability. The excitable state seems to depend primarily on the operation of active transport systems within the cell membrane that bring about a separation of electrically charged atoms (ions) according to both type and amount. Active transport systems involve a molecule (carrier) that can attach to a particular ion and move it one direction or the other through the membrane. Attachment of the ion usually requires an *energy source* (adenosine triphosphate [ATP]) and *enzymes.* All three components of such a system involve work by the cell to synthesize the materials, hence the designation as an active process.

In a neuron, active transport tends to deposit large amounts of sodium and chloride ion *outside* the cell, while keeping potassium ion *inside* the cell. At rest, that is, when the neuron is *not* forming or transmitting impulses, the separation of sodium (Na), chloride (Cl), and potassium (K) ions is as follows:

	Outside cell	Inside cell
Sodium	145*	12
Chloride	120	4
Potassium	4	155

Also present *inside* the cell is about 155 mEq of large (usually protein) molecules that also are negatively ionized and too large to move through the membrane by any method. If we make

*Concentrations in mEq/L.

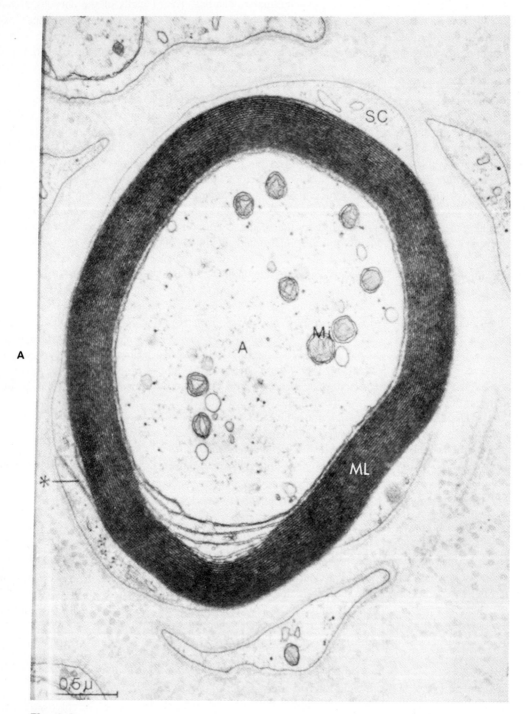

Fig. 9-6
Biopsies of human sural nerve. **A,** Myelinated axon.

From Babel, J., Bischoff, A., and Spoendin, H.: Ultrastructure of the peripheral nerve system and sense organs: atlas of normal and pathologic anatomy, St. Louis, 1970, The C.V. Mosby Co.

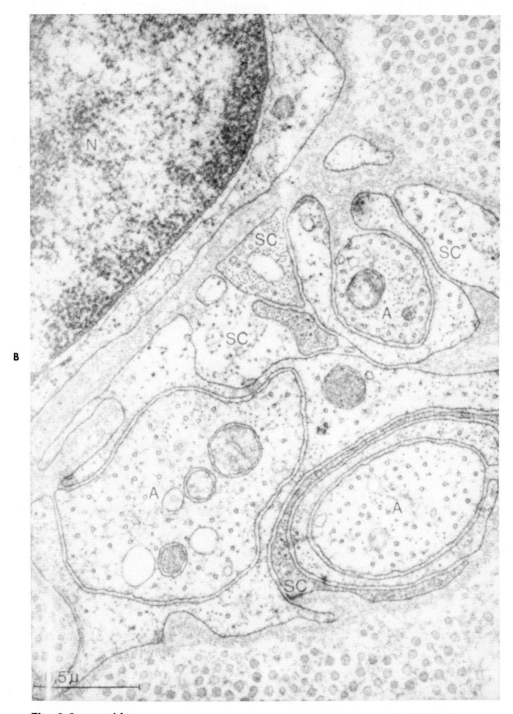

Fig. 9-6, cont'd
B, Nonmyelinated axons, several in group surrounded by single Schwann's cell. *A,* Axon; *M,* mitochondria; *SC,* Schwann's cell cytoplasm; *ML,* myelin lamellae; *N,* nucleus of Schwann's cell. *Indicates overlapping of Schwann's cell cytoplasm.

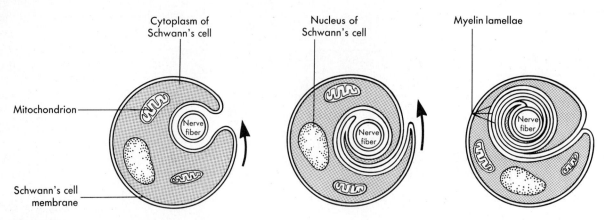

Fig. 9-7
Formation of myelin lamellae around axon in peripheral nervous system. Schwann's cell is believed to "wrap around" axon.

a simple calculation of charges inside and outside, it becomes apparent that there are *more* positively charged substances outside the cell than inside. For example:

Outside	Inside
Na + K = 149$^+$	Na + K = 167$^+$
Cl = 120$^-$	Cl + Protein = 159$^-$
∴ 29$^+$	∴ 8$^+$

In this condition, with an excess of positive charge outside the cell, the membrane is said to be *polarized*, with the outside of the cell electrically positive to the inside.

Formation of a nerve impulse. When stimulated, the cell loses its ability to keep the sodium outside the fiber, as if the transport system had failed. Permeability (ease of passage) to sodium ion increases about 500 times, that to potassium about 40 times. Although both ions can move more freely, potassium ions tend to be held inside the cell by the negatively charged proteins, and so large amounts of sodium ion diffuse into the cell, making for an excess number of positive charges inside the cell. In this state the cell is said to be *depolarized*.

A depolarized area now lies adjacent to a still polarized area:

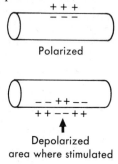

Using the ion-laden fluids inside and outisde the cell, a current flows between the polarized and depolarized areas like batteries connected with wires:

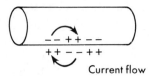

This current flow creates an electrical field that is the *nerve impulse*.

Transmission of the impulse. To cause the electrical field to move along the membrane, we have only to realize that the current acts as a stimulus to cause depolarization of the adjacent polarized membrane—it repeats the events just described, current flows that depolarizes the *next* segment of membrane, and the whole disturbance moves down the membrane. After moving away from a point of stimulation, the electrical field no longer inhibits the transport system; it pumps the sodium out of the cell, returning the membrane to its original state; that is, the membrane is *repolarized*. From time of stimulation to repolarization, most neurons take 1 millisecond (msec, $1/1000$ second) so that the changes occur very rapidly.

Properties of the neuron. The sequence of events described confers certain properties on neurons that enable them to more efficiently carry out their tasks. Several of the more important properties are:

They follow the all-or-none law. If a stimulus is strong enough to cause depolarization, it causes complete depolarization *(all)*; if it is not strong enough, there is no depolarization *(-or-none)*. There is no halfway activ-

Fig. 9-8
Some types of glia in central nervous system. *Above,* Ependyma and neuroglia in region of central canal of child's spinal cord: *A,* ependymal cells; *B* and *D,* fibrous astrocytes; *C,* protoplasmic astrocytes. Golgi method. *Below,* Interstitial cells of central nervous system: *A,* protoplasmic astrocyte; *B,* fibrous astrocyte; *C,* microglia; *D,* oligodendroglia.

ity here; an impulse strong enough to cause something to happen later on will result.

Recovery time after stimulation is very short. This enables a neuron to be ready to transmit more information very quickly after being stimulated once before. Further transmission is quickly needed to ensure proper function.

The strength of the conducted impulse does not decrease as it passes along a nerve fiber. Therefore the impulse will not "fizzle out" before it gets where it is going (often many feet along a nerve pathway).

A myelin sheath on a nerve fiber enables a more rapid speed of conduction (by 10 to 100 times) than if the sheath is absent. The term **saltatory** conduction is applied to conduction along a myelinated fiber. The events are as described previously, but current flow can only occur between the nodes that are placed several millimeters apart in the sheath. The impulse "jumps" from node to node instead of depolarizing small lengths of membrane. This may be likened to covering a given distance with long strides instead of tiny steps—a person will get there faster.

Glial cells

Glial cells outnumber neurons in the nervous system by about five to one. They have processes but do not have axons or dendrites and do not conduct nerve impulses. They do possess the ability to divide throughout life and possess nuclei and the organelles of all actively metabolizing cells (mitochondria, ER, ribosomes, and so forth). Seven types of glial elements are found in the nervous system (Fig. 9-8; see also Fig. 9-5). The seven types are astrocytes—protoplasmic (mossy) and fibrous types, oligodendroglia, microglia, satellite cells, ependyma, and Schwann's cells.

Astrocytes are star-shaped cells that form transversely oriented networks supporting nerve fibers of the CNS. Astrocytes also form coverings on about 85% of the surfaces of cerebral capillaries and may be important in transferring nutrients to neurons. The protoplasmic astrocyte has many highly branched processes, whereas the fibrous astrocyte has fewer and less highly branched processes.

Oligodendroglia resemble "plump pillows" with a major process from each corner of the pillow. These cells are oriented longitudinally along fibers in the CNS and produce the myelin sheaths of CNS fibers.

Microglia are small cells with a flattened body, and several branched processes, studded with spinelike *gemmules,* exit from the cell body. These cells develop from mesenchyme (mesodermal) cells that invade the CNS. When injury or infection in the CNS occurs, these cells become active (ameboid) and phagocytic, aid in fighting inflammation, and help wound healing.

Satellite cells form capsulelike coverings around peripheral ganglions and may also transfer nutrients to neurons.

Ependyma are wedge-shaped ciliated cells that line the cavities (ventricles and central canal) of the CNS and that produce small amounts of cerebrospinal fluid. The nervous system develops as a tube, hence the epithelial lining.

Schwann's cells have already been mentioned as the peripheral myelin-producing cells of the neurilemma and myelin sheath.

Synapse

Where one neuron connects to another neuron, the junction is termed a *synapse*. The neuron that carries an impulse *to* the synapse is termed a **presynaptic neuron,** whereas the one carrying the impulse *away* from the synapse is a **postsynaptic neuron.** The synapse is an area of functional but *not* anatomical continuity between the two neurons. The two parts of the synapse are separated by a cleft about 150 angstrom units (Å) wide.

The junction may take several forms (Fig. 9-9). It may occur between the axon terminations of the presynaptic neuron and the dendrites, body, or axon of the postsynaptic neuron. Conduction across the synapse is chemical, with the chemical being released by the axons of the presynaptic neuron to act on the membranes of the postsynaptic neuron.

The synaptic transmitter chemicals are synthesized by the axon terminals of the presynaptic neuron and are stored as membrane-surrounded **vesicles** in the axon endings. A nerve impulse causes the release of the chemical that diffuses across the gap (cleft) be-

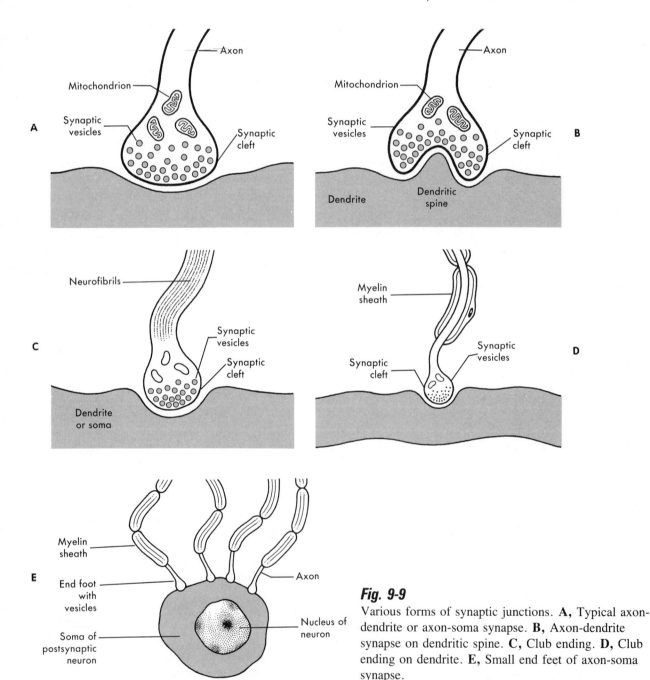

Fig. 9-9
Various forms of synaptic junctions. **A,** Typical axon-dendrite or axon-soma synapse. **B,** Axon-dendrite synapse on dendritic spine. **C,** Club ending. **D,** Club ending on dendrite. **E,** Small end feet of axon-soma synapse.

tween the two neurons and binds to receptor sites on the postsynaptic neuron. The binding serves as a stimulus to cause depolarization of the postsynaptic neuron's membranes. A chemical matched to the transmitter substance is always present to destroy the transmitter after it has exerted its effect. This prevents continued stimulation by the transmitter and allows more messages to cross the synapse.

This type of transmission endows the synapse with different properties from the neurons that compose it. Several of these properties are:

Synapses conduct in only one direction. The transmitter chemical is released from only the presynaptic axon. Thus there are no "short circuits" or wrong-way transmission of information. According to the chemical produced, transmission may be enhanced *(facilitation)* or reduced *(inhibition).* The synapse thus is a point at which *control* may be established on the impulse's further progress.

Synaptic transmission is slower than that along a nerve fiber. It takes a measurable time per synapse (about 0.5 to 1 msec) for chemical release, diffusion, and binding before the impulse can continue its course.

Synapses are very sensitive to their chemical environment. Oxygen lack, change in pH (acidity), and many drugs may alter synaptic function. Pain-killing drugs, anesthetics, and hypnotics (sedatives) reduce transmission ("downers"); others increase transmission ("uppers").

Reflex arc

To achieve a relationship between a change in the internal or external environments of the body and a response to those changes, reflex arcs (Fig. 9-10) have been established. An arc may be characterized as the simplest *controlling* unit of the body, and it requires at least two neurons. The parts of the arc include the following:

receptor responds best to some particular form of energy or change, such as light, sound, pressure, or temperature.

afferent (sensory) neuron is one whose peripheral process is connected to the receptor and whose central process carries the impulse to the CNS.

center or *synapse* occurs within the CNS, providing a connection that conveys the impulse to an efferent neuron.

efferent neuron has its cell body in the CNS and may reach to the organ that is to respond or to a peripheral synapse, whence a second neuron reaches the effector.

effector is usually a type of muscular or glandular tissue that responds in a manner appropriate to maintaining body homeostasis.

Any reflex, or reflex act, to be of value to the body, must possess certain characteristics:

The arc must be complete with no breaks in it; otherwise the pathway is interrupted, and no response will occur.

Reflex activity is involuntary but can often be terminated voluntarily. We would not want or be able to voluntarily monitor the blood pressure, heart and breathing rates, gut activity, and so forth. Vital activities are handled continuously and automatically without our thinking about it.

Reflex activity is stereotyped; that is, a given reflex always has the same result. There are different reflexes for different results, but the same thing happens if a particular reflex is triggered. This obviously results in constancy of a particular activity.

Reflex response is always predictable, a fact we can use to *test* for integrity of a reflex arc.

Reflex response is always purposeful, designed for a useful response that is of value to the body.

Since many examples of reflex control of body function will be cited in later chapters, it is necessary to understand the structure and function of the arc and its result.

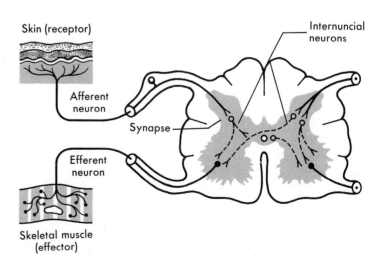

Fig. 9-10
Components of reflex arc.

Nerves

The axons or dendrites of individual neurons are microscopic structures. Bundles of these nerve fibers are bound together into larger peripheral structures that are visible to the naked eye. Such bundles are referred to as *nerves* (Fig. 9-11).

A fine sheath of reticular and collagenous fibrils surrounds each nerve fiber and binds them to one another. This tissue constitutes the *endoneurium*. In large nerves groups of fibers called *fascicles* are bound together by collagenous tissue called the *perineurium*. Several fascicles may be bound into the whole nerve by the collagenous tissue of the *epineurium*. These various levels of connective tissue are continuous with one another and are distinguished by *position* within the nerve, not by type of tissue. The endoneurium and perineurium may also contain blood vessels and lymphatic vessels that supply nutrients or remove fluids and metabolic wastes from the nerve. The epineurium, as the outer heavy sheath of connective tissue, may pass these structures but does not have networks of vessels in it.

Although mature neurons cannot undergo mitosis, it is possible for damaged axons or dendrites to regenerate. Peripheral regeneration is most common; regeneration in the CNS seems possible, but the chance of reestablishing anything resembling the original connection is remote because of the lack of a neurilemma and because of the millions of fibers in the brain and spinal cord. The comments to follow are therefore directed to regeneration in the PNS.

Damage to a nerve fiber brings about changes in the cell body (the *axon reaction*) and in the fiber distal to the point of injury *(wallerian degeneration)*. The axon reaction results in swelling of the cell body and fragmentation of the Nissl bodies. In wallerian degeneration the distal portion of the fiber swells, the myelin sheath undergoes dissolution, and the axon degenerates, leaving only the outer neurilemma or other glial elements as a sheath. Blood cells and glial cells remove the products of degeneration. These processes take about 3 weeks to complete. Next the proximal portion of the axon develops an ameboid, bulbous tip that grows through the neurilemmal tube at about 2 mm/day. The axon can eventually reach the area it formerly served and reestablish innervation of the organ. The new connections seem further apart or more at random than originally, so control of function is not as precise as it was.

Multiple sclerosis is one example of a group of *demyelinating diseases* characterized by loss of myelin sheaths in the brain and spinal cord. Since the myelin sheaths not only enable faster conduction but also largely insulate one fiber from its neighbor, loss of sheaths allows impulses to short out neighboring fibers. Disturbances in motor function, sensory function, and reflex activity are common.

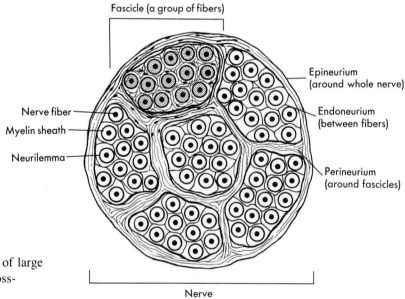

Fig. 9-11
Connective tissue components of large peripheral nerve as seen in cross-section.

Central nervous system

The CNS consists of the brain and spinal cord. These organs form the areas for ultimate reception of stimuli and their interpretation or analysis and serve as the initiators for responses that maintain whole body homeostasis.

Brain

The brain, defined as that part of the CNS enclosed within the skull, has three major portions. The *cerebrum* accounts for the greater portion of the 1.4-kg weight of the brain. The *cerebellum* is the next smaller part, and the *brainstem* is the smallest portion (Fig. 9-12).

Cerebrum. The cerebrum consists of two halves or *hemispheres* and presents a convoluted or folded appearance. The upfolds are designated as *gyri* (sing., *gyrus*) and the shallow downfolds are named *sulci* (sing., *sulcus*). In certain areas of the cerebrum, deep downfolds called *fissures* separate the cerebrum into its hemispheric *lobes*.

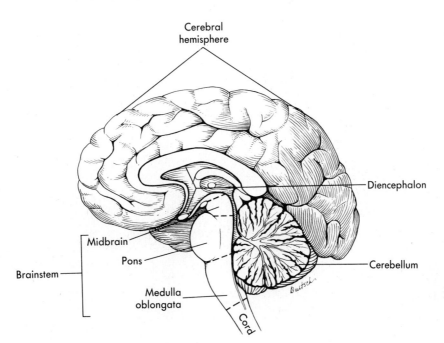

Fig. 9-12
Major parts of brain as seen in midsagittal section.

Major fissures and sulci (Fig. 9-13). The *longitudinal fissure* runs from anterior to posterior in the midline and separates the right cerebral hemisphere from the left. Within a given hemisphere there are several fissures. The *lateral fissure (of Sylvius)* runs from anterior to posterior on the lateral side of the hemisphere. A *central fissure (central sulcus* or *Rolando's fissure)* runs in the coronal plane about halfway from the anterior edge of each hemisphere. The *parietooccipital fissure* runs coronally near the posterior pole of the hemisphere and is more pronounced on the medial side of the hemisphere than on the lateral side. The *calcarine fissure* runs roughly from anterior to posterior on the medial side of the posterior part of a hemisphere. The *transverse fissure,* although not part of the cerebrum, may be noted as the fissure lying between the cerebral hemispheres above and the cerebellum, brainstem, and diencephalon below.

Lobes of the cerebrum. The extent of the lobes of the cerebrum can be appreciated by viewing the lateral, medial, and basal aspects of the cerebral hemispheres.

The *frontal lobe* is that portion of the cerebral hemisphere anterior to the central fissure. The lobe rests in the anterior cranial fossa of the skull. The *parietal lobe* lies between the central fissure anteriorly and the parietooccipital fissure posteriorly, with the lateral fissure forming its lateral boundary. This lobe occupies the superior two thirds of the middle cranial fossa. The *occipital lobe* lies posterior to the parietooccipital fissure and occupies the posterior part of the middle cranial fossa. The *temporal lobe* is set off by the lateral fissure. Its posterior limit is determined by an arbitrary line drawn along the plane of the parietooccipital fissure and its intersection with a similar line extending from the posterior limit of the lateral fissure. The *insula (island of Reil)* is considered by some to be an additional lobe. It is attached to the inferior surface of the frontal lobe and is hidden from view by the anterior third of the temporal lobe.

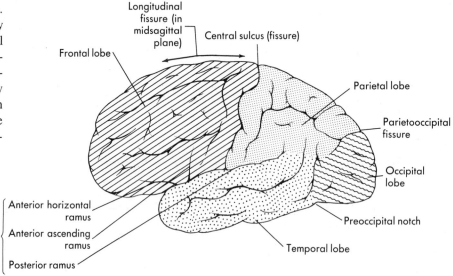

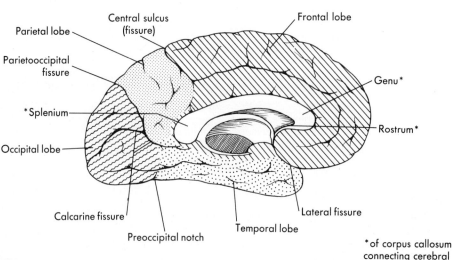

Fig. 9-13
Major fissures, sulci, and lobes of cerebrum.

Gyri and secondary sulci of the cerebrum (Fig. 9-14). The convolutions of the cerebrum form rather well-defined gyri that may in many cases reflect functional areas of the cerebrum.

The precentral gyrus occupies the dorsolateral aspect of the frontal lobe ahead of the central fissure. Anterior to this gyrus is the **precentral sulcus** that sets off the rest of the frontal lobe. This remaining frontal lobe area is subdivided into *superior, middle,* and *inferior frontal gyri* by the *superior* and

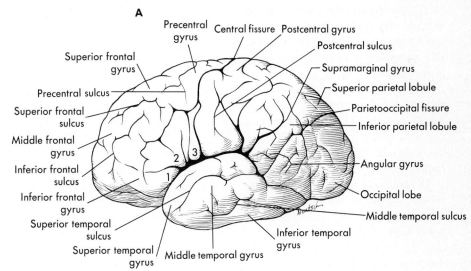

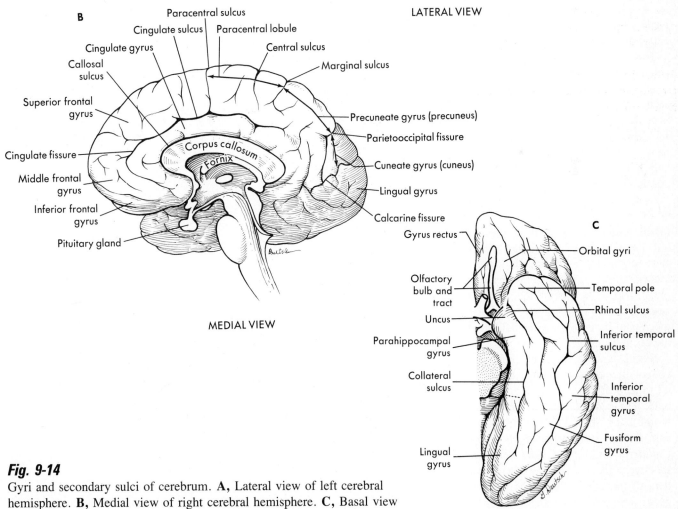

Fig. 9-14
Gyri and secondary sulci of cerebrum. **A,** Lateral view of left cerebral hemisphere. **B,** Medial view of right cerebral hemisphere. **C,** Basal view of left cerebral hemisphere. *1, 2,* and *3* indicate orbital, triangular, and opercular portions of inferior frontal gyrus respectively.

inferior frontal sulci. The inferior frontal gyrus is divided into three parts by the ***anterior horizontal*** and ***anterior vertical rami*** of the lateral fissure. The three parts of this gyrus are designated the *opercular, triangular,* and *orbital portions* from posterior to anterior.

The parietal lobe contains the ***postcentral gyrus*** immediately behind the central fissure, and the ***postcentral sulcus*** sets off the remainder of the lobe that contains ***superior*** and ***inferior parietal lobules***.

The temporal lobe contains ***superior, middle,*** and ***inferior temporal gyri,*** separated by ***superior*** and ***inferior temporal sulci***. The inferior gyrus extends onto the basal surface of the lobe. On the upper surface of the superior temporal gyrus, several ***transverse temporal gyri*** extend to the floor of the lateral fissure.

The occipital lobe has no gyri or sulci of special significance.

On the medial and basal surfaces of a hemisphere, there are several prominent gyri and sulci. The ***cingulate gyrus*** lies over the *corpus callosum,* a band of fibers connecting the cerebral hemispheres across the midline. The ***cingulate sulcus*** separates the gyrus of the same name from the superior frontal gyrus anteriorly and the ***paracentral lobule*** posteriorly. Behind the paracentral lobule lies the ***marginal sulcus***. Between the latter sulcus and the parietooccipital fissure lies the *precuneate gyrus,* part of the parietal lobe. The ***cuneate gyrus*** lies posterior to the parietooccipital fissure on the medial side of the occipital lobe.

Basally the hemisphere presents a convolution extending from the occipital pole nearly as far as the temporal pole. The posterior part of this convolution is the ***lingual gyrus;*** the anterior part is the ***parahippocampal gyrus***. ***Collateral*** and ***rhinal sulci*** define respectively the medial and lateral borders of the convolution. The ***fusiform gyrus*** lies between the lingual and parahippocampal gyri and the inferior temporal gyrus. The ***gyrus rectus*** lies on the medial and basal surface of the frontal lobe, and the ***orbital gyri*** form the lateral and basal portions of the frontal lobe.

Functional areas of the cerebrum. Brodmann's maps of the cerebral hemispheres (Fig. 9-15) reflect the fact that the neuron-containing portion of the cerebrum lies in a thin external covering of gray matter called the cerebral cortex. Brodmann recognized that the cellular arrangement in the cortex was different in different parts of the cerebrum and, noting that injuries to different parts of the cerebrum produced different symptoms, came to the conclusion that there were different functional regions of the organ. Although today it is recognized that there are not always indisputable structural-functional relationships, it *is* certain that there are particular regions of the cerebrum that are concerned with one particular aspect of function. The discussion to follow emphasizes general and important functional areas of the cerebrum, using Brodmann's numbered regions to outline their extent, and mentions some disorders that may occur as a result of lesions in specific areas.

FRONTAL LOBES. *Area 4* occupies the precentral gyrus and is designated as the ***primary somatic motor area.*** Using low-voltage electric current to stimulate various parts of the region, it may be demonstrated that contractions of specific muscles or small groups of muscles will occur. Stronger stimulation

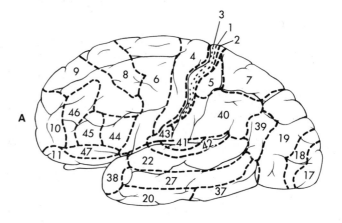

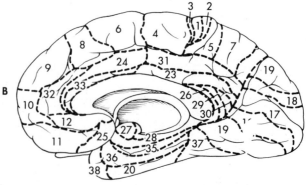

Fig. 9-15

Locations of cytoarchitectural areas according to Brodmann. Within certain numbered regions are specific functional areas (see text). **A,** Lateral aspect of cerebrum. **B,** Medial aspect of cerebrum.

causes a larger number of muscles to contract, often causing a coordinated motion such as closing the hand. The body is also shown to be represented upside down in area 4, with the legs and feet hanging over the superior border of the gyrus and the head near the lateral fissure. Those areas of the body where complex movements (hands, face, and head) occur have a larger area of representation. Thus the area for the back is smaller than that for the face. A *supplementary motor area* occupies areas **24** and **31** on the medial aspect of each hemisphere and representation of movement in this region is grosser or less specific than in area 4. A given side of either motor region (primary or supplementary) primarily controls muscles on the opposite side of the body.

Injury to these motor areas results in *paralysis,* or inability to voluntarily contract a muscle or group of muscles. The outgoing fibers from these motor areas to the spinal cord constitute the *upper motor neurons.* As the fibers pass downward through the brainstem, they undergo about an 85% crossing-over *(decussation)* in the medulla. Thus if injury is to one side of the cerebrum, paralysis is on the opposite side of the body. Also, the paralysis is of the *spastic type,* in which the muscles are in a state of contraction while incapable of voluntary movement; this type of paralysis is typical of an *upper motor neuron lesion.*

Area 6 is called the *premotor area* and when stimulated causes contractions of specific muscles if area 4 is intact, and movements of head, trunk, and neck will occur. It may be that *learned* motor activity lies in this area, because lesions here cause a loss of *performance* but not a loss of the ability to contract muscles.

Area 8 is called the *frontal eye field* and is responsible for eye movements that scan the visual field. Lesions here interfere with the ability to direct the eyes to objects in the visual field.

The large region of frontal lobe anterior to area 8 is called the *prefrontal cortex* and includes *areas 9 to 12.* Stimulation here produces no specific movements or sensation, and the areas receive fibers from other brain areas. The regions appear to inhibit emotional reactions and to house our moral and social senses, our intelligence, and many everyday experiences.

PARIETAL LOBES. *Areas 3, 1, and 2,* from front to back, lie in the *postcentral gyrus.* These areas are called the *general sensory* or *somesthetic areas.* Skin sensations (pain, heat, cold, touch, pressure), limb position, and sense of tension in muscles (deep sensibility) are received here. The body is also represented upside down in areas 3, 1, and 2, and lesions will affect the ability to discriminate such sensations as listed previously in particular body areas.

Parietal association areas are found in regions **5** and **7** and provide for recognition of textures, shapes, and degrees of change, such as heat and cold. Lesions here render a person incapable of recognizing familiar objects by touch alone.

OCCIPITAL LOBES. *Area 17* is the *visual area* of the cerebrum, receiving the termination of the visual pathways from the retina. Lesions here will result in defects of vision because even though the retina will respond to light, the brain connection is lost. The *visual association areas* lie anterior to area 17 and include *areas 18* and *19.* These regions relate past visual experience (memory?) to present experience and aid in integrating eye movements that fix the eyes on objects.

TEMPORAL LOBES. The *auditory area* occupies *area 41* and receives the nerve fibers from the ear. Lesions here result in inability to hear specific frequencies of sound; that is, they create *central deafness.* *Area 42* is the *auditory association area* and is important in language functions as well as in storing memories of auditory *and* visual experiences. *Broca's speech areas* lie in *areas 44* and *45* and are concerned with coordinated and appropriate use of the vocal apparatus.

Lesions of the temporal lobes usually produce *agnosia* (loss of ability to recognize objects), *apraxia* (loss of ability to speak properly), and *aphasia* (inability to use or understand the spoken or written word).

The reader should attempt to relate these functional areas of the cerebrum to the individual gyri of the organ.

Histology of the cerebrum. The gray matter of the cerebrum forms an outer covering, the cerebral *cortex,* that varies between 1.5 and 4 mm in thickness. It is thickest in the motor area and thinnest over the visual area. It has a surface area of about 2000 cm^2, a volume of about 600 cm^3, and is estimated to contain at least 12 to 15 billion neurons. It forms about 40% of the total weight of the human brain.

According to phylogenetic, histological, and functional considerations, several types of cortex are recognized:

paleocortex contains only three layers of cells and is restricted to the olfactory regions of the frontal lobe.
archicortex is typical of the several parts of the temporal lobe that are involved in the limbic system of the brain. The latter is the system that is concerned with emotional expression.
neocortex is typical of those areas concerned with motor functions and reception of sensory data for general sensation, vision, hearing, and taste. In humans, 90% of the cerebral cortex is neocortex, and it contributes also to our intellectual, social, and cultural lives.

Neocortex contains five characteristic types of cells, arranged in six ill-defined layers (Fig. 9-16). The five types of cortical neurons have the following characteristics:

pyramidal cells range in height from 10 to 100 μm. They have a shape suggestive of a slim pyramid and occur in small, medium, large, and giant sizes. A heavy *apical dendrite* rises from the top of the cell and is directed toward the surface of the cortex. Several *basal dendrites* project laterally from the base of the cell, and its *axon* rises from the center of the base and is directed away from the cortical surface.

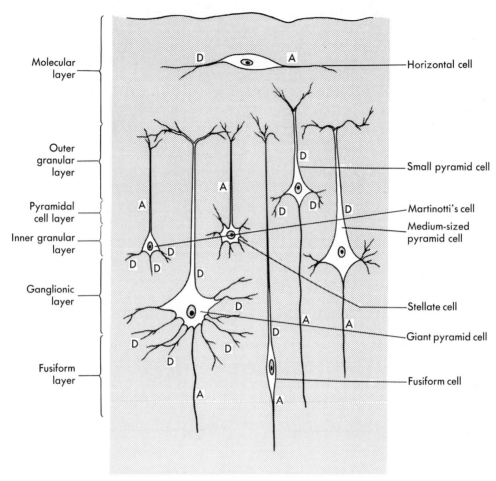

Fig. 9-16
Cellular arrangement of cerebral cortex. (See text for explanation.) *A*, Axon; *D*, dendrite.

stellate (granule) cells are star-shaped cells about 8 μm in diameter. Several short dendrites and a short axon connect with neighboring neurons.

fusiform cells have a tapered and vertically oriented cell body with dendrites rising from both ends. The axon comes off the side of the cell and is directed away from the cortical surface.

Martinotti's cells have a rhomboid-shaped body, a superficially directed axon, and several deeply directed dendrites.

horizontal cells (Cajal's cells) resemble fusiform cells but lie parallel to the cortical surface. These cells connect the pyramidal cell dendrites to one another.

The names of the layers of neocortex from the surface inward and the cells characteristic of each layer are as follows:

molecular (plexiform) layer contains the branches of the dendrites of pyramidal and fusiform cells and the horizontal cells.

outer granular layer contains many small pyramidal and stellate cells.

pyramidal cell layer contains medium and large pyramidal cells.

inner granular layer contains stellate cells and some Martinotti's cells.

ganglionic layer contains large and giant pyramidal cells (Betz's cells), stellate cells, and Martinotti's cells.

fusiform (polymorph) layer contains mainly fusiform cells.

In paleocortex and archicortex only, layers one, five, and six of the neocortex are present.

Medullary body of the cerebrum. The internal white matter of the cerebrum forms its **medullary body.** Within this material are bundles of axons and dendrites of neurons organized into three kinds of fibers (Fig. 9-17):

association fibers connect various parts of one cerebral hemisphere.

commissural fibers connect one hemisphere with the other via the *corpus callosum* and the *anterior* and *posterior commissures.*

projection fibers consist of fibers entering and leaving the hemispheres as motor and sensory pathways. These fibers enable the brain to "keep in touch" with the peripheral body areas. The **internal capsule** is an area composed mainly of efferent motor fibers leaving the cerebrum.

Interruption of the association fibers of the medullary body results in lack of

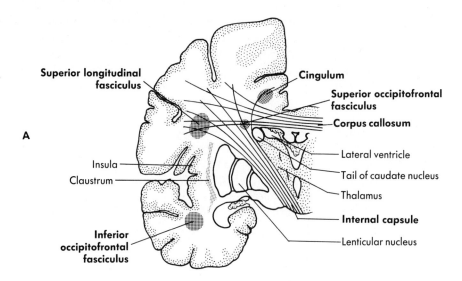
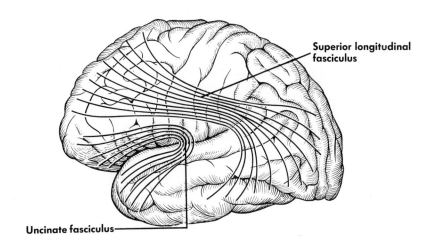
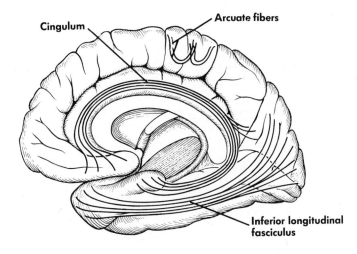

Fig. 9-17
Some cerebral association, projection, and commissural fibers (boldface). **A,** Coronal section. **B,** *Upper,* lateral view; *lower,* medial view.

ability to integrate activities within one hemisphere. Response to sensory input may not occur; that is, sensorimotor integration is defective.

Interruption of commissural fibers creates a "split brain" in which information is not directly passed between hemispheres. Experiments on animals in which the corpus callosum has been sectioned has led to the suggestion that in humans (by extrapolation from animals) the left cerebral hemisphere controls the "routine" activities of the body, whereas the right hemisphere is the more "creative" and "intellectual" hemisphere. In any event the two hemispheres are not identical in the jobs they have.

Deep within the medullary body are the **basal ganglions** (or *nuclei*), which are sometimes called the "central gray matter" of the cerebrum (Fig. 9-18). Specifically, the ganglions are the *caudate nucleus, putamen, globus pallidus* (the latter two collectively forming the *lenticular nucleus*), **claustrum**, and *amygdaloid nucleus*. Several **subthalamic nuclei** *(red nucleus, substantia nigra)* are often included as basal ganglions. These nuclei deal with involuntary motor functions such as control of tremor and tone and suppression of purposeless movements.

Lesions of the basal ganglions are associated with the *assumption* of purposeless movements. Among such movements are *choreiform movements*, which are jerky, rapid, and uncoordinated *(St. Vitus' dance)*; athetoid

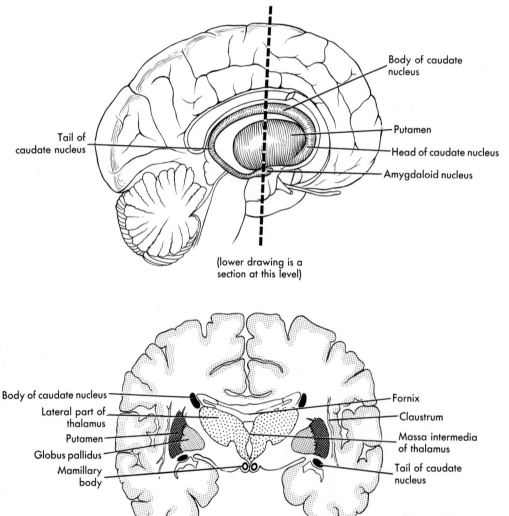

Fig. 9-18
Basal ganglions and associated structures. *Above,* Several major ganglions projected on cerebral hemisphere. *Below,* Frontal section of cerebrum to show positions of ganglions.

movements, which are slow and sinuous; and *Parkinson's disease* (shaking palsy), a disorder most often seen in the aged, which is associated with tremors, rigid facial expression, and postural abnormalities. *L-dopa,* a chemical similar to synaptic transmitters of the CNS, has been used with some success in reducing the severity of the disease's symptoms, suggesting that it may in part be due to lack of such materials in the brain.

Cerebellum

Gross anatomy. The cerebellum weighs 140 to 150 g in the adult. A central narrow **vermis** supports two laterally placed **cerebellar hemispheres.** Arrangement of gray and white matter is as in the cerebrum, with an outer **cortex** of gray matter surrounding an inner mass of white matter, the **central white matter** or *medullary body.* The white matter has a treelike arrangement within the cerebellum and is designated as the *arbor vitae.* The surface of the cerebellum is thrown into numerous parallel ridges called **folia** that are separated by **cerebellar fissures.** In several areas the fissures are quite deep and divide the organ into lobes. Fig. 9-19 indicates the names and positions of the fissures and lobes.

Attaching the cerebellum to the brainstem are three pairs of **cerebellar peduncles** *(inferior, middle,* and *superior)* that carry afferent and efferent fibers to and from the organ.

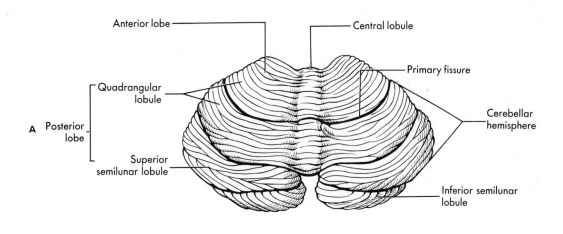

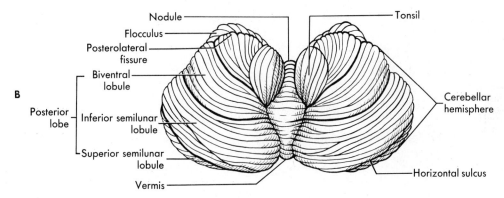

Fig. 9-19
Cerebellum. **A,** Superior surface. **B,** Inferior surface.

Four pairs of **central** (*deep*) **nuclei** are found in the medullary body (Fig. 9-20). The nearly spherical *fastigial nucleus* lies nearest the midline. The *globose nucleus* contains two or three small masses of cells and is placed laterally to the fastigial nucleus. The oval *emboliform nucleus* is next most laterally placed, and the largest and most lateral of the nuclei, the **dentate nucleus,** has a crumpled shape.

Histology of the cerebellum. Structure of the cerebellar cortex is the same in all parts. There are three portions to the cerebellum, reflecting phylogenetic age. The oldest portion or **archicerebellum** consists of the *flocculonodular lobe* and the *uvula* and receives afferent fibers from the inner ear equilibrial structures (maculae, semicircular canals). It projects to neurons that pass to skeletal muscles and control their activities in posture maintenance. The **paleocerebellum,** consisting of most of the *vermis* and adjacent portions of the hemispheres, receives input from general sensory receptors in muscle and skin and also projects to skeletal muscle. The **neocerebellum** includes the greater portion of the *hemispheres* and receives fibers primarily from the cerebral cortex. Cerebellar functions are associated with coordinating movements; controlling their rate, direction, and force of progression; and stopping them at the proper place.

More specifically, the cerebellum continually monitors the motor impulses that originate elsewhere in the brain ("intent") with the movements themselves as detected by receptors in skin, muscles, and tendons ("performance") and adjusts the rate, direction, and force of the movement. The activities are considered under three interrelated headings:

Error control ensures that a movement goes where it is intended, with appropriate force.
Damping cancels the tendency for oscillation at the end of a motion.
Prediction stops the movement at the appropriate *time*.

Lesions of the organ are obviously associated with lack of these three actions; motions become uncoordinated, slow, and tremorous.

The cerebellar cortex is organized into three layers (Fig. 9-21). The outer layer is termed the **molecular layer** and contains the dendritic branches of *Purkinje's cells* and some *stellate cells*. The middle layer is the **Purkinje's cell layer** and contains the characteristic Purkinje's cells. The latter have a flask-shaped cell body, apical dendrites that branch profusely in the molecular layer, and deep-directed axons. The inner layer is the **granular layer** and is composed of tightly packed *granule cells*. Incoming fibers, carried in the white matter, are of two types: **mossy fibers** terminate on the granule cells; **climbing fibers** terminate on Purkinje's cells.

The connections between the several types of cells in the cerebellum create "functional units" (see Fig. 9-21) capable of analyzing incoming data and controlling muscular movement. Axons of Purkinje's cells are directed to the deep nuclei of the cerebellum before impulses are passed to the skeletal muscles.

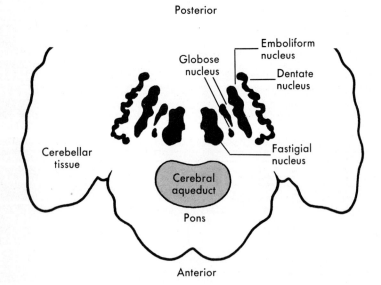

Fig. 9-20
Deep nuclei of cerebellum as seen in cross-section of pons.

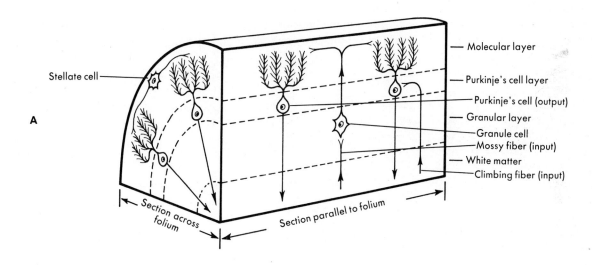

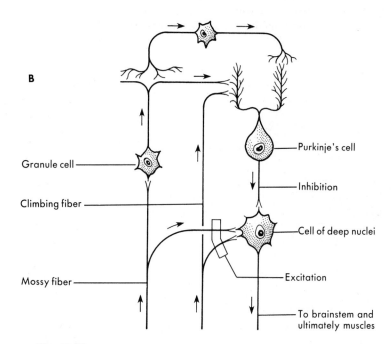

Fig. 9-21
A, Basic histology of cerebellum. **B,** Functional unit of cerebellum.

Diencephalon. The diencephalon (Fig. 9-22) is derived from the same area that gave rise to the cerebrum (the "old" forebrain). It is positioned atop the brainstem between the inferior portions of the cerebral hemispheres and consists of the *epithalamus,* the *dorsal thalamus* (sometimes called just *thalamus*), the *ventral thalamus,* and the *hypothalamus.*

Epithalamus. The epithalamus consists of the **pineal gland** (or *pineal body*) and the **habenular nuclei.** The pineal gland is cone shaped, measures about 7 × 5 mm, and may be concerned with timing the onset of puberty and the establishment of circadian rhythms. The habenular nuclei receive impulses from the olfactory system and project to the limbic system that in turn controls emotional expression. Thus a connection between the sense of smell and sex drive, rage reactions, and visceral changes associated with these factors is established.

Dorsal and ventral thalamus. The dorsal and ventral thalamus make up the bulk of the diencephalon and consist of many groups of cell bodies or **nuclei** that act primarily as receiving

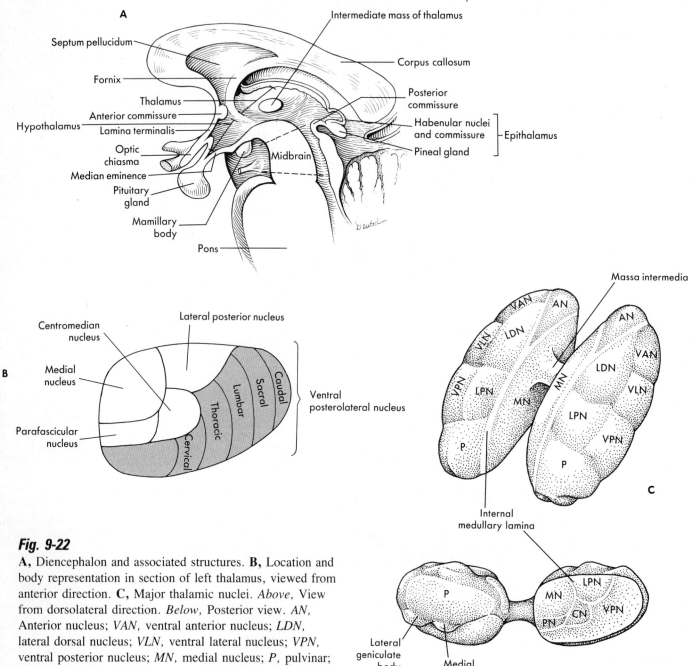

Fig. 9-22
A, Diencephalon and associated structures. **B,** Location and body representation in section of left thalamus, viewed from anterior direction. **C,** Major thalamic nuclei. *Above,* View from dorsolateral direction. *Below,* Posterior view. *AN,* Anterior nucleus; *VAN,* ventral anterior nucleus; *LDN,* lateral dorsal nucleus; *VLN,* ventral lateral nucleus; *VPN,* ventral posterior nucleus; *MN,* medial nucleus; *P,* pulvinar; *PN,* parafascicular nucleus; *CN,* centromedian nucleus.

areas for general sensations from the skin and from the organs of hearing and vision. The more important nuclei of the thalamus are as follows:

reticular nucleus receives fibers via the reticular formation from general sensory receptors and projects to the cerebral cortex. It is part of the system that maintains the waking state of the organism.

midline nuclei receive fibers from the viscera and taste buds and send fibers to other parts of the diencephalon.

medial geniculate body receives fibers from the cochlea (organ of hearing) and sends fibers to the temporal lobe.

lateral geniculate body receives fibers from the retina and sends fibers to the visual portion of the cerebral cortex.

ventral posterior nucleus receives sensory fibers concerned with pain, heat, cold, touch, and pressure. Outgoing fibers pass to the sensory portions of the cerebral cortex. **The ventral lateral** and **ventral anterior nuclei** receive fibers from the basal ganglions and cerebellum and project to the motor areas of the frontal lobes. Voluntary motor activity may thus be influenced.

Remaining nuclei include the *pulvinar, lateral posterior, lateral dorsal, medial,* and *anterior nuclei.* Incoming fibers are sensory in nature, and outgoing fibers pass to the interpretive areas of the cerebral cortex where memory storage, intellect, social sense, and "feelings" may occur.

Lesions of the thalamus produce the **thalamic syndrome,** which is characterized by intractable pain and lack of appreciation of general sensation.

Hypothalamus. The hypothalamus (Fig. 9-23) weighs about 4 g and contains groups of cells that constitute nuclei. The names and functions of these nuclei are presented in Fig. 9-23. The hypothalamus is concerned with homeostasis of several vital functions necessary for organism survival, for example, temperature regulation, feeding, and water balance. The hypothalamus also produces nine polypeptides that regulate the activity of the pituitary gland of the endocrine system. Thus nervous activity may be coupled to hormone secretion or suppression.

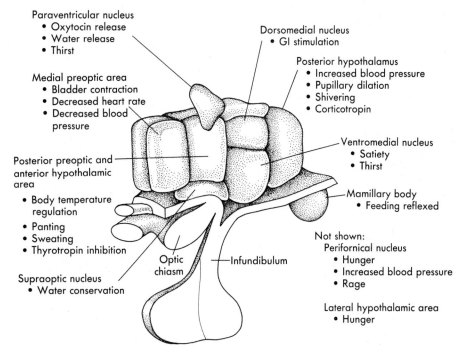

Fig. 9-23
Nuclei of hypothalamus.

Brainstem. The brainstem consists of three parts. Listed from the inferior portion to the superior portion, they are the **medulla oblongata** (or simply *medulla*), **pons**, and **midbrain** (Fig. 9-24). These regions contain *nuclei* that are involved in specific motor and sensory functions and also transmit *tracts*, bundles of fibers carrying particular kinds of nerve impulses.

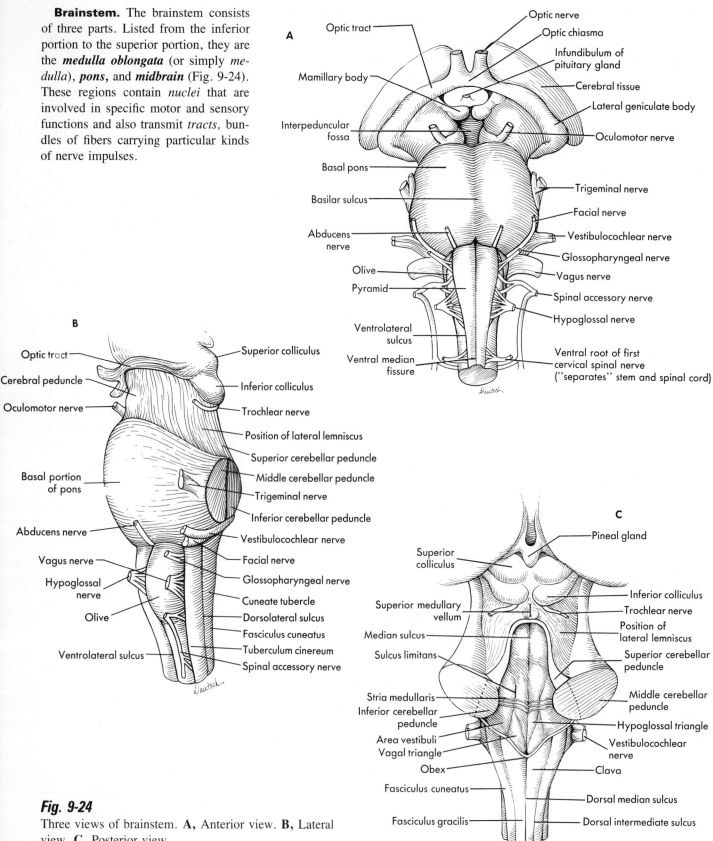

Fig. 9-24
Three views of brainstem. **A**, Anterior view. **B**, Lateral view. **C**, Posterior view.

Medulla. The medulla forms the inferior 3 cm or so of the brainstem. It is slightly wedge shaped, being wider at the superior end. Within the medulla there is no clear-cut organization of gray and white matter, because extensive rearrangement of fiber tracts is occurring in this area. A section of the medulla about halfway along its length presents the features shown in Fig. 9-25. Conspicuous are the ascending fibers derived from the dorsal area of the spinal cord white matter that form the **gracile** and **cuneate tracts**. The tracts convey sensations of fine touch, pressure, and body position. These tracts terminate in the **gracile** and **cuneate nuclei** located in the upper portion of the medulla. From these nuclei, axons sweep toward the middle of the medulla as the **internal arcuate fibers** that *cross* the midline as the **decussation** (crossing) *of the medial lemniscus.* After crossing, the fibers form the **medial lemniscus** that progresses superiorly to terminate within the thalamus. The cord tracts for pain, thermal sensations, and crude touch and pressure (spinothalamics) and the spinotectal tract for relaying data to the midbrain form the **spinal lemniscus,** an obvious feature in the medullary section. The spinoreticular tracts, carrying impulses derived from the skin and viscera, terminate on the lateral reticular nuclei, a prominent area in the ventrolateral region of the medulla. Spinal tracts carrying information from receptors in muscles and tendons (spinocerebellar tracts) pass through the medulla in its lateral portions to be distributed to the cerebellum via the cerebellar peduncles.

Nuclei in the medulla, in addition to the gracile and cuneate nuclei, include the following:

olivary nuclei, composed of *inferior, medial accessory,* and *dorsal accessory* subdivisions, are located in the ventrolateral portion of the medulla. The largest of these is the inferior nucleus, which looks like a crumpled purse. These nuclei receive fibers from the basal ganglions and cerebral cortex and send fibers to the cerebellum via the *olivocerebellar tract*. The nuclei are thus part of the motor system of the body.

arcuate nucleus lies on the external portion of the ventral side of the medulla and receives fibers from the cerebral cortex. Fibers pass from the nucleus to the cerebellum.

Another set of motor tracts that pass through the medulla are the **corticospinal fibers**. About 40% of the fibers of these tracts originate in area 4 of the cerebral cortex, the remainder from other cerebral motor areas. The fibers are gathered in the ventral region of the medulla to form the **pyramid.** A bit below the middle of the medulla, and thus *not* appearing in Fig. 9-25, the corticospinal fibers undergo a *decussation* or crossing. About 85% of the fibers cross the midline and are continued down the spinal cord as the *lateral corticospinal tracts;* the remainder do not cross the midline, and they form the *ventral corticospinal tracts*. These tracts deal with impulses leading to voluntary contraction of skeletal muscles. The **rubrospinal tract,** an involuntary motor pathway, originates in the midbrain and forms a band of fibers in the lateral medulla.

In addition to the nuclei described, the medulla contains the nuclei of cranial nerves nine through twelve (glossopharyngeal, vagus, accessory, hypoglossal). Our section shows the nuclei of nerves ten and twelve. The nuclei of the eleventh and ninth nerves lie superior to the level of the section.

More information on cranial nerve nuclei will be provided later in this chapter.

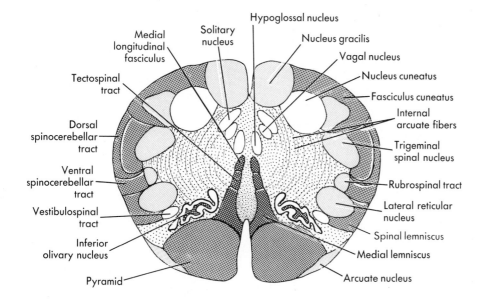

Fig. 9-25

Section through medulla depicting extent of structures named.

Pons. The pons (Fig. 9-26) is about 2.5 cm long and is best distinguished grossly by the conspicuous bulge on its ventral surface. The bulge, forming the **basal portion** of the pons, consists of longitudinal and transversely oriented fiber bundles receiving motor fibers from one cerebral hemisphere and relaying them to the *opposite* cerebellar hemisphere via the **cerebellar peduncles.** The peduncles are three masses of fibers *(superior, middle,* and *inferior peduncles)* making the connection described. **Pontine nuclei** are found in the basal portion. They send fibers to the cerebellum. The dorsal pons, called the **tegmentum,** contains ascending fiber tracts (continuations of the medial and spinal lemnisci) that are passing to the thalamus and the descending corticospinal and rubrospinal tracts.

Additionally, the nuclei of cranial nerves five through eight (trigeminal, abducent, facial, vestibulocochlear) lie in the tegmentum. The **lateral lemniscus** is derived from the fibers of the eighth nerve. Again, further information as to location of these nuclei will be provided later in this chapter.

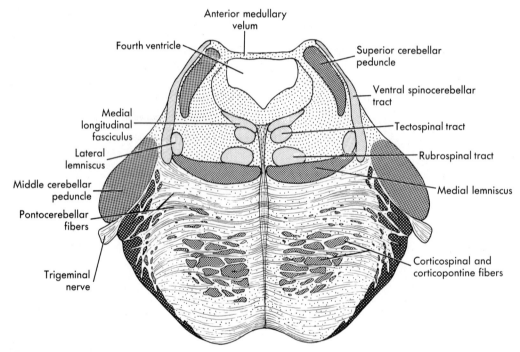

Fig. 9-26
Section through pons depicting extent of structures named.

Midbrain. The midbrain is a wedge-shaped (from anterior to posterior) section of the stem superior to the pons and is about 1.5 cm in length. Grossly, its most characteristic features are the paired *superior* and *inferior colliculi* (collectively called corpora quadrigemina). On the ventral surface or base are the *cerebral peduncles* that form the bulk of the midbrain. A section of the midbrain at the level of the superior colliculi shows the features of Fig. 9-27.

The inferior colliculi contain neuron cell bodies that receive fibers from the cochlea and send fibers to the medial geniculate body of the thalamus. The superior colliculi receive fibers from the visual cortex and send fibers to eye and neck muscles to orient the head to follow moving stimuli or to avoid objects seen.

The medial and spinal lemnisci pass through the midbrain. The *red nucleus* forms large masses on either side of the midline. It is a motor nucleus concerned with muscle tone and posture.

The *basis pedunculi* are obvious ventrally placed bands of fibers carrying motor impulses from the cerebrum to cerebellum and spinal cord.

Nuclei of cranial nerves three and four (oculomotor and trochlear) are located in the midbrain. Fibers from these nuclei pass to extrinsic eye muscles.

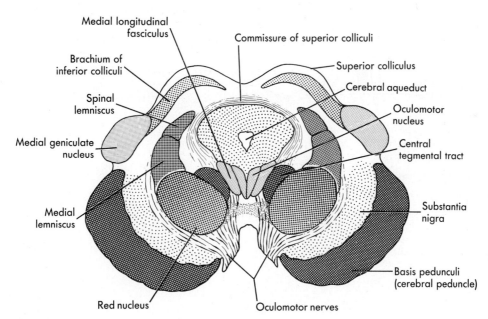

Fig. 9-27
Section through midbrain depicting extent of structures named.

Reticular formation. Extending through the brainstem is a column of gray matter separate from the nuclei already described. It is called the *reticular formation* (Fig. 9-28). Specific nuclei are named in various parts of the stem (see Fig. 9-28). The formation receives fibers from the sensory lemnisci and sends fibers to the cerebellum, spinal cord, and cerebral cortex. The first two connections are concerned with involuntary motor function. Cerebral connections enable sensory input to activate or awaken the cortex. Thus the formation and its cortical connections are part of the *reticular activating system* (RAS) that contributes to the waking state of the organism.

Limbic system. Combining several of the cerebral and diencephalic structures mentioned previously, the limbic system is concerned with expression of emotion. The amygdaloid nuclei, habenular nuclei, hypothalamus, and tracts connecting these areas, plus the hippocampal gyri that floor the lateral ventricles, are the anatomical parts of the system. Stimulation of the amygdala produces aggressive behavior, whereas their destruction renders an animal passive. The hypothalamus is required for the visceral component of emotional expression such as motility of the tract and changes in heart rate and blood pressure. It is the limbic system that is most greatly influenced by psychosomatic disorders, that is, when psychic activity is translated into physical symptoms. Lest the system be regarded only as a system of punishment, there appear to be "pleasure centers" in the hypothalamus and thalamus that give feelings of affection and wellbeing when stimulated.

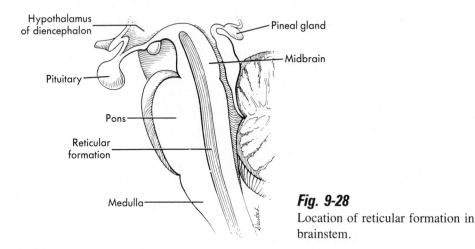

Fig. 9-28
Location of reticular formation in brainstem.

Spinal cord

The *spinal cord* (Fig. 9-29) is an elongated structure 42 to 45 cm long having an average diameter of 1 cm. It weighs about 30 g and in the adult extends from the rim of the foramen magnum to the level of the upper border of the second lumbar vertebra. In the newborn it reaches the third lumbar vertebra but recedes during the growth of the spinal column to assume its final position in the adult.

Externally, the cord presents two enlargements. One, the *cervical enlargement*, extends from the third cervical to second thoracic vertebrae and shows a maximal diameter of about 38 mm. The *lumbar enlargement* begins at the level

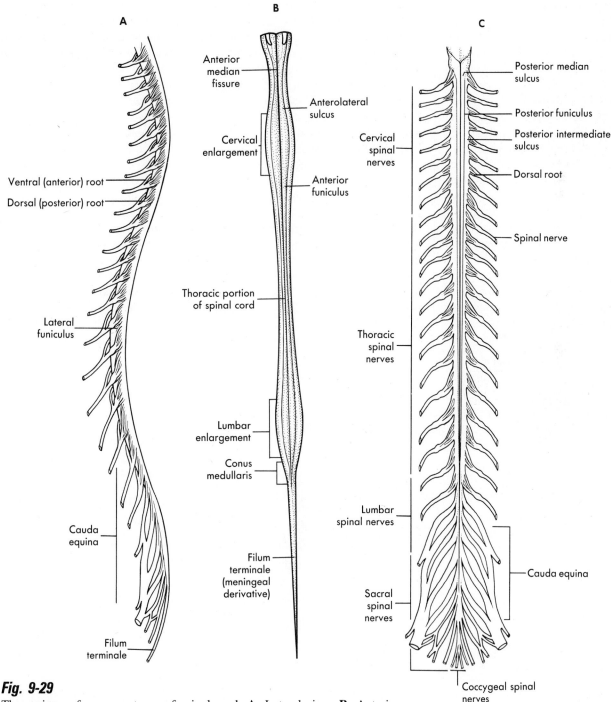

Fig. 9-29
Three views of gross anatomy of spinal cord. **A,** Lateral view. **B,** Anterior view. **C,** Posterior view.

of the ninth thoracic vertebra, reaches a maximal diameter of about 33 mm opposite the twelfth thoracic vertebra, and then tapers rapidly to a point to form the **conus medullaris**. From the conus, fibrous tissue forms the **filum terminale** that extends through the vertebral canal to the first coccygeal vertebra where it is attached. The enlargements are associated with the provision of fibers to and from the appendages.

A fissure and several sulci groove the length of the cord. The **anterior median fissure** is about 3 mm in depth and, as its name suggests, grooves the midline of the anterior length of the cord. The **posterior median sulcus** lies opposite the fissure and marks the position of the **posterior median septum**, a partition of glial cells that penetrates about halfway into the cord, dividing it into right and left portions. The **posterolateral sulcus** is a longitudinal groove marking the position of the attachments of the posterior roots of the spinal nerves. In the cervical and upper thoracic portions of the cord a **posterior intermediate sulcus** lies between the posterior and posterolateral sulci. An **anterolateral sulcus** marks the exit from the cord of the anterior roots of the spinal nerves.

Thirty-one pairs of spinal nerves originate from the cord, each with a posterior and anterior root. The 31 pairs consist of 8 cervical, 12 thoracic, 5 lumbar, 5 sacral, and 1 coccygeal nerve. The origins of the nerves also mark **spinal segments** or **neuromeres**. The length of these segments varies in the cord, being about 13 mm in the cervical cord and 26 mm in the midthoracic region, and then diminishes rapidly in the lumbar region (15 mm) to about 4 mm in the sacral cord. Since the cord is shorter than the spine, spinal nerves from the thoracic region downward take progressively more oblique courses to exit from their appropriate intervertebral foramina. Beyond the conus medullaris, large numbers of nerve roots run vertically within the vertebral canal to form the **cauda equina**, so named because of the resemblance to the hairs in the tail of a horse. Posterior (dorsal) roots of the spinal nerves have their cell bodies in a **posterior root ganglion** that forms an enlargement a few millimeters from the cord itself and enter the cord in a continuous series at the posterolateral sulcus. Anterior (ventral) roots exit from the cord along the anterolateral sulcus in groups of 6 to 8 rootlets.

Internally (Fig. 9-30) the cord may be seen to be composed of an inner core of gray matter arranged roughly in the form of a letter H and an outer layer of white matter consisting primarily of myelinated nerve fibers. The gray matter consists of four regions. Projecting toward the posterolateral sulcus is the **posterior column**, to the side is the **lateral column**, and toward the anterolateral sulcus projects the **anterior column**. The columns extend the length of the cord but because of their appearance in a cross-section are often called *horns*. The fourth part of the gray matter is the **gray commissure**, the bar of the H, divided into *anterior* and *posterior gray commissures* by the **central canal**. Between the posterior and lateral gray columns lies the cord **reticular formation** distinguished by a netlike arrangement of fibers and glia. About halfway up the posterior column is a

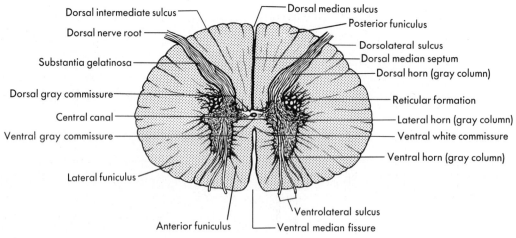

Fig. 9-30
Generalized cross-section of spinal cord.

rather clear area called the **substantia gelatinosa**. It consists mostly of glial cells.

There are several columns of *cells* within the gray columns (Fig. 9-31). The major ones are as follows:

lateral column contains bodies for autonomic fibers.
posterolateral column, anterolateral column, and *medial column* are all found in the ventral gray matter and house the cell bodies of neurons that supply the skeletal muscles.
posterior column cells are located in the posterior gray column and form the substantia gelatinosa, nuleus proprius, and nucleus dorsalis.

The white matter of the cord is composed of millions of myelinated and unmyelinated nerve fibers and is divided by the gray matter into three funiculi, the **posterior funiculus, lateral funiculus,** and **anterior funiculus** (see Fig. 9-30). The funiculi contain functional groupings of fibers that form the **tracts** or **fasciculi** of the cord. The tracts may be designated as *ascending* (sensory) or *descending* (motor) and serve as the re-

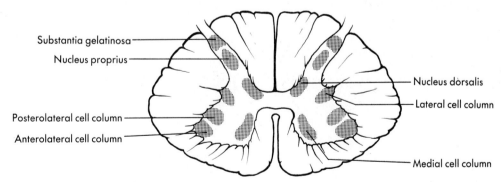

Fig. 9-31
Cell columns of spinal cord gray matter.

lay pathways for sensory and motor impulses to and from the brain. Fig. 9-32 depicts the positions and names of these tracts, and Table 9-1 indicates the type of information each tract carries, its origin, and termination.

Clinically, it becomes obvious that if fibers in a particular tract are damaged, there will be a loss of function or sensation in a particular body area as determined by the level or segment at which the damage occurs.

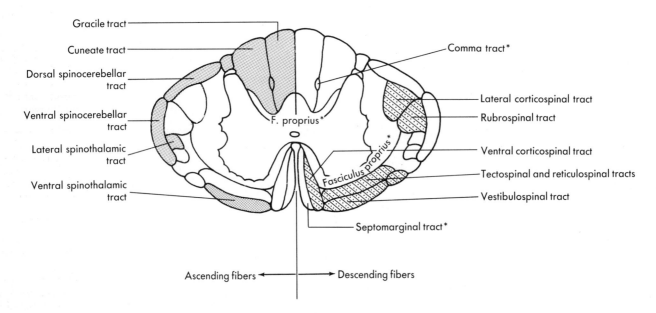

Fig. 9-32
Major spinal tracts. *Left*, Sensory or ascending tracts. *Right*, Motor or descending tracts. *Indicates intersegmental tracts.

Table 9-1. Summary of major spinal tracts

Name of tract and ascending (sensory) or descending (motor)	Origin of the tract or area from which fibers come	Termination of the tract	Crossed or uncrossed	Function or type of impulses carried
Lateral corticospinal (descending)	Motor areas of cerebral cortex	On motor neurons of spinal cord	Crossed in medulla; constitutes 80%-85% of corticospinal fibers	Voluntary motor impulses to skeletal muscles
Reticulospinal (descending)	Reticular formation of brainstem	On motor neurons of spinal cord	Crossed in brainstem	Involuntary motor impulses to skeletal muscles to increase their tone
Rubrospinal (descending)	Red nucleus of midbrain	On motor neurons of spinal cord	Crossed in brainstem	Involuntary motor impulses to skeletal muscles (for tone, posture)
Ventral corticospinal (descending)	Motor areas of cerebral cortex	On motor neurons of spinal cord	Uncrossed; constitutes 15%-20% of corticospinal fibers	Voluntary motor impulses to skeletal muscles
Vestibulospinal (descending)	Vestibular nuclei of brainstem	On motor neurons of spinal cord	Uncrossed	Involuntary motor impulses to skeletal muscles for maintenance of equilibrium
Cuneate (ascending)	Skin, muscles	Medulla	Uncrossed	Touch, pressure, two-point discrimination, vibrational sense, appreciation of body position
Dorsal spinocerebellar (ascending)	Muscles, joints, tendons	Cerebellum	Uncrossed	Proprioceptive
Gracile (ascending)	Skin, muscles	Medulla	Uncrossed	Touch, pressure, two-point discrimination, vibrational sense, appreciation of body position
Lateral spinothalamic (ascending)	Skin	Thalamus	Crossed in cord	Pain, heat, cold
Ventral spinocerebellar (ascending)	Muscles, joints, tendons	Cerebellum	Uncrossed	Proprioceptive
Ventral spinothalamic (ascending)	Skin	Thalamus	Crossed in cord	Pain, heat, cold

Adapted from McClintic, J.R.: Basic anatomy and physiology of the human body, ed. 2, 1980. Copyright © 1980. Reprinted by permission of John Wiley & Sons, Inc.

Functions of spinal cord. In addition to acting merely as a transmitter of nerve impulses over its tracts, the spinal cord can mediate a variety of reflex activities that do not involve the brain.

Myotatic reflexes are elicited when a muscle or its tendon is subjected to a tension or stretch. Tiny sensory structures within the muscle (muscle spindles) or around tendons (Golgi tendon organs) send information to the cord that indicates degree of tension and to some extent direction of its application. The incoming sensory neuron connects directly to an outgoing motor neuron that causes the stretched muscle to reflexly contract. The purpose of such activity is usually to maintain posture in the face of a force tending to throw the body out of balance.

Flexor reflexes (*withdrawal reflexes*) result when a body part contacts a painful stimulus, such as cutting a finger, touching a hot object, or the like. The receptors here are generally in the skin, and the reflex contraction of flexor muscles removes the body part from harm's way. This reflex usually involves at least three neurons, an *internuncial neuron* being inserted between sensory input and motor output. By making connections between different cord levels possible, such neurons can involve cord levels and body areas other than those directly concerned with stimulus and response.

Crossed-extension reflexes combine a flexor reflex on one side of the body with muscular extension on the other. If a hurt limb is withdrawn from a stimulus, the opposite limb must usually be extended to maintain posture. This type of reflex obviously involves information being transmitted *across* the cord and also requires at least three neurons.

Clinical considerations. Specific symptom development with spinal cord lesions depends on several things:

Extent of damage
Location of damage
Level of the lesion

Complete transection (cutting through) of the cord causes loss of all sensation and results in paralysis in those parts of the body served by the cord below the level of the cut. This is because of the interruption of the tracts of the cord. At the same time **spinal shock** will develop. It is a loss of reflex activity believed to be caused by lack of excitatory brain impulses to cord neurons. However, recovery of spinal reflex activity will occur, usually after several weeks or months have passed. Hemisection (cutting the cord halfway through) produces the **Brown-Séquard syndrome**. This condition is characterized by paralysis on the *same* side of the body as the damage to the cord, as well as *same* side loss of touch and pressure. *Opposite* side loss of pain and thermal sensations occurs because the fibers carrying these sensations cross to the opposite side of the cord after entering it.

Tabes dorsalis involves syphilitic destruction of the posterior funiculi with consequent loss of touch, pressure, and muscle sense.

Multiple sclerosis involves loss of myelin sheaths on cord tracts, with "short-circuiting" of cord pathways.

Poliovirus seems to exhibit a predilection for anterior gray column neurons, those that form the *lower motor neurons* to skeletal muscle. Destruction of these cells causes a *flaccid* paralysis, the muscle in a relaxed and paralyzed state that characterizes a *lower motor neuron lesion.*

Meninges

The brain and spinal cord are surrounded by three membranes, collectively referred to as the **meninges** (sing., *meninx*). They are the **dura mater**, the **arachnoid**, and the **pia mater**.

The dura mater (pachymeninx) is a thick membrane composed of collagenous connective tissue and is the outer tough protective meninx. The *cranial dura mater* encloses the brain and acts not only as a brain covering but also as the periosteum for the interior of the cranial cavity. It is thus composed of an outer *periosteal layer* and an inner *meningeal layer* that are separated in certain places to provide space for the venous sinuses (p. 287). The inner layer also forms separating partitions between the major portions of the brain. For example, the *falx cerebri* (Fig. 9-33) is found extending between the cerebral hemispheres, the *tentorium cerebelli* (Fig. 9-34) separates the cerebellum from both cerebral hemispheres, and the *falx cerebelli* projects between the two cerebellar hemispheres. **Meningeal blood vessels** are found running through the dura.

The *cranial arachnoid* is a delicate avascular membrane that is separated from the dura by a narrow **subdural space**. A thin film of fluid lies in the subdural space. The arachnoid follows the falx and tentorium but does not extend into the sulci or fissures of the brain. Delicate strands of connective tissue extend through the **subarachnoid space**, which is filled with cerebrospinal fluid, and connect to the *cranial pia mater* that invests the entire cerebral and cerebellar surface, following the sulci and fissures. The cranial pia contains blood vessels derived from the cerebral and cerebellar arteries that form the major portion of the blood supply to the brain.

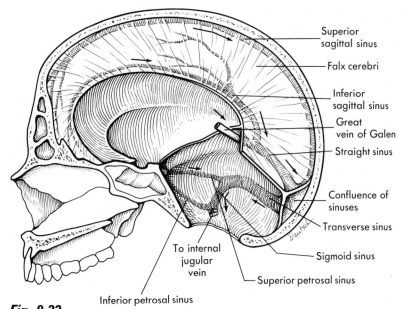

Fig. 9-33
Falx cerebri and associated structures. Arrows indicate direction of blood flow.

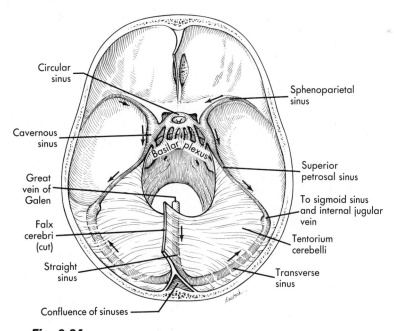

Fig. 9-34
Tentorium cerebelli and associated structures. Arrows indicate direction of blood flow.

The *spinal dura mater* (Fig. 9-35) forms a loose tubular sheath around the spinal cord and connects to the cranial dura at the foramen magnum. It is composed of a single layer corresponding to the inner cranial dura layer. An **epidural space** separates the spinal dura from the walls of the vertebral canal; the space contains loose connective tissue, a venous plexus, and branches of the spinal nerves as they enter or exit the cord. Again, a subdural space separates the spinal dura from the *spinal arachnoid*. A large subarachnoid space (see Fig. 9-35) containing cerebrospinal fluid separates the spinal arachnoid from the spinal *pia mater*. The latter is thicker than the cranial pia and consists

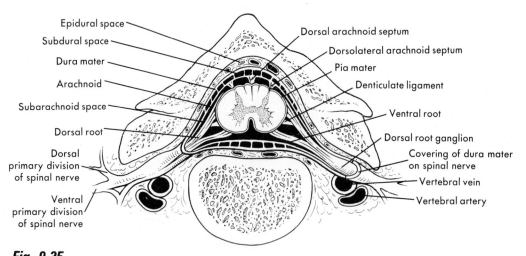

Fig. 9-35
Location of spinal meninges.
After Everett, N.B.: Functional neuroanatomy, ed. 6, Philadelphia, 1971, Lea & Febiger.

of an outer layer of collagenous tissue and a delicate inner layer than adheres to the cord surface and projects into the anterior median fissure. At three points on the posterior aspect of the cord, pial fibers extend to the arachnoid to form the *arachnoid septae* (see Fig. 9-35). Laterally, the pia forms **denticulate ligaments** (Fig. 9-36) extending along the entire cord. The lateral margins of the ligaments are scalloped and attach to the dura. The septae and denticulate ligaments support the cord and maintain its position within the dural sheath. As the spinal nerves exit from the vertebral canal, they are covered by elongations of the spinal dura (see Fig. 9-35).

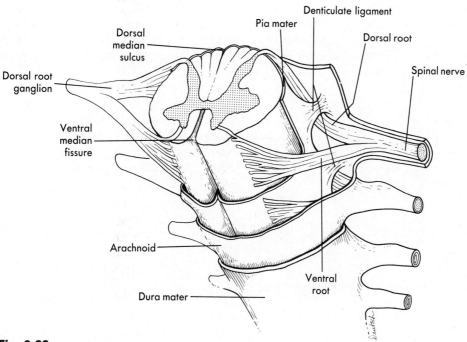

Fig. 9-36
Location of denticulate ligament of spinal cord.

Ventricles and cerebrospinal fluid

Recalling that the CNS develops as a tube, we might expect it to contain structures derived from the cavity of the original tube. The brain contains several *ventricles* and the *cerebral aqueduct;* the spinal cord has the *central canal* as the remnant of the original cavity. All spaces have an ependymal cell lining.

The two *lateral ventricles* lie within the cerebral hemispheres (Fig. 9-37). Each ventricle has a *body* from which arise three *horns.* The *anterior horn* extends into the frontal lobe, the *posterior horn* into the occipital lobe, and the *inferior horn* into the temporal lobe. Lying in the body and part of the inferior horn of each ventricle is a *choroid plexus,* a vascular structure responsible for the formation of the bulk of the cerebrospinal fluid (CSF). An *interventricular foramen* connects each lateral ventricle with the *third ventricle,* a slitlike single cavity lying between the thalami. A choroid plexus lies in the roof of this ventricle. The *cerebral aqueduct (of Sylvius)* runs through the midbrain. It is very small and connects the third ventricle with the *fourth ventricle* that lies between the cerebellum posteriorly and the pons and upper medulla anteriorly. The fourth ventricle is a flattened and diamond-shaped cavity containing a choroid plexus in its roof. Three openings in the roof of the ventricle permit CSF to pass from the fourth ventricle into the subarachnoid space. They are a single *median aperture (foramen of Magendie)* and two *lateral apertures (foramina of Luschka).* CSF is produced by ependymal cells of the central canal of the spinal cord. This fluid drains upward into the fourth ventricle within the skull. The subarachnoid space is enlarged in several areas to form what are called *cisternae* (Fig. 9-38).

Cerebrospinal fluid produced by the four choroid plexuses drains from the lateral to third to fourth ventricles and enters both the cranial and spinal subarachnoid spaces. From the cranial space, *arachnoid granulations* (Fig. 9-

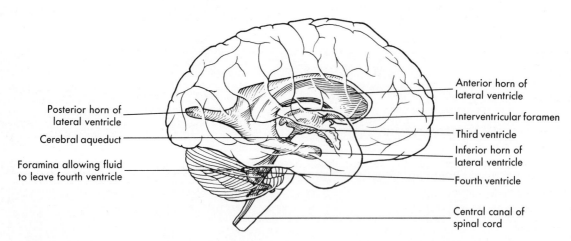

Fig. 9-37
Ventricles of brain viewed as if cast of system were superimposed on brain.

39), berrylike tufts of arachnoid, project into the superior sagittal sinus, a venous sinus of the brain. CSF may thus pass via these granulations into the circulation. The bulk of both cranial and spinal CSF is absorbed in this manner, although some is drained into spinal venous vessels. About 23 ml of CSF is in the ventricles and 117 ml in cranial and spinal subarachnoid spaces.

Since the adult spinal cord does not extend beyond the second lumbar vertebra and a CSF-filled subarachnoid space continues beyond this level, it is possible to insert a needle into the space without damaging the cord. Such a procedure is called **spinal tap** or **puncture**. The needle is inserted at **L4** in a direction perpendicular to the back to avoid placing the needle tip into the cauda equina. Samples of CSF may be withdrawn for analysis, and if a volume of fluid is replaced with an equal volume of anesthetic, a *spinal block* results. The block stops both sensory and motor impulses from traversing the spinal nerves. The anesthetic is gradually absorbed by the blood vessels in the meninges and will be excreted.

Subdural hematoma refers to bleeding into the subdural space following trauma to the head that ruptures the dural blood vessels. The clot often creates pressure that can close blood vessels or damage brain cells.

Hydrocephalus is a condition resulting from a block to CSF circulation, excess production of fluid, or slow absorption of fluid. The common denominator is increased CSF pressure that can inflate the ventricles from within or compress the brain from without. Blockage is most common in the cerebral aqueduct because of its small size—the lateral and third ventricles will become enlarged. Surgical intervention is the usual treatment with a bypass tube being extended from a lateral ventricle to the veins of the neck.

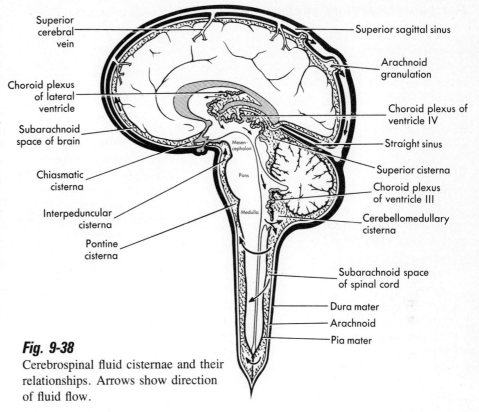

Fig. 9-38
Cerebrospinal fluid cisternae and their relationships. Arrows show direction of fluid flow.

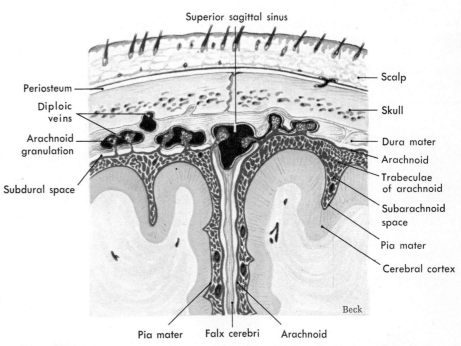

Fig. 9-39
Subarachnoid villi (arachnoid granulations) and their relationship to brain.

Peripheral nervous system

The PNS consists of the nervous tissue outside the limits of the brain and spinal cord. Its parts conduct impulses to and from the CNS. The PNS consists of nerve fibers, plexuses, ganglions, and specialized end organs that are derived from *cranial nerves* and *spinal nerves.*

Cranial nerves

The brain gives rise to 12 pairs of *cranial nerves.* They are numbered using Roman numerals and are named from anterior to posterior as:

I. Olfactory
II. Optic
III. Oculomotor
IV. Trochlear
V. Trigeminal
VI. Abducent (abducens)
VII. Facial
VIII. Vestibulocochlear (auditory, statoacoustic, acoustic)
IX. Glossopharyngeal
X. Vagus
XI. Accessory (spinal accessory)
XII. Hypoglossal

The origins of the cranial nerves are shown in Fig. 9-40) and their characteristics and functions are summarized in Table 9-2. Locations in the brainstem of the nuclei of the cranial nerves are presented in Fig. 9-41.

A discussion of each nerve is presented next.

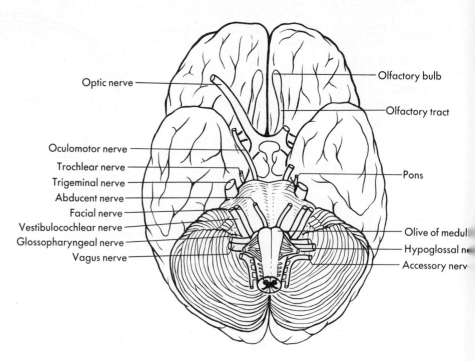

Fig. 9-40
Basal view of brain showing cranial nerves.

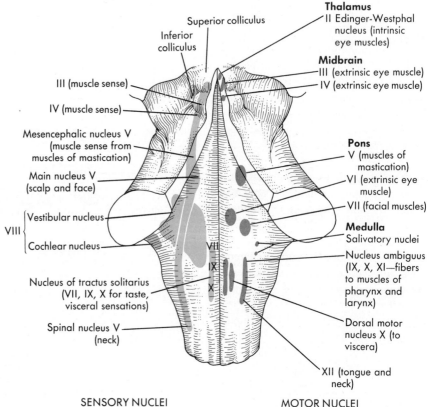

Fig. 9-41
Positions of cranial nerve nuclei in brainstem and diencephalon.

Table 9-2. Summary of the cranial nerves

Nerve	Composition*	Origin	Connection with brain or peripheral distribution	Function
I. Olfactory	S	Nasal olfactory area	Olfactory bulb	Smell
II. Optic	S	Ganglionic layer of retina	Optic tract	Sight
III. Oculomotor	MS	M—midbrain	Four of six extrinsic eye muscles (rectus superior oculi, rectus medialis oculi, rectus inferior oculi, obliquus inferior oculi)	Eye movement
		S—ciliary body of eye	Nucleus of nerve in midbrain	Focusing, pupil changes, muscle sense
IV. Trochlear	MS	M—midbrain	One extrinsic eye muscle (obliquus superior oculi)	Eye movement
		S—eye muscle	Nucleus of nerve in midbrain	Muscle sense
V. Trigeminal	MS	M—pons	Muscles of mastication	Chewing
		S—scalp and face	Nucleus of nerve in pons	Sensation from head
VI. Abducent	MS	M—nucleus of nerve in pons	One extrinsic eye muscle (rectus lateralis oculi)	Eye movement
		S—one extrinsic eye muscle	Nucleus of nerve in pons	Muscle sense
VII. Facial	MS	M—nucleus of nerve in lower pons	Muscles of facial expression	Facial expression
		S—tongue (anterior two thirds)	Nucleus of nerve in lower pons	Taste
VIII. Vestibulocochlear (statoacoustic, acoustic, auditory)	S	Internal ear: balance organs, cochlea	Vestibular nucleus, cochlear nucleus	Posture, hearing
IX. Glossopharyngeal	MS	M—nucleus of nerve in lower pons	Muscles of pharynx	Swallowing
		S—tongue (posterior two thirds), pharynx	Nucleus of nerve in lower pons	Taste, general sensation
X. Vagus	MS	M—nucleus of nerve in medulla	Viscera	Visceral muscle movement
		S—viscera	Nucleus of nerve in medulla	Visceral sensation
XI. Accessory	M	Nucleus of nerve in medulla	Muscles of throat, larynx, soft palate, sternocleidomastoideus trapezius	Swallowing, head movement
XII. Hypoglossal	M	Nucleus of nerve in medulla	Muscles of tongue and infrahyoid area	Speech, swallowing

From McClintic, J.R. Basic anatomy and physiology of the human body, ed. 2. Copyright © 1980. Reprinted by permission of John Wiley & Sons, Inc.
*M, Motor; S, sensory.

Olfactory nerve. The olfactory nerve (Fig. 9-42) is a purely sensory nerve that originates within the olfactory epithelium lining the apexes of the nasal cavities. Neuroepithelial cells of the olfactory area send many bundles of nerve fibers through the cribriform plate of the ethmoid bone into the cranial cavity, where they synapse with cells in the olfactory bulb. The nerve proper stops at this point. The remaining parts are considered as parts of the brain and will be discussed with the special sense organs.

Optic nerve. The optic nerve (Fig. 9-43) is a purely sensory nerve formed by axons of retinal cells, and it leaves the eyeball on the nasal side of the posterior pole. It passes through the optic foramen. At the *optic chiasma,* which lies on the tuberculum sellae, the fibers derived from the nasal half of each retina undergo a crossing and enter the *optic tract* that passes to the thalamus. The distribution of the remaining parts of the visual system will be considered in the chapter on the eye.

Oculomotor nerve. The oculomotor nerve (Fig. 9-44) is a mixed nerve carrying sensory impulses from the ocular muscles and motor impulses leading to movements of the eyelid and eyeball that cause changes in the size of the pupil. Specifically, the nerve supplies the levator palpebrae superioris, the rectus inferior oculi, rectus superior oculi, rectus medialis oculi, obliquus inferior oculi, and the sphincter pupillae iris muscle. It passes through the superior orbital fissure.

Trochlear nerve. The smallest of the cranial nerves, the trochlear nerve (see Fig. 9-44), transmits sensory and motor impulses from and to the obliquus superior oculi. It also uses the superior orbital fissure.

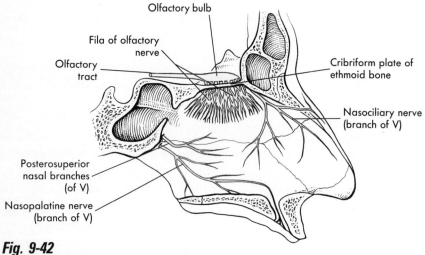

Fig. 9-42
Olfactory nerve and branches of trigeminal nerve on nasal septum.

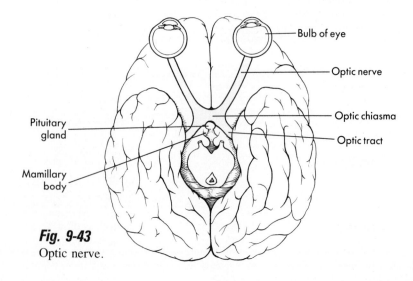

Fig. 9-43
Optic nerve.

Trigeminal nerve. The largest of the cranial nerves, the trigeminal nerve (Fig. 9-45) has a large sensory and a smaller motor component. Two roots, corresponding to the sensory and motor components, attach to the brainstem. The sensory root has the large *semilunar ganglion* on it and receives sensory data via three large branches that form nerves: the ophthalmic, maxillary, and mandibular nerves.

The *ophthalmic nerve* is derived from the eyeball, the conjunctiva, the mucous membranes of the nose, the paranasal sinuses, and the skin of the forehead, eyes, and nose. It passes to the brain through the superior orbital fissure.

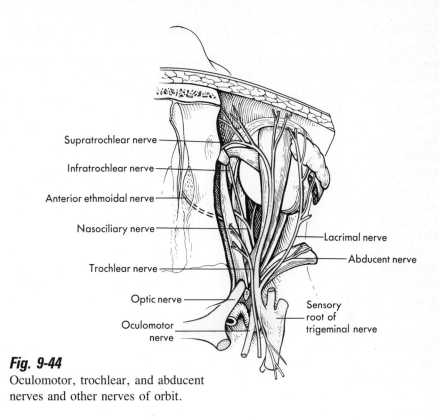

Fig. 9-44
Oculomotor, trochlear, and abducent nerves and other nerves of orbit.

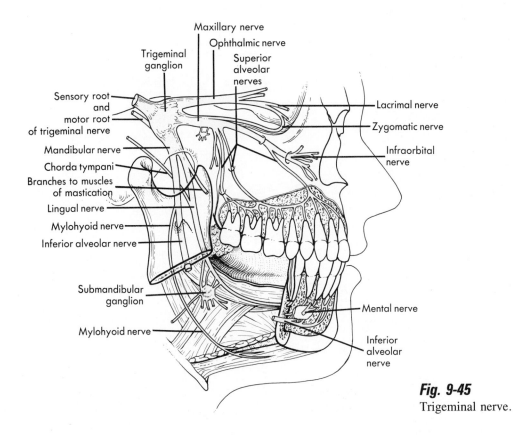

Fig. 9-45
Trigeminal nerve.

The *maxillary nerve* is derived from the skin of the middle portion of the face, lower eyelid, side of nose and upper lip, and mucous membranes of the nasal pharynx (soft palate), roof of mouth, and upper gums and teeth. It passes to the brain through the foramen rotundum. The nerve also receives fibers from the cranial dura mater.

The *mandibular nerve* is itself mixed in composition. The sensory portion is derived from the skin of the temporal area anterior to the ear, lower face, cheek, lower lip, mucous membranes of cheek, tongue, lower teeth and gums, mandible, and part of the cranial dura mater. The motor portion includes the entire motor division of nerve V. The motor fibers pass to the muscles of mastication (masseter, temporalis, pterygoidei) and to the mylohyoideus, anterior digastricus, tensor tympani, and soft palate muscles. The foramen ovale serves as the cranial passage for the mandibular nerve.

Abducent nerve. The abducent nerve (see Fig. 9-44) carries sensory impulses from and motor impulses to the rectus lateralis oculi muscle. It also uses the superior orbital fissure to reach the orbit. We thus see that three cranial nerves (III, IV, VI) innervate the extrinsic muscles of the eyeball.

Facial nerve. The facial nerve (Fig. 9-46) is a mixed nerve with a large motor and a much smaller sensory component.

The motor component supplies the muscles of facial expression, the scalp muscles, the stylohyoideus muscle, and the posterior digastricus muscle. It arises from the pons, passes into the internal auditory meatus, and ultimately exits from the stylomastoid foramen. It then courses through the substance of the parotid salivary gland over the ramus of the mandible. At this point the nerve is exposed to external damage as from a blow or at childbirth from a forceps-assisted delivery.

The sensory portion of the nerve takes its origin from taste buds on the anterior two thirds of the tongue. These fibers are initially found in the lingual nerve, a part of the trigeminal nerve, but later enter the seventh nerve.

Vestibulocochlear nerve. The purely sensory vestibulocochlear nerve (Fig. 9-47) consists of two distinct sets of fibers derived from the cochlea and called the **cochlear nerve** and from the vestibular ganglion that receives fibers from the organs of balance and equilibrium of the inner ear (semicircular canals and maculae). The latter set of fibers is designated the *vestibular nerve*.

Both branches pass through the internal auditory meatus to their central connections.

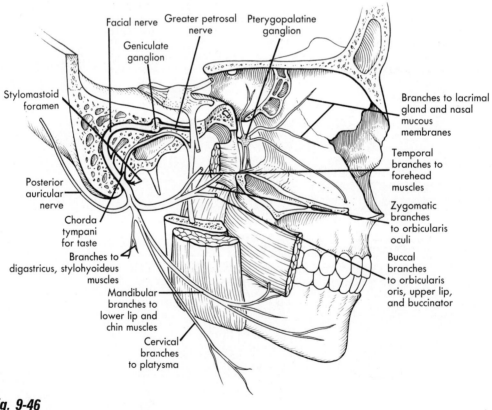

Fig. 9-46
Facial nerve.

Glossopharyngeal nerve. The glossopharyngeal nerve (Fig. 9-48) is a mixed nerve. The motor portion supplies secretory fibers to the parotid salivary gland and small glands in the tongue and pharynx, plus a branch to the stylopharyngeus muscle. The sensory portion of the nerve is derived from the posterior one third of the tongue and serves the sense of taste for that area.

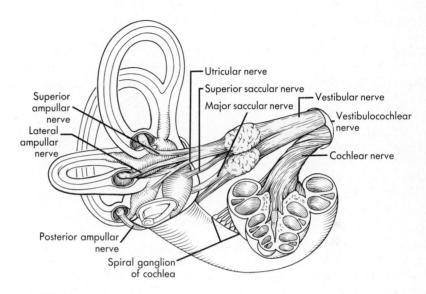

Fig. 9-47
Vestibulocochlear nerve.

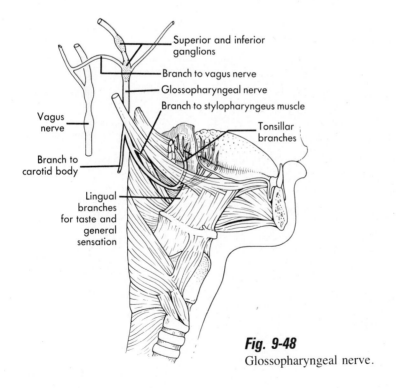

Fig. 9-48
Glossopharyngeal nerve.

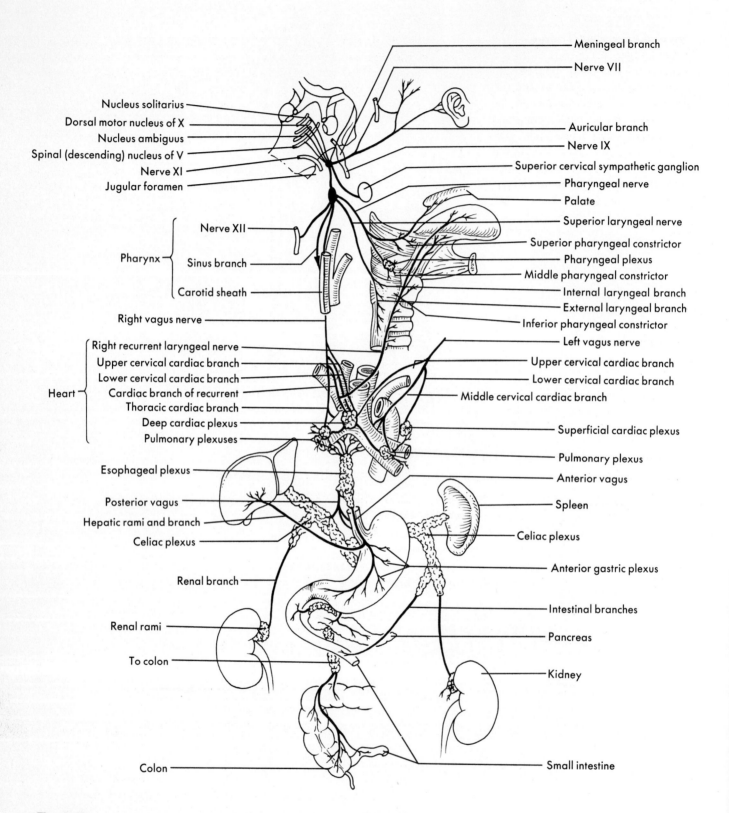

Fig. 9-49
Distribution of tenth cranial nerve (vagus).

Vagus nerve. The name of the vagus nerve (Fig. 9-49) means *a wanderer*. The name is appropriate, for this longest of the cranial nerves is distributed to viscera of the thoracic and abdominopelvic cavities. It uses the jugular foramen to pass from the skull, then progresses down the neck and is bound with the common carotid artery and internal jugular vein by a common connective tissue sheath.

The nerve carries sensory fibers from the viscera and motor fibers to cardiac and smooth muscle and many digestive system glands. This nerve will be mentioned again as an important part of the autonomic nervous system.

Accessory nerve. The accessory nerve (Fig. 9-50) is motor in function and consists of two parts. The *cranial part* is derived by four or five rootlets from the sides of the medulla. The *spinal part* is derived from motor cells in the anterior gray column of the first five cervical segments of the spinal cord. The spinal rootlets pass into the skull through the foramen magnum and join with the cranial rootlets. Both parts exit the skull through the jugular foramen and innervate the sternocleidomastoideus and trapezius muscles.

Hypoglossal nerve. Arising from the medulla, the hypoglossal nerve (Fig. 9-51) exits the skull through the hypoglossal canal and passes anteriorly to innervate the extrinsic and intrinsic muscles of the tongue.

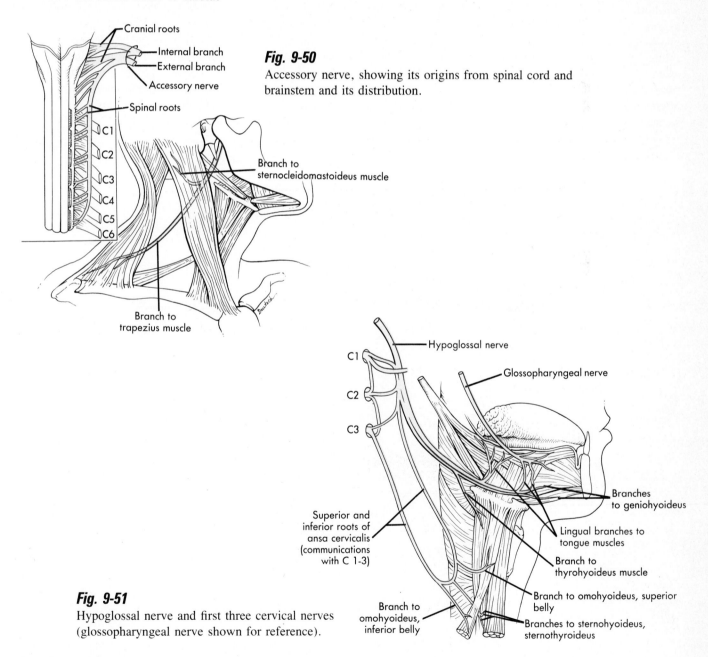

Fig. 9-50
Accessory nerve, showing its origins from spinal cord and brainstem and its distribution.

Fig. 9-51
Hypoglossal nerve and first three cervical nerves (glossopharyngeal nerve shown for reference).

Spinal nerves and plexuses

Spinal nerves. Thirty-one pairs of spinal nerves arise from the spinal cord. They are divided into eight cervical, twelve thoracic, five lumbar, five sacral, and one coccygeal. Each nerve is attached to the cord by a *posterior (sensory)* and *anterior (motor) root*. A *spinal (posterior root) ganglion* containing the cell bodies of the sensory neurons is usually placed at the level of the intervertebral foramen.

White communicating rami provide a pathway for autonomic efferent fibers to join the chain of sympathetic ganglions that lie alongside the vertebral column or for autonomic afferents to reach the cord. *Gray communicating rami* provide a pathway for neurons originating in these ganglia to rejoin the spinal nerve (Fig. 9-52).

Cervical plexus. Cervical nerves 1 to 4 form the cervical plexus (Fig. 9-53). A plexus is a "coming together" of divisions of these nerves before being distributed to their respective organs. The plexuses described in this section are paired. The cervical plexus is located deep to the sternocleidomastoideus muscle of the neck. The plexus supplies nerves to the skin of the head behind and in front of the ear, the skin of the neck, upper chest, shoulder, and many of the hyoid and neck muscles. Fig. 9-53 shows the finer branches from the plexus; the reader should pay attention to the names of the branches, since they indicate the muscles supplied, and to the phrenic nerve. It is an important, although not the only nerve to the diaphragm. It is a derivative of the third cervical nerve.

Brachial plexus. The brachial plexus (Fig. 9-54) supplies nerves to the upper limb. The spinal nerves from cervical five through thoracic one form the plexus. The *roots* of cervical nerves five and six join to form the **superior trunk,** the seventh cervical nerve alone forms the **middle trunk,** and the **inferior trunk** is formed by the joining of roots from the eighth cervical and first thoracic nerves. These trunks progress peripherally for a short distance and then split into **anterior** and **posterior divisions**. These divisions next form **cords,** designated as *lateral, medial,* and *posterior* with reference to the axillary artery. From the cords a number of nerves arise that pass to the skin and muscles of the shoulder girdle and upper appendage.

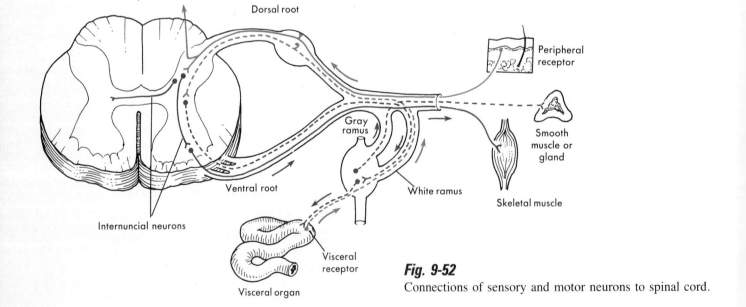

Fig. 9-52
Connections of sensory and motor neurons to spinal cord.

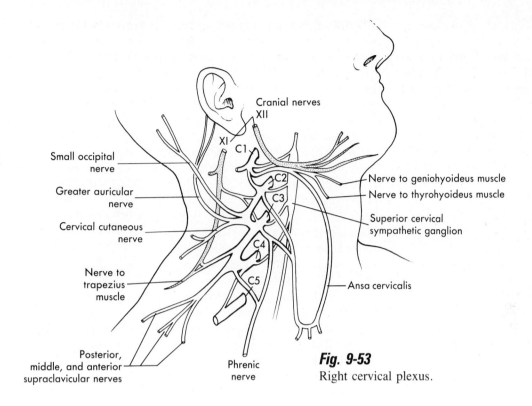

Fig. 9-53
Right cervical plexus.

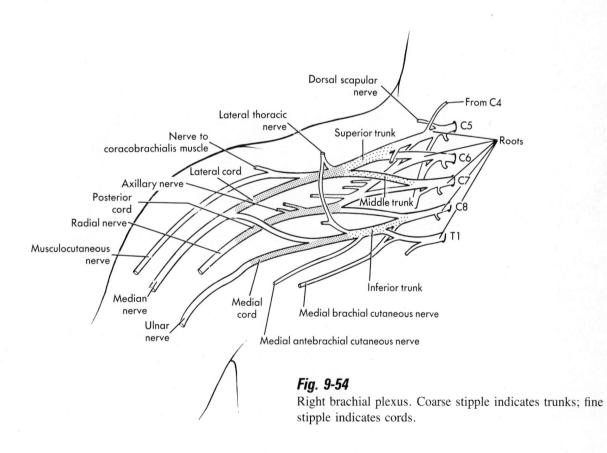

Fig. 9-54
Right brachial plexus. Coarse stipple indicates trunks; fine stipple indicates cords.

Many of the names of the nerves arising from the brachial plexus suggest their distribution. We should concentrate our attention on five of the brachial plexus nerves to the upper appendage:

axillary nerve supplies the *skin of the shoulder* area and the *deltoideus muscle*.

musculocutaneous nerve supplies primarily the skin and muscles of the *anterior upper arm*. Specific muscles innervated include the biceps, brachialis, and coracobrachialis.

median nerve supplies the *radial side of the anterior forearm and hand*. Thus skin and *flexor muscles* of the wrist and digits are innervated.

ulnar nerve serves the *ulnar side of the anterior forearm and hand* and thus also supplies flexor muscles of the wrist and fingers.

radial nerve supplies the *skin and muscles of the posterior arm, forearm, and hand*. The triceps brachii and the extensor muscles of the wrist and digits are muscles supplied.

Thoracic nerves. Thoracic nerves two through twelve do not form a named plexus. They are designated as the *intercostal nerves* and supply the skin of the thorax, the intercostal muscles, and the abdominal and back muscles. Fig. 9-55 indicates the distribution of a typical intercostal nerve. Note that the posterior division of the nerve passes to the muscles of the spine, whereas the anterior division supplies the thoracic and abdominal muscles.

Lumbosacral plexus. The name *lumbosacral plexus* is applied to the combination of the lumbar, sacral, and coccygeal nerves that supply the skin and muscles of the lower limb, perineum, and sacrum. Lumbar nerves one through four are considered to form a *lumbar plexus,* whereas lumbar nerves four and five and the sacral and coccygeal nerves form the *sacral* and *coccygeal plexuses.* Overlap and intermingling of branches of these nerves leads some anatomists to view the whole (lumbosacral) plexus rather than its parts. Fig. 9-56 indicates the plan of these plexuses.

Again, I call attention to several of the nerves that exit from the lumbosacral plexus:

iliohypogastric nerve supplies the skin of the buttocks and hypogastric area and the obliquus internus abdominis and transversus abdominis muscles.

ilioinguinal and *genitofemoral nerves* supply the skin of the external genitalia and inner side of the thigh.

lateral femoral cutaneous nerve supplies the skin of the lateral, anterior, and posterior parts of the thigh as far as the knee.

obturator nerve is the motor nerve for the obturator and adductor muscles of the thigh.

femoral nerve supplies the skin and muscles of the anterior thigh. Thus the vastus, sartorius, rectus femoris, and pectineus are among the muscles supplied.

The **sacral plexus** is concerned primarily with the nerve supply to the posterior thigh and the leg.

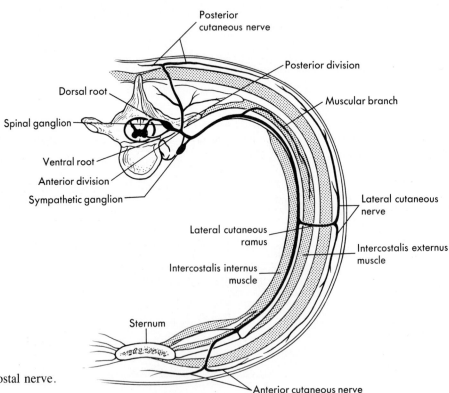

Fig. 9-55
Distribution of intercostal nerve.

Refer to Fig. 9-56 for the plan of the plexus. The major nerves exiting from this plexus are the following:

superior gluteal nerve supplies the gluteus minimus and medius.
inferior gluteal nerve supplies the gluteus maximus.
sciatic nerve is the name given to the combined *tibial* and *common peroneal nerves*. From the nerve, before it branches, is the *nerve to the hamstring muscles* and skin of the same region. The *tibial nerve*, derived from the sciatic nerve above the knee, supplies the posterior calf muscles and the skin of the same area. Branches continue onto the sole of the foot.
common peroneal nerve supplies the skin and muscles of the remainder of the leg.

Pudendal and coccygeal plexuses. Fibers derived from the pudendal and coccygeal plexuses (see Fig. 9-56) supply the external genitalia, bladder, prostate gland, seminal vesicles, urethra, and skin of the coccyx area.

Autonomic nervous system

The term *autonomic nervous system* (ANS) encompasses the whole system of nerves, ganglions, plexuses, and receptors that convey sensory and motor impulses between the body viscera and the CNS. To establish reflex controlling mechanisms for the viscera, it is necessary that the system include *visceral afferent* (sensory) and *visceral efferent* (motor) divisions.

Visceral afferent fibers. Fibers conducting impulses *from* viscera *to* the CNS cannot be separated into an anatomical system different from other fibers because they have their cell bodies in the posterior root ganglion, along with fibers from skin and skeletal muscles (the somatic system). The types of sensations that the receptors of the visceral afferents detect are, however, different from those detected by receptors in the somatic system. They detect pressure changes, chemical differences in body fluids, and tension in the viscera in which they are located.

Some of the more important sources of visceral afferent fibers follow:

In the *head*, afferent fibers are derived from blood vessels and meninges. The ninth and tenth cranial nerves serve as the major route for entry into the CNS.

In the *neck*, fibers are derived from the larynx, trachea, and esophagus and are involved in coughing and swallowing reflexes. Nerves from the carotid sinus convey information from receptors monitoring blood pressure and gas content (O_2, CO_2) of the blood. Most such fibers enter the CNS via the vagus nerve.

In the *thorax*, fibers are derived from the lungs, aorta, and pleurae and convey impulses relating to lung expansion and pressure and gas content in the aorta.

From the *heart* are derived fibers that are concerned with blood volume and that lead to alterations in heart rate.

From the *abdomen*, fibers are derived from intestines, stomach, liver, gallbladder, kidneys, and reproductive organs that enter the CNS through the spinal nerves. Tension, pressure, and pain are the main types of sensations these nerves carry.

In the *extremities*, the afferent fibers are derived from blood vessels and convey impulses associated with pressure in those vessels.

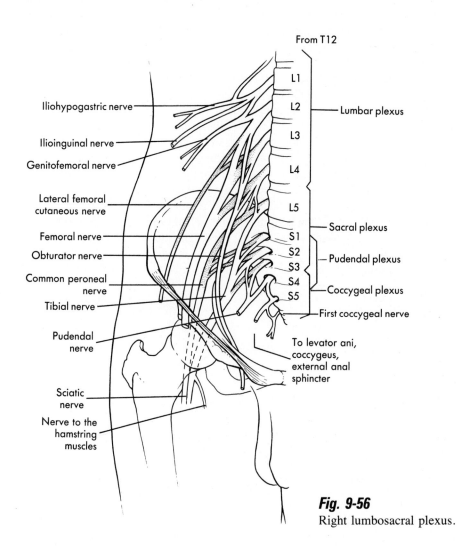

Fig. 9-56
Right lumbosacral plexus.

Visceral efferent fibers. The visceral efferent fibers *do* form separate anatomical and functional subdivisions. The difference here arises from the fact that two neurons are usually required to transmit motor impulses to the viscera and from the fact that a given organ may be caused to either elevate or depress its level of activity; thus two different functional systems are required to achieve such effects.

On the efferent side the neuron that carries impulses from the CNS to a ganglion in the periphery is called a **preganglionic neuron;** the one traveling from the ganglion to the organ is the **postganglionic neuron.** Different effects on a given organ are mediated by the production of different chemicals at the terminal ends of postganglionic neurons.

Parasympathetic (craniosacral) division. Autonomic *cranial outflow* of visceral efferents is carried in the third, seventh, ninth, and tenth cranial nerves. The oculomotor nerve supplies impulses to the sphincter pupillae of the iris (for light control) and to the ciliary muscle (for accommodation of near vision). The facial nerves supply fibers to the lacrimal gland (tear secretion) and to submandibular and sublingual salivary glands (saliva secretion). The glossopharyngeal nerve supplies fibers to the parotid salivary gland (saliva secretion). The vagus, as the major nerve of the cranial parasympathetic outflow, supplies fibers to the heart, most of the smooth muscle in the respiratory and digestive systems, and the muscular blood vessels associated with these organs.

Also forming part of the parasympathetic system is the **sacral outflow,** derived from sacral nerves two through four. Smooth muscle of the colon, bladder, external genitalia, uterus, vagina, and pelvic organs of the male reproductive system are supplied.

The postganglionic neurons of this system secrete acetylcholine (ACh), which has the general effect of maintaining normal levels of organ function (for example, stimulate gut muscle, depress heart rate, and cause vasodilation).

Sympathetic (thoracolumbar) system. Nerves making up the sympathetic system are derived from the thoracic and lumbar spinal nerves. Fibers pass to the organs mentioned previously and provide the *only* innervation to blood vessels in the skeletal muscles, skin, and spleen and to the sweat glands. Their postganglionic fibers secrete norepinephrine (nor-E), which, in general, is associated with elevation of activity (heart and breathing rates elevated). However, gut smooth muscle is inhibited and vessels in viscera are constricted, but those in skeletal muscle, heart, and lungs are dilated. All in all, the responses fit the body to resist sudden stress (the "fight-or-flight" reaction).

The distribution of fibers of the two systems is presented in Fig. 9-57.

Autonomic ganglions and plexuses. The term *ganglion* is generally used to refer to peripheral masses of nervous tissue that contain cell bodies or synapses (junctions between neurons). The term *plexus* usually refers to a network of nerve fibers or blood vessels without cell bodies being present. The terms are often used synonymously.

In the autonomic system there are many ganglions that act as the junction areas between pre- and postganglionic neurons. The **lateral** or **vertebral ganglions,** also called the *sympathetic ganglions,* form paired chains of 22 ganglions that lie alongside the vertebral bodies in the thoracic and abdominal cavities (see Fig. 9-57). They receive preganglionic fibers from the thoracic and lumbar spinal nerves. Four of the ganglions (superior, middle, and inferior cervical ganglions and stellate ganglion) receive special names; the rest do not. **Collateral** or **prevertebral ganglions** are part of the sympathetic nervous system and form groups of nerve cells close to organs in the neck, thorax, and abdomen. The ciliary, sphenopalatine, otic, cardiac, celiac, and mesenteric ganglions are examples of such ganglions (see Fig. 9-57). **Terminal ganglions** are parasympathetic and lie next to or within the organ supplied. The plexuses within the walls of the intestine (submucosal, myenteric) are examples of terminal plexuses.

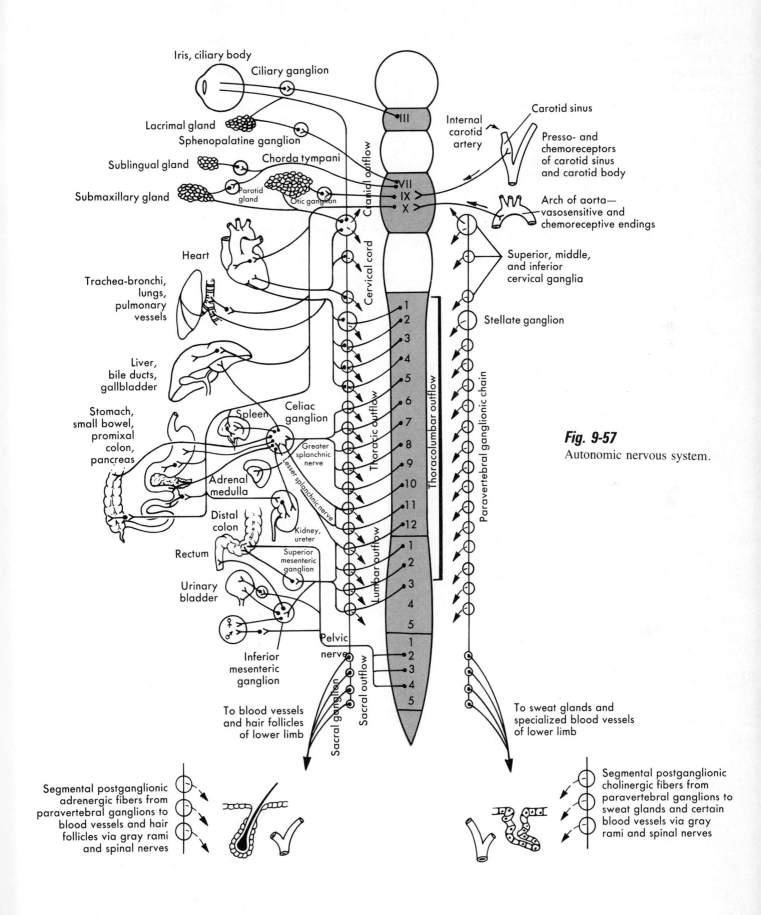

Fig. 9-57
Autonomic nervous system.

Summary

1. The nervous system develops from the germ layer known as ectoderm.
   a. The embryo develops a neural plate, a neural groove, neural folds, and a neural tube. Neural crest cells separate from the tube.
   b. The head end of the tube develops an enlargement that is subdivided into a fore-, mid-, and hindbrain. From these parts, the several sections of the brain will form.
   c. The remainder of the tube forms the spinal cord.
   d. Three layers of cells develop in the walls of the neural tube and form the gray and white matter of the CNS.
   e. Peripheral nervous structures develop from neural crest material.
2. The nervous system is organized into several parts.
   a. The CNS consists of the brain and spinal cord.
   b. The PNS includes all other nervous tissue. It is subdivided into spinal nerves, cranial nerves, and the ANS.
3. Neurons (from neuroblasts) and glial cells (from spongioblasts) form the cells of the nervous system.
   a. Neurons may be uni-, bi-, or multipolar in form; they may have processes (axons and dendrites) that convey nerve impulses, and these processes may be surrounded by sheaths (myelin sheath and neurilemma).
   b. The excitable state in neurons is achieved actively by separating both different kinds and amounts of ions. A nerve impulse is formed when the transport mechanism is interrupted, sodium ion flows into the cell, and an electrical field is created by current flow. The impulse is transmitted by depolarization of the next segment of the cell by the impulse, with recovery of the original state occurring behind the transmitted impulse.
   c. Neurons follow the all-or-none law, recover very rapidly, do not allow the strength of the impulse to decrease as it is transmitted, and, if possessing a myelin sheath, exhibit saltatory conduction (a faster rate of conduction).
4. Glial cells occur in several varieties (astrocytes, oligodendrocytes, microglia, satellite cells, ependyma, Schwann's cells) and act as nutritive, protective, and supportive cells for neurons and their processes.
5. A synapse is an area of functional but not anatomical continuity between two neurons.
   a. Transmission of impulses across a synapse is by chemical means, the chemical being destroyed after exerting its effect.
   b. A synapse conducts in only an axon-dendrite direction, determines if the impulse passage will be enhanced, blocked, or changed, transmits more slowly than a neuron, and is very sensitive to the characteristics of its environment.
6. Reflex arcs enable automatic response to changes in internal or external environments. An arc always has five parts:
   a. A receptor to detect change
   b. An afferent neuron to carry impulses *to* the CNS
   c. A center to integrate information
   d. An efferent neuron to carry impulses *away from* the CNS
   e. An effector to respond and maintain homeostasis
7. If a nerve cell body is damaged, the entire unit dies. Axons or dendrites can regenerate if damaged peripherally.
   a. Fiber damage results in changes in the cell body (the axon reaction) and in the fiber itself (wallerian degeneration). The fiber degenerates, and a new tip develops on the part proximal to the injury. It grows to its former area of innervation by passing through the neurilemma tube that remains.
   b. Demyelinating diseases result in loss of myelin sheaths and loss of function.
8. Nerves are individual neuron processes bound together by connective tissue (endo-, peri-, and epineurium) to form the large visible (with the naked eye) conduction pathways of the PNS.
9. The brain consists of four major parts:
   a. The cerebrum is the largest portion.
   b. The cerebellum is next largest in size.
   c. The brainstem supports the other two parts.
   d. The diencephalon lies above the brainstem.

10. The cerebrum has the following characteristics:
    a. It is convoluted, with gyri (upfolds), sulci (shallow downfolds), and fissures (deep clefts) that separate it into hemispheres and lobes.
    b. Fissures include longitudinal, central, parietooccipital, calcarine, and lateral.
    c. Lobes include frontal, parietal, occipital, and temporal.
    d. Within the lobes, major gyri and sulci are named.
    e. Functional areas of the lobes are numbered, and a discussion of the major functional regions is presented (pp. 266-267).
    f. The cerebral cortex histology shows (in most areas) five types of cells and six layers.
    g. The white matter of the cerebrum constitutes the medullary body that consists of association, projection, and commissural fibers.
    h. The basal ganglions lie within the cerebral white matter. They are motor in function.
11. The cerebellum is characterized by the following:
    a. It has a vermis and two hemispheres.
    b. It has an outer cortex organized in three layers and an inner medullary body.
    c. It contains several deep nuclei.
    d. It coordinates muscular activity by error control, damping, and predictive functions.
12. The diencephalon is composed of the thalamus and hypothalamus.
    a. The thalamus contains many named nuclei (see Fig. 9-22, C) that receive and organize sensory information.
    b. The hypothalamus contains many nuclei (see Fig. 9-23) that control body homeostasis (food intake, temperature, water balance).
13. The brainstem consists of the medulla, pons, and midbrain.
    a. The medulla contains fiber tracts and nuclei of the ninth through twelfth cranial nerves.
    b. The pons contains the cerebellar peduncles and nuclei of the fifth through eighth cranial nerves.
    c. The midbrain contains the colliculi and nuclei of the third and fourth cranial nerves.
    d. The reticular formation is a column of gray matter extending through the whole brainstem.
    e. The limbic system includes parts of the diencephalon and stem and functions in the expression of emotions.
14. The spinal cord has the following characteristics:
    a. It extends through the vertebral canal.
    b. It has two enlargements and a tapering conus.
    c. It is grooved by several sulci and a single fissure.
    d. It gives rise to 31 pairs of spinal nerves.
    e. It possesses (internally) centrally placed gray matter and peripherally placed white matter. The gray matter is divided into posterior, lateral, and anterior columns; the white matter is divided into corresponding funiculi.
    f. It has functional areas designated as tracts (see Fig. 9-33).
    g. It controls myotatic, flexor, and crossed-extension reflexes on its own.
15. Membranes collectively known as the meninges surround the CNS.
    a. An outer dura mater is tough and protective.
    b. A middle arachnoid encloses a space filled with fluid.
    c. An inner pia mater is a vascular membrane.
    d. The dura forms membranes that lie between brain portions.
16. Ventricles within the brain contain cerebrospinal fluid produced by choroid plexuses. The fluid also fills the subarachnoid spaces around the brain and spinal cord.
17. The PNS consists of all nervous tissue outside the CNS.
    a. There are 12 pairs of cranial nerves. Names and functions are presented in Table 9-2.
    b. Thirty-one pairs of spinal nerves supply peripheral organs.
    c. Cervical, brachial, lumbar, and sacral plexuses are formed by the spinal nerves. Major nerves exiting from these plexuses and their areas supplied are described.
18. The autonomic nervous system supplies body viscera with sensory and motor nerves.
    a. The parasympathetic division is composed of cranial and sacral spinal nerves.
    b. The sympathetic division is composed of thoracic and lumbar spinal nerves.
    c. Two neurons usually are present in the motor pathways, one in the sensory pathway.
    d. Several locations of autonomic ganglions are described.

Questions

1. What are the major steps in the early development of the nervous system?
2. How is the nervous system organized?
3. Describe the structure of a multipolar neuron and the sheaths that are present on its axon. Account for formation and transmission of a nerve impulse.
4. What are the several types of glial cells? Where are they found? What functions do they have?
5. How is a synapse constructed? How does an impulse cross it? What are some of the properties of the synapse?
6. Describe a reflex arc and how it operates. Give some characteristics of a reflex.
7. For each of the following parts of the brain, give an outline of major anatomical features and the functions each has.
   a. Cerebral hemisphere
   b. Cerebellum
   c. Brainstem
   d. Diencephalon
8. What are the meninges of the CNS? Where is each located? What function(s) does each serve?
9. Describe the external and internal structure of the spinal cord.
10. Describe the form and location of the four ventricles of the brain.
11. Name the cranial nerves and the function of each.
12. Describe the manner of attachment of the spinal nerves to the spinal cord and the major plexuses they form.
13. Name *at least* two major nerves rising from each plexus (named in 12) and the area of the body supplied.
14. Name the two divisions of the autonomic nervous system, their components, and effects on the body.

Readings

Austin, G.: The spinal cord, Springfield, Ill., 1971, Charles C Thomas, Publisher.

Barr, M.L.: The human nervous system, an anatomic viewpoint, ed. 3, New York, 1979, Harper & Row, Publishers, Inc.

Brown, D.R.: Neurosciences for allied health therapies, St. Louis, 1980, The C.V. Mosby Co.

Bunge, R.P.: Glial cells and the central myelin sheath, Physiol. Rev. **48:**197, 1968.

Galaburda, A.M., and others: Right-left asymmetries in the brain, Science **199:**852, 1978.

Gluhbegovic, N., and Williams, T.H.: The human brain: a photographic guide, New York, 1980, Harper & Row, Publishers, Inc.

Kostyuk, P.G., and Vasilenko, D.: Spinal interneurons, Ann. Rev. Physiol. **41:**115, 1979.

Moore, K.L., Bertram, E.G., and Barr, M.L.: Study guide and review manual of the human nervous system, Philadelphia, 1978, W.B. Saunders Co.

Rapoport, S.I.: Blood brain barrier in physiology and medicine, New York, 1976, Raven Press.

Reinis, S., and Goldman, J.M.: The development of the brain, Springfield, Ill., 1980, Charles C Thomas, Publisher.

Routtenberg, A.: The reward system of the brain, Sci. Am. **239:**154, Nov. 1978.

10

Organs of sensation

Objectives
Some characteristics of receptors
Classification of receptors
Cutaneous and membrane sensation
    Location and structure of receptors
    Pathways to the central nervous system
Visceral sensation
    Location and structure of receptors
    Pathways to the central nervous system
Muscle, tendon, and joint sensation
    Location and structure of receptors
    Pathways to the central nervous system
Eye
    Development
    Size, shape, and landmarks
    Tunics of the eye
    Visual pathways
    Extrinsic eye muscles
    Eyelids and lacrimal apparatus
    Selected disorders of the eye and its associated structures
Ear
    Development
    Structure
    Auditory and equilibrial pathways
    Selected disorders of hearing and equilibrium
Sense of taste
    Distribution and structure of taste buds
    Taste pathways
Sense of smell
    Location and structure of the olfactory epithelium
    Olfactory pathways
Summary
Questions
Readings

OBJECTIVES After studying this chapter, the reader should be able to:

Give some of the basic characteristics of receptors.
Define a receptor, and classify receptors according to location and adequate stimulus.
Give the names, locations, structures, and neural pathways for all receptors other than those of the special senses.
Describe the landmarks of the eye, its tunics, and the structure of each subdivision of the tunics.
Outline the course of the visual pathways to the brain.
Describe the lacrimal apparatus.
Name the extrinsic eye muscles, their innervation, and the eye movements each muscle causes.
List the three parts of the ear, the structures included in each part, and the structure of the specialized sense organs of the ear.
Describe the neural pathways for the senses of equilibrium and hearing.
Compare the locations, structure, and mode of operation of the senses of taste and smell.
Trace the neural pathways for the senses of taste and smell.

Maintaining body homeostasis requires detection of changes that occur both within and outside the body. *Receptors* of diverse types are the structures that respond to changes and may be considered the starting point for reflexly controlled effector response. The fact that many types of receptors are found within the body is the result of specialization of response to particular kinds of stimuli that constitute the adequate stimulus for each receptor. All receptors transduce (change) a form of energy into a nerve impulse for transmission to the CNS. Receptors may have an extremely complicated structure, as in the eye and ear, or be as simple as naked nerve endings within the skin serving the sense of pain.

The object of this chapter is to discuss the types of stimuli that each receptor responds best to, to indicate the pathways a particular sensation traverses to its termination in the brain, and to indicate the value of each receptor to the overall operation of the body.

Some characteristics of receptors

Specialization of receptor response is embodied in the *law of adequate stimulus,* which states that a particular receptor will respond best although not exclusively to a particular form of energy. Thus the eye is stimulated best by light but gives a sensation of light flashes if pressure is applied to it. **Specificity** to some degree follows as a result of the law. Once a receptor has been stimulated by a particular form of energy, the nerve impulse that is carried from it is no different from that carried from an entirely different receptor (transduction). What we appreciate as a particular sensation depends on the central connections the nerves ultimately make. These tenets are embodied in the *law of specific nerve energies*. A receptor can often **signal intensity** of a stimulus by altering the frequency of its discharge. In general, higher frequency of receptor discharge results from stronger stimuli and is interpreted by the brain as greater intensity of stimulation. With continuing stimulation at a given strength some receptors slow their frequency of firing; others do not. The former are called **rapidly adapting receptors** and seem designed to communicate qualitative data. That is, they notify of a stimulus, then cease or diminish greatly their transmission of information. Touch and pressure receptors are examples of this type. **Poorly adapting receptors** do not adapt or do so very slowly and continue to transmit information as long as the stimulus continues. Examples of the second type of receptors are the muscle and joint organs that must continue to supply the brain with data on body position, movement, and posture.

This list of receptor characteristics emphasizes that receptors provide much information about the nature of the stimulus as well as its mere presence.

Classification of receptors

One method of classifying receptors uses their *positions* within the body. According to this scheme we may define the following:

exteroceptors are located at or near the external body surface and react primarily to stimuli originating in the external environment. Included here are receptors sensitive to touch and pressure, heat, cold, and pain and the eye and ear.
enteroceptors lie within the walls of internal body organs and sense pressure, tension, tastes, odors, and changes in composition of body fluids. This group may include the same type of receptors as in the first group, such as pressure, and includes several of the *special sense organs*, the taste buds, olfactory epithelium, and equilibrial structures of the ear.
proprioceptors lie within muscles and tendons and give information as to body position and rate and direction of muscle-caused movements.

Another means of classifying receptors is by their adequate stimulus. Thus we may speak of the following:

chemoreceptors are structures monitoring the chemical composition of body fluids. A good example is the receptors in certain blood vessels that measure oxygen and carbon dioxide levels of the blood and the cells of taste buds and olfaction.
mechanoreceptors measure pressure, bending, or tension. They are found in the skin (touch, pressure), internal viscera (pressure, tension), in the inner ear (hearing, equilibrium), and in muscles, joints, and tendons.
  baroreceptors are a type of mechanoreceptor responding to pressure.
photoreceptors, exemplified by the rods and cones of the retina, are stimulated by light.
thermal receptors respond to small changes in skin temperature brought about by activity or alterations in environmental temperature.

It seems appropriate to consider receptors from the point of view of location and to point out what types of stimuli each best responds to.

Cutaneous and membrane sensation

Location and structure of receptors

Naked nerve endings. The simplest of the receptors located in the skin and mucous membranes of the body are **naked nerve endings** serving the senses of pain and temperature (Fig. 10-1, *A*). The nerve endings lie between the cells of the lower epidermal layers and in the epithelium of the cornea, mouth, nasal cavities, pharynx, anal canal, and other areas. Naked nerve endings for temperature are especially abundant in the cornea and mucous membranes.

Four varieties of pain are generally recognized:

"Bright" *pain* is exemplified by the sensation experienced when cutting a finger. It may be intense but usually does not last long and is easily localized.
"Dull" *pain* of the sort experienced when having a toothache is slower to develop, lasts longer, and has a diffuse quality. It is often difficult to "take."
Deep pain results from tension or pressure within internal body viscera and often produces autonomic symptoms such as nausea.
Referred pain occurs when there is a source of pain in one organ, but the pain is localized as coming from some other body area. For example, the pain of angina (from insufficient blood flow to the *heart*) seems to be coming from the shoulders or arm. This is explained as the result of intense sensory information from the heart reaching neurons in the cord and "spreading" to neighboring neurons that receive information from the appendage. The brain interprets the signal as though it originated from the limb.

Thermal receptors. The *end bulb of Krause (Krause's corpuscle)* (Fig. 10-1, *B*) serves the cutaneous and mucous membrane sensation of cold. Actually monitoring the skin or membrane temperature at its particular dermal or laminal location, the bulb will respond to both a change in temperature (a fall of temperature of as little as 0.003° C) and to the *rate* of fall of temperature

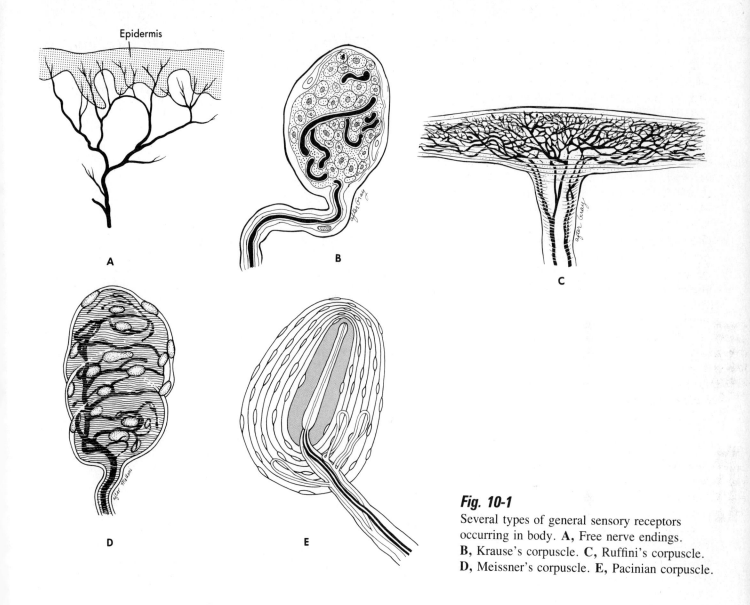

Fig. 10-1
Several types of general sensory receptors occurring in body. **A,** Free nerve endings. **B,** Krause's corpuscle. **C,** Ruffini's corpuscle. **D,** Meissner's corpuscle. **E,** Pacinian corpuscle.

(0.001° C/sec). Krause bulbs normally form impulses between 10° C and 41° C, with a peak rate of discharge at 10° C to 20° C. They again discharge strongly at 46° C to 50° C. Thus when we step into a hot shower, we may feel cold and shiver.

The *corpuscles (organs) of Ruffini* (Fig. 10-1, *C*) are found at the junction of the dermis and hypodermis and are considered to be receptors for warmth. These receptors form impulses between 20° C and 45° C, with a peak at 37.5° C to 40° C. As small a change as 0.001° C/second (in an increasing direction) will suffice to cause response.

Numbers of cold receptors exceed warmth receptors by between 4 and 10:1. Cold thus seems to be a greater threat to the body than warmth, if numbers of receptors indicate anything.

Both heat and cold receptors send information to the heat gain and heat loss centers of the hypothalamus. If the data suggest the temperature outside is cold, the heat gain center initiates an increased degree of muscular contraction (tone), a cutaneous vasoconstriction, and diminished sweating, measures that conserve or increase body heat. Reverse reactions occur if the heat loss center is activated. The receptors are therefore a very important part of the body's temperature-regulating mechanism.

Tactile sensations. *Meissner's corpuscles* (Fig. 10-1, *D*) are delicate receptors found in the dermal papillae of the skin. They form impulses when the surface of the skin is deformed by light pressure (touch).

Pacinian corpuscles (Fig. 10-1, *E*) are multilayered deeper lying structures that respond to more intense deformation of the skin. These receptors are found in some unlikely places in the body: for example, the pancreas and the mesenteries supporting the abdominal viscera.

Pathways to the central nervous system

Pain fibers from all areas except those served by cranial nerve V pass over the afferent neurons of the spinal nerves and the posterior root and enter the spinal cord. A synapse occurs in the posterior gray column of the side of entry with a second-order neuron that crosses to the opposite side of the cord and enters the lateral spinothalamic tract. This tract ascends to the posterolateral ventral nucleus of the thalamus, where synapses are made with third-order neurons. The third-order neurons ascend to areas 3, 1, and 2 of the parietal lobe, where final analysis is provided.

Thermal sensations also use the lateral spinothalamic tract, the same thalamic nucleus as pain fibers, and terminate, as third-order neurons, in areas 3, 1, and 2 of the cerebral cortex. The pathways are shown in Fig. 10-2.

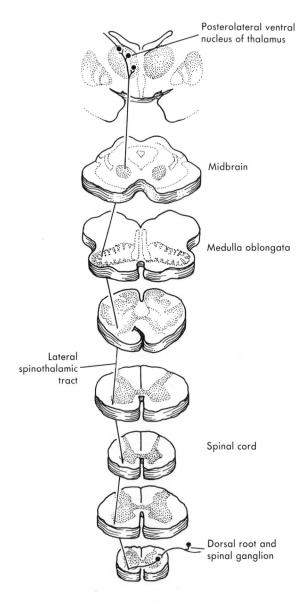

Fig. 10-2
Spinal tract for pain, heat, and cold (lateral spinothalamic tract).

Tactile sensations enter the spinal cord over the posterior roots. A synapse may occur in the posterior gray column, in which case the second-order neuron crosses to the opposite side of the cord and enters the ventral spinothalamic tract ("crude" touch and pressure). If no synapse occurs in the posterior columns, the fibers stay on the same side of the cord and ascend via the gracile and cuneate tracts to the gracile and cuneate nuclei of the brainstem. Fibers pass next to the thalamus and from there to areas 3, 1, and 2 of the cerebral cortex. These pathways are shown in Fig. 10-3.

Anesthesia refers to partial or complete loss of sensation from a body part. It may be caused by receptor destruction, as when the skin is badly burned, or by interruption of the nervous pathway to the brain, as in cord section. *Paresthesia* is an abnormal or heightened sensation without demonstrable objective cause. It is usually attributed to a peripheral or central nerve lesion. An interesting type of abnormal sensation is that of the "phantom limb." Amputees may have a sensation referred to a missing body part. Nerves that are severed by surgery may recover to the point that irritation will cause impulses to be transmitted to the brain, and that organ then localizes the sensation as though it originated in the missing part. In other words, the pathway beyond the amputation still remains.

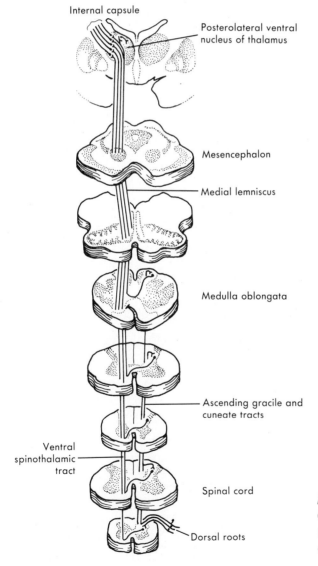

Fig. 10-3

Cord pathways for touch and pressure (ventral spinothalamic and gracile and cuneate tracts).

Visceral sensation

Location and structure of receptors

The primary stimuli visceral receptors respond to appear to be pressure or tension and certain chemical changes in the blood. Baroreceptors responding to pressure or tension are found in the blood vessels, alimentary tract, uterus, and urinary bladder, to name some obvious sites. They form receptors for reflex arcs that control cardiovascular function, gut motility, and micturition, to name a few examples.

The structure of baroreceptors is exemplified by the organs found in the carotid sinus and aorta that sense elevation of blood pressure (Fig. 10-4). In these regions, lamellated oval bodies about 4 μm long may be found. They resemble pacinian corpuscles. Distention of the vessel wall or elongation of the vessel causes impulse formation by the receptor.

Chemoreceptor structure may be exemplified by the aortic and carotid bodies that sense elevation of carbon dioxide or depression of oxygen levels in the bloodstream. In these bodies are *glomus cells* that are surrounded or partially enclosed by *support cells* (Fig. 10-5). The latter cells lie between the glomus cells and the nerve fiber supplying the assembly. Chemoreceptors appear to be involved primarily in control of rate and depth of breathing to ensure adequate carbon dioxide elimination and oxygen intake.

Specific nerves for pain appear to be lacking in most internal viscera. Distention or stretching of an organ may result in a sensation interpreted as pain.

Pathways to the central nervous system

The general pathway taken by visceral sensation involves the ninth and tenth cranial nerves and the thoracic, lumbar, and sacral spinal nerves. Those fibers from spinal nerves are carried in the lateral funiculus to terminations in the thalamus and hypothalamus. Those fibers found in cranial nerves terminate in the sensory nucleus of the involved nerve and from there pass to the hypothalamus and to the cardiac, respiratory, and vasomotor centers in the medulla. Reflex arcs are thus established that control rate and depth of breathing, heart rate, and blood pressure.

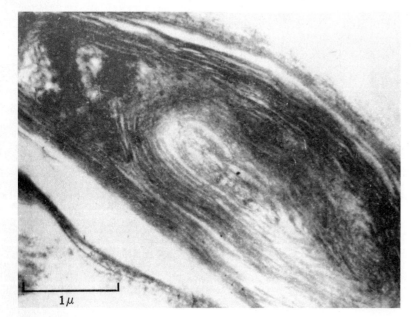

Fig. 10-4
Structure of baroreceptor (carotid sinus).

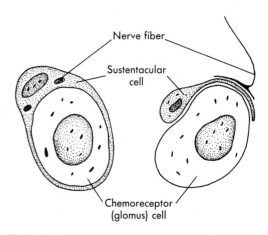

Fig. 10-5
Two theories of structure of carotid body chemoreceptor. *Left,* Sustentacular (support) cell may envelop glomus cell. *Right,* Support cell is intermediate between nerve and glomus cell.

Muscle, tendon, and joint sensation

Location and structure of receptors

Receptors in muscles, tendons, and joints provide information about pressure, tension, and degree of muscle contraction as movements occur.

Joint sensation. Pacinian corpuscles are found in the ligamentous structures of joint capsules, where they provide awareness of position and movement of body parts (the *kinesthetic sense*). If a joint is injured and swells, pacinian corpuscles contribute to appreciation of that swelling.

Neuromuscular spindles. Spindles (Fig. 10-6) are important in the reflex control of muscle tension and are the most important receptors for the myotatic (stretch) reflexes. Spindles are about 1 mm in diameter and about 6 mm long and are oriented parallel to the fibers of the muscles in which they are found. Each spindle consists of a fusiform-shaped connective tissue capsule that contains two to ten tiny **intrafusal fibers**. These fibers lack central striations and have their nuclei within the fiber rather than at their periphery. Some fibers, known as *nuclear bag fibers*, have an expansion in their center in which the nuclei are concentrated; in others, known as *nuclear chain fibers*, no central expansion is present, and nuclei are arranged in a row in the fiber.

Two types of sensory nerve endings terminate on both types of fibers. The **annulospiral endings** form spirally wrapped structures around the midregion of the intrafusal fibers, whereas *flower spray endings* are derived from other sensory neurons and terminate by expansions on the ends of the intrafusal fibers. Extension of the intrafusal fibers stimulates both types of sensory endings. Motor fibers are also supplied to the intrafusal fibers, terminating as typical neuromuscular junctions on the contractile ends of the intrafusal fibers.

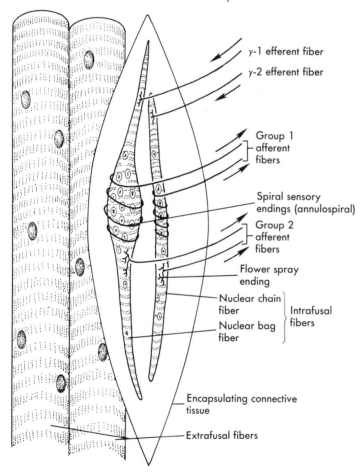

Fig. 10-6
Structure and innervation of muscle spindle. Group 1 and 2 afferent fibers supply annulospiral and flower spray endings respectively. γ-1 and γ-2 efferent fibers supply contractile portions of nuclear bag and nuclear chain fibers respectively.

Thus the amount of tension required to elongate the fibers may be adjusted. The spindles also convey impulses to the spinal cord, where connections are made via efferent fibers to the fibers of the muscle itself (the extrafusal fibers), resulting in reflex contraction of the whole muscle that maintains body posture.

Neurotendinous spindles. Also known as Golgi tendon organs, neurotendinous spindles (Fig. 10-7) are quite numerous in the area of junction of a muscle with its tendon. A thin connective tissue capsule surrounds a few collagenous fibers of the tendon on which the branches of the sensory nerve terminate. More tension is required to stimulate these receptors, because of the limited ability of tendons to elongate. These receptors function in the same manner as the muscle spindles.

Naked nerve endings. Naked nerve endings terminate in joints and apparently serve the sense of pain if the joint is injured, becomes inflamed, or becomes swollen. Some pain fibers also supply the bursae associated with some joints and may signal the pain of bursitis.

Pathways to the central nervous system

Afferent fibers from muscle, joint, and tendon receptors enter the CNS over spinal posterior roots on which their cell bodies are located or via cranial nerves. Those fibers entering via spinal nerves use primarily the spinocerebellar tracts as their pathways to the cerebellum. Information concerning rate of movement of body parts, strength of contraction of muscles, and direction of movement is integrated in the cerebellum to provide coordinated and directed motion via motor fibers that eventually terminate on the muscles themselves.

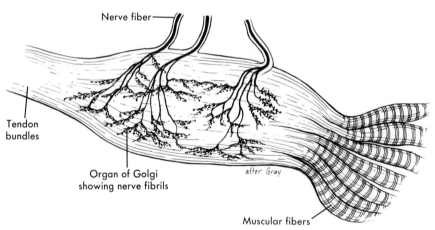

Fig. 10-7
Golgi tendon receptors.

Eye

The eye is the organ of vision and includes devices enabling control over amount of light entering the organ and focusing of images on the retina.

Development

Eye development (Fig. 10-8) is first noticeable early in the fourth week with the appearance of the *optic grooves* in the neural folds at the head end of the embryo. An outpocketing from the groove results in the formation of hollow *optic vesicles* whose outer ends expand to form *optic cups.* The surface ectoderm next to the cups is thickened by the presence of the cup and gives rise to the *lens placodes* that will form the lenses of the eyes. The placodes fold inward to form a *lens vesicle,* and the "lips" of the optic cup grow around the lens vesicle, enclosing it within the developing eye. The outer and inner layers of the optic cup form the retina, whereas the uveal and scleral coats form from mesenchyme that lies outside the optic cup.

The eye muscles, six in all, develop from three myotomes, each served by a single cranial nerve. One of the myotomes forms four of the muscles of each eye while retaining the single cranial nerve innervation. The other two myotomes give rise to a single muscle, each served by a single cranial nerve.

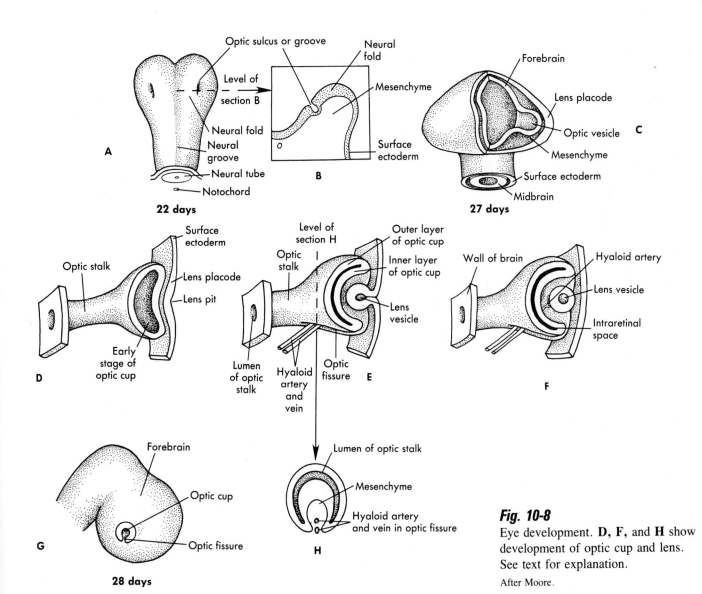

Fig. 10-8

Eye development. **D, F,** and **H** show development of optic cup and lens. See text for explanation.

After Moore.

Size, shape, and landmarks
(Fig. 10-9)

The adult human eyeball is a roughly spherical organ about 24 mm in diameter and weighing 6 to 8 g. The center of the cornea is the **anterior pole,** whereas the **posterior pole** lies opposite at the greatest diameter in the anteroposterior dimension; it lies between the *fovea* (a depression of the retina) and the **optic papilla,** where the optic nerve exits from the eye. A line drawn between the two poles establishes the **anatomical** *(optical) axis* of the eye. The **visual axis** is a line drawn between the fovea and the apparent center of the pupil; it establishes the path of light rays when a person is looking at a distant object. The **equator** is a line drawn at right angles to the visual axis through the greatest lateral expansion of the eyeball.

The anterior transparent portion of the eyeball (cornea) has a radius of curvature of about 7.8 mm, whereas the remainder of the eyeball has a radius of curvature of about 13 mm. Thus segments of *two* spheres form the eyeball.

Tunics of the eye

Three tunics or tissue layers form the eyeball. The outer *fibrous tunic* includes the *sclera* (white of the eye) and the transparent *cornea.* The centrally placed *uvea* includes the anterior *ciliary body* and the posterior *chorioid* (choroid). The inner *retina* contains the light-sensitive cells of the eye.

Fibrous tunic

Sclera. The sclera forms about the posterior five sixths of the eyeball. It is about 1 mm thick at the posterior pole, 0.3 mm thick at the equator, and 0.6 mm thick at the sclerocorneal junction. It consists of flat collagenous bundles running in various directions parallel to the surface of the eyeball. Elastic fibers occupy the spaces between the collagenous bundles, and fibrocytes also lie between collagenous bundles. Melanocytes (pigment cells) occur in the deeper layers of the sclera and are especially abundant near the exit of the optic nerve. The superficial layers of the sclera are more compact and form what is called Tenon's capsule; it provides a firm layer for attachment of the tendons of the muscles that move the eyeball.

Cornea. The cornea (Fig. 10-10) is 0.8 to 0.9 mm thick at its center and about 1 mm thick at its edges. It, plus the *aqueous fluid*–filled anterior cham-

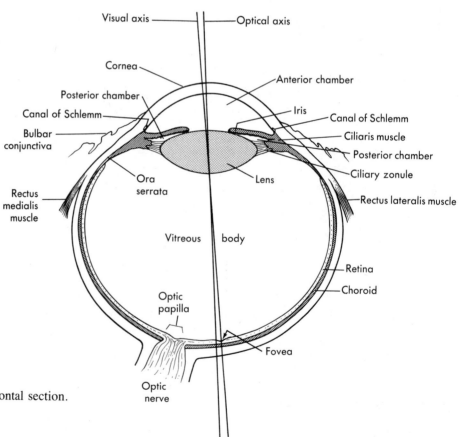

Fig. 10-9
Right eye in horizontal section.

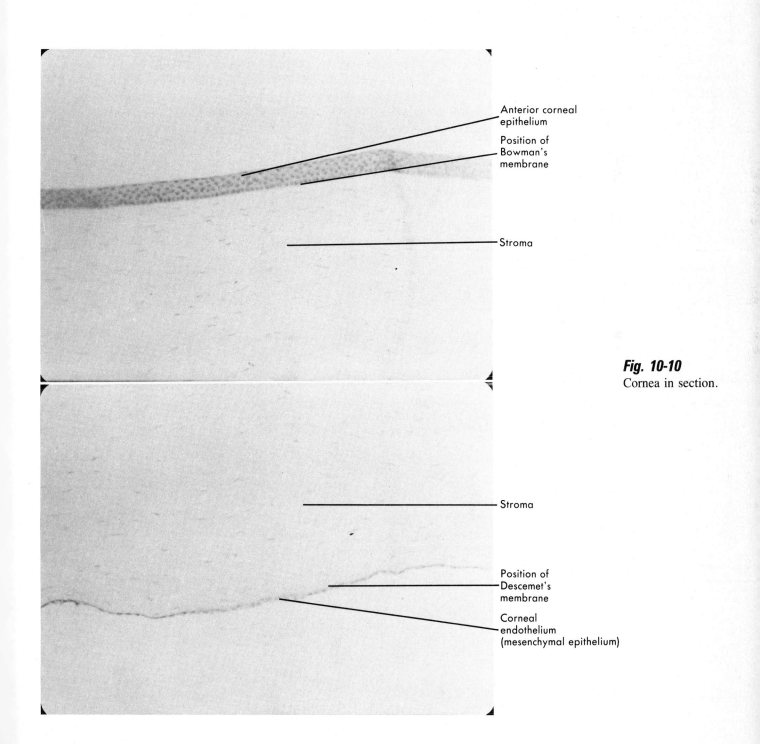

Fig. 10-10
Cornea in section.

ber (see Fig. 10-9), forms a lens of fixed curvature that initially refracts (bends) light rays as they enter the eye. Five layers of tissue compose the cornea:

anterior corneal epithelium is a stratified squamous type, 50 μm thick, and usually consisting of five layers of cells. Many free nerve endings are found between the cells and give the cornea senses of heat, cold, touch, and pain. The epithelium has great regenerative capacity; minor injuries are healed by adjacent cells gliding to fill the defect.

Bowman's membrane is a 6- to 9-μm thick *basement membrane* for the epithelium. Actually the superficial layer of the stroma, the membrane consists of randomly ordered collagen fibrils.

stroma (substantia propria) forms 90% of the thickness of the cornea and consists of transparent collagenous fibrils.

Descemet's membrane is a 5- to 10-μm thick basement membrane for the innermost layer.

corneal endothelium (mesenchymal epithelium) is a simple squamous epithelium forming the posterior corneal surface.

At the sclerocorneal junction is a flattened circumferentially arranged vessel, Schlemm's canal. Aqueous humor produced by the ciliary processes enters the posterior chamber, flows through the pupil to the anterior chamber, and then flows through a loosely arranged mass of fibers at the cornea-iris junction to be drained by the canal. A circulation of fluid is thus established.

Uvea. The uvea is the vascular tunic of the eyeball. Posteriorly, the uvea is composed of the **chorioid;** anteriorly, it consists of the **ciliary body,** which includes the *iris, lens, ciliary zonule, ciliary processes,* and *ciliary muscle.*

Chorioid. Reflecting its vascular nature, the outer two layers of the chorioid (see Fig. 10-9) are called the *vessel layer,* containing small arteries and veins, and the *choriocapillaris,* containing capillary networks. These layers contain many melanocytes, and the entire chorioid has a brown to black color. *Bruch's* (glassy) *membrane* is about 2 μm thick and acts as a basement membrane for the *pigment epithelium,* a layer considered to be part of the chorioid, although listed as the first of the 10 layers of the retina.

Ciliary body. The main mass of the ciliary body (Fig. 10-11; see also Fig. 10-9) is composed of the *ciliary muscle.* It is smooth muscle that is innervated and that can exhibit variable degrees of contraction for focusing. The muscle has an anterior origin in the area of the base of the iris and a posterior insertion on the chorioid. Its contraction pulls the chorioid forward, relaxing the tension on the *ciliary zonule,* the circularly arranged ligaments that attach to the lens. The lens then "rounds up" to change its curvature and bring light rays to a proper focus on the retina. The *ciliary processes* have been described as the site of aqueous humor production.

The *iris* is a ringlike curtain of tissue positioned ahead of the lens. The opening in its center is the *pupil.* The iris is

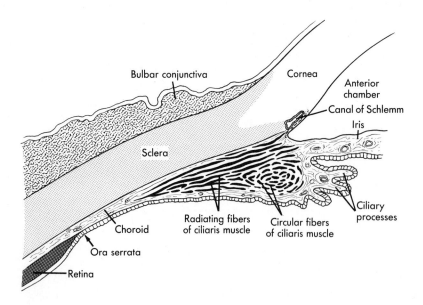

Fig. 10-11
Ciliary body.

composed of loose, melanocyte-rich connective tissue (the pigments of the melanocytes give "eye color"). Anteriorly it is covered by a layer of fibroblasts and melanocytes and posteriorly by an extension of the pigment layer of the chorioid. The iris can adjust the amount of light entering the eye. The *sphincter pupillae* is a mass of circularly arranged smooth muscle close to the pupillary margin. The *dilator pupillae* consists of radially arranged columns of smooth muscle in the outer portion of the iris. Both muscles are served by nerves.

The *lens* is a transparent, biconvex structure located behind the pupil. It is surrounded by an *elastic capsule* that affords attachment for the ciliary zonule. Anteriorly, there is an *epithelium* whose cells form the *lens fibers* that compose the bulk of the lens. These fibers are normally pliable and can be "molded" by the elastic capsule according to the tension on the ciliary zonule. As we age, the plasticity of the lens diminishes, and the lens tends to remain in a more flattened state.

Retina. The retina contains the photosensitive elements of the eye, the rods and cones. In most parts of the retina, 10 layers of tissue compose the tunic (Fig. 10-12):

external pigment epithelium appears as a single layer of cuboidal cells whose inner margins are filled with pigment granules. The pigment absorbs "stray" photons of light, and the cells themselves may phagocytose the debris that results from shedding of rod and cone discs.

bacillary layer is the second layer, consisting of the photosensitive and supportive portions (actually highly modified dendrites) of the rods and cones.

outer limiting membrane is perforated and supports the rod and cone segments.

outer nuclear layer contains the nuclei of the rods and cones.

outer plexiform layer consists of the axons of the rod and cone cells.

inner nuclear layer contains the dendrites and cell bodies and some axons of the bipolar cells that are one of the neuron types in the retina.

inner plexiform layer consists of axons of the bipolar cells and dendrites of the next neuron in the "chain."

ganglion cell layer consists of multipolar neuron cell bodies.

layer of optic nerve fibers is composed of the axons of the ganglion cells that exit from the eyeball as the optic nerve.

inner limiting membrane is a layer of glial cells forming the innermost retinal layer

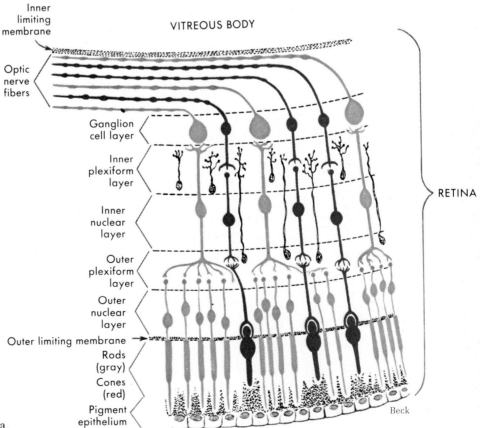

Fig. 10-12
Organization of retina.

Rods and cones. The photosensitive elements of the retina are represented by *rods* and *cones* (Fig. 10-13). Rods are slender elongated cells whose outer lamellated segments contain a visual pigment called *rhodopsin*. Cones are shorter and broader and also have an outer lamellated segment. One theory suggests that there are three types of cones, distinguished on the basis of their content of a red-sensitive, green-sensitive, or blue-sensitive pigment. Light quanta are believed to break down a pigment molecule, and the products create an electrical potential that leads to rod or cone depolarization. Rods are absent from the fovea, cones only being present; both receptors are lacking on the optic disc (Fig. 10-14). Rods are low-level light intensity receptors and convey no sense of color; cones are high-level light intensity receptors and also confer color vision.

Outer segments of both the rods and cones contain disclike layers of membranes that are theorized to contain the photosensitive pigments. These "discs" appear to be shed or to be phagocytosed by the pigment epithelium cells and may be renewed by the inner segment of the rod or cone.

Vitreous body. The interior of the eyeball between the lens and retina is filled by the vitreous body (see Fig. 10-9). It adheres to the retina and, in fresh condition, is a colorless, structureless, gelatinous mass with glasslike transparency. It aids in maintaining the shape of the eyeball and in transmitting light rays to the retina.

Visual pathways (Fig. 10-15)

From the previous discussion it should be clear that three neurons (rod or cone, bipolar cell, ganglion cell) lie *within* the eye. The **optic nerves** pass from each eye and proceed to the **optic chiasma**, where the nerve fibers from the nasal half of each retina cross to the opposite side. Beyond the chiasma, the fibers form the **optic tracts** that terminate in the **lateral geniculate** bodies of the thalamus. From the geniculate bodies, the **geniculocalcarine tracts** *(optic radiation)* convey the visual impulses to the visual cortex (area 17).

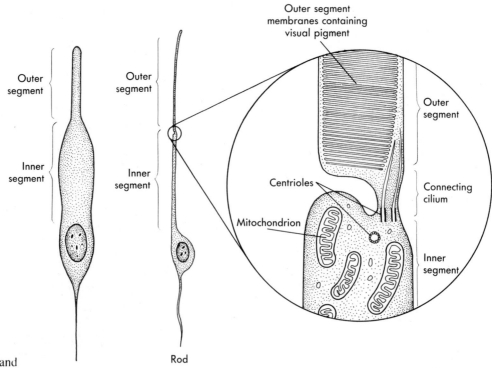

Fig. 10-13
Morphology of rods and cones.

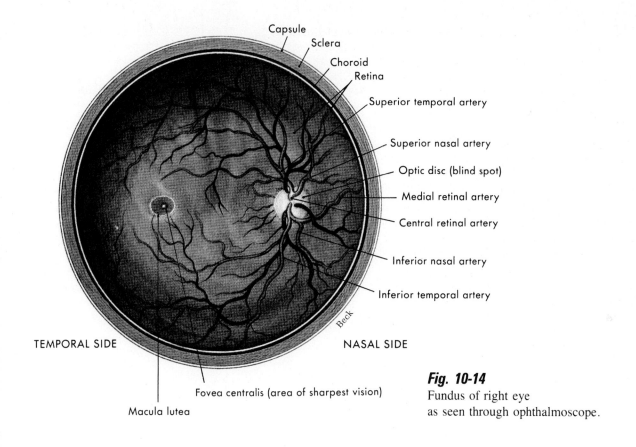

Fig. 10-14
Fundus of right eye as seen through ophthalmoscope.

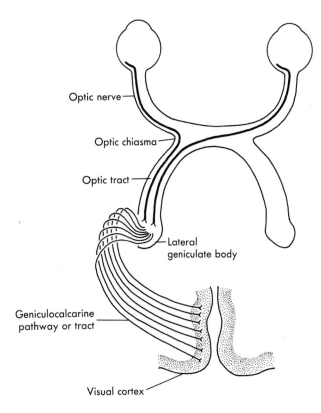

Fig. 10-15
Neural pathways for sense of sight.

Extrinsic eye muscles

Each eye is supplied with six muscles (Fig. 10-16) that take their origins from various areas of the orbit and that insert into the sclera. The muscles are described in the following paragraphs and are summarized in Table 10-1.

The four rectus muscles take their origins from a fibrous ring that attaches to the rim of the optic foramen. The *rectus lateralis oculi* muscle inserts on the lateral aspect of the eyeball, the *rectus medialis oculi* muscle on the medial aspect of the eyeball, and the *rectus superior oculi* and *rectus inferior oculi* muscles on the corresponding aspects of the eyeball.

The *obliquus superior oculi* muscle rises from the orbit above the fibrous ring, and its tendon passes through a fibrocartilaginous ring attached to the anterior, upper, and inner aspect of the orbit. The tendon then passes *posteriorly* to insert on the sclera at the superior and medial aspect. Thus the *direction of pull* of the muscle is converted to an angle different from that expected.

The *obliquus inferior oculi* muscle rises from the orbital surface of the maxillary bone and passes directly to the inferior and medial aspect of the eyeball.

The nerve supply of the six muscles and their action on the eyeball are also presented in Table 10-1.

Eyelids and lacrimal apparatus

Eyelids. The eyelids (see Fig. 10-17) provide protection for the eyeball from light and trauma approaching from the front. The outer surface is formed by *thin skin*. The dermis contains pigment cells that have varying concentrations of yellow to brown pigments ("eye shadow" is designed to obscure these colors). Along the edge of each lid are several rows of *eyelashes* that are replaced every 3½ to 5 months. Inner to the skin are *skeletal muscle* fibers belonging to the orbicularis oculi. The next layer is called the **palpebral fascia** and in the upper lid is a continuation of

Table 10-1. Extrinsic eye muscles

Name	Innervation (cranial nerve)	Action on eyeball
Obliquus inferior oculi	III Oculomotor	Upward and outward
Obliquus superior oculi	IV Trochlear	Downward and outward
Rectus inferior oculi	III Oculomotor	Downward and inward
Rectus lateralis oculi	VI Abducent	Outward
Rectus medialis oculi	III Oculomotor	Inward
Rectus superior oculi	III Oculomotor	Upward and inward

Adapted from McClintic, J.R.: Basic anatomy and physiology of the human body, ed. 2. Copyright © 1980. Reprinted by permission of John Wiley & Sons, Inc.

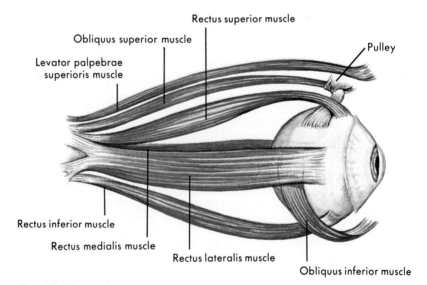

Fig. 10-16

Six extrinsic muscles of right eye. Note how pulley converts direction of pull of obliquus superior oculi.

the tendon of the levator palpebrae muscle that raises the upper lid. A *tarsal plate*, as the next layer, is composed of dense connective tissue and stiffens the lid. *Meibomian glands*, sebaceous glands, are found next to the tarsal plate and produce part of the material that is often found in the eyes on awakening. The innermost layer of the lid is the *conjunctiva*, an epithelium that is usually pseudostratified in type. The conjunctiva of the lid (palpebral conjunctiva) is continuous with the corneal epithelium, changing to a stratified squamous epithelium (bulbar conjunctiva) at the connection to the eyeball.

Lacrimal apparatus (Fig. 10-17). On the superior temporal aspect of each eyeball is located a two-lobed **lacrimal gland.** Cells producing a watery fluid (tears or lacrimal fluid) empty through a number of short ducts onto the superior temporal surface of the eyeball. Blinking and eye movements move the fluid across the surface of the eye, preventing it from drying. The secretion also contains a bacteriolytic enzyme (lysozyme) that acts to minimize microorganism growth on the eye. The tears are collected at the inner margins of the eyelids and pass through tiny openings, the **puncta,** into the eyelid. From here they enter the **lacrimal ducts.** The ducts empty into the **lacrimal sac,** which drains by way of the **nasolacrimal duct** into the nasal cavity. The nose may "run" if excess fluid passes through the duct system (crying).

Selected disorders of the eye and its associated structures

The purpose of this section is to indicate the anatomical basis for some of the more common disorders related to the eye and its accompanying structures.

Disorders of image formation.

The cornea, lens, and iris function similarly to a camera in that to achieve a sharp image of the world the focal powers of the refracting media of the eye must correspond to the anatomical dimensions of the eye. The normal or **emmetropic** eye can bring to a sharp focus on the retina images from near infinity to about 20 cm from the face. If, at given object distance from the eye, the image of that object is focused *in front* of the retina, a blurred image will result. In this case the eye is said to be **myopic** (nearsighted), and the eyeball is deemed to be *too long* for the focal power of the lens systems. What is needed to correct this condition is a lens placed ahead of the eye itself that can *diverge* the incoming light rays and cause the cornea and lens to focus them further back in the eyeball. If the image is focused *behind* the retina, the eye is said to be **hypermetropic** (farsighted). Again, a blurred image will result, the eyeball is *too short* for the focal system, and a lens that *converges* the incoming light rays will aid the eye to focus the image on the retina. These conditions are diagrammatically presented in Fig. 10-18.

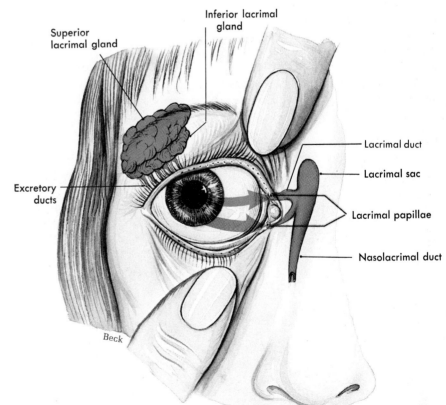

Fig. 10-17
Lacrimal apparatus.

Astigmatism (*astigmia*) results when the lateral and vertical directions of the cornea do not have equal curvatures but have curvatures like a teaspoon. In the astigmatic eye each dimension acts as a separate lens of different curvature, and two images are formed (double vision). To correct this condition obviously requires an artificial lens that also has different curvatures in its horizontal and vertical planes to compensate for the defects of the cornea.

Presbyopia ("old eye") occurs as the lens hardens with age. Proper focusing of an image on the retina requires that the lens become more rounded the closer the object is to the eye. Hardening prevents "rounding up," and objects must be further from the eye to be seen clearly. The condition is treated by a biconvex lens that aids the bending of light rays as they enter the eye.

Cataract refers to opacity of the lens, its capsule, or both structures. The result will be failure to transmit light rays to the retina. Removal of the lens followed by fitting of glasses will preserve vision.

Other disorders. *Color blindness* may be interpreted as loss of a cone or a particular pigment in a cone. Thus insensitivity to particular wavelengths (and thus colors) of light may result. *Protanopia* (red-blindness), *deuteranopia* (green-blindness), and *tritanopia* (blue-blindness) are described. Red-

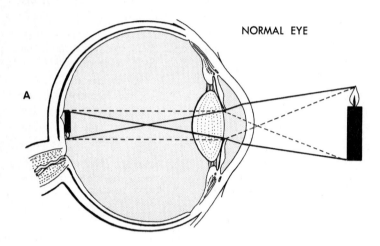

Fig. 10-18
A, Emmetropic or normal eye. **B,** Myopic (nearsighted) eye. **C,** Hypermetropic (farsighted) eye. Diagram indicates type of lens required to correct abnormal situation.

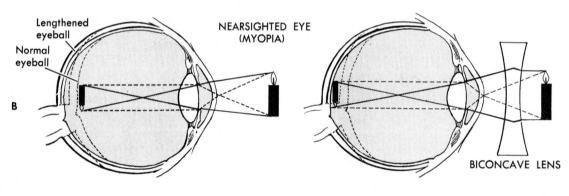

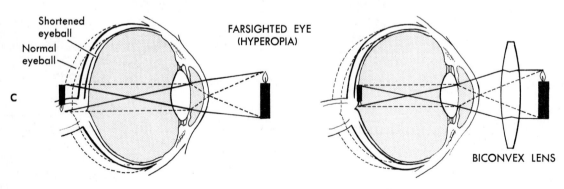

green blindness is the most common type, is inherited as a sex-linked recessive trait, and is perhaps of greatest bother in reading traffic lights; a person who cannot tell the color of the lighted bulb on the traffic light must learn its position.

Glaucoma is a condition in which intraocular pressure rises above its normal value of about 15 mm Hg (as tested with a tonometer rested on the cornea). Pressure of the aqueous humor is what is tested, and if it rises, this may be caused by overproduction by the ciliary processes or blockage of Schlemm's canal (the drainage route). Rising pressure can cause compression of retinal blood vessels, death of visual receptors, and blindness.

Conjunctivitis ("pink eye") refers to inflammation of the conjunctiva, usually caused by bacteria (staphylococcus, streptococcus). Tears contain lysozyme, a bacteriolytic enzyme, and if tear production decreases, conjuntivitis may ensue. *Trachoma* is bilateral *viral* conjunctivitis affecting many individuals in the so-called emerging nations, where sanitary practices and disease-control programs are substandard.

Strabismus refers to the inability of both eyes to be directed at objects. One or more of the extrinsic eye muscles may be too short or too long, resulting in an unbalanced pull on the eyeball, or nerve supply to the muscles may be damaged (recall that *three* cranial nerves supply the eye muscles).

Last, as shown in Fig. 10-19, damage in various portions of the visual pathways produces characteristic types of visual losses. Keep in mind that the lens reverses the image of the world, so that loss of a particular retinal area results in a visual field loss that is opposite to the retinal loss.

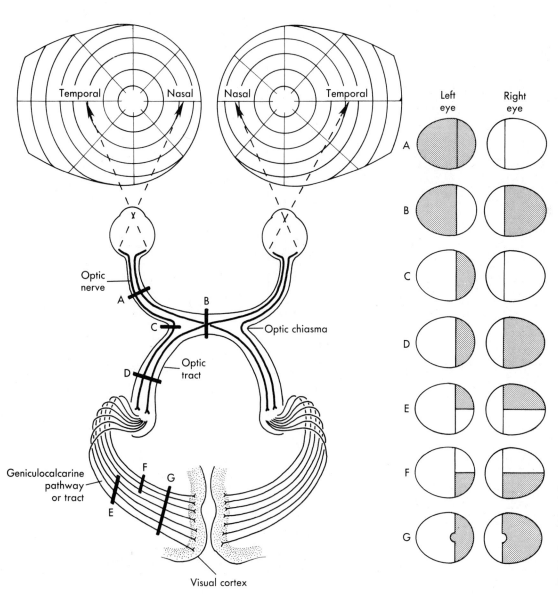

Fig. 10-19
Types of visual loss resulting from lesions in various parts of visual pathways. Shaded areas indicate deficits of vision.

Ear

Development

Of the three anatomical subdivisions of the ear (outer, middle, and inner), the inner ear begins its development first. Early in the fourth week, an *otic placode* appears on each side of the hindbrain (Fig. 10-20). The placodes invaginate to form an *otic pit* whose edges come together to form an *otic vesicle*. The vesicle is the forerunner of the cellular portion of the inner ear called the *membranous labyrinth*. Mesenchyme forms the bony tissue around the cellular part, creating the *osseous labyrinth*.

The middle ear cavity develops at 4 to 5 weeks as an outgrowth of the primitive pharynx (Fig. 10-21). The pharyngeal connection remains as the **eustachian tube**. The **ear ossicles** develop from cartilaginous structures that form supports for the pharyngeal wall (Fig. 10-22).

The external ear includes the **pinna** and the **external auditory meatus**. The pinna is derived at 5 weeks from six mesenchymal swellings or **auricular hillocks** (Fig. 10-23) that surround a groove, the **branchial groove,** on the upper neck of the embryo. The developing pinna then migrates to a position on the head. The meatus rises from a **meatal plug** that is a derivative of the surface ectoderm (see Fig. 10-21). The plug is canalized to create the meatus, and the **eardrum** remains as the membrane between the middle ear cavity and the meatus.

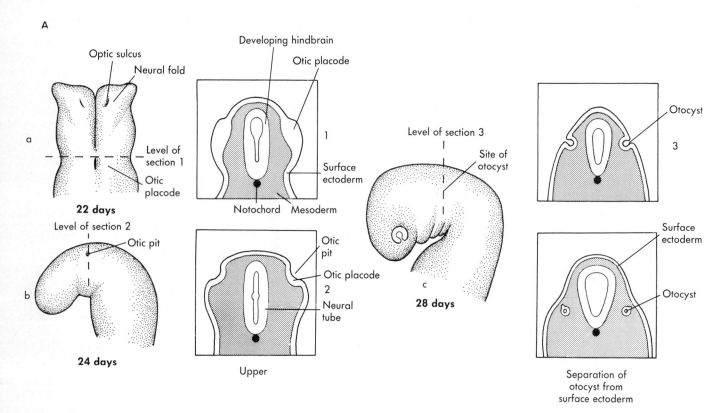

Fig. 10-20

Development of inner ear. **A,** Early development (22 to 28 days) of inner ear. **B,** *a* to *e* show development of membranous labyrinth from weeks 5 to 8; *f* to *i* show development of organ of Corti from weeks 8 to 20.

10 Organs of sensation

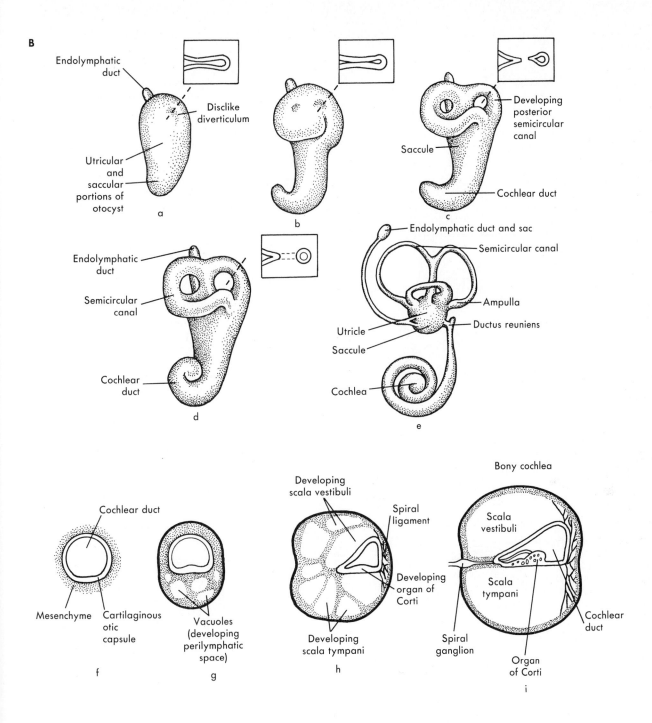

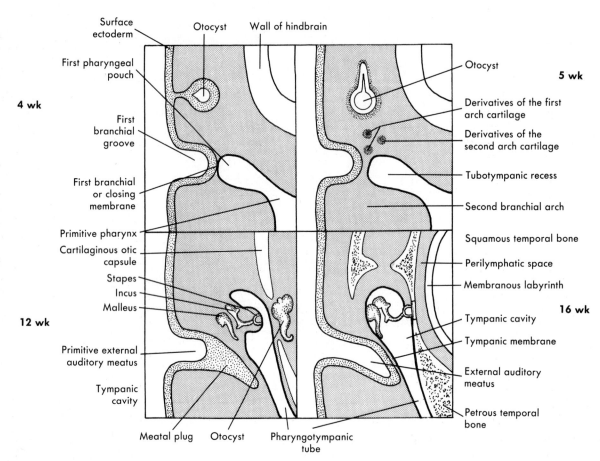

Fig. 10-21
Development of middle ear.
After Moore.

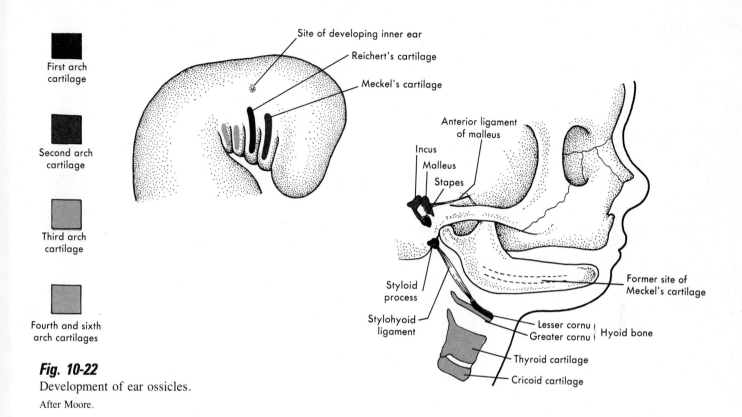

Fig. 10-22
Development of ear ossicles.
After Moore.

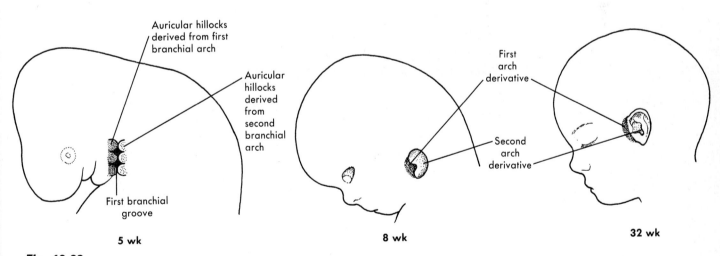

Fig. 10-23
Development of pinna of external ear.
After Moore.

Structure

The ear (Fig. 10-24) is conventionally divided into three portions designated as the *outer (external) ear*, *middle ear (tympanic cavity)*, and *inner (internal) ear*.

Outer ear. The outer ear consists of the *auricle (pinna)* and the *external auditory meatus (ear canal)*. The auricle (Fig. 10-25) has several parts and consists of an irregular plate of elastic cartilage covered by thin skin. The auricular muscles, mentioned in Chapter 8, attach to and move the auricle; they are of no importance in the human. The external auditory meatus is slightly S shaped and pursues a medial and inferior course to terminate at the eardrum. Its outer half is supported by a continuation of the auricular cartilage, whereas the inner half runs through the temporal bone. The meatus is lined with thin skin that is firmly attached to the perichondrium or periosteum of the canal wall. The meatus contains the brown waxy *cerumen* secreted by sebaceous and *ceruminous glands* (a type of apocrine sweat gland). The cerumen prevents the skin from drying and is bitter to the taste, presumably discouraging the entry of insects into the canal.

Middle ear. The middle ear consists of the *tympanic cavity*, the *auditory ossicles* (ear bones), the *pharyngotympanic (eustachian) tube*, the *tympanic membrane* (eardrum), and the tendons of two small *muscles* that attach to two of the ear ossicles.

Tympanic cavity and eustachian tube. The tympanic cavity (see Fig. 10-24) is an irregular air-filled space in the temporal bone. Its lateral wall is formed mainly by the tympanic membrane and the other limits by bony tis-

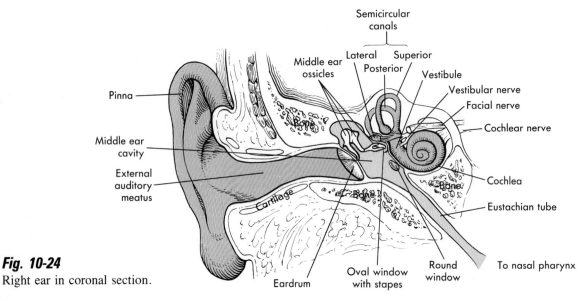

Fig. 10-24
Right ear in coronal section.

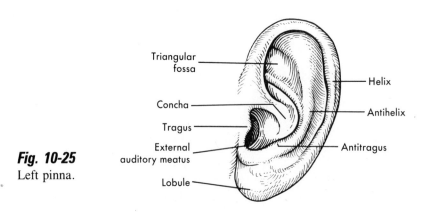

Fig. 10-25
Left pinna.

sue. Anteriorly, the cavity communicates with the pharynx by way of the pharyngotympanic (eustachian) tube; the tube allows equalization of air pressure on the two sides of the eardrum (our ears "pop" on ascent or descent in altitude; the pop is the tube's opening). Posteriorly, communication is made with the mastoid air cells. Thus bacteria can pass from the throat to the mastoid process (mastoiditis). The lining of all parts of the cavity is a simple squamous epithelium underlaid by a delicate lamina propria. Glands are not visible in the lining, although it is called a mucous membrane. The tube has a simple cuboidal or columnar epithelium provided with cilia.

Auditory ossicles (Fig. 10-26). Three small bones, the *malleus (hammer)*, *incus (anvil)*, and *stapes (stirrup)*, form a system of levers that transfer vibrations of the eardrum to the organ of hearing. The handle of the malleus attaches to the eardrum, whereas the footplate of the stapes fits into the oval window of the inner ear. The lining of the middle ear cavity is reflected over the ossicles and is firmly attached to their periosteums.

Tympanic membrane. The eardrum is oval in shape, has a conelike apex that projects medially and forms at the same time the medial wall of the external auditory meatus and the lateral wall of the tympanic cavity. Its apex is maintained by its attachment to the handle of the malleus. It consists of two layers of collagenous tissue (except in its superior temporal quadrant) and is covered externally by the thin skin of the meatus and internally by the mucous membrane of the tympanic cavity. The membrane vibrates as sound waves strike it.

Middle ear muscles. The *tensor tympani* lies above the lateral end of the pharyngotympanic tube and sends

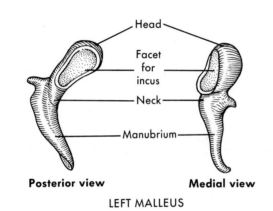

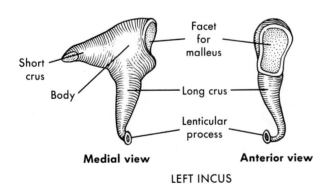

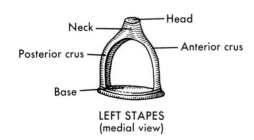

Fig. 10-26
Ear ossicles.

its tendon to attach to the handle of the malleus. The **stapedius** arises from the posterior wall of the tympanic cavity and inserts on the stapes. These muscles form the effectors of a reflex arc that causes the movement of the ossicles to be decreased with high-intensity sounds; this prevents damage to the eardrum and internal ear.

Inner ear. The inner ear (Fig. 10-27) consists of two parts: the **osseous labyrinth**, a series of channels located within the petrous portion of the temporal bone, and the **membranous labyrinth**, a series of membranous fluid-filled sacs containing complex cellular structures that follow the tortuosities of the osseous channels.

The labyrinths may be subdivided into three major portions: the **vestibule**, the **semicircular canals**, and the **cochlea**.

Vestibule. The vestibule is the centrally placed cavity of the osseous labyrinth and lies medial to the tympanic cavity. Membranous partitions divide the vestibule into a superior and larger portion called the **utriculus** and a smaller inferior portion called the **sacculus**. Within each of these divisions lies a sensory structure called a **macula** (see Fig. 10-28); thus there are two maculae in each inner ear. Two types of cells, *hair cells* and *supporting cells*, are found in the maculae. The hair cells each have a single *cilium* and numerous *microvilli* that are embedded in a gelatinous *matrix* that is located above the epithelium. Numerous concretions called *otoliths* (ear stones) are found in the matrix and increase its mass. Support cells are columnar in shape and support the hair cells. They have no cilia. The maculae form nerve impulses when head position is changed or when linear acceleration or deceleration occurs and the mass above the epithelium shifts and bends the cilia. Thus notification of head position is produced.

Semicircular canals. The fluid-filled semicircular canals (see Fig. 10-27) in each ear are designated as the *superior*, *posterior*, and *lateral* canals. They are set in three mutually perpendicular planes to one another. Each canal has next to the utricle an enlarged **ampulla** that contains perpendicularly to the axis of the canal itself a structure called the **crista ampullaris** (Fig. 10-28). Each crista is composed of hair and supporting cells, and the hair cell cilia are here embedded in a gelatinous **cupula** above the epithelium. Movements of the head in a nonlinear direction are most effective in causing the fluid in the canals to move toward or away from the cupula, bending it. Nerve impulses are thus generated in the hair cells.

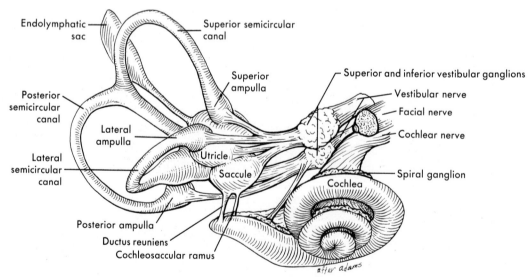

Fig. 10-27
Membranous labyrinth of inner ear. (*Right side*, Anterior view.)

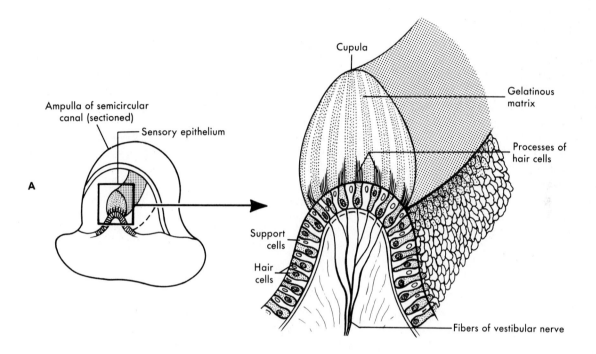
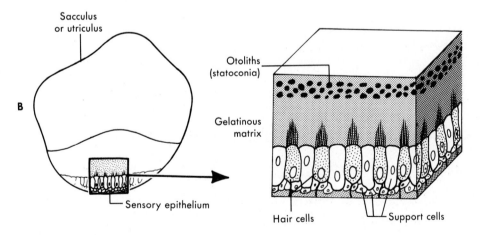

Fig. 10-28
Location and morphology of equilibrium structures of inner ear. **A,** Crista ampullaris of semicircular canal. **B,** Macula of sacculus or utriculus.

Cochlea. The cochlea (see Fig. 10-27) consists of a channel coursing 2¾ turns around a central supporting pillar of bone called the **modiolus.** Shaped like a snail shell, the apex of the cochlea is directed downward and outward when the head is held erect. The membranous labyrinth is structured to divide the bony channel into three portions (Fig. 10-29). The *scala vestibuli* lies above the **vestibular membrane**, the *scala media* between the vestibular and **basilar membranes**, and the *scala tympani* below the basilar membrane. Resting on the basilar membrane is the **organ of Corti**, the organ of hearing (Fig. 10-30). The sensory elements of the organ of Corti are again hair cells. The stapes pushes in and out on the oval window, setting up shock waves in the cochlear fluid. The basilar membrane vibrates more or less selectively to the frequency of the shock waves, and this movement is translated into bending of the hair cells' cilia. Nerve impulses are generated by bending of the cilia.

Auditory and equilibrial pathways

The structures of the inner ear are served by the eighth **vestibulocochlear** or cranial nerve. The nerve clearly has two major subdivisions: the *vestibular nerve* serves the sensory receptors of the maculae and the cristae; the *cochlear nerve* serves the organ of Corti.

Fibers of the vestibular nerve proceed from the equilibrial organs to the **vestibular ganglion** located within the internal auditory meatus, where a synapse occurs. Axons from the ganglion enter the medulla oblongata and pass to the vestibular nuclei of the pons and to the deep nuclei of the cerebellum. A pathway enabling muscular response designed to maintain balance and equilibrium is afforded.

Dendrites of the cochlear hair cells pass to the **spiral ganglion** located in the modiolus of the cochlea. Axons of the ganglion cells form the cochlear nerve that passes to the medulla, where synapses are made in the **cochlear nuclei.** Fibers pass from the medullary nuclei to the medial geniculate body of the thalamus, where a synapse is made, and from there to the auditory areas of the temporal lobe (areas 41 and 42).

Selected disorders of hearing and equilibrium

Deafness is the main concern when dealing with disorders of the inner ear. It may be defined as partial or complete loss of the ability to hear. According to where the point of origin of the deafness lies, there are several types of deafness.

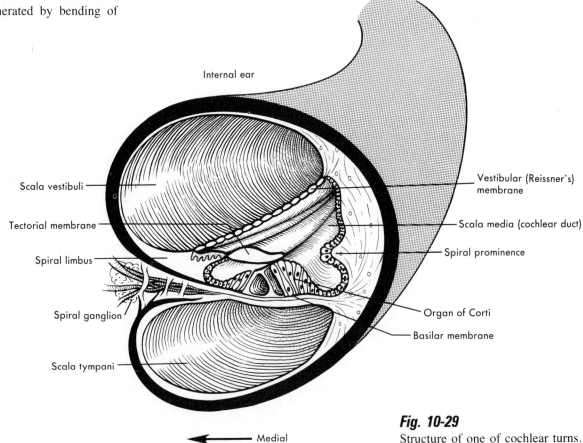

Fig. 10-29
Structure of one of cochlear turns.

Transmission (conduction) deafness occurs when sound waves are prevented from reaching the eardrum or when the ossicles fail to transmit the sound waves to the cochlea. Obstruction of the external auditory meatus by foreign objects or accumulation of cerumen may reduce or block sound waves from reaching the eardrum. Ear plugs used in noisy situations employ this principle. Inflammation of the middle ear cavity (for example, *otitis media*) may result in fluid production that limits ossicle movement or may cause the ossicles to *fuse* with one another to prevent movement. ***Fixation*** of the stapes footplate in the oval window (*otosclerosis*) may prevent shock waves from being created in the cochlear fluids.

Nerve deafness results from damage to the organ of Corti or to the neural pathways. Drugs, exposure for a long time to high-intensity sounds, or blows to the head are several causes of nerve damage. If one cochlea is affected, loss occurs in only one ear; lesions in the brainstem affect both ears because of connections made to the cochlear nuclei by both eighth cranial nerves; cerebral lesions usually produce inability to comprehend the spoken word (*aphasia*).

Disturbances in the ability to maintain equilibrium, or dizziness, may reflect damage to the equilibrial structures or pathways. Infections, antibiotic administration, or trauma to the head may injure these structures and can result in confinement to bed because of inability to locomote properly.

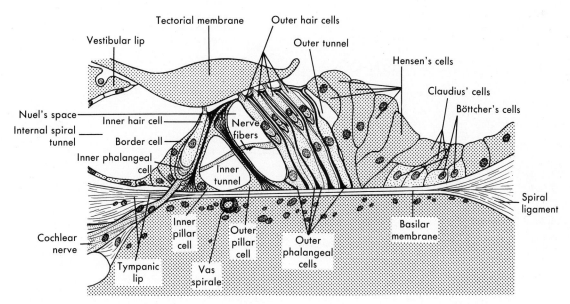

Fig. 10-30
Organ of Corti. Sensory elements are hair cells.

Sense of taste

Distribution and structure of taste buds

The receptors for the sense of taste are the *taste buds*, located primarily on the tongue (a few may be found on the soft palate or pharynx wall). They are found in the walls of several of the types of papillae of the tongue and are missing on the upper surface of the tongue (nervous structures are very delicate and would be damaged by the use of the tongue in swallowing and chewing). Each bud (Fig. 10-31) is an oval structure containing neurons or *taste cells* and *supporting cells*. Both types of cells have microvilli on their apexes that project through a *taste pore* to the epithelial surface. It has been suggested that a substance to be tasted may fit into a suitably shaped receptor on the microvilli of the nerve cells. There are only four taste sensations *(salt, sweet, sour,* and *bitter)*, which are distributed differently on the tongue (Fig. 10-32). No difference in structure has been discerned between buds that are sensitive to the different tastes.

Taste pathways (Fig. 10-33)

Buds located on the anterior two thirds of the tongue are served by the seventh cranial nerve (facial) and those on the posterior one third by the ninth cranial nerve (glossopharyngeal). Buds elsewhere are supplied by the tenth cranial nerve (vagus). The *tractus solitarius* of the brainstem receives the nerves, and fibers then pass to the thalamus and ultimately to the parietal sensory areas (areas 3, 1, and 2).

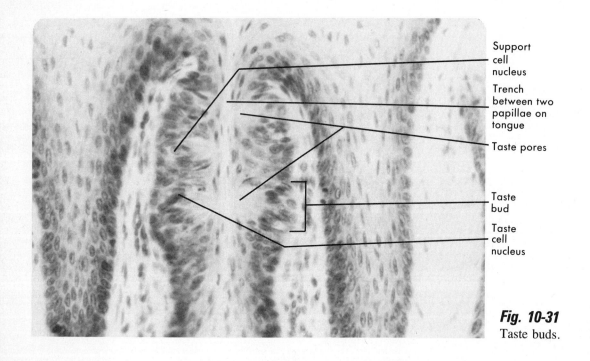

Fig. 10-31
Taste buds.

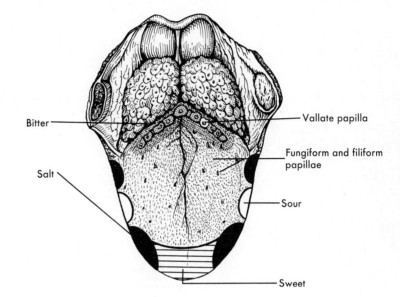

Fig. 10-32
Distribution of sense of taste on tongue.

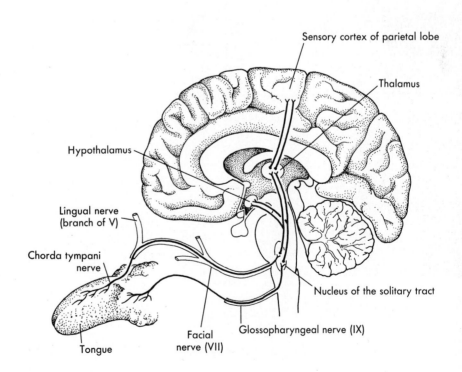

Fig. 10-33
Basic neural pathways for sense of taste.

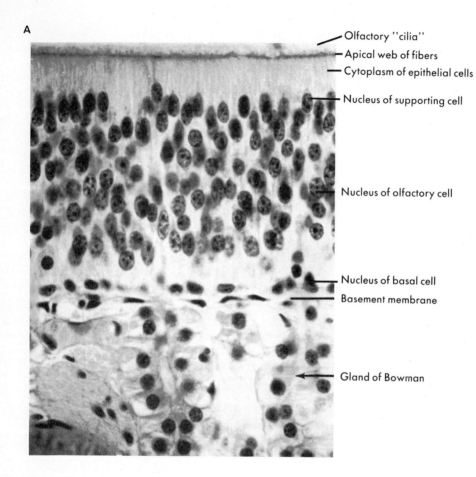

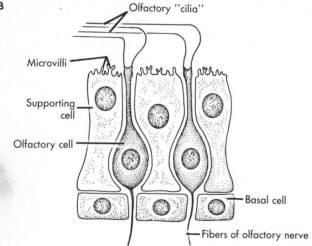

Fig. 10-34
Olfactory epithelium. **A,** Light microscope view. **B,** Structure as revealed under electron microscope.

Sense of smell

Location and structure of the olfactory epithelium

The olfactory epithelium extends 8 to 10 mm down the nasal septum in the apexes of the nasal cavities. Total surface area in both nasal cavities for olfaction is about 500 mm^2. The epithelium is of a pseudostratified type and contains three types of cells: supporting cells, basal cells, and olfactory cells (Fig. 10-34). Under the electron microscope (see Fig. 10-34) the organization and morphology of these cells may be appreciated.

Supporting cells are generally columnar elements having a stiffening apical web of fibrils that give support to the olfactory cilia. Their nuclei are located in the upper portion of the epithelium.

Basal cells are short, rounded elements whose nuclei form a single row at the base of the epithelium.

The olfactory cells have their nuclei in the intermediate area of the epithelium and are elongated bipolar neurons. The apex of the cell bears a rounded expansion from which six to eight long, nonmotile *olfactory cilia* (highly modified dendrites) extend to lie parallel to the surface of the epithelium. Again, it is believed that a substance to be smelled must achieve a "fit" with receptor sites on the cilia. According to one theory, there are seven basic odors: *camphoraceous, musky, floral, ethereal, pungent, putrid,* and *pepperminty.* The sense of smell is, on the average, about 25,000 times more sensitive than the sense of taste, as judged by concentrations that can be detected. The sense of smell complements the sense of taste; as anyone who has a cold knows, food becomes tasteless (actually, "smell-less").

Olfactory pathways (Fig. 10-35)

Axons of the olfactory cells pass through the cribriform plate of the ethmoid bone and synapse with *mitral cells* in the olfactory bulb. Fibers from the mitral cells form the *olfactory tract* that leads to the *olfactory areas* of the brain. The latter are located in the frontal lobes (intermediate and medial olfactory areas) and in the temporal lobe (lateral olfactory area and amygdaloid nucleus). The amygdaloid nucleus, as part of the limbic system, affords a connection that allows olfactory stimuli to influence behavior, notably food-seeking and sexual behavior.

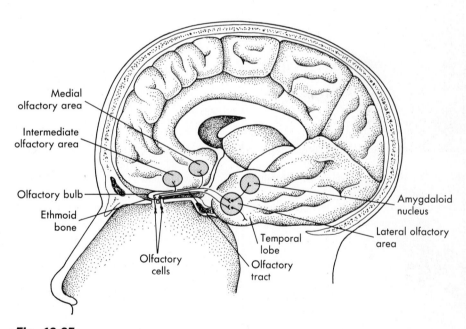

Fig. 10-35
Basic neural pathways for sense of smell.

Summary

1. Receptors in general exhibit certain characteristics:
   a. They follow the law of adequate stimulus, responding best to one particular form of energy.
   b. The transmitted impulse from a receptor does not differ in nerves from different receptors. The law of specific nerve energies states that the CNS is responsible for determining the appreciation and location of received stimuli.
   c. Receptors can signal intensity of a stimulus by adjusting their rate of firing.
   d. Adaptation, or a decreased rate of firing with continuation of a stimulus of a certain strength, is a feature of certain receptors. Others continue a high rate of firing to ensure continued supply of information to the CNS.
2. Receptors of diverse types serve as the organs to detect changes in the internal and external environments of the body, and their activity leads to maintenance of homeostasis.
   a. Each receptor responds best, although not exclusively, to a particular type of stimulus, its adequate stimulus.
   b. Receptors are classified by their location (extero-, entero-, and proprioceptors) or by their adequate stimulus (chemicals, pressure or tension, thermal changes, or light and sound waves).
3. Sensations in the skin and body membranes, receptors, types of sensation served, and major neural pathways are as follows:
   a. Naked nerve endings serve the sense of pain and "crude" temperature; four varieties of pain are described (bright, dull, deep, referred); spinothalamic tracts to the parietal sensory cortex.
   b. End bulbs of Krause serve the sense of cold; spinothalamic tracts to parietal sensory cortex.
   c. Corpuscles of Ruffini serve the sense of warmth; spinothalamic tracts to parietal sensory cortex (3b and 3c are involved in the body temperature regulation).
   d. Meissner's corpuscles serve the sense of touch; gracile and cuneate tracts to parietal sensory cortex.
   e. Pacinian corpuscles serve the sense of pressure; gracile and cuneate tracts to the parietal sensory cortex.
4. Visceral sensations have their receptors in internal body organs, except muscles, tendons, and joints. Their receptors, type of sensation served, and major neural pathways are as follows:
   a. Baroreceptors (sensitive to pressure or stretching) are specialized organs resembling pacinian corpuscles; they control blood pressure and lung expansion by reflex arcs; pathways use mainly the vagus nerve and brainstem nuclei.
   b. Chemoreceptors respond to changes in P_{O_2}, P_{CO_2}, or pH and are involved in control of respiratory and circulatory phenomena; pathways use mainly the vagus nerve and brainstem nuclei.
5. Muscle, tendon, and joint sensations, type of stimulus detected, and major neural pathways may be described as follows:
   a. Pacinian corpuscles in joint capsules provide awareness of position and movement; spinocerebellar tracts to the cerebellum.
   b. Neuromuscular spindles in the fleshy part of a muscle respond to tension or stretch applied to the muscle; spinocerebellar tracts to the cerebellum.
   c. Neurotendinous spindles are located at muscle-tendon junctions and also respond to tension or stretch; spinocerebellar tracts to the cerebellum.
   d. Naked nerve fibers supply joint capsules for the sense of pain; spinothalmic tracts to the parietal sensory cortex.
6. The eye is the organ of vision. It develops in the fourth week from optic grooves, vesicles, and cups.
   a. It has several gross landmarks, including anterior and posterior poles, fovea, optic papilla, equator, and optical and visual axes.
   b. Three tunics compose the eye: fibrous, uvea, and retina.
   c. The fibrous tunic includes the protective sclera (white of the eye) and the anterior transparent cornea. The cornea has five layers in it.
   d. The uvea is composed of a posterior vascular chorioid and the anterior ciliary body that includes the iris, lens, lens ligaments, and ciliary muscle, all involved in image focusing and light control.
   e. The retina contains the visual receptors, rods, and cones and is the visual tunic of the eye. Ten layers of neurons and their supporting structures compose the retina. Rods are night vision receptors; cones are for color and daylight vision.
   f. Aqueous humor is a watery fluid filling the anterior part of the eye. It is drained by Schlemm's canal at the sclerocorneal junction.
   g. Vitreous humor is a gelatinous substance filling the posterior part of the eye and helping to maintain its shape.
   h. Visual pathways include the optic nerve, chiasma, tract, lateral geniculate body, and geniculocalcarine pathway to the occipital lobe.
7. Each eye has six muscles that move it. Names, nerve supply, and movements occurring are presented in Fig. 10-16 and Table 10-1.
8. The eyes are supplied with eyelids and lacrimal apparatuses.
   a. The lids protect the eyes and help spread the tears over the eyes to lubricate and cleanse them. Structure of the lids is described.
   b. The lacrimal gland secretes a fluid that is drained into the nasal cavities by the lacrimal ducts, lacrimal sacs, and nasolacrimal ducts.
9. Several common disorders associated with the eye are discussed.
10. The ear is the organ of hearing and equilibrium. It consists of outer, middle, and inner portions.
    a. The outer ear includes the auricle and external meatus. It develops at 4 to 5 weeks.
    b. The middle ear includes the tympanic cavity, ossicles, eustachian tube, eardrum, and two muscles. It develops at 4 to 5 weeks from a pharyngeal outgrowth.

c. The inner ear includes the vestibule, its subdivisions, the utriculus and sacculus, the semicircular canals, and the cochlea. It develops at 4 weeks from an otic placode.
d. The utriculus and sacculus contain maculae for detection of position of the head; the semicircular canals contain cristae that respond to movement of the head. The vestibular portion of the eighth cranial nerve carries impulses ultimately to the cerebellum.
e. The cochlea contains the organ of Corti, the organ of hearing. The cochlear portion of the eighth cranial nerve carries impulses ultimately to the temporal lobe.
f. Selected disorders of the inner ear are described.

11. The sense of taste is served by taste buds.
    a. Buds are located primarily on the tongue.
    b. Buds contain taste cells, support cells, a taste pore, and microvilli on the taste cells that have receptor sites on them.
    c. There are four taste sensations that are distributed unequally on the tongue.
    d. Taste pathways include the seventh and ninth cranial nerves, with the parietal lobe as ultimate termination.

12. The sense of smell is served by the olfactory epithelium in the roofs of the nasal cavities.
    a. The epithelium contains basal, support, and nerve cells, the latter with olfactory cilia that bear receptor sites.
    b. There are seven basic odors.
    c. Olfactory pathways include the olfactory nerves, bulb, and tract, with terminations in the frontal and temporal lobes.

Questions

1. In classifying receptors, what criteria may be employed in naming them? Include mention of receptor properties.
2. To what sort of stimuli do the following receptors respond?
   a. Pacinian corpuscles
   b. Meissner's corpuscles
   c. Krause's corpuscles (end bulbs of Krause)
   d. Ruffini's organs (corpuscles)
   e. Muscle spindles
   f. Baroreceptors
3. For each of the preceding receptors, name the spinal pathway(s) that transmit(s) the sensation, and where in the brain the pathway terminates.
4. What structures of the eye are involved in focusing an image? Describe the structure of each.
5. What structures of the eye help to maintain its shape?
6. What parts of the eye are included in the uvea?
7. Describe the structure of the retina, tracing the pathway an impulse would take from the retina to the occipital lobes of the brain.
8. Name the extrinsic muscles of the eye, their nerve supply, and the movement of the eyeball each muscle causes.
9. Where are tears produced, and how do they find their way to the nasal cavity?
10. What are the parts of the ear? For each, briefly describe the structure of the subdivisions of each major part, and indicate their role in the body.
11. Compare the location, structure, and mode of operation of the receptors for taste and smell.

Readings

Apple, D.J., and Rabb, M.F.: Clinicopathologic correlation of ocular disease: a text and stereoscopic atlas, St. Louis, 1978, The C.V. Mosby Co.

Botelho, S.Y.: Tears and the lacrimal gland, Sci. Am. **211**:78, Oct. 1974.

Parker, D.E.: The vestibular apparatus, Sci. Am. **243**:118, Nov. 1980.

Parr, J.: Introduction to ophthalmology, New York, 1978, Oxford University Press.

Singh, R.P.: Anatomy of hearing and speech, New York, 1980, Oxford University Press.

11

Circulatory system

Objectives

Development of the blood

Blood vascular system
   Blood
   Heart
   Development of blood and lymph vessels
   Structure of the blood vessels
   Major systemic arteries
   Major systemic veins
   Special features of the circulation
   Fetal circulation

Lymph vascular system
   Lymph
   Lymph vessels
   Lymph organs

Summary

Questions

Readings

OBJECTIVES After studying this chapter the reader should be able to:

List the parts of the blood and lymph vascular systems and describe their basic development.

Give the functions of the blood.

Describe the formed elements of the blood as to size, numbers, types, and general functions.

Describe the location, membranes, and structure of the heart.

List, locate, and describe the structure of cardiac muscle and nodal tissue.

Show how the heart is supplied by blood vessels.

Describe the structure of the three types of blood vessels in the body.

Name the major arteries and veins of the body and the areas supplied or drained.

Show how lymph differs from blood in location and content of formed elements.

Describe the pattern of the lymph vessels in the body, naming the major vessels.

Describe the location and structure of the lymph organs, including nodules, nodes, tonsils, spleen, and thymus.

The circulatory system of the human may be subdivided into a ***blood vascular system*** and a ***lymph vascular system***. Both subdivisions have a *fluid*, blood or lymph, which circulates through blood or lymphatic *vessels* and lymph *organs* (nodes, tonsils, spleen, thymus). The blood vascular system also has a pump, the *heart,* which provides the work necessary to circulate the blood. The system "ties together" all body structures by carrying essential nutrients to all body organs, removing their wastes of activity, and carrying products of cellular activity to other cells or to appropriate organs of excretion.

Development of the blood
(Fig. 11-1)

Growth of the embryo depends on adequate supplies of oxygen and nutrients, and thus blood and blood vessels are among the earliest body structures to develop. At about $2^{1}/_{2}$ weeks of development (when the embryo is only 1.5 mm long) dense masses of cells called **blood islands** appear in the mesoderm of the embryo and its surrounding tissues (yolk sac, chorion). Cavities appear within the islands, and the cells around these cavities flatten to form the linings (endothelium) of what are to become primitive blood vessels. Some mesodermal cells are shed into the cavities, and they differentiate into primitive blood cells. The first cells to be formed are nucleated *red blood cells,* capable of transporting oxygen and carbon dioxide. The liver, spleen, lymph glands, and bone marrow form blood cells beginning at about 6 weeks of development and continue until shortly before birth, when bone marrow and lymph organs remain as the only important sites of blood cell formation.

Blood vascular system

Blood

Blood is a tissue with a liquid intercellular material, the ***plasma,*** which contains free cells or cell-like structures, the ***formed elements.*** It accounts for about 7% of the body weight, and amounts to 5 to 6 L in the typical adult.

Functions of the blood. Functions of the blood are oriented toward two major activities.

Since over half the blood volume is the liquid plasma, **transport** is an important function of the blood. Materials are dissolved in the plasma water, suspended in it, or carried attached to certain plasma components. Amino acids, small lipids, carbohydrates, minerals, vitamins, and enzymes, to name a few, are dissolved or suspended in the plasma as they are absorbed from areas like the gut. To some extent dissolved in the plasma but to a greater degree carried in red blood cells are oxygen and carbon dioxide, the major respiratory gases. Heat is absorbed by plasma water and is carried primarily to the skin for elimination from the body. Hormones are largely carried attached to plasma proteins or white blood cells for distribution through the body.

Regulation of homeostasis is the major function involving the blood. As the immediate *source of water* for the body generally, the bloodstream must have its content regulated within narrow limits. The plasma contains chemicals that *regulate body pH* (acidity). *Regulation of body temperature,* mentioned previously in connection with heat transport, depends on blood water content. The blood contains chemicals (antibodies) that *protect against disease* and *blood loss* (clotting).

All in all, blood may be considered the most dynamic body fluid we possess.

Plasma. Plasma constitutes 53% to 57% of whole blood and is a histologically homogeneous fluid that is 90% water, 7% to 9% protein, 0.1% glucose, about 1% inorganic substances, and the remainder other transported substances. The proteins are separable into three categories: ***Albumins,*** the smallest (molecular weight 70,000) and most plentiful type (4 to 5 g/100 ml), create an osmotic pressure in the blood that helps to draw water into the bloodstream to maintain its fluid levels, and they also bind barbiturates and certain hormones (for example, thyroxin). ***Globulins*** are present to the extent of 2 to 3 g/100 ml and are notable as the fraction containing the antibodies of the blood. ***Fibrinogen*** is present to the extent of about 0.25 g/100 ml and is an important protein in blood clotting.

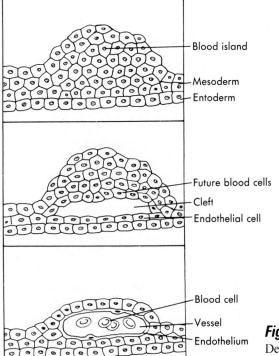

Fig. 11-1
Development of blood islands, with vessel and blood cell formation.

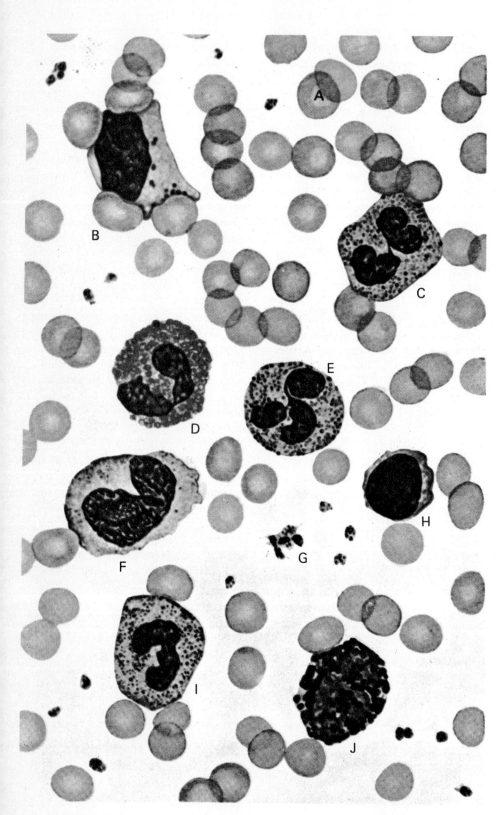

Fig. 11-2

Morphology of blood elements. Cell types found in smears of peripheral blood from normal individuals. Arrangement is arbitrary, and number of leukocytes in relation to erythrocytes and platelets (thrombocytes) is greater than what would occur in actual microscopic field. *A*, Erythrocytes; *B*, large lymphocyte with azurophilic granules and deeply indented by adjacent erythrocytes; *C*, neutrophilic segmented cell (neutrophil); *D*, eosinophil; *E*, neutrophilic segmented cell (neutrophil); *F*, monocyte with blue-gray cytoplasm, coarse linear chromatin, and blunt pseudopods; *G*, platelets (thrombocytes); *H*, lymphocyte; *I*, neutrophilic band; and *J*, basophil.

Formed elements. The formed elements (Fig. 11-2) constitute 43% to 47% of the volume of whole blood. Three distinct types of formed elements are recognized: *erythrocytes* (red blood cells, red blood corpuscles), *leukocytes* (white blood cells), and *platelets* (thrombocytes).

Erythrocytes. The most numerous of the formed elements are erythrocytes, averaging about 5 million/mm^3 of blood. To get that many units in such a small volume, they must be very small; in the blood itself they measure about 8.6 μm in diameter, and on a dried smear, because of dehydration and shrinkage, they are about 7.5 μm in diameter. As they appear in the circulation, normal erythrocytes are biconcave (indented on both sides) nonnucleated discs. The nucleus has been lost during the maturation process that, after birth, occurs in the red bone marrow of the body. The biconcave shape of the unit gives the greatest possible surface-volume ratio, a fact of importance as the cells take on oxygen or carbon dioxide. As suggested, the cells function as the transporters of most of the body's oxygen and for about one third of the carbon dioxide transport of the body. Hemoglobin, an iron-based respiratory pigment within the erythrocytes, is the medium of this oxygen and carbon dioxide transport. Normal blood contains 13.5 to 16.0 g of hemoglobin per 100 ml of blood.

Erythrocytes are produced by bone marrow at rates estimated to be about 3.5 million cells/kg of body weight per day (the "average" adult weighs 70 kg; therefore 3.5 million × 70 = 245 million cells produced per day). They maintain their function for 90 to 120 days in the circulation and then are phagocytosed, primarily by cells of the spleen and liver. If numbers in the bloodstream remain essentially constant, rates of destruction must be balanced to production.

Leukocytes. The least numerous of the formed elements are leukocytes, having a normal range of 5000 to 9000, or an average of about 7500 cells/mm^3. According to site of production, leukocytes may be grouped as *myeloid* (bone marrow) or *lymphoid* (lymphatic organs) in origin.

Myeloid leukocytes are of three types, all of which have *lobed* or *segmented nuclei* and contain *specific granules* in their cytoplasm. The latter are structures produced by each type of cell and therefore have characteristic sizes, shapes, and staining reaction in each type of cell.

Neutrophils (*heterophils*) are the most plentiful leukocytes, forming 60% to 70% of total leukocyte numbers. They usually have two or three lobes in their nuclei (which may sometimes appear as several nuclei; lobes are connected by sometimes nearly invisible nuclear membranes), and their cytoplasm contains small blue- *and* red-staining granules.

Eosinophils constitute 2% to 4% of all leucocytes, usually have a two-lobed nucleus, and have large, shiny, orange- or yellow-staining granules in their cytoplasm.

Basophils are rare, constituting 1% or less of all leukocytes. Their nucleus is S shaped and difficult to see, and the cytoplasm contains large purple-staining granules.

Note that all myeloid leukocytes have a lobed nucleus in common and that the differentiating characteristic is the granules in the cytoplasm.

Lymphoid leukocytes are of two varieties. **Lymphocytes** constitute 20% to 25% of the leukocyte population. They are the smallest leukocytes and have a rounded nucleus that nearly fills the cytoplasm.

Monocytes constitute 3% to 8% of the leukocytes, are the largest leukocyte, and commonly have a highly indented or horseshoe-shaped nucleus.

Leukocytes are all ameboid phagocytic cells that function primarily in providing protection against microorganisms or their products. Neutrophils, monocytes, and lymphocytes are the most important cells in phagocytosis of microorganisms, and the lymphocytes can become *plasma cells,* the major source of circulating antibodies. Eosinophils may be involved in the detoxification of foreign proteins such as those entering the body as allergens or on parasites. In any event, eosinophil numbers increase in allergies and parasitic infections. Basophils seem mostly to migrate from the bloodstream to take up residence in connective tissues as "tissue basophils."

Platelets. Platelets are the smallest of the formed elements, average 250,000/mm^3, and are fragments of a large bone marrow cell called a megakaryocyte. They contain several factors important in blood clotting. The facts mentioned previously and others related to the formed elements are summarized in Table 11-1.

Table 11-1. Summary of formed elements

Element	Normal numbers	Origin (area of)	Diameter (μm)	Morphology	Function(s)
Erythrocytes	4.5-5.5 million/mm^3 (highest in males)	Myeloid (marrow)	8.5 (fresh) 7.5 (dry smear)	Biconcave, nonnucleated disc; flexible envelope and interior	Transports O_2 and CO_2 by presence of hemoglobin; buffering by hemoglobin
Leukocytes	6000-9000/mm^3		9-25		
Neutrophils	60%-70% of total number	Myeloid	12-14	Lobed nucleus, fine heterophilic specific granules in cytoplasm	Phagocytosis of particles, wound healing. Granules contain peroxidase for destruction of microorganisms.
Eosinophils	2%-4% of total number	Myeloid	12	Lobed nucleus; large, shiny red or yellow specific granules in cytoplasm	Detoxification of foreign proteins? Granules contain peroxidases, oxidases, trypsin, phosphatases. Numbers increase in autoimmune states, allergy, and in parasitic infection (schistosomiasis, trichinosis, strongyloidiasis).
Basophils	0.15% of total number	Myeloid	9	Obscure nucleus; large, dull, purple specific granules in cytoplasm	Control viscosity of connective tissue ground substance? Granules contain heparin (liquefies ground substance), serotonin (vasoconconstrictor), histamine (vasodilator).
Lymphocytes	20%-25% of total number	Lymphoid (lymph organs)			
Small			9	Nearly round nucleus filling cell, cytoplasm a clear staining, thin rim	Phagocytosis of particles, globulin production.
Large			12-14	Nucleus nearly round more cytoplasm	
Monocytes	3%-8% of total number	Lymphoid	20-25	Nucleus kidney- or horseshoe-shaped, cytoplasm grayish	Phagocytosis. globulin production.
Platelets (Thrombocytes)	250,000-350,000/mm^3	Myeloid	2-4	Chromomere and hyalomere	Carries a clotting factor.

Adapted from McClintic, J.R.: Basic anatomy and physiology of the human body, ed. 2. Copyright © 1980. Reprinted by permission of John Wiley & Sons, Inc.

11 Circulatory system

Heart

The heart is basically a hollow muscular pump that creates a pressure on the blood, causing it to circulate through the body's blood vessels.

Development of the heart (Fig. 11-3). At approximately 2 weeks of embryonic life, a ***cardiogenic plate*** of mesoderm develops in the anterior end of the embryo. In the next 2 to 4 days, the plate organizes into two elongated masses called the ***heart cords.*** Within the cords cavities will develop, creating hollow ***heart tubes.*** Starting with the anterior end, the paired heart tubes

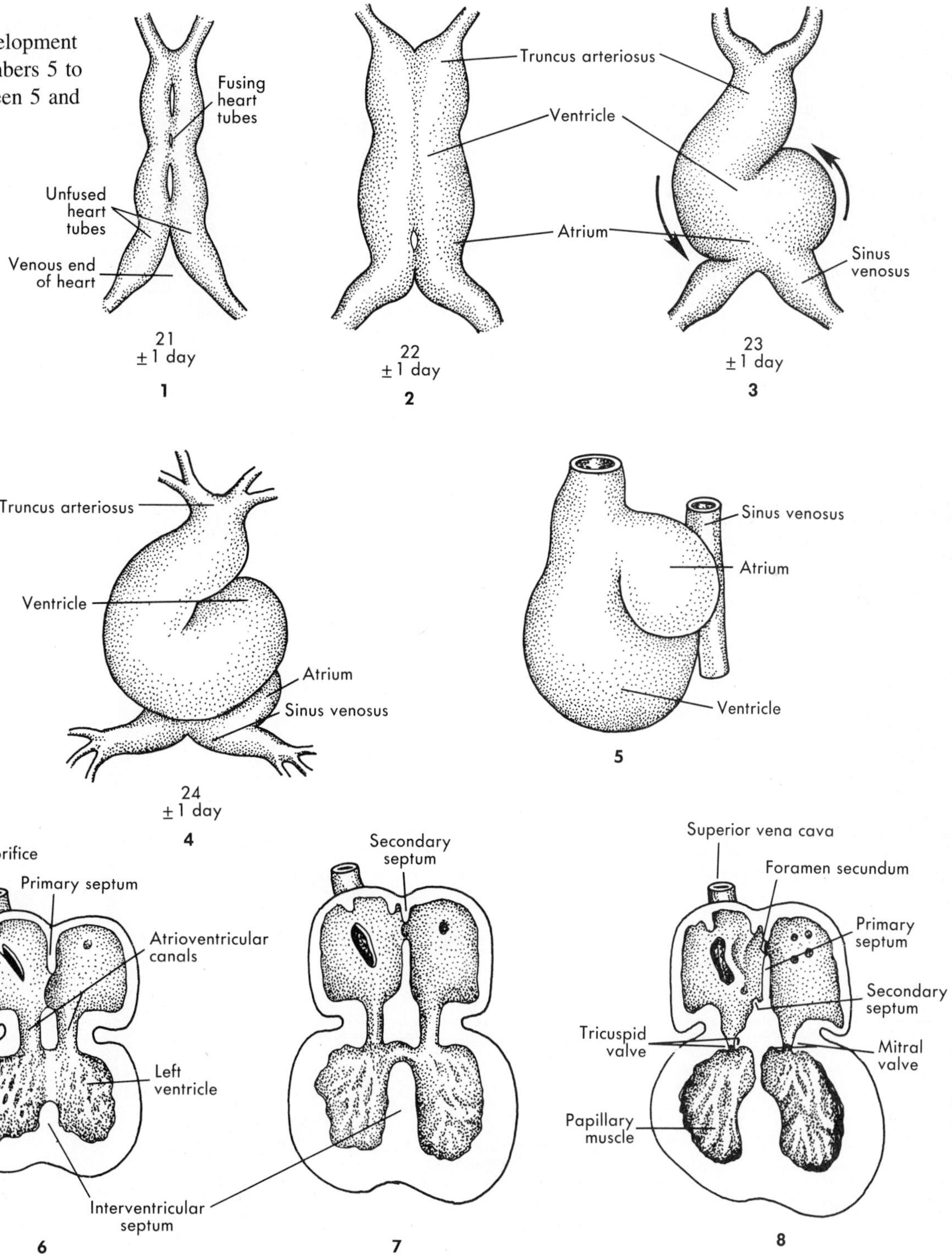

Fig. 11-3
Stages in development of heart. Numbers 5 to 8 occur between 5 and 8 weeks.
After Moore.

fuse, forming by day 22 a single tube. By 28 days, the single tube develops a series of constrictions that separate the tube into several parts named, from anterior to posterior, the *bulbus arteriosus, ventricle, atrium,* and *sinus venosus.* Next, rapid growth of the tube occurs, and the ventricle and sinus venosus fold over the atrium to lie posterior to it, the normal fetal position of the chambers. Between weeks 5 and 7, the interior of the atrium and ventricle are partitioned to form a four-chambered organ.

The single atrium is divided by what is called the **primary septum.** This septum is perforated by an *ostium* or opening. A **secondary septum** next forms to the right of the primary septum but does not join centrally; the two flaps overlap one another. The opening that remains forms the *foramen ovale,* which allows blood from the right atrium to pass to the left atrium, bypassing the nonfunctional fetal lungs.

The ventricle is partitioned by an *interventricular septum* that grows upward from the apex of the common ventricle.

The bulbus arteriosus divides into two vessels, the *aorta,* associated with the left ventricle, and the *pulmonary trunk* that leaves the right ventricle.

As these changes are occurring, mesoderm is forming the three **layers** of the heart wall and the specialized **nodal tissue** is differentiating. The latter behaves like nervous tissue, initiating and distributing impulses that cause the heart's muscle to contract.

Connections of the vascular system are established by the placenta for nutrient, gas, and waste exchange with the uterus, and the circulation shows the features depicted in Fig. 11-4.

The time during which the atrium and ventricle are being subdivided and the bulbus arteriosus is dividing is critical in terms of interference with these processes. Measles virus, if the mother contracts the disease at this time, can lead to atrial septal defects (ASD), ventricular septal defects (VSD), or malformations of the great vessels.

At birth, pressure differentials close the foramen ovale, and the ductus arteriosus, which carries blood from pulmonary trunk to aorta, shuts down. The blood now must pass through the lungs.

Fig. 11-5 depicts several of the more

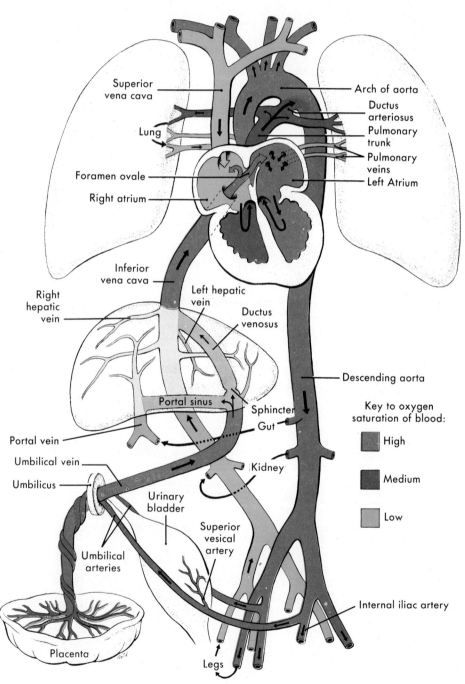

Fig. 11-4

Plan of fetal circulation. Colors indicate state of oxygenation of blood.

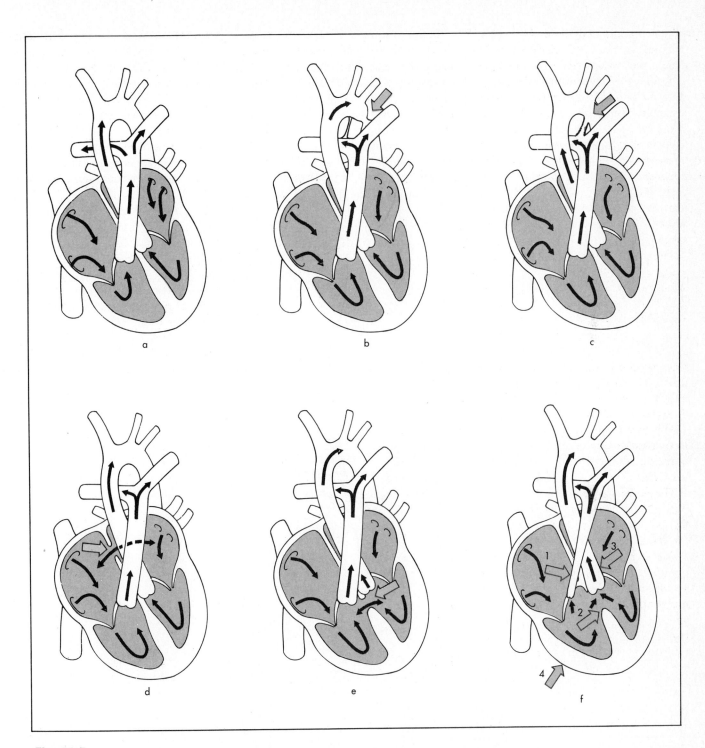

Fig. 11-5
Some more common congenital anomalies of heart. *a*, Normal heart; *b*, coarctation of aorta (aortic lumen is narrowed); *c*, patent ductus arteriosus (blood tends to bypass lungs); *d*, atrial septal defect (oxygenated and unoxygenated blood mix); *e*, ventricular septal defect (blood passes from left to right ventricle); *f*, tetralogy of Fallot with *(1)* narrowing of pulmonary artery or valve; *(2)* ventricular septal defect; *(3)* overriding aorta; and *(4)* hypertrophied right ventricle.

common congenital (born with) disorders of the heart. Each diagram includes a short description of the disorder. Valvular problems may develop as a result of malformation, disease (for example, rheumatic fever), or age. A *stenosed* valve is one that fails to open completely. Stretched valves or those that fail to close properly are said to be *incompetent.* Both conditions leave a larger than normal volume of blood in the ventricle to be ejected, and the heart may enlarge as a result of this extra volume.

Size. At almost any age after birth, the size of the human heart approximates the size of the clenched fist of the owner. At birth the heart measures about 25 mm in length, width, and depth, and weighs about 20 g. In the adult it measures about 12.5 cm in length, 9 cm in width, 5 cm in depth, and weighs about 300 g. It grows most rapidly between 8 and 12 years of age and again between 18 and 25 years of age.

Location and description (Fig. 11-6). The heart rests on the diaphragm in the lower part of the mediastinum, at the level of the fifth to eighth thoracic vertebrae. It is enclosed in a double-walled sac known as the *pericardium,* about which more will be said in the next paragraph. The broad upper end of the organ is the *base,* and it is directed toward the right shoulder. The more pointed lower end or *apex* is directed toward the seventh to eighth left ribs. A line drawn from the center of the base to the apex defines the *axis* of the heart and further emphasizes that the organ is not placed vertically within the chest. About one third of the organ lies to the right of the midsternal line and two thirds to the left of the line, demonstrating that it is not centered within the chest.

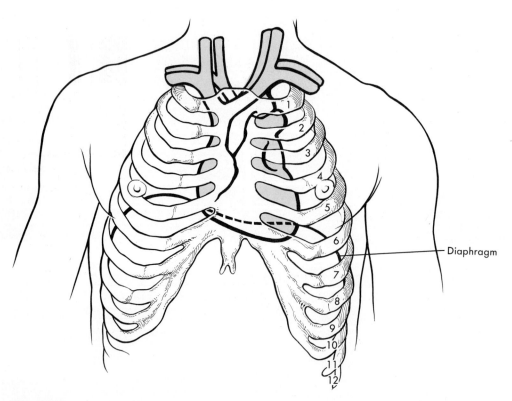

Fig. 11-6
Location of heart in thorax.

Pericardium. The pericardium (Fig. 11-7) is a double-walled sac composed of two layers of tissue designated as the *serous pericardium* and the *fibrous pericardium*. The serous pericardium lines the inside of the pericardial sac as the parietal pericardium *and is reflected onto the surface of the heart itself as the* visceral pericardium *or epicardium* of the heart wall. The layer consists of a *mesothelium* (simple squamous epithelium) beneath which is a thin connective tissue layer known as the *submesothelial layer.* The two layers of the serous pericardium normally lie close to one another and are lubricated by a film of fluid that allows nearly frictionless movement of the heart within the pericardium as it beats. The potential space between the serous membranes is the *pericardial cavity.*

The fibrous pericardium forms a flask-shaped sac that attaches superiorly to the bases of the aorta and pulmonary trunk and is adherent to the parietal pericardium. Ventrally, it attaches to the manubrium of the sternum and xiphoid process, inferiorly to the diaphragm, and anteriorly and posteriorly it is opposed to the parietal pleurae of the lungs. As we breathe, the fibrous pericardium and heart are pulled downward within the chest on inspiration and are elevated during expiration.

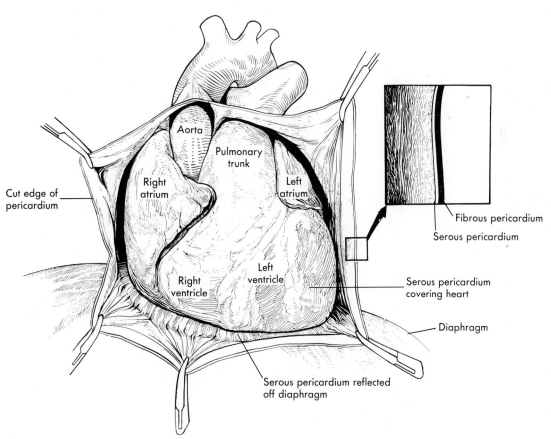

Fig. 11-7
Relationship of pericardium to heart.

Heart wall. As the heart develops, three basic layers of tissue form its walls (Fig. 11-8). An inner **endocardium** is composed of a lining *endothelial layer* and several *layers of connective tissue and smooth muscle*. The valves of the heart are formed from the endocardium. In some parts of the heart a **subendocardial layer** contains the specialized nodal tissue that causes the contraction of the heart muscle. The centrally placed **myocardium** is composed of *cardiac muscle* and accounts for about three fourths of the total width of the heart wall. The outer layer of the heart wall is the **epicardium** or serous pericardium described earlier. The epicardium carries the larger vessels of the coronary circulation and may contain significant quantities of adipose tissue.

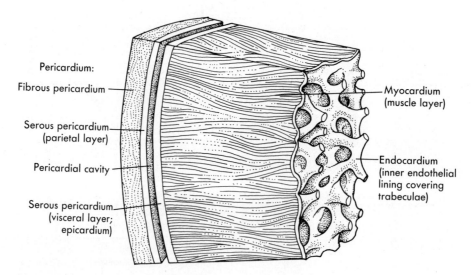

Fig. 11-8
Structure of heart wall and its relationship to pericardium.

Chambers and valves. The heart is actually two pumps in one (Fig. 11-9). The right side of the heart receives blood from the body in general and sends it to the lungs for gas exchange. The left side of the heart receives blood from the lungs and sends it to the body generally.

The upper receiving chambers of the heart are the *atria,* and the lower pumping chambers are the *ventricles.* There are thus two atria, right and left, and a right and left ventricle for a total of four chambers. Atria have ridges of muscle in their walls, the *pectinati,* and are separated from the ventricles by an externally located *coronary sulcus.* Internally, they are separated from one another by an *interatrial septum.* An *interventricular septum* separates the two ventricles. Externally, the two ventricles are separated by an *anterior* and *posterior interventricular sulcus.*

Blood that enters the right atrium passes through the *right atrioventricular orifice* into the right ventricle. The orifice is guarded by the right atrioventricular or *tricuspid valve.* This valve consists of three flaps or cusps that are attached by their bases to a fibrous ring that surrounds the orifice; the free edges are scalloped and attach to numerous strands of collagenous tissue called the *chordae tendineae.* The chordae are in turn attached to *papillary muscles* in the wall of the ventricle. The muscles are elevated portions of the myocardium covered with endocardium, and the whole assemblage prevents the valve from reversing as the ventricle contracts. Ridges of muscular tissue in the ventricle wall are the *trabeculae carneae.* Blood passes next

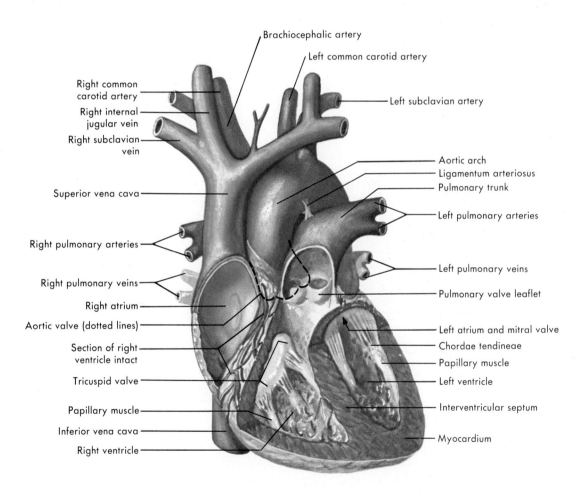

Fig. 11-9
Heart in frontal section, with associated organs.

through the **pulmonary valve** *(pulmonary semilunar valve)* (Fig. 11-10) in the base of the pulmonary trunk that exits from the right ventricle. This valve consists of three cuplike pockets of tissue. As the ventricle relaxes, blood attempts to return to the ventricle, fills the pockets of the valve, and closes it. After oxygenation, blood returns to the left atrium through four pulmonary veins (two from each lung) and passes through the *left atrioventricular orifice* to the left ventricle. This orifice is guarded by the **mitral** *(left atrioventricular* or *bicuspid) valve.* It is constructed like the tricuspid valve (but has only two cusps) and possesses chordae tendineae and two large papillary muscles. The wall of the left ventricle has trabeculae carneae like those of the right ventricle. From the left ventricle, blood passes through the **aortic valve** into the aorta. This valve is built like the pulmonary valve.

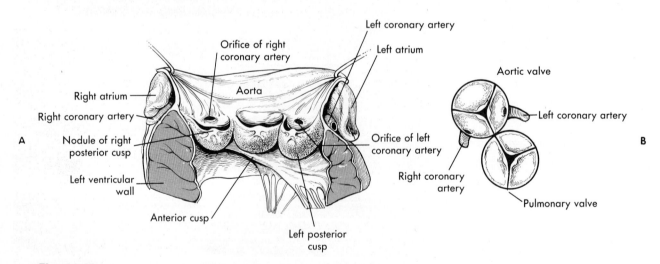

Fig. 11-10
Structure of semilunar valve. **A,** Valve cusps seen as if aorta were cut lengthwise and flattened. **B,** Two valves as viewed from above.

Conductile and contractile tissues of the heart. Epithelial tissue lines the internal and external surfaces of the heart; connective tissue forms a *cardiac skeleton* that supports the valves of the heart and sets the general shape of the organ. For pressure to be generated on the blood, **cardiac muscle** is present in the heart as its contractile tissue, and for the muscle to be stimulated to contract, there is specialized **nodal tissue** that originates and distributes nerve impulses.

Cardiac muscle (Fig. 11-11) is a type of striated muscle found only in the heart. It consists of branching *fibers* about 100 μm long by 15 μm wide. Each fiber is separated from its neighbor by a dark-staining **intercalated disc** that occurs at the I bands of the muscle. A *sarcolemma* is present, as are *sarcoplasm* and *myofibrils*. The tubules of the **sarcoplasmic reticulum** and the **transverse (T) tubules** are present but are less well developed than in skeletal muscle. They are larger and do not form triads as in skeletal muscle; the T-tubules lie on Z lines in cardiac muscle. *Nuclei* are centrally placed in the fibers, and the *myofibrils* sweep around the nuclei. Functionally, the tissue behaves as though it were one continuous mass or *syncytium;* that is, stimulation of any part causes spread of a wave of depolarization throughout the entire mass. The intercalated discs are anatomical but not functional limits to a fiber.

Other properties of cardiac muscle include the fact that it *follows the all-or-none law*. This law says that if a stimulus is strong enough to cause depolarization, it causes a complete depolarization. The value of this property to continued circulation of blood should be obvious—it is desirable to have the full contraction to develop pressure on the blood. Also, cardiac muscle *follows the law of the heart*, which indicates that a greater degree of stretch on the muscle, as by the chambers filling to a greater degree, results in a stronger contraction. Thus an automatic mechanism for emptying the greater quantity of blood and for matching demand to output is available. Cardiac muscle *cannot be tetanized*, or caused to sustain a contraction. Again, filling and emptying is the only way to achieve continual circulation.

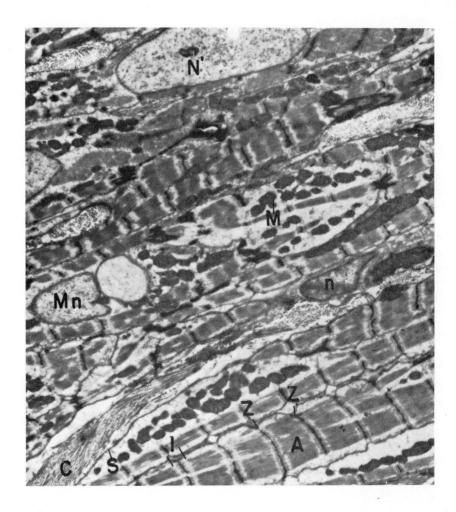

Fig. 11-11
Cardiac muscle histology.
N, Nucleolus; *M,* mitochondrion; *Z,* Z line; *A,* A line; *I,* I line; *S,* sarcolemma; *n,* nucleus of fibroblast; *C,* collagen fibrillae; *Mn,* muscle cell nucleus. (×6000.)

The cardiac muscle is disposed as two main masses, one for the atria and one for the ventricles. In the ventricles the fibers have a distinct, rather spiral orientation resembling a twisted towel (Fig. 11-12). When the ventricles contract, the blood is more "wrung" from the chambers than "pushed," causing the apex of the heart to rotate and to "tap" the chest wall as the *apex beat*.

The nodal tissue develops from cardiac muscle but has largely lost its power of contractility while developing the powers of spontaneous depolarization and conductility to a high degree. Fibers of the nodal tissue are two to four times larger in diameter than cardiac muscle fibers and are rich in sarcoplasm and poor in myofibrils. The few myofibrils that are present are generally peripherally arranged in the fibers. In the heart the nodal tissue is disposed in several discrete masses, called nodes or bundles, within the subendocardial layer of the heart (Fig. 11-13).

The *sinoatrial (SA) node* measures about 18 by 6 mm and lies in the right atrium at its junction with the superior vena cava. Because of its highest rate of spontaneous depolarization, the SA node acts as the *pacemaker* of the heart beat. Impulses that originate in the SA node pass through the atrial musculature, stimulating it to contract, and eventually arrive at the *atrioventricular (AV) node*. This is a node that is a bit smaller than the SA node and that is located in the right atrium near the top of the interventricular septum. Because

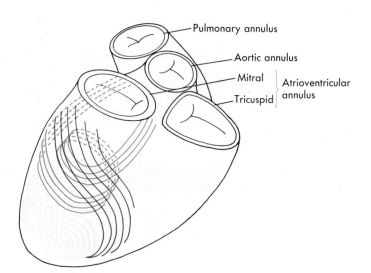

Fig. 11-12
Courses of major cardiac muscle fiber bundles in heart.

of the arrangement of the node, impulses are slowed in passing through it. This allows completion of atrial contraction to occur. Next in line and attached to the AV node is the *AV bundle (of His)* that lies in the top portion of the interventricular septum. From the bundle arise two **bundle branches** that pass toward the apex of the heart and fan out over the respective ventricle to terminate as fine **Purkinje's fibers** "inserted" into the innermost layer or two of myocardial muscle. The system from the AV node onward is a fast-conducting transmission system designed to bring impulses nearly simultaneously to all parts of the ventricular muscle. Because of the functional syncytial nature of the myocardium, the system brings a stimulating impulse that spreads throughout the ventricular mass to create a coordinated squeeze and maximal pressure on the blood within the ventricular cavities.

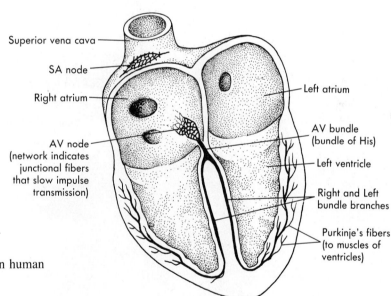

Fig. 11-13
Locations of nodal tissue in human heart.

Blood vessels of the heart (Fig. 11-14). Blood supply to the heart originates as two *coronary arteries* whose orifices are located behind the flaps of the aortic valve. The *right coronary artery* passes to the right and enters the coronary sulcus, following it around the right margin of the heart to its posterior surface, giving off branches to the right ventricle as it goes. On the posterior aspect of the heart, a *right interventricular artery* branches from the coronary artery and passes toward the apex of the heart in the posterior interventricular (longitudinal) sulcus, giving vessels to both ventricles. A large *marginal branch* supplies vessels to the right atrium and SA node and terminates on the posterior surface of the right ventricle to which it supplies vessels. The *left coronary artery* follows a course beneath the left atrium and forms an *anterior descending branch* and the *circumflex artery*. The former passes to the left to the anterior surface of the heart, follows the anterior interventricular (longitudinal) sulcus between right and left ventricles, and supplies branches to both ventricles. The circumflex artery passes around the heart to the left in the coronary sulcus and reaches nearly to the posterior interventricular sulcus. It supplies vessels to the left atrium and ventricle.

The veins of the heart (see Fig. 11-14) collect primarily into the *coronary sinus* that runs in the posterior coronary sulcus and empties into the right atrium between the openings of the superior vena cava and the tricuspid valve. Its major tributaries include:

great cardiac vein begins at the apex of the heart, ascends along the anterior interventricular sulcus, and joins the coronary sinus at its left extremity.
middle cardiac vein begins at the apex of the heart, ascends in the posterior interventricular sulcus, and joins the sinus at its right extremity.
posterior vein of the left ventricle accompanies the circumflex branch of the left coronary artery to the sinus.

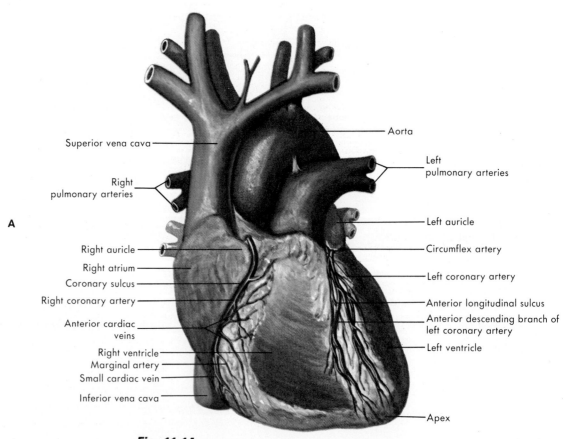

Fig. 11-14
Coronary vessels and associated structures. **A,** Anterior view. **B,** Posterior view.

Alternative pathways for venous return include three or four **anterior cardiac veins** that collect blood from the anterior aspect of the right ventricle and the **thebesian veins.** Both these sets of vessels empty their blood directly into the atria and ventricles, bypassing the coronary sinus.

Diminished blood flow to the myocardium *(ischemia)* is usually associated with the development of *angina*, a pressing or dully painful sensation commonly referred to the chest or left arm. It is believed to be due to metabolic changes occurring in ischemia that result in nerve stimulation by the abnormal metabolites. The ischemia itself may result from the deposition of fatty plaques *(atherosclerosis)* in arterial walls, spasm of the arteries, or clot formation *(thrombosis)* in the arteries. If a vessel is plugged and tissue beyond the plug dies, an **infarction** has occurred. If deprived of nutrients for more than a few minutes, the muscle's excitability may be altered so that the heart *fibrillates*. This term refers to a series of very rapid weak quivering contractions (200 to 400/min) that are completely ineffective in pumping blood. *Defibrillation* involves shocking the heart to render all the tissue electrically the same and hoping that the SA node can then resume control of the heart. *Bypass* operations involve grafting a piece of vein between the aorta and the heart wall to create blood channels that circumvent narrowed or plugged coronary arteries.

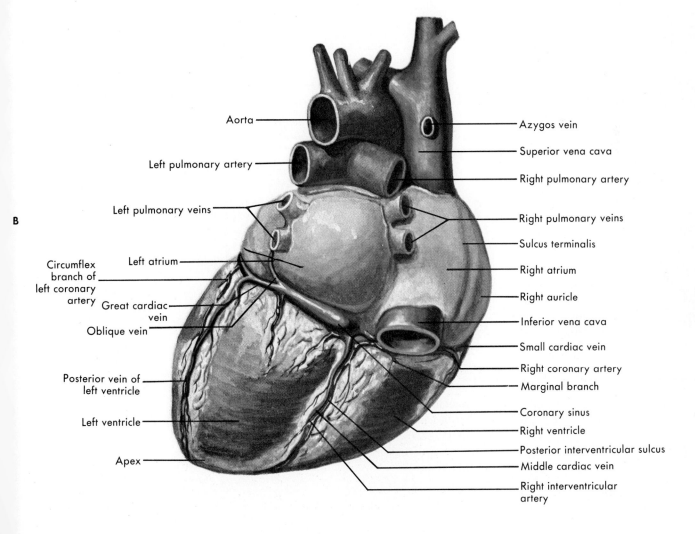

Nerves to the heart (Fig. 11-15). The heart receives motor fibers from both divisions of the autonomic nervous system. Parasympathetic fibers are carried in the *vagus nerves* and supply the SA and AV nodes and muscular arteries of the coronary system. Their effect is to slow heart rate and nodal conduction and to constrict coronary vessels. Sympathetic fibers are carried in the *cardiac nerves*, derived from the cervical spinal nerves. These fibers also terminate on the nodal masses and muscular arteries of the heart. Their effect is to increase heart rate and conduction and to dilate muscular arteries of the heart.

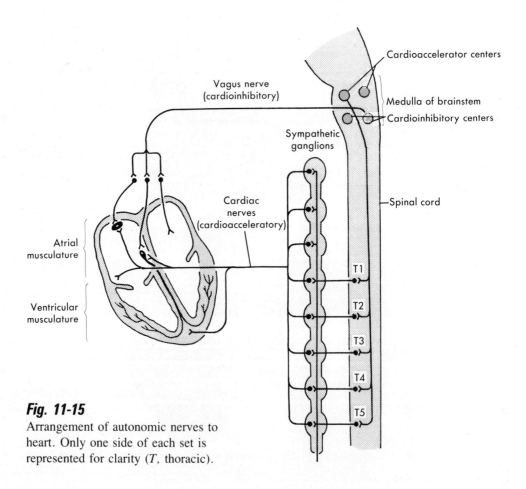

Fig. 11-15
Arrangement of autonomic nerves to heart. Only one side of each set is represented for clarity (*T*, thoracic).

Development of blood and lymph vessels (Figs. 11-16 and 11-17)

Blood vessels develop (see Fig. 11-1) from masses of mesoderm called **blood islands** that form first in the yolk sac and then in the embryo. At about 2 weeks of embryonic development, cavities appear in these islands and some of the mesodermal cells are found within the cavities. The cells lining the cavities flatten to form an endothelial lining for the cavities. Growth and joining of these cavities create networks of blood channels or what may now be called vessels. The network is extended by "budding" from the original vessels. The mesodermal cells within the vessels differentiate to form first red blood cells and somewhat later the various types of white blood cells. Other mesodermal cells outside the primitive vessels will give rise to the muscular and connective tissue coats around all vessels except those that will become capillaries. By about 3 weeks of development, the vessels indicated in Fig. 11-17 have formed, and a rudimentary circulation is established between the embryo and the developing placenta.

Lymph vessels begin to develop in the embryo at about 5 weeks. They develop in a manner similar to that of the blood vessels and make connections with those blood vessels that will become veins. Initial development of vessels occur from large areas known as

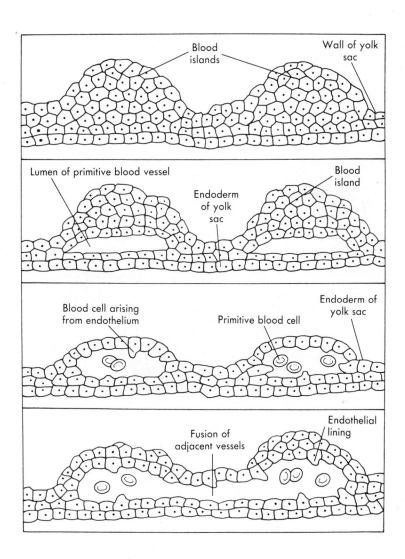

Fig. 11-16
Early development of blood vessels.

lymph sacs. There are six sacs that initially form: two in the neck region *(jugular lymph sacs)*, two in the pelvic area *(iliac lymph sacs)*, one against the posterior abdominal wall *(retroperitoneal lymph sac)*, and one dorsal to the retroperitoneal sac *(cisterna chyli)*. Lymph vessels (or lymphatics) grow from these sacs following the courses of the major veins to form networks in the developing body organs and muscles. Head, neck, upper trunk, and arms are supplied with lymphatic vessels from the jugular sacs, the lower trunk and legs from the iliac sacs, and the gut and its derivatives from the retroperitoneal sac and cisterna chyli.

Further information on development of blood vessels and lymphatic vessels may be found in the readings at the end of this chapter.

Structure of the blood vessels

Blood vessels serve as the "plumbing" of the body, carrying blood to and from the tissues and organs. There are three general categories of blood vessels distinguished by their relationship to the heart and body tissues and organs.

Arteries are vessels that carry blood *away from the heart*, regardless of the degree of oxygenation of that blood.
Capillaries permeate body tissues and organs and act as the *exchange vessels* between the bloodstream and the tissue fluids that surround body cells.
Sinusoids are vessels that have a function and structure similar to capillaries. They are found in the liver, brain, pancreas, and adrenal glands.
Veins are vessels carrying blood *to the heart*, again regardless of the degree of oxygenation of that blood.

The general arrangement of the body's blood vessels is depicted in Fig. 11-18. Two main circulatory pathways are evident:

A *pulmonary circulation* begins with the right ventricle, carrying blood to the lungs and back to the left atrium.
A *systemic circulation* begins with the left ventricle, carrying blood to the body generally and returning it to the right atrium.

The comments to follow are directed primarily to the vessels of the systemic circulation. Any major differences in vessels in other parts of the body will be noted in the section on circulation in special regions later in this chapter.

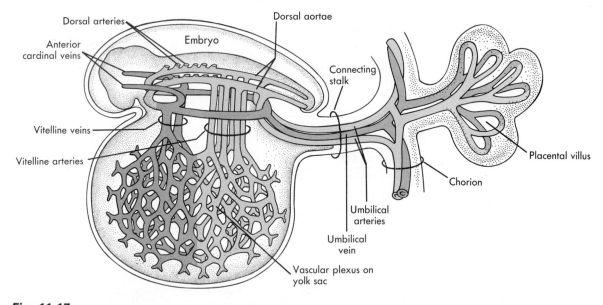

Fig. 11-17
Later development of blood vessels. Degree of development shown occurs at about 3 weeks.
After Moore.

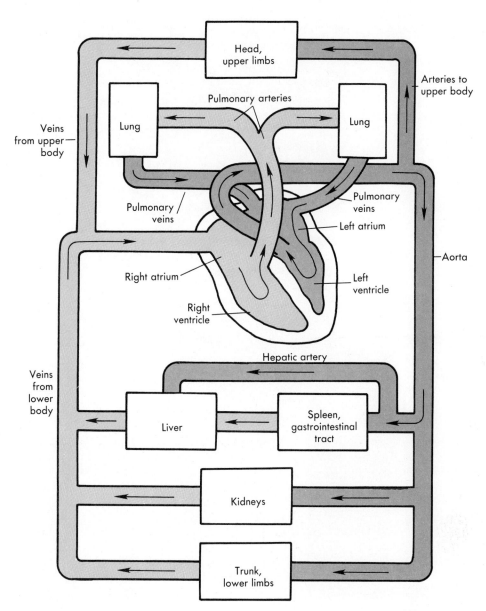

Fig. 11-18
Schematic representation of circulatory system.

Arteries. Arteries (Fig. 11-19; Table 11-2) are usually designated as *large*, *medium*, *small* (arterioles), and *metarterioles* primarily on the basis of size. The terms also imply structural differences in the vessels. In general, the smaller the artery, the greater the number of vessels.

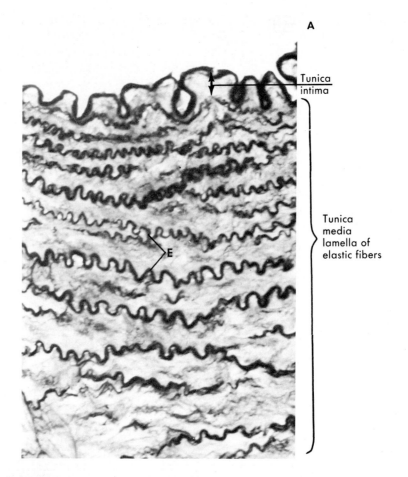

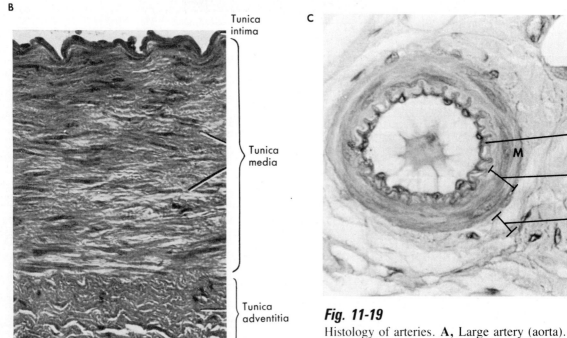

Fig. 11-19
Histology of arteries. **A,** Large artery (aorta). *E,* Elastic laminae. **B,** Medium artery. **C,** Small artery (arteriole). *M,* Smooth muscle.

Large arteries, also known as *elastic arteries,* are exemplified by the aorta, the pulmonary trunk, and the initial branches of these vessels. Such vessels have a three-layered structure in their walls:

tunica intima (tunica interna) consists of an *endothelial lining,* a layer of *subendothelial connective tissue,* and an elastic membrane called the *internal elastic lamina.*

tunica media consists of 50 or more lamellae (or layers) of elastic membranes separated by small amounts of reticular and collagenous fibers and a few smooth muscle cells.

tunica adventitia (tunica externa) is thinner than the media and is composed of loose connective tissue. This layer commonly carries the *vasa vasorum,* a network of small blood vessels that nourish the tissues of the vessel wall.

Elastic arteries expand passively as the ventricular volume is ejected into them and recoil to help keep the blood moving when the ventricles relax. Stiffening of elastic artery walls results in a greater pressure during ventricular contraction (systole) and a lower pressure during ventricular relaxation (diastole).

Medium arteries are a type of *muscular artery.* They are exemplified by the named arteries of the appendages and viscera, and as in the large arteries, three layers of tissue form their walls:

tunica intima is composed of an *endothelial lining,* a *subendothelial connective tissue layer,* and an *internal elastic lamina.*

tunica media contains about 12 to 40 layers of *smooth muscle* that is circularly disposed around the vessel. Contraction of the muscle diminishes and relaxation increases the diameter of such a vessel.

tunica adventitia is thinner than the media and consists of a loosely woven network of *collagenous and elastic fibers.* A vasa vasorum may be present.

Small arteries *(arterioles)* are unnamed and are the very numerous branches of the medium arteries passing to body tissues and organs. They also have a three-layered wall:

tunica intima consists of the *endothelial layer.* There is little connective tissue and no elastic lamina.

tunica media contains from two to about a dozen layers of circular *smooth muscle.*

tunica adventitia is about as wide as the media and consists of loose networks of longitudinal *collagenous and elastic fibers.* A vasa vasorum is not present.

Arterioles, because of their great numbers, act as stopcocks to control the flow of blood into capillary net-

Table 11-2. Summary of artery structure

Type of vessel	Average size	Example(s) or location	Tunica intima	Tunica media	Tunica adventitia	Comments
Large artery	20-25 mm	Aorta, pulmonary trunk	Endothelium and ct (connective tissue)	Elastic ct—60 to 70 layers	Loose ct	Elastic vessels expand to contain cardiac output; recoil to push blood onward
Medium artery	2-10 mm	Named vessels of arms, legs, viscera	As above	25-40 layers of smooth muscle	As above	Muscle makes them important in terms of changing size; not too numerous
Small artery (arteriole)	2 mm to about 20 μm	Unnamed; close to or in tissues and organs	As above	5-10 layers of smooth muscle	As above	Large numbers make these the primary controllers of blood flow and pressure
Metarteriole	10-20 μm	At entrances to capillary beds	Vessel wall consists only of endothelium and a layer of smooth muscle			Act as sphincters to control filling of capillary beds

Adapted from McClintic, J.R.: Basic anatomy and physiology of the human body, ed. 2. Copyright © 1980. Reprinted by permission of John Wiley & Sons, Inc.

works and cause redistribution of blood from inactive to active areas as their diameters change.

Metarterioles are located at the entry to capillary beds. An *endothelial lining* is surrounded by one cell or one layer of *smooth muscle cells*. These vessels act as precapillary sphincters and control the blood flow into particular portions of a capillary network.

Muscular arteries are usually provided with two types of **vasomotor nerves.** Nerves that cause the muscle to contract are termed *vasoconstrictor nerves* and are derived from the sympathetic division of the autonomic nervous system. At their endings they produce norepinephrine, which is the actual stimulus for smooth muscle contraction. *Vasodilator nerves* are derived from the parasympathetic division of the autonomic system and secrete acetylcholine, which inhibits contraction and causes dilation of the vessels. The nerves are connected to vasomotor centers in the medulla of the brainstem, and by this connection the brain can influence muscular vessel diameter. The importance of all this is that blood vessel size will determine blood pressure by changing the capacity of the vessels relative to a rather fixed volume of blood. Thus vasodilation is associated with a fall and vasoconstriction with a rise of blood pressure.

Persistent vasoconstriction is one cause of **hypertension,** or *high blood pressure*. It can be life threatening by placing strain on blood vessel walls, especially in the brain, and on the heart that has to work muscle harder to circulate blood against the high pressure in the vessels.

Hypotension, or *low* blood pressure, is a desirable state as long as the pressure is not so low as to fail to adequately perfuse such organs as the kidney and brain. Less work by the heart is required to circulate blood, and vessel walls are not subjected to as much strain.

Normal values for the blood pressure are given as 110 to 135 mm Hg systolic and 70 to 90 mm Hg diastolic.

Capillaries. *Capillaries* (Fig. 11-20), as the areas where exchange of materials occurs between bloodstream and (ultimately) cells, must have very thin walls. They are also the most numerous vessels (it has been estimated that there is 60,000 miles of capillaries in the average adult body). An *endothelial tube* some 7 to 8 μm in diameter is surrounded by a few collagenous fibrils. Thus only one layer of cells, the endothelium, must be traversed by any material entering or leaving the capillary. Transfer of substance across a capillary wall is by diffusion, osmosis, and filtration, aided by pinocytosis, as evidenced by the many pinocytic vesicles found in capillary walls (see Fig. 11-20).

Adipose tissue is rich in capillaries. It is claimed that 1 pound of fat adds 3 miles of capillaries to the circulation and creates that much more tube that takes a pressure to perfuse. Heavy people therefore often have higher blood pressures than lean ones.

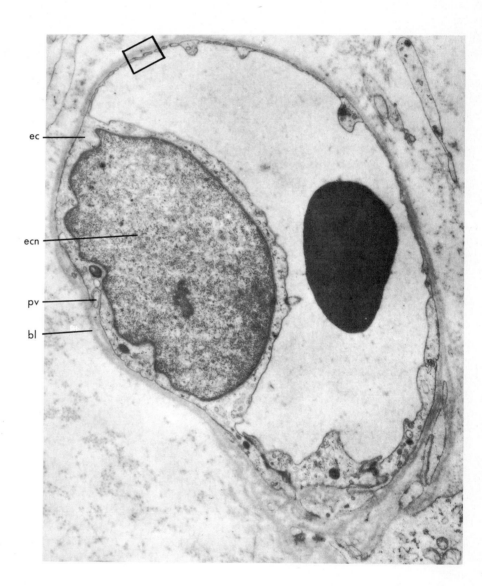

Fig. 11-20
Histology of capillary as seen under electron microscope (×11,000). *pv*, Pinocytic vesicles; *ec*, endothelial cell cytoplasm; *ecn*, endothelial cell nucleus; *bl*, basement lamina (membrane). Enclosed in the box are several capillary "pores."

From Bevelander, G., and Ramaley, J.A.: Essentials of histology, ed. 8, St. Louis, 1979, The C.V. Mosby Co.

Veins. Veins (Fig. 11-21; Table 11-3) are subdivided into three types on the basis of size and structure. In order of the blood flow as it would pass through them in the body they are *small veins (venules)*, ***medium veins***, and ***large veins***.

Venules drain the capillary beds. They are about as numerous as the arterioles to which they correspond. Size of venules varies greatly, from about 20 μm to 0.2 mm. ***Venules*** are typically two layered in structure: A lining *endothelium* is surrounded by a layer of *collagenous* connective tissue. No subdivision into tunics is evident, and a vasa vasorum is not present. In the smaller venules the connective tissue alone forms the second layer of the wall. When the vessel reaches 50 to 60 μm in diameter, smooth muscle cells appear, scattered in the connective tissue. Numbers of muscle cells increase as the size of the vein increases, but a *densely* packed muscle layer such as composes artery media is not formed.

Medium veins are three-layered

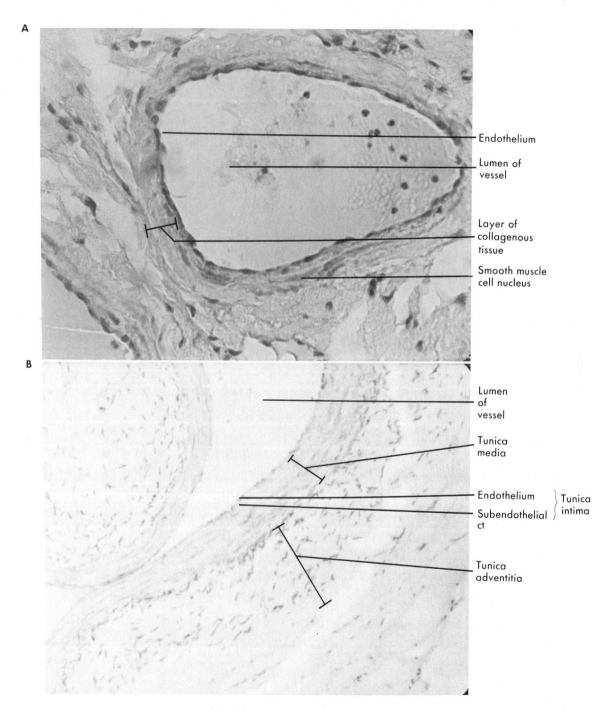

Table 11-3. Summary of vein structure

Type of vessel	Average size (mm)	Example(s) or location	Tissue in			Comments
			Tunica intima	Tunica media	Tunica adventitia	
Large vein	20-30	Vena cava	Endothelium and ct	Ct and scattered smooth muscle cells	Loose ct	Few in number, little change in size
Medium vein	2-20	Named veins of appendages and viscera	As above	As above	As above	As above
Small vein (venule)	<2	In tissues and organs	As above	Collagenous ct and a few smooth muscle cells	Lacking	Most numerous; can significantly alter venous volume by change in size

From McClintic, J.R.: Basic anatomy and physiology of the human body, ed. 2. Copyright © 1980. Reprinted by permission of John Wiley & Sons, Inc.

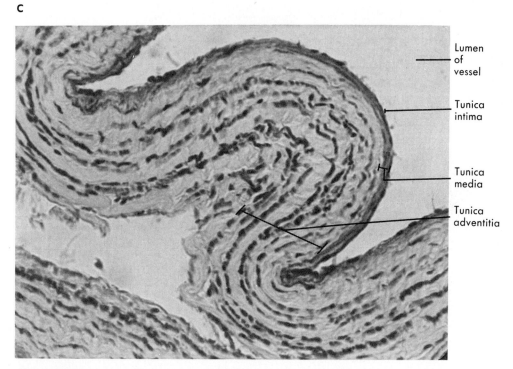

Fig. 11-21
Histology of veins. (Venule at twice magnification of other two.)
A, Venule (small vein). **B,** Medium vein. **C,** Large vein.

structures that are exemplified by the named veins of the appendages and viscera:

tunica intima is composed of an *endothelium* lying on a thin *subendothelial connective tissue layer*.

tunica media is composed of several layers of loosely woven *smooth muscle and connective tissue*.

tunica adventitia is two to five times thicker than the media and is composed of *loose connective tissue*.

Large veins are exemplified by those closest to the heart, such as the venae cavae. They are three layered also, with the media extremely thin; these vessels are almost two layered. Thus the adventitia becomes the primary layer of the vessel wall.

Medium veins and the larger venules are provided with valves that ensure a movement of blood *toward* the heart through the vessels and usually carry a vasa vasorum.

Veins contain blood at very low pressure, the pressure having been lost in traversing capillary beds. Veins are volume, not pressure, vessels and normally contain about 60% of the blood volume.

Clinical considerations. Structure of the wall of an artery may be altered by processes that stiffen, thicken, or lead to the deposition of substances in the vessel wall. The term **arteriosclerosis** is a generalized term used to describe the results of a thickening or hardening process. The term **atherosclerosis** (Fig. 11-22) refers to the changes that occur when certain types of lipids enter and are retained within artery walls. Proliferation of arterial smooth muscle, as well as the presence of the lipid itself, decreases vessel diameter, leading to diminished blood

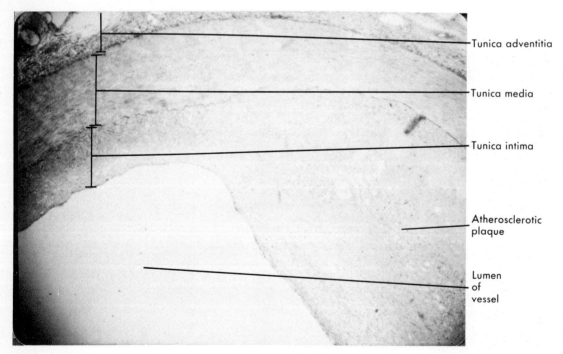

Fig. 11-22
Section of artery that has become atherosclerosed. (Note thickened intima and plaque.)

flow to vital organs and to rise of blood pressure. Smoking, high-fat diets, and lack of exercise have been shown to accelerate the atherosclerotic process.

Aneurysms (Fig. 11-23) are thin blisterlike areas in a blood vessel. They may be due to failure of the vessel wall to develop properly, or by high blood pressure. Rupture with consequent hemorrhage is a constant threat.

Aneurysms may take several forms:

A *berry aneurysm* is a small rounded or saccular dilation that is common in the cerebral circle of Willis. Vessels here are thin, and in persons having hypertension they pose a constant danger of rupture.

Fusiform aneurysms are spindle shaped and are usually due to atherosclerosis.

Dissecting aneurysms are common in the aorta and are the result of blood entering the wall of the vessel and splitting its media. The wall is weakened and separates, allowing ballooning to occur. High pressure in the aorta speeds the process, and death usually occurs from hemorrhage as the vessel comes apart.

Varicosities (Fig. 11-24) occur in veins, primarily the superficial medium-sized vessels of the lower appendage. Veins are thin walled, and if blood accumulates in them, as in the legs when a person is standing still for long periods of time, the veins may become stretched. Their valves do not close, blood stretches the vessels distally, and a vicious circle ensues. Surgical removal of the vessels may be required, or the wearing of supportive hosiery may slow the enlargement of the vessels.

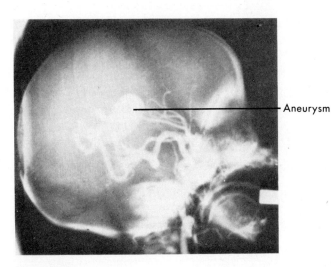

Fig. 11-23
Aneurysm of cerebral vessel.

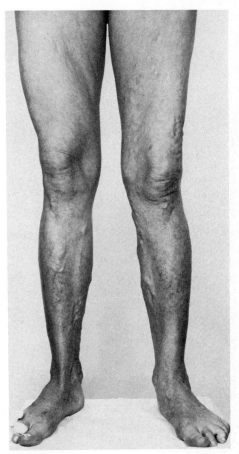

Fig. 11-24
Varicose veins. Note dilation and tortuosity of superficial venous plexuses.

Major systemic arteries

Aorta. The main trunk of the systemic arterial system is the **aorta**, which consists of three portions: the **ascending aorta** rises from the left ventricle and ascends toward the head for about 5 cm; the **arch** of the aorta curves to the left in the normal individual; and the **descending aorta** begins where the arch pursues a clearly inferior course. It passes through the chest as the *thoracic aorta,* through the abdomen as the *abdominal aorta,* and terminates in the lower abdomen by branching to give off the common iliac arteries.

In the sections to follow, the major branches from the aorta will first be described. Then a more or less regional approach will be adopted that will trace the aortic branches into the head and neck, appendages, and thoracic and abdominal cavities.

Branches of the ascending aorta. The ascending aorta (Fig. 11-25; Table 11-4) gives rise to the **right** and **left coronary arteries,** whose distribution to the heart has been described in a previous section.

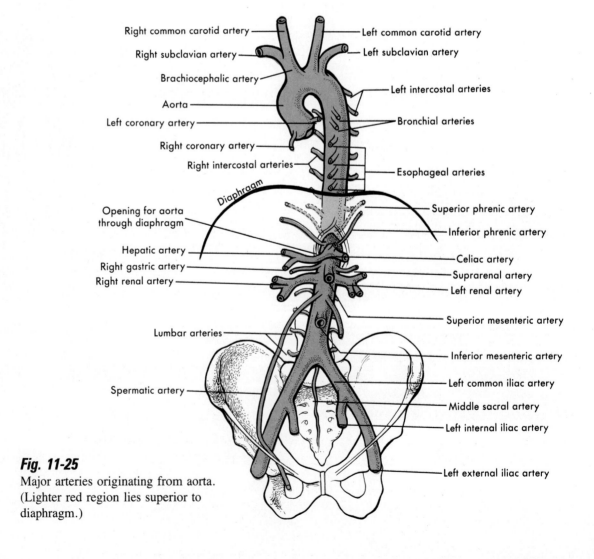

Fig. 11-25
Major arteries originating from aorta. (Lighter red region lies superior to diaphragm.)

Branches of the arch of the aorta. Three vessels arise from the aortic arch. From proximal (the heart) to distal, they are as follows:

brachiocephalic artery terminates by dividing into the *right subclavian* and *right common carotid* arteries. The brachiocephalic artery gives branches to the thyroid gland and sometimes supplies branches to the thymus gland.

left common carotid artery rises separately from the aortic arch.

left subclavian artery rises separately from the aortic arch.

The courses of all these vessels will be described in a section to follow.

Branches of the thoracic descending aorta. As the aorta passes through the chest, it gives off branches that are named according to the structures supplied (see Fig. 11-25). They include the following:

pericardial arteries supply the pericardium.
bronchial arteries follow the tubes of the respiratory system into the lungs.
esophageal arteries supply the esophagus.
phrenic (superior and inferior) *arteries* supply the diaphragm
intercostal arteries usually form nine pairs

Table 11-4. Major branches of the aorta and area(s) supplied

Arising from	Branch	Area supplied by branch
Ascending aorta	Coronary arteries	Heart
Arch of aorta	Brachiocephalic artery gives rise to: Right common carotid artery Right subclavian artery	Right side head and neck Right arm
	Left common carotid artery	Left side head and neck
	Left subclavian artery	Left arm
Descending aorta Thoracic portion	Pericardial arteries	Pericardium
	Intercostal arteries	Intercostal muscles, chest muscles, pleurae
	Superior phrenic arteries	Posterior and superior surfaces of the diaphragm
	Bronchial arteries	Bronchi of lungs
	Esophageal arteries	Esophagus
	Inferior phrenic arteries	Inferior surface of diaphragm
	Spinal arteries	Spinal cord meninges
Abdominal portion	Celiac artery gives rise to: Hepatic artery Left gastric artery Splenic artery	Liver, stomach, duodenum Stomach and esophagus Spleen, part of pancreas and stomach
	Superior mesenteric artery	Small intestine, cecum, ascending and part of transverse colon
	Suprarenal arteries	Adrenal glands
	Renal arteries	Kidneys
	Spermatic (male) or ovarian (female) arteries	Testes Ovaries
	Inferior mesenteric artery	Part of transverse colon and descending colon (left colic), sigmoid colon (sigmoid), most of rectum (superior rectal)
	Common iliac arteries, which give rise to: External iliac arteries Internal iliac arteries Midsacral artery	Terminal branches of aorta Lower limbs Uterus, prostate gland, buttock muscles Coccyx

Adapted from McClintic, J.R.: Basic anatomy and physiology of the human body, ed. 2. Copyright © 1980. Reprinted by permission of John Wiley & Sons, Inc.

of vessels and are distributed to the intercostal muscles, pectoral muscles, serratus anterior, mammary glands, and from which *spinal branches* pass through the intervertebral foramina to the spinal cord and its membranes.

Branches of the abdominal descending aorta. The abdominal aorta gives off branches that supply most of the organs located within the abdominopelvic cavity (see Fig. 11-25). The abdominal aorta begins at the diaphragm and gives off, in order from the diaphragm inferiorly, the following major branches:

celiac artery (Fig. 11-26), an unpaired vessel, rises just inferior to the aortic hiatus of the diaphragm. In turn it usually divides into three branches: the *left gastric* that supplies the anterior and posterior aspects of the stomach; the *hepatic artery* that mainly supplies the liver but also sends branches to the stomach and duodenum (gastroduodenal artery); and the *splenic artery* (lineal artery) that sends branches to the pancreas, stomach, and spleen.

superior mesenteric artery (Fig. 11-27), an unpaired vessel, rises about 1.25 cm inferior to the celiac artery and is responsible for the blood supply of the entire small intestine and the large intestine as far as the middle of the transverse colon. Its major branches include the *ileocolic, right colic,* and *middle colic* arteries for the large intestine and the *intestinal* arteries for the small intestine.

suprarenal arteries are paired and pass to the adrenal glands.

renal arteries are paired and pass from the aorta to the kidneys. Their distribution will be considered in Chapter 14.

gonadal arteries, known as *ovarian* or *spermatic* arteries according to sex, are small paired vessels that proceed to ovaries or testes.

inferior mesenteric artery (Fig. 11-28), a single vessel, branches to form the *left colic* artery, which supplies the distal half of the transverse colon and the descending colon; the *sigmoid* artery, which supplies the sigmoid colon; and the *superior rectal* (hemorrhoidal) artery, which supplies part of the rectal vasculature.

middle sacral artery rises from the aorta just before its bifurcation into the common iliac arteries and passes to the rectum.

Completing the blood supply of the abdominal organs is the branch of the external iliac artery called the **internal iliac (hypogastric)** artery. A host (to 10) of branches are formed by this vessel. The more important ones are:

vesical arteries supply the urinary bladder, vas deferens, seminal vesicle, and prostate gland.

middle rectal artery supplies the rectum.

uterine artery supplies the uterus.

vaginal artery supplies the vagina.

obturator artery supplies the muscles of the posterior pelvis.

internal pudendal artery supplies the penis or the labia majora, clitoris, and erectile tissue of the vagina.

gluteal arteries supply the muscles of the buttocks and posterior sacrum.

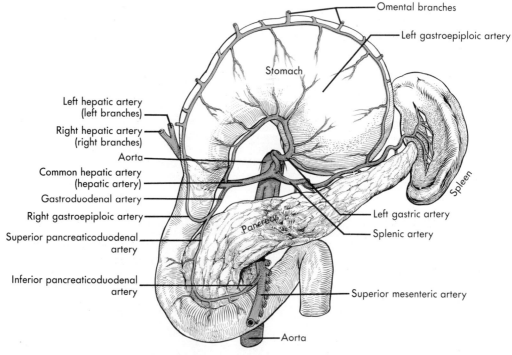

Fig. 11-26
Distribution of branches of celiac artery. (Stomach has been reflected superiorly.)

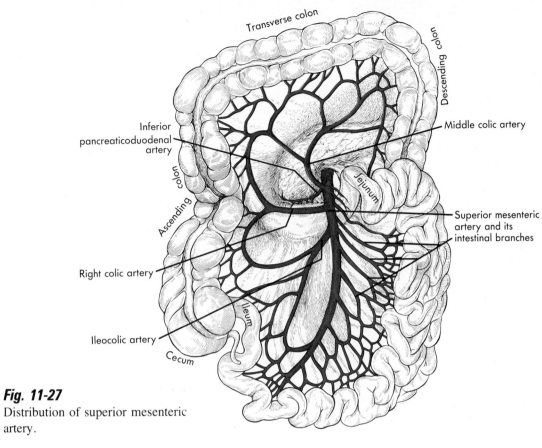

Fig. 11-27
Distribution of superior mesenteric artery.

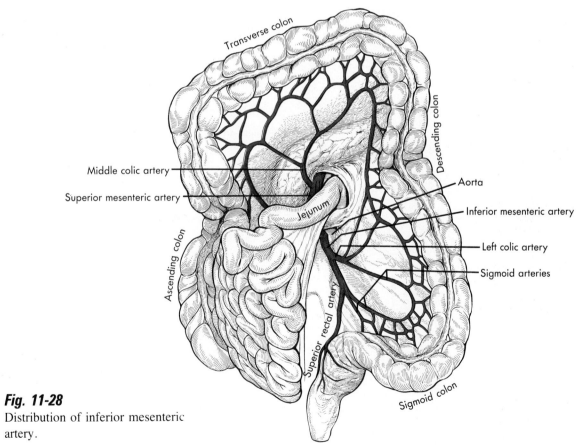

Fig. 11-28
Distribution of inferior mesenteric artery.

Arteries of the head and neck

(Fig. 11-29). The *common carotid* arteries pass upward along the neck medial to the sternocleidomastoideus muscles and at about the level of the superior border of the thyroid cartilage each bifurcates into an *internal carotid* and an *external carotid* artery. The internal carotid artery passes through the carotid canal into the cranial cavity to form a major part of the cerebral circulation. The external carotid artery forms the circulation for the structures of the head and neck not within the cranial cavity.

As the external carotid artery ascends toward the skull it gives off, in order, the following important branches:

superior thyroidal artery forms a large portion of the blood supply to the thyroid gland and to muscles and skin in the laryngeal area.
ascending pharyngeal artery rises from the posterior aspect of the external carotid artery and supplies muscles of the pharynx and soft palate and anterior muscles of the neck.
lingual artery branches at the level of the greater horn of the hyoid bone and supplies the tongue and adjacent structures.
facial (external maxillary) artery branches at the level of the mandibular angle and supplies most of the face below the level of the nose.
occipital artery branches at the level of the facial artery and supplies the posterior part of the skull.
superficial temporal artery branches at the level of the mandibular joint, passes upward in front of the ear, and supplies the structures on the sides of the skull.
maxillary artery is a large branch from the previous vessel and passes anteriorly to give off branches to the structures of the face (such as masseter muscle and cheeks), the lower jaw and teeth (passing into the bone through the mandibular foramen), and the upper jaw and teeth.

The vessels serving the brain will be considered in a later section.

Arteries supplying the upper appendage

(Fig. 11-30; Table 11-5). The *right subclavian artery* rises from the brachiocephalic artery, the left subclavian artery as a separate branch from the aorta. They pass laterally beneath the clavicles, and as they go, each vessel gives off the following branches:

vertebral artery is the largest branch. It passes up the neck through the transverse foramina of the cervical vertebrae and enters the foramen magnum to form part of the blood supply to the brain.
thyrocervical artery is short and immediately branches to send vessels to the thyroid gland, muscles and skin of the anterior neck, and muscles of the upper back.
internal thoracic (internal mammary) artery proceeds downward over the chest, giving branches to the intercostal muscles, mammary gland, sternum, and anterior abdominal muscles.
costocervical artery is the most distal branch and gives vessels to the posterior neck muscles.

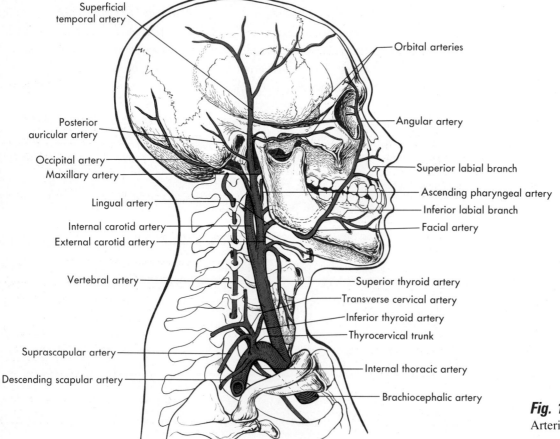

Fig. 11-29
Arteries of head and neck.

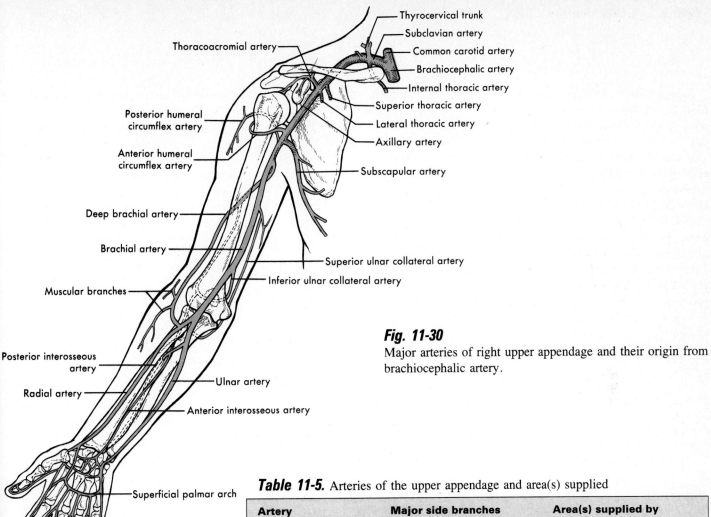

Fig. 11-30
Major arteries of right upper appendage and their origin from brachiocephalic artery.

Table 11-5. Arteries of the upper appendage and area(s) supplied

Artery	Major side branches	Area(s) supplied by branches
Subclavian (beneath clavicle)	Vertebral	Brain and cervical spinal cord
	Internal thoracic (mammary)	Mammary glands, diaphragm, pericardium
	Thyrocervical Costocervical	Muscles, organs, and skin of neck and upper chest
Axillary (armpit)	Long thoracic Ventral thoracic Subscapular	Muscles and skin of shoulder, chest and scapula
Brachial (upper arm)	Muscular	Biceps muscle
	Deep brachial	Triceps muscle; both also supply skin and other tissues of upper arm
Radial Ulnar (forearm)	Muscular Medial	Muscles and skin of the forearm
Palmar (volar) arch	Metacarpals	Muscles and skin of the hand
Digital	—	Muscles and skin of fingers

Adapted from McClintic, J.R.: Basic anatomy and physiology of the human body, ed. 2. Copyright © 1980. Reprinted by permission of John Wiley & Sons, Inc.

The **axillary artery** (Fig. 11-31) is the continuation of the subclavian artery across the axilla (armpit). It begins at the first rib and ends by the tendon of the teres major. It gives off several branches, the largest of which is the **subscapular artery**. This vessel supplies the scapular muscles.

The **brachial artery** begins at the teres major tendon and terminates by branching just below the elbow into radial and ulnar arteries. The brachial artery supplies branches named *muscular* branches and the *deep brachial* branch, to the muscles of the arm and terminates by forming the next two vessels.

The **radial artery** pursues a course along the radius to the wrist. It is superficial at the wrist lateral to the flexor carpi radialis tendon and serves as a common site for taking the pulse. Many *muscular branches* supply the muscles of the radial half of the forearm.

The **ulnar artery** pursues a course along the ulna. It gives off a large *medial* (interosseous) *artery* that along with the ulnar branches supplies the muscles on the ulnar side of the forearm.

At the wrist, branches of the radial artery and ulnar artery form an archlike anterior vessel known as the **volar arch**. From the arch, metacarpal vessels follow the metacarpal bones and terminate by branching into digital arteries to the fingers, and branches pass also to the thumb.

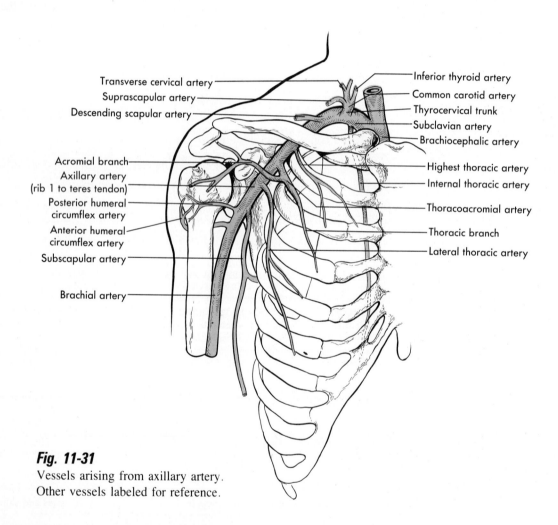

Fig. 11-31
Vessels arising from axillary artery. Other vessels labeled for reference.

Arteries supplying the lower appendage (Fig. 11-32). As described earlier, the aorta terminates by branching into the common iliac arteries. These in turn form internal and external branches. The *internal iliac* artery supplies the pelvic viscera, and the *external iliac artery* passes toward the thigh. As it crosses the rim of the acetabulum onto the thigh, the vessel becomes the *femoral artery*. It passes along the femur to terminate above the knee as the popliteal artery. Three large branches of the femoral artery are the following:

lateral circumflex femoral artery supplies muscles and skin of the hip joint region. A *descending branch* supplies the lateral thigh muscles.
deep femoral artery supplies posterior and medial thigh muscles.
descending genicular artery supplies medial and anterior thigh muscles and structures associated with the knee joint.

The *popliteal artery* extends from its origin on the lower thigh to the level of the fibular head. It gives off numerous branches to the knee joint and to the proximal ends of the leg muscles.

The popliteal artery branches to form the *anterior* and *posterior tibial arteries* (see Fig. 11-32) that follow along the respective aspects of the tibia. The anterior vessel gives off branches to the muscles and skin of the anterior leg. The posterior vessel gives off branches to the posterior leg muscles and also gives rise to the *peroneal artery* that passes along the fibula supplying the lateral leg muscles.

At the ankle the anterior tibial artery becomes the *dorsalis pedis artery,* and from this, a *plantar arch* and a *deep plantar artery* arise. The dorsalis pedis artery gives branches to the structures on the top of the foot, the deep plantar artery to the sole of the foot; from the plantar arch, *metatarsal* and *digital arteries* supply the metatarsal structures and toes.

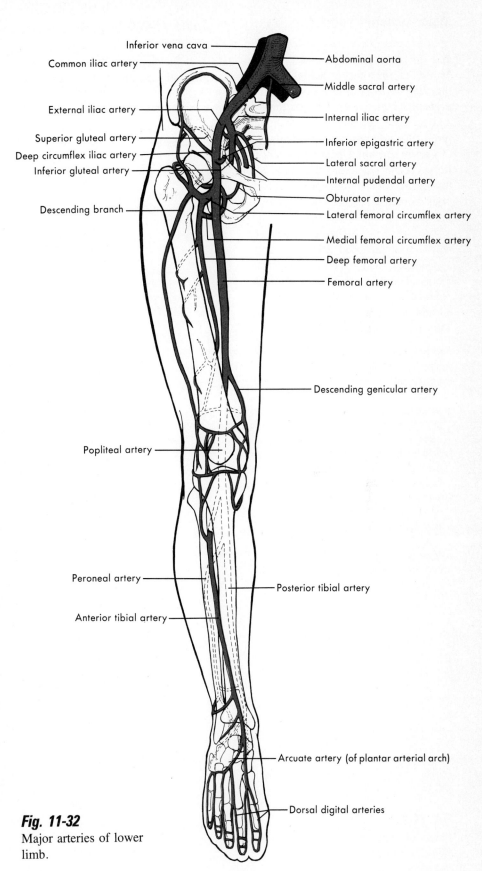

Fig. 11-32
Major arteries of lower limb.

Major systemic veins

In general, veins accompany arteries and are named in the same fashion. Those that meet this criterion are called the *deep veins*. In the appendages, *superficial veins* lie just beneath the skin and are often used as sites of venipuncture or for administration of fluids.

Venae cavae. The *superior* and *inferior venae cavae* are the major systemic veins that return blood from the body generally to the right atrium of the heart.

The superior vena cava collects blood from the head and neck, upper appendages, and upper portion of the thorax. The major tributaries of the superior vena cava are shown in Fig. 11-33 and are summarized in Table 11-6. They may be listed as follows:

internal jugular vein drains blood from the structures within the cranial cavity.
external jugular vein drains blood from the external surfaces of the skull and has as its major tributaries the *superficial* and *middle temporal veins*, the *facial vein*, and the *mandibular vein*.
subclavian veins and the internal and external jugular veins from both sides of the body combine to form two
innominate (brachiocephalic) veins that in turn form the superior vena cava. The *vertebral vein* empties into the subclavian vein, where it joins the innominate vein.
azygos vein empties into the superior vena cava just before it pierces the pericardium to empty into the right atrium.

Table 11-6. Veins forming the superior vena cava

Vein	Formed from	Area(s) drained
Internal jugular	Dural sinuses	Inside of skull and brain
External jugular	Veins of face	Muscles and skin of scalp and face
Subclavian	Axillary, cephalic,* basilic,* and their tributaries, scapular, and thoracic veins	Upper appendage, chest, mammary glands
Innominate (brachiocephalic)	Internal jugular, external jugular, and subclavian	None
Azygos	Veins along the spine	Muscles of spine; can substitute for inferior vena cava

Adapted from McClintic, J.R.: Basic anatomy and physiology of the human body, ed. 2. Copyright © 1980. Reprinted by permission of John Wiley & Sons, Inc.
*These are the superficial veins of the extremities.

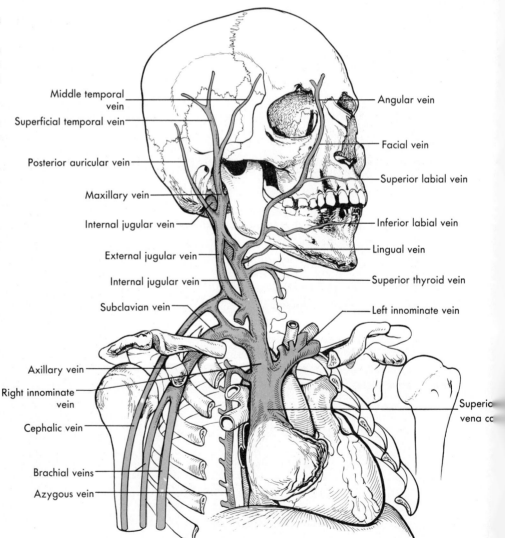

Fig. 11-33
Veins forming superior vena cava and their contributing vessels.

The inferior vena cava corresponds to the descending aorta and collects blood generally from the body parts that lie below the level of the midthorax. It is formed inferiorly by the fusion of the common iliac veins and receives several tributaries that are described as follows (Fig. 11-34; Table 11-7):

hepatic veins (several) enter the inferior vena cava just below the diaphragm. They drain the liver.
renal veins (a pair) enter the vena cava about halfway along its length. The left renal vein receives *gonadal veins* and *veins from the adrenal glands*.
small veins from the back muscles enter along the course of the vessel.

Notably missing as tributaries to the inferior vena cava are the vessels corresponding to the superior and inferior mesenteric arteries and their tributaries. Such vessels *do* exist and will be described in a later section under the heading of the *hepatic portal system*.

Table 11-7. Veins forming the inferior vena cava

Vein	Formed from	Area(s) drained
Hepatics	Sinusoids of liver	Liver
Renals	Veins of kidney	Kidney
Gonadals	Veins of gonads	Gonads: testes, and ovaries
Common iliac	External iliac	Lower appendage
	Internal iliac	Organs of lower abdomen
	Saphenous*	Superficial structures of lower appendages

Adapted from McClintic, J.R.: Basic anatomy and physiology of the human body, ed. 2. Copyright © 1980. Reprinted by permission of John Wiley & Sons, Inc.
*These are the superficial veins of the extremities.

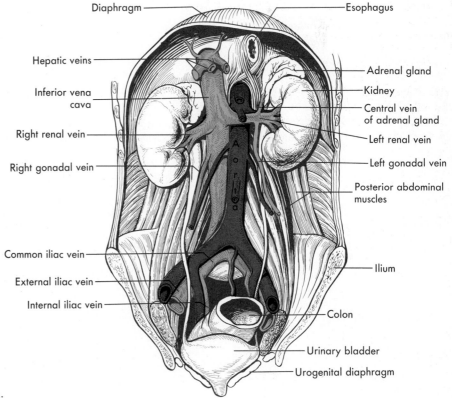

Fig. 11-34
Veins forming inferior vena cava.

Veins of the upper appendage.
It is perhaps most logical to consider the veins of the upper appendage (Fig. 11-35) in the order the blood flows through them, that is, to begin with the hand and proceed proximally. The vessels form a superficial and deep set of veins as listed in the following.

Digital veins empty into three *dorsal metacarpal veins* that in turn form a superficial *dorsal venous network* visible on the back of the hand.

From the dorsal venous network, three large vessels rise from the muscles of the forearm and arm. The *cephalic vein* runs superficially along the lateral aspect of the upper appendage to empty into the axillary vein. The *basilic vein* is a superficial vein that courses along the medial aspect of the upper appendage to empty into the brachial vein. The *median antebrachial vein* drains the anterior forearm and joins the basilic vein just below the elbow. Across the anterior aspect of the elbow joint, the cephalic and basilic veins are connected by the *median cubital vein,* a common site for venipuncture.

Deep veins include a palmar venous arch that gives rise to *radial* and *ulnar veins* that course alongside the corresponding arteries. The *brachial vein* runs along the humerus and at the armpit becomes the *axillary vein.* The latter vessel becomes the *subclavian vein* at the first rib and joins the innominate vein.

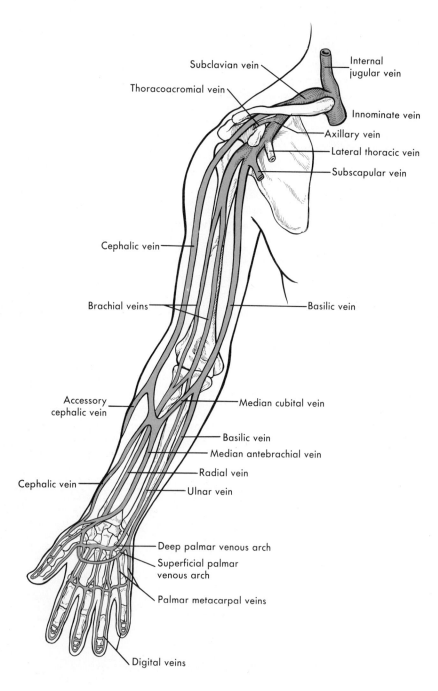

Fig. 11-35
Major veins of right upper limb in anterior view. (Dorsal veins not depicted.)

11 Circulatory system 385

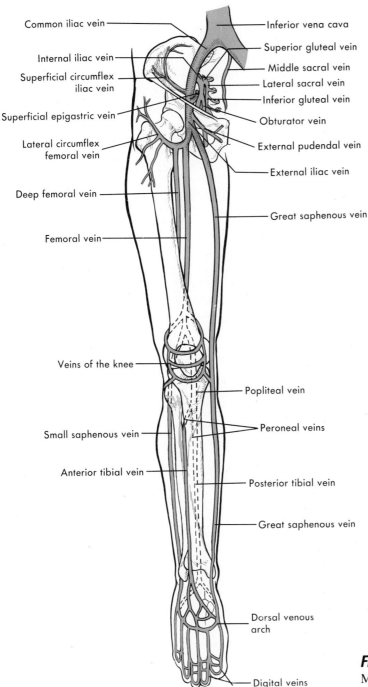

Veins of the lower appendage.
As in the upper appendage, in the lower appendage also there is a superficial and a deep set of veins (Fig. 11-36), with the deep vessels following alongside of and named in the same fashion as the arteries. Again, proceeding from distal to proximal the superficial vessels are as follows:

digital veins form a *plantar venous arch.* From the medial side of the arch rises the *great saphenous vein* that courses along the medial aspect of the lower appendage to end in the femoral vein high in the thigh.

small saphenous vein rises from the lateral margin of the plantar arch and courses up the posterior side of the leg to empty into the popliteal vein behind the knee joint.

The deep veins include:

plantar digital veins form four *metatarsal veins* that join a *plantar arch.*

posterior tibial vein, rising from the plantar arch, is joined by the *peroneal veins* below the knee.

anterior tibial veins unite with the posterior tibial vein to form the popliteal vein.

popliteal vein courses across the knee joint, receiving small veins from the knee structures.

femoral vein arises above the knee from the popliteal vein and courses alongside the femur, receiving the *deep femoral vein.*

The femoral vein becomes the *external iliac vein* that begins at the groin. The iliac vein receives branches from the upper thigh and sacral areas.

The external iliac vein is joined by the *internal iliac vein* that drains the gluteal area, uterus, vagina, penis, and rectum.

A *common iliac vein* is formed by the other two iliac veins and joins with its fellow to form the inferior vena cava.

Fig. 11-36
Major veins of right lower limb.

Special features of the circulation

Hepatic portal system. The hepatic portal system (Fig. 11-37) includes all the veins draining the abdominal portion of the alimentary tract (stomach, intestines) plus the spleen, pancreas, and gallbladder. In traversing this route the veins collect to form the *hepatic portal vein* that conveys blood to the *sinusoids* of the liver. From the latter, blood is carried to the inferior vena cava by the previously described hepatic veins. The major veins that form the portal vein are as follows:

splenic vein receives the *inferior mesenteric vein* (from the last half of the large intestine and rectum), several *gastric veins* (from the stomach), and the *pancreatic veins* (from the pancreas).

superior mesenteric vein drains blood from the small intestine and the first half of the large intestine.

coronary vein (also known as the gastric vein) also drains the stomach.

pyloric and *cystic veins* drain the distal end of the stomach and the gallbladder.

The value of the portal circulation lies in the fact that nutrients absorbed from the alimentary tract (except large lipids) are passed first to the liver, where that organ removes the building blocks it requires for its activity and places into the outflowing blood its products and wastes.

Blood supply of the brain

Arteries. The brain receives its blood via two major arterial routes (Fig. 11-38).

The *internal carotid artery* originates at the level of the hyoid bone and passes up the neck to enter the skull through the carotid canal. After entering the skull, branches are given off to the meninges, pituitary gland, and eye *(ophthalmic artery),* branches of which supply the retina, eyelids, and several of the extrinsic eye muscles. The *anterior cerebral arteries* rise from the carotid artery and supply most of the medial side of the anterior part of the cerebral hemispheres. They are connected by the **anterior communicating artery.**

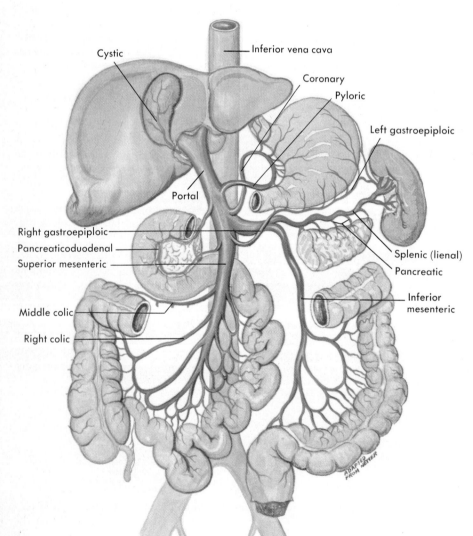

Fig. 11-37
Hepatic portal system.

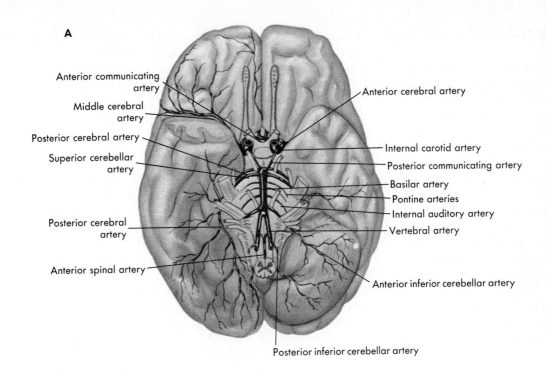
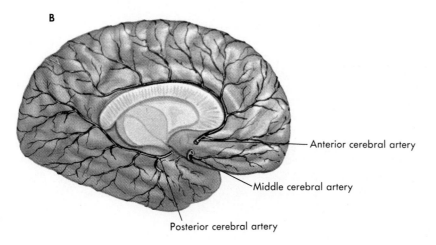
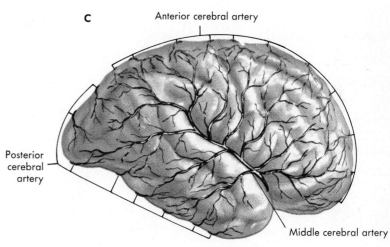

Fig. 11-38
Arteries of brain and their distributions. **A,** Basal view. **B,** Medial view. **C,** Lateral view.

The *middle cerebral artery* is a large vessel rising from the carotid artery and supplying most of the central part of the cerebral hemispheres. The *posterior communicating artery* rises from the carotid artery and passes posteriorly to join with the *posterior cerebral artery*.

The *vertebral arteries*, rising from the subclavian arteries, pass up the neck through the transverse foramina of the cervical vertebrae, enter the skull through the foramen magnum, and at the level of the pons join to form the **basilar artery**. The **anterior spinal artery** is given off in a Y-shaped manner just before the fusion of the vertebral arteries, as is a **posterior inferior cerebellar artery** from each vertebral artery. The area supplied by each of these vessels is inherent in its name.

The basilar artery lies on the ventral surface of the brainstem and gives off (in ascending order) the ***anterior inferior cerebellar arteries***, the ***labyrinthine arteries*** to the inner ear, the ***pontine arteries*** to the pons, the ***superior cerebellar arteries***, and the ***posterior cerebral arteries*** that supply the occipital and temporal lobes of the cerebrum. Joining with the posterior cerebral artery is the previously mentioned posterior communicating artery.

A moment's look at Fig. 11-38 will indicate that the posterior cerebral, posterior communicating, internal carotid, anterior cerebral, and anterior communicating arteries create a system of vessels allowing blood to flow from one vessel or one side to the other. This circular ring of vessels is called the **circle of Willis** and allows continued flow of blood to the brain if one of the major vessels (carotid or vertebral) is narrowed or blocked.

Veins (Fig. 11-39). Venous drainage from the brain is accomplished by vessels that empty into the **dural sinuses** of the brain. These vessels are actually spaces between the two layers of the dura mater that act as conduits for the blood.

The *superior cerebral veins* drain the upper, medial, and lateral surfaces of the cerebral hemispheres, opening into the ***superior sagittal sinus***. *Inferior cerebral veins* drain the inferior surfaces of the hemispheres, and most join the **cavernous** and **petrosal sinuses**.

The *superior sagittal sinus* lies beneath the sagittal suture of the skull and continues in the midline of the skull to the internal occipital protuberance. The ***inferior sagittal sinus*** is carried in the lower margin of the falx cerebri. It is joined by the **great cerebral vein** that collects the blood from cerebral veins,

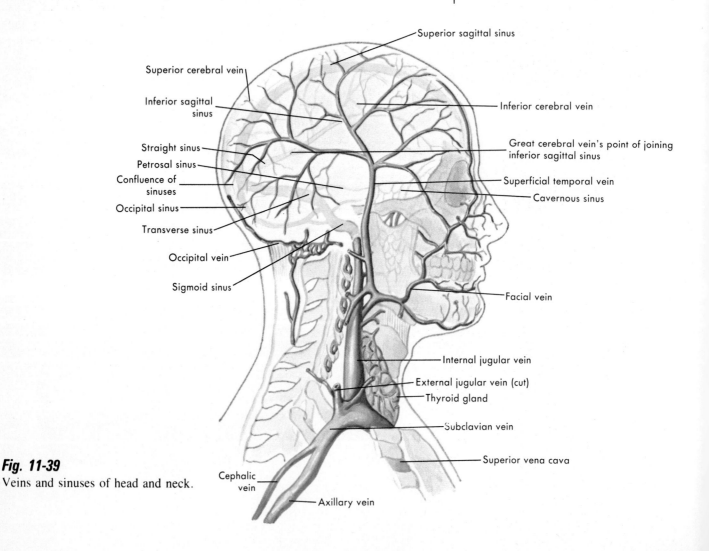

Fig. 11-39
Veins and sinuses of head and neck.

and together the sinus and great vein form the **straight sinus,** the latter carried in the posterior falx. Blood flows posteriorly in the sinuses mentioned. The **occipital sinus,** situated in the falx cerebelli, collects blood from the area of the foramen magnum. Blood in this sinus flows upward and posteriorly. The several sinuses form the **confluence of the sinuses** at the internal occipital protuberance. Blood reaching the confluence from the superior sagittal sinus is directed into the **right transverse sinus.** Blood from the straight and occipital sinuses is directed into the **left transverse sinus.** The transverse sinuses proceed horizontally along the lateral margins of the occipital bone and at the junction with the temporal bone become the **sigmoid sinus,** so named because of the S-shaped bend it undergoes to reach the jugular foramen. The sigmoid sinus becomes the *internal jugular vein* at the jugular foramen.

Diploic veins (Fig. 11-40) occupy the spongy bone (diploë) of the cranial bones. They collect blood from the diploë and end in the dural sinuses.

Emissary veins receive blood from the scalp, pass through the cranial bones, and empty into the dural sinuses. They provide a route by which scalp infections may reach the brain.

Another interesting feature of the cerebral circulation is the presence of what is called the **blood-brain barrier.** The term refers to the fact that certain substances, such as protein molecules, small lipids, and many antibiotics do not pass through the walls of cerebral capillaries at all, whereas urea, chloride ion, and sucrose suffer severe restriction of their passage. Other substances, such as glucose, potassium ion, and sodium ion pass freely. It has been suggested that restriction of passage of certain substances "protects" the brain from materials that could harm it.

There appears to be an anatomical basis for such a barrier in that:

Cerebral capillaries lack or have fewer and smaller "pores" in their walls.
The lining endothelial cells overlap one another and are sealed by tight junctions that restrict passage of materials between cells.
Astrocytes cover about 85% of the capillary surface of the brain, causing substances to pass two cells (endothelial and glial) before reaching neurons.

Last, the cerebral circulation maintains a remarkable constancy of flow in spite of exercise, chemical environment, and systemic blood pressure. Flow averages 750 ml/min. The brain is enclosed in the rigid skull, and increase in flow would mean an increased volume of the cranial contents that could compress the brain. The major factor controlling flow appears to be the baroreceptors in the aorta and carotid sinus that monitor pressure to the brain.

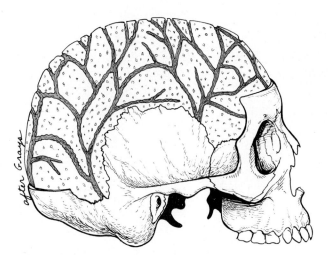

Fig. 11-40
Diploic veins in cranial bones.

Pulmonary vessels. The pulmonary circulation (Fig. 11-41) begins with the *pulmonary trunk* that rises from the right ventricle of the heart. It is about 5 cm long and branches to form right and left pulmonary arteries directed toward the respective lung. These vessels form a number of *segmental arteries* that pass to the bronchopulmonary segments of the lungs. The vessels ultimately form capillary networks around the alveoli of the lungs.

From the alveolar capillary beds the vessels form *segmental veins* that empty into the four *pulmonary veins*, two from each lung, that in turn empty into the left atrium of the heart.

Vessels in the pulmonary circulation are generally shorter and thinner walled than systemic vessels. Blood is pumped through the pulmonary circulation at pressures about one fourth that of the systemic vessels, and the vessels therefore do not have to be constructed with the thickness of systemic vessels.

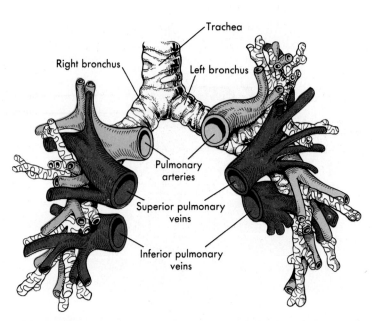

Fig. 11-41
Major pulmonary vessels.

Fetal circulation (Fig. 11-42)

The circulation of the fetus incorporates a number of devices to shunt blood past the nonfunctional lungs and to establish connections with the placenta, which serves as the area of exchange of nutrients and wastes between the fetus and its mother.

The **umbilical arteries** are two vessels derived from the internal iliac arteries. They carry oxygen-poor and waste-rich blood to the placenta where exchange of material occurs. A single **umbilical vein** drains the placenta. The vein, carrying blood rich in oxygen and nutrients, passes to the liver where some blood is distributed to the liver cells, and a large amount passes via the **ductus venosus** to the inferior vena cava.

Entering the right atrium of the heart, blood passes via the **foramen ovale**, in the interatrial septum, directly to the left atrium. The blood that does enter the right ventricle passes out the pulmonary trunk to the **ductus arteriosus**, a vessel connecting the trunk to the aorta. At birth the foramen, ductus, and umbilical arteries are closed, and the adult pattern of circulation through the lungs is established.

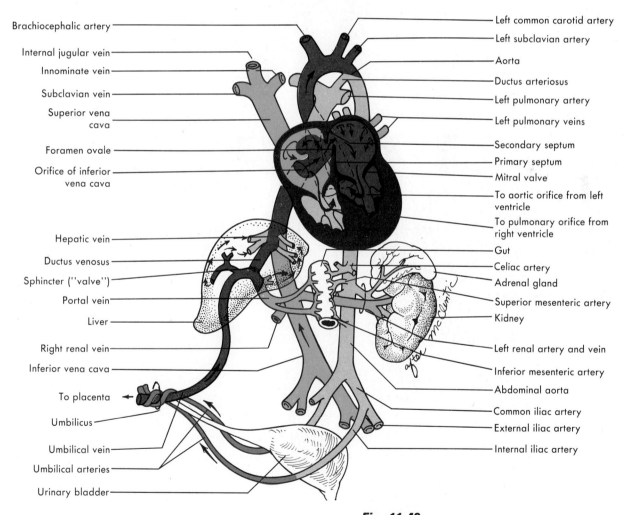

Fig. 11-42

Plan of fetal circulation. Colors indicate state of oxygenation of blood. (Blue, lowest; red, highest; gray, intermediate.)

Lymph vascular system

The lymph vascular system consists of a fluid, the **lymph,** that is carried in lymph vessels or **lymphatics,** and a series of **lymph organs** including the *lymph nodes, tonsils, spleen,* and *thymus.*

Lymph

The term *lymph* is applied to the fluid that circulates through the lymphatics of the body. It is formed from the tissue fluid, the fluid lying around the body cells. Lymph has no red cells, few platelets, and little protein but may contain from 500 to 50,000 nongranular leukocytes per mm^3 and has the same composition as plasma for small

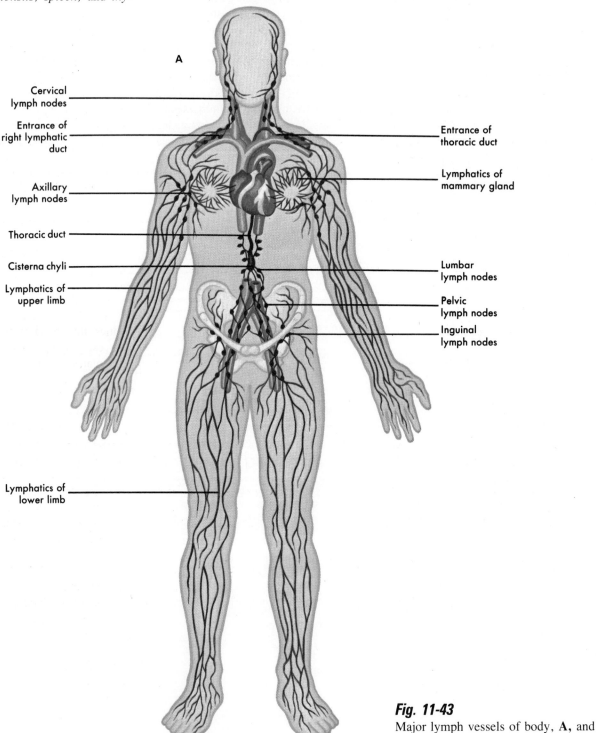

Fig. 11-43
Major lymph vessels of body, **A,** and their areas of drainage, **B.**

molecules. Large molecules are low (except in the abdominal lymphatics) because they do not filter from the blood capillaries into the tissue fluid. The same reasoning applies to red cells and platelets, whereas more nongranular leukocytes are added to the fluid as it passes through certain of the lymph organs.

Lymph vessels (Fig. 11-43)

The system of vessels that collects tissue fluid permeates nearly all body tissues and organs* and begins as **lymph capillaries.** These are blind-ended vessels that vary in caliber but are several times larger than blood capillaries. They are lined with endothelium, freely branch and rejoin with one another, and possess much larger "pores" or openings in their walls than do blood capillaries. The term *plexuses* is often applied to the networks of lymph capillaries that permeate body tissues and organs. Draining the lymph capillary networks are fewer and larger vessels called **collecting vessels.** These vessels enter *lymph nodes* as *afferent lymphatics* and leave the nodes as *efferent lymphatics.* Collecting vessels usually accompany the blood vessels of the body. Collecting vessels have a structure resembling that of very thin-walled veins; that is, an intima, media, and adventitia are present but are very thin, and these vessels usually possess *valves.*

*No capillaries have been found in the central nervous system, meninges, eyeball, internal ear, cartilage, and epidermis.

Collecting vessels ultimately form two large **lymph ducts** that drain specific areas of the body (see Fig. 11-43).

The **thoracic duct** is about 40 cm in length and begins with an enlargement, the **cisterna chyli,** that is located on the anterior surface of the second lumbar vertebra. It progresses upward, receiving tributaries from the musculature of the back, and empties into the left subclavian vein at its junction with the left internal jugular vein. The **right lymphatic duct** is a short vessel that empties into the right subclavian vein at *its* junction with the right internal jugular vein.

The lymphatic vessels serve to return some 10% of the fluid filtered from the blood vessels to the blood vascular system, carry large lipid molecules absorbed from the intestines to the blood circulation, return protein (filtered or produced) from tissue spaces, and provide a route for lymphocytes and monocytes produced in lymph organs to enter the blood circulation. Those lymphatics that connect to lymph nodes also carry lymph that is cleansed of particulates as it passes through the nodes.

Lymph organs

Lymph nodules. The term *lymph nodule* is generally used to describe the *solitary* unencapsulated (not having a connective tissue capsule) masses of lymphocytes and macrophages that are found in the walls of organs in systems that contact the external environment (for example, respiratory and digestive). *Aggregated* lymph nodules form the Peyer's patches of the ileum, the last portion of the small intestine. From such nodules, lymphocytes may migrate into the connective tissue and epithelial surfaces of these organs to combat microorganisms or other substances that may enter the body through the lining tissues.

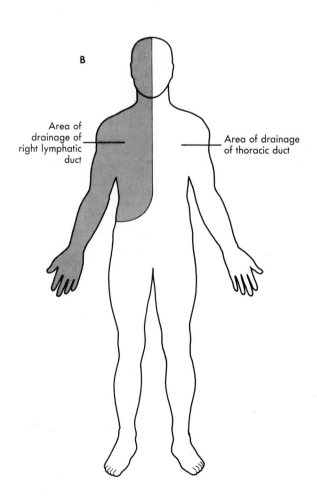

Lymph nodes. Lymph nodes (Fig. 11-44) are capsulated organs measuring from a few millimeters to 1 cm in diameter. They are placed in the course of collecting lymphatic vessels and any particulate material in the lymph flow is removed by the lymphocytes and macrophages within the node. Cancer cells shed from a tumor and entering the lymph stream may also be trapped in nodes, hence the practice of examining the nodes in the area of a tumor. *Afferent lymphatic vessels* penetrate a collagenous tissue *capsule* around the organ carrying the lymph into a *cortical sinus*. Lymph may flow through the sinus or pass through the interior of the node to be carried out of the node by *efferent lymphatic vessels*. Both sets of vessels have valves to ensure a one-way flow through the node. The outer region of the node, or *cortex*, consists of densely packed masses of lymphocytes and macrophages. The masses are sometimes called *nodules*. Within the dense masses, lighter areas called *germinal centers* give evidence of active proliferation of cells. The inner region or *medulla* of the node consists of cordlike masses of lymph cells between which are lymph-filled spaces called the *medullary sinuses*. Connective tissue *septa* penetrate the node from the capsule and carry blood vessels into and out of the node. The interior of the node is supported by a network of reticular connective tissue fibers.

Distribution of lymph nodes in the body is widespread (see Fig. 11-43). Nevertheless, we can see that there are aggregations of nodes in particular body areas that seem to be strategically placed to guard the approaches to the interior of the body.

- *cervical lymph nodes* lie in the neck and cleanse the lymph from the head and neck. These enlarge when one has a cold.
- *axillary lymph nodes* are found in the area of the armpit and chest and cleanse the lymph flowing from the upper appendages.
- *thoracic* and *abdominal nodes* are located alongside the spine, the trachea, and the iliac vessels; *mesenteric nodes* are located in the membranes suspending the abdominal organs in their cavity. These nodes cleanse lymph derived from the organs that may contact the environment.
- *inguinal nodes* are concentrated in the groin and cleanse lymph from the lower appendages.

The nodes provide a very important line of defense against organisms that may enter through epithelial wounds or be absorbed through organ walls.

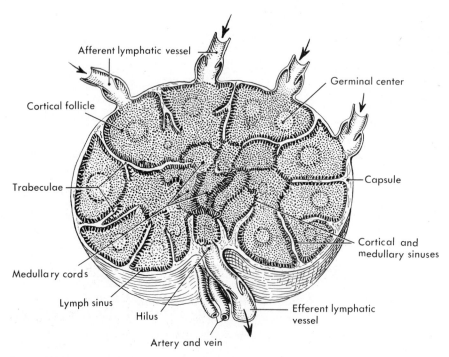

Fig. 11-44
Lymph node. Cortical and medullary areas contain both lymphocytes and macrophages. Arrows indicate direction of lymph flow. Note valves in afferent and efferent vessels.

Tonsils. Three sets of tonsils are present in the body. The *palatine tonsils* are located at each side of the base of the soft palate, the *pharyngeal tonsils* (adenoids) are found in the posterior wall of the upper pharynx (throat), and the *lingual tonsils* are in the base of the tongue. All have a capsule beneath the organ and are covered with epithelium. Internally, masses of lymphatic tissue are present, but no cortex or medulla exists. Germinal centers are common, and the cells are placed into efferent lymphatic vessels. No afferent vessels are present. Thus, functionally the tonsils seem only to produce cells.

Viewed from the mouth, the tonsils form a ringlike arrangement around the openings of the respiratory and digestive systems into the body (Fig. 11-45). The name *Waldeyer's ring* is given to this arrangement, and it has been suggested that no part of the ring be removed unless it interferes with normal body function.

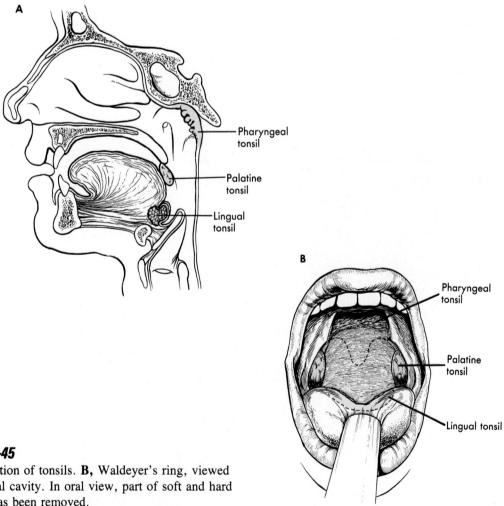

Fig. 11-45
A, Location of tonsils. **B,** Waldeyer's ring, viewed from oral cavity. In oral view, part of soft and hard palate has been removed.

Spleen. The spleen (Fig. 11-46) is situated in the upper abdomen to the left of the stomach. It is the largest lymph organ, measuring about 12 by 7 by 3 cm in the adult. It is placed in the course of blood vessels and serves functions related to the blood as well as producing nongranular white blood cells.

It has a **capsule** composed mainly of elastic connective tissue and smooth muscle. **Trabeculae** *(septa)* extend from the capsule into the organ and attach to reticular fibers that form the supporting internal framework of the organ. The trabeculae also carry the blood vessels into and out of the organ.

Internally, the tissue of the spleen is composed of **splenic pulp** that is of two types: *white pulp* and *red pulp*. The white pulp consists of dense masses of lymphatic tissue (splenic nodules) containing an arteriole (the *central artery*, which usually is not centrally placed). Red pulp consists of blood in the venous sinuses of the organ. White pulp produces nongranular leukocytes; red pulp (at least the cells in the blood of it) are subjected to phagocytosis. The spleen is thus an organ where significant numbers of old or abnormal red cells are destroyed. Efferent lymphatic vessels (only) drain the white cells into the circulation.

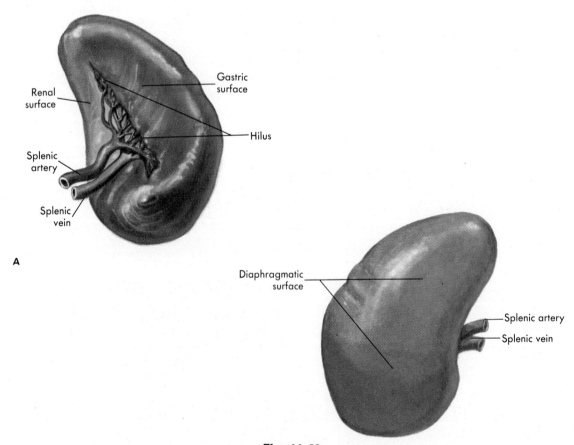

Fig. 11-46
A, Gross and, B and C, microscopic structure of spleen.

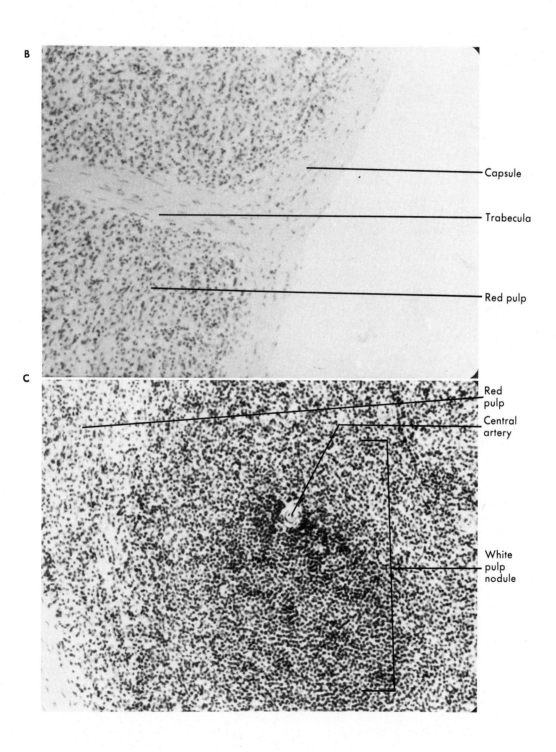

Thymus. The *thymus* (Fig. 11-47) is a two-lobed organ located behind the sternum in the central part of the mediastinal cavity. It has a collagenous capsule and is divided into **lobules** by connective tissue septa. Each lobule has an outer dense area of lymph tissue, the **cortex,** and an inner loosely arranged mass of lymph tissue, the **medulla.** Germinal centers are not present, and only efferent lymphatic vessels are found. The organ is an important part of the body's immune system, producing cells that combat foreign cells (for example, transplants) that may be introduced into the body.

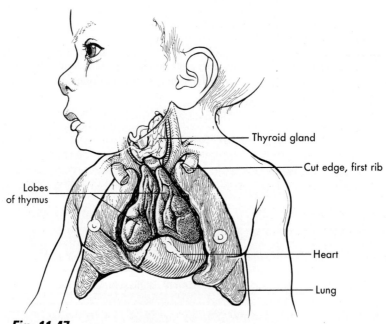

Fig. 11-47
Location and gross anatomy of thymus.

Summary

1. The circulatory system consists of a blood vascular system and a lymph vascular system.
   a. The blood vascular system consists of the blood, the heart, and the blood vessels.
   b. The lymph vascular system consists of the lymph, lymphatic vessels, and the lymph organs such as the lymph nodes, tonsils, spleen, and thymus.
2. Blood develops at about $2\frac{1}{2}$ weeks from blood islands of mesodermal origin. Red cells are formed first to carry oxygen for development.
3. The blood is composed of plasma and formed elements.
   a. Functions of the blood are related to transport and maintenance of homeostasis.
   b. Plasma forms over half of whole blood and is a complex solution of chemicals. Plasma proteins create osmotic pressure, carry antibodies, and are important in blood coagulation.
   c. The formed elements include erythrocytes (red blood cells), five types of leukocytes (white blood cells), and platelets.
4. The heart develops from a cardiogenic plate that forms heart cords. These develop into heart tubes that form the heart and its parts.
5. The heart provides the pressure necessary to circulate the blood.
   a. It approximates the size of the fist of the owner regardless of age.
   b. It is located in the mediastinum; is enclosed in a double-walled pericardial sac; has a base, apex, and axis; and is located in a slanted and left-of-center position in the chest.
   c. Its walls consist of three main layers of tissue: an inner endocardium (in places where nodal tissue is present there is also a subendocardial layer), a central myocardium of cardiac muscle, and an outer epicardium.
   d. It has four chambers: two upper atria and two lower ventricles. Other features are an external coronary sulcus between atria and ventricles and two interventricular sulci; internally, there are interatrial and interventricular septa, valves (tricuspid and mitral), chordae tendineae, and papillary muscles.
   e. Valves are present in the bases of the arteries leaving the ventricles.
6. Two types of tissue are essential for the heart to beat.
   a. Cardiac muscle forms the myocardium, has branching fibers separated by intercalated discs, is striated, and has a system of T-tubules and sarcoplasmic reticulum. It follows the all-or-none law and the law of the heart and cannot be tetanized.
   b. Nodal tissue in the SA and AV nodes, AV bundle, bundle branches, and Purkinje's fibers initiates and distributes the impulses that cause cardiac muscle contraction.
7. The heart itself is supplied by the coronary vessels.
   a. Two coronary arteries originate from the aorta. Their major branches lie in coronary and interventricular sulci. The major branches are listed.
   b. Coronary veins accompany the arteries, and most empty into the coronary sinus that passes blood to the right atrium. The major veins are listed.
   c. Diminished blood flow to the heart can lead to angina, heart attacks, and death of muscle.
8. Nerves to the heart are parasympathetic (slow heart rate and constrict muscular vessels) and sympathetic (speed heart rate and dilate muscular vessels).
9. Blood vessels develop as cavities in blood islands at 2 weeks. Lymph vessels develop from lymph sacs at about 5 weeks.
10. There are three general types of blood vessels that have a characteristic structure.
    a. Arteries carry blood from the heart, are three-layered (intima, media, adventitia), and are classed as elastic and muscular arteries.
    b. Capillaries are one-layered (endothelium) and serve as the exchange vessels in the body tissues and organs.
    c. Veins carry blood to the heart. Small vessels are two-layered (endothelium and connective tissue), whereas large vessels have the same three layers as do arteries.
11. The major systemic arteries and veins are presented in a regional fashion. Branches of the aorta, arteries of head and neck, upper appendages, and lower appendages, veins forming the venae cavae, and the veins of the head and neck and appendages and the areas supplied or drained are discussed and presented in pictorial and tabular form.
12. Circulation of the portal system, brain, lungs, and differences in fetal and adult circulations are described.
13. The lymph is the fluid circulating through the lymph vessels of the body.
    a. It is derived from tissue fluid around cells.
    b. It lacks red cells and platelets, is deficient in large molecules, and contains variable numbers of nongranular leukocytes.
14. Lymph vessels begin in the body tissues and organs and connect ultimately to the subclavian veins.
    a. Lymph capillaries form plexuses in tissues and organs.
    b. Collecting vessels accompany blood vessels and are larger than capillaries.
    c. Two lymph ducts (thoracic, right lymphatic) collect lymph from the other vessels and carry it to the subclavian veins.
15. Lymph organs are of several types. All produce nongranular leukocytes.
    a. Nodules are unencapsulated masses of lymphatic tissue common in the walls of respiratory and digestive systems.
    b. Lymph nodes are found in the course of collecting lymphatics and cleanse the lymph as it passes through the nodes. Node structure is described. Nodes are especially numerous in the neck, axilla, groin, and abdomen.
    c. There are three sets of tonsils: palatine, pharyngeal, and lingual. They guard the entrances of respiratory and digestive systems into the body.
    d. The spleen is set in the course of blood vessels and plays a role in the destruction of aged red cells.
    e. The thymus is an important part of the immune system.

Questions

1. Of what parts do the blood and lymph vascular systems consist? Can you find any major differences between the two systems? If so, what are they?
2. What are the formed elements of the blood? From what does each originate, and what jobs does each formed element have?
3. Give the orientation and relationships of the heart in the thorax.
4. What are the layers in the wall of the heart, and of what type of tissue does each layer consist?
5. Name the chambers of the heart, describe the nature of any valves separating the chambers, and describe the devices present to ensure a one-way flow of blood through the heart.
6. What is nodal tissue, where is it found, and what tasks does it have in the heart?
7. Describe the structure of cardiac muscle.
8. How is the heart supplied with blood? Name the major arteries and veins involved and describe their locations.
9. Compare the structure of arteries, capillaries, and veins. What is the predominant layer and type of tissue in each type of vessel?
10. Trace a drop of blood from the aorta to the following areas, and return it to the right atrium. List the major vessels through which the drop would pass.
    a. Left anterior cerebral hemisphere
    b. Right little finger (if two return pathways are possible list vessels in both paths)
    c. Liver
    d. Small intestine
    e. Left great toe
11. What is the circle of Willis and what is its functional significance?
12. How does the fetal circulation differ from that of the adult? *Why* is it different?
13. What are the following?
    a. Collecting vessel
    b. Lymphatic plexus
    c. Thoracic duct (What body area[s] does it drain?)
14. Compare the structure of a lymph node and the spleen.
15. What is Waldeyer's ring?

Readings

Adolph, E.F.: The heart's pacemakers, Sci. Am. **213**:32, March 1967.

Benditt, E.P.: The origin of atherosclerosis, Sci. Am. **236**:74, Feb. 1977.

Erslev, A.J.: Pathophysiology of blood, Philadelphia, 1975, W.B. Saunders Co.

James, D.G.: Circulation of the blood, Baltimore, 1978, University Park Press.

Jarvik, R.K.: The total artificial heart, Sci. Am. **244**:74, Jan. 1981.

Schwartz, C.J., Warthesses, N.T., and Wolf, S., editors: Structure and function of the circulation, New York, 1980, Plenum Publishing Corp.

Weiss, H.J.: Platelets: physiology and abnormality of function, N. Engl. J. Med. **293**:531, 1975.

Wood, J.E.: The venous system, Sci. Am. **218**:86, Jan. 1968.

12

Respiratory system

Objectives
Development of the system
Organs of the system
Nose and nasal cavities
   Nose
   Nasal cavities
Pharynx
Larynx
   Cartilages
   Muscles
Trachea and bronchi
Lungs
   Relationships of the lungs to the thoracic cavity
   Gross anatomy
   Bronchioles
   Respiratory portion of the system
   Bronchopulmonary segments and lobules
   Blood supply of the lungs
   Nerves of the lungs
   Breathing
Summary
Questions
Readings

OBJECTIVES After studying this chapter, the reader should be able to:

List the organs making up the respiratory system, divide them into conducting and respiratory divisions, and state what organs compose upper and lower portions of the system.

Describe the structure of the nose and nasal cavities, the histology of these organs, and the functions served by the organs.

Show the relationships of the paranasal sinuses to the nasal cavities.

Describe the structure, gross and microscopic, of the pharynx.

Outline the structure of the larynx, including its cartilages and muscles, and the histology of its wall.

Describe the structure of the trachea and bronchi and their relationships to the thorax and lungs.

Describe the lungs in terms of their relationships to the pleural cavities and pleurae, their gross and microscopic anatomy, the bronchopulmonary segments and lobules, and their blood and nerve supply.

List the changes occurring in the linings of the respiratory system, with particular attention to the protective and exchange functions of the epithelium.

The organs of the respiratory system provide a means of bringing oxygen-rich, carbon dioxide–poor atmospheric air into contact with a diffusing surface, the alveoli of the lungs. The blood vessels of the lungs bring to this same diffusing surface blood that contains much carbon dioxide and less oxygen. Oxygen, because it is in higher concentration in the alveoli than in the bloodstream, passes into the blood, and carbon dioxide, higher in concentration in the blood, passes into the lungs. Essential for the intake of air into and expulsion of air from the lungs are the muscles of respiration, that is, the diaphragm, intercostales muscles, and other muscles of the neck and thorax. It is appropriate to review these muscles now.

Since the respiratory system contains air that is derived from the external environment, air that may contain pollutants of many varieties, the system is provided with numerous devices to warm, moisten, and cleanse inhaled air. The system, together with the mouth, lips, and tongue, also is involved in phonation and word forming.

Development of the system
(Fig. 12-1)

At about $3^{1}/_{2}$ to 4 weeks of embryonic life, the nose and nasal cavities begin their development. The first evi-

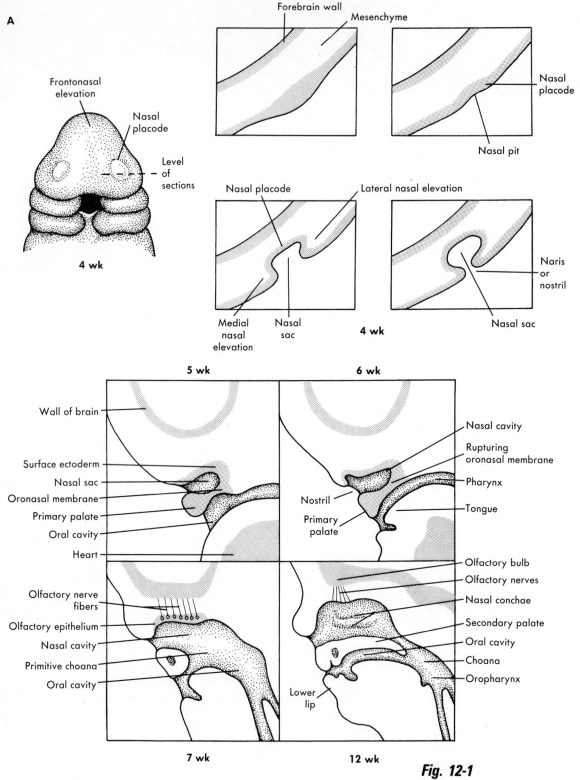

Fig. 12-1
Development of respiratory system.
A, Development of nasal region.
B, Development of system in first 12 weeks.
Continued.

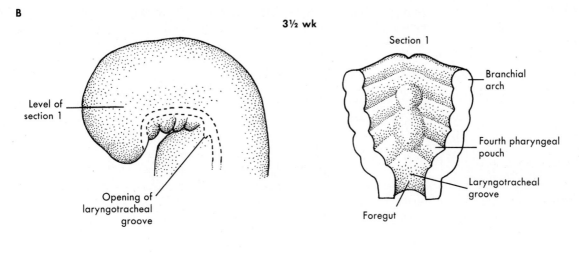

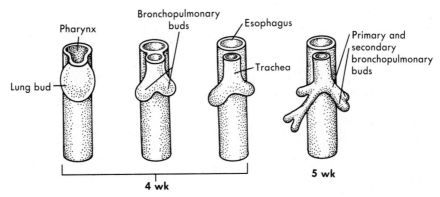

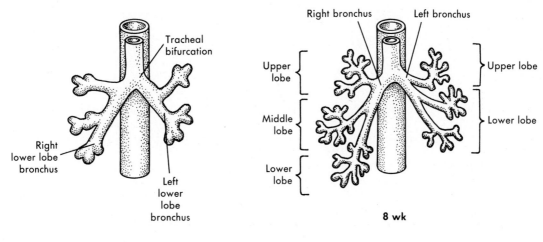

Continued.

dence of formation is provided by a thickened area of ectoderm known as the **olfactory placode** that appears on the front and lower part of the head. In the placode an **olfactory pit** that is extended backward to fuse with the anterior part of the foregut appears next; the foregut will form the **pharynx**. The posterior extension of the pit is separated from the mouth cavity by a thin membrane that ruptures at about 7 to 8 weeks of development to form a single **nasal-oral cavity**. Next two bony plates begin to grow horizontally across the cavity, and a third vertical plate grows downward from the roof of the nasal portion of the cavity. The three plates meet by 3 months and form respectively the **hard palate** and the **nasal septum**. Thus two nasal cavities are separated from the mouth cavity.

Failure of the horizontal plates to fuse in the midline creates a **cleft palate** that makes it difficult for an infant to develop the pressures required for sucking and swallowing.

Development of the lower portion of the system begins at about 26 days with the formation of the **laryngotracheal groove** in the floor of the foregut. The groove forms a tube from which at its lower end a **lung bud** develops. This bud elongates and undergoes a series of branchings that form the tubes of the system and their air sacs (alveoli). About 24 generations of branchings occur as the bud grows. Muscle, cartilage, and connective tissue coats of the system are formed from mesoderm from about 10 weeks onward.

No alveoli are found in the lungs until about 26 weeks of age, their number increasing beyond this time. A child born prematurely cannot survive until enough alveoli have formed to permit gas exchange—this usually occurs at 26 to 28 weeks of age.

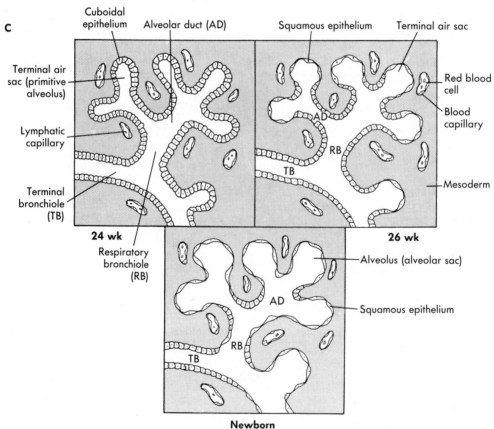

Fig. 12-1, cont'd
C, Development of alveoli.
After Moore.

Organs of the system

The organs of the respiratory system (Fig. 12-2) may be divided into a **conducting division** whose walls are too thick to permit any significant exchange of gases and a **respiratory division** containing *alveoli* whose walls are thin enough to allow gas exchange. The conducting division consists of the *nose* and *nasal cavities, pharynx, larynx, trachea, bronchi, bronchioles,* and *terminal bronchioles*. The respiratory division is composed of the *respiratory bronchioles, alveolar ducts,* and *alveolar sacs*. All parts of the system beyond the bronchi are contained within the lungs.

Another way of subdividing the system is to distinguish an **upper respiratory tract** or **system** (URT or URS) consisting of the *nose, nasal cavities,* and *pharynx* and a **lower respiratory tract** or **system** (LRT or LRS) consisting of the remaining organs. This type of subdivision is often used by medical personnel because of the tendency for infections to involve the nose and throat (an *upper respiratory infection* or URI) as opposed to the lower tract (a *lower respiratory infection* or LRI).

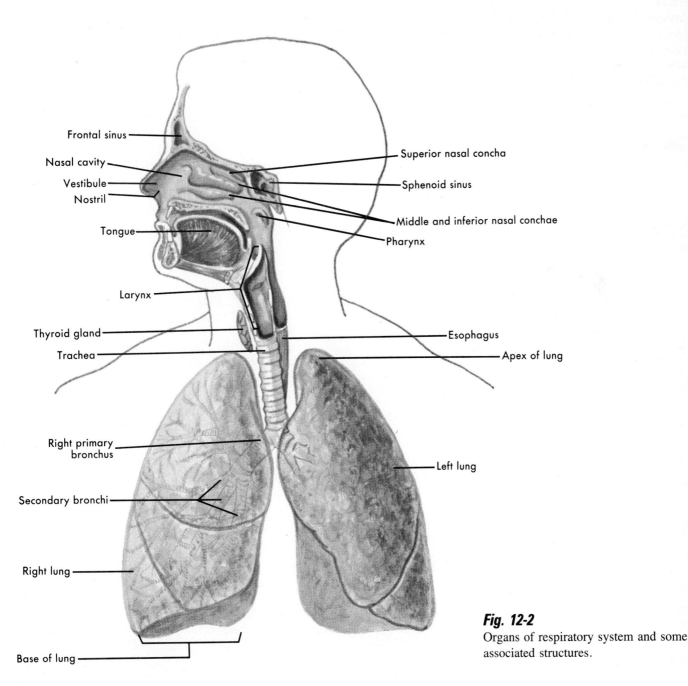

Fig. 12-2

Organs of respiratory system and some associated structures.

Nose and nasal cavities

Nose

The nose has a pyramidal shape, with its "point" termed the *apex*. The orifices of the nose, termed the *nostrils* or *nares*, are separated from one another by a median septum called the *columna*. The first 0.5 cm or so inside the nares is called the *vestibule* and is provided with a stratified squamous epithelium containing large hairs or *vibrissae* and sebaceous glands. The hairs act as a coarse filter to block the entry of large foreign particles that might be contained in the inhaled air.

The framework of the nose (Fig. 12-3) is formed of both bone and cartilage.

The *nasal bones* form the "bridge" of the nose that lies between the orbits. The *frontal processes of the maxillae* form the posterolateral support of the nose. The cartilaginous framework consists of the *septal cartilage* lying in the midline and connecting posteriorly to the nasal septum; the two *lateral nasal cartilages* situated inferior to the nasal bones; the two *greater alar cartilages* that form the apex and inferior and lateral portions of the framework; and several small pieces of cartilage, the *lesser alar cartilages,* which lie between the greater alar cartilages and the *frontal processes* of the maxillae.

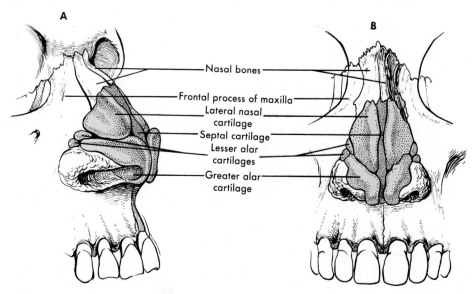

Fig. 12-3
Structure of nose. **A,** Anterolateral view. **B,** Anterior view.

Nasal cavities

The boundaries of the nasal cavities (Fig. 12-4) are formed by the cartilages mentioned previously and by the **conchae** (superior, middle, inferior), beneath each of which lies a **meatus,** the **maxillae** and **palatine** bones, the perpendicular plate of the **ethmoid bone,** and the **vomer.** The posterior openings of the cavities into the nasal pharynx are called the **choanae.** Each nasal cavity is divided into an upper **olfactory portion** and a **respiratory portion.** The structure of the olfactory portion was described in Chapter 10 and should be reviewed now.

The respiratory portion of the nasal cavities is lined by a mucous membrane that adheres closely to the underlying cartilaginous or bony framework. The epithelium of the membrane is of a *pseudostratified ciliated* type and contains many *goblet cells* that produce mucus. The mucus traps much particulate matter in the inhaled air, and the cilia move the mucus coating toward the pharynx, where it is usually swallowed. This obviously provides cleansing of incoming air. A layer of connective tissue called the *lamina propria* lies beneath the epithelium; its lowest layers of fibers act as a perichondrium or periosteum for the supporting framework. The lamina is heavily infiltrated with lymphocytes that afford additional protection (through phagocytosis and antibody production) against microorganisms and particles. Large thin-walled veins in the lamina bring blood close to the surface of the membrane and allow heat to radiate into the nasal cavities to warm the inhaled air. *Seromucous glands* in the lamina produce a watery and mucous secretion that moistens inhaled air and provides additional trapping of particles. This secretion also contains an immunoglobulin *(IgA)* that lyses (breaks down) many types of microorganisms.

The entire membrane is spongy and commonly swells when inflamed to give a "plugged nose." *Rhinitis* refers to inflammation of the nasal mucosa.

Emptying into the nasal cavities are the openings of the paranasal sinuses (see Fig. 12-4). These sinuses are the *frontal, ethmoidal, maxillary,* and *sphenoidal.* Their linings are the same as those of the nasal cavities, and because of the small size of the orifices, inflammation may close them easily, often creating pain and an alteration in the quality of the voice. Also emptying into the nasal cavities are the **nasolacrimal ducts** that carry the tears from the eyes.

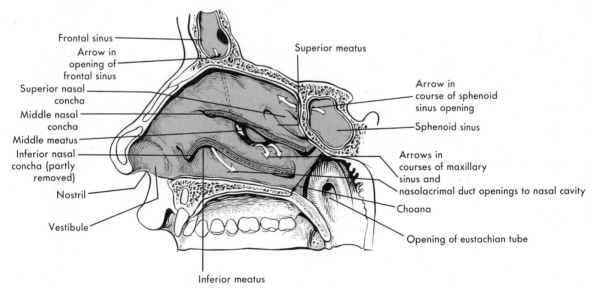

Fig. 12-4
Medial view of right nasal cavity.

Pharynx

The pharynx or throat (Fig. 12-5) extends from behind the nasal cavities to the level of the epiglottis. It is subdivided into three parts: The **nasal pharynx** forms about the upper one third, has the choanae opening into it, and receives in its lateral walls the openings of the *eustachian tubes* from the middle ear cavities; the *pharyngeal tonsil* (adenoid) is located in its posterior wall. The **oral pharynx** forms the middle one third of the throat and lies behind the oral cavity. This portion serves both the respiratory and digestive systems. The **laryngeal pharynx** forms the lower one third of the throat, and in its lower region the larynx and esophagus form separate passageways for air and food.

The wall of the pharynx is basically three layered. The epithelium is pseudostratified ciliated with goblet cells in the upper part of the nasal pharynx, a continuation of the nasal cavity epithelium. A transition to stratified squamous epithelium occurs in the lower nasal pharynx, and this type is found in the remainder of the pharynx. The underlying lamina propria contains many lymphocytes, and its deeper layers, called the **pharyngobasilar fascia,** act as a covering for the **pharyngeal constrictors** (*superior, middle,* and *inferior*) that form the third layer. The constrictors are composed of skeletal muscle and are important in swallowing.

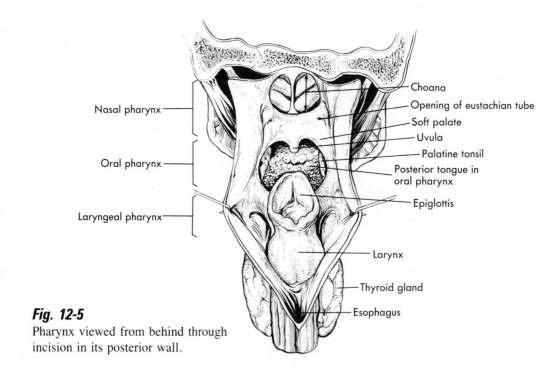

Fig. 12-5
Pharynx viewed from behind through incision in its posterior wall.

Larynx

Cartilages

The larynx (Fig. 12-6; see also Fig. 12-7) is basically a cartilage–formed and cartilage–supported boxlike organ that conveys air from the throat to the trachea, acts as the source of the sound that is shaped into speech, and blocks the respiratory passageways during swallowing to prevent aspiration of food and drink.

Three major cartilages, the *thyroid cartilage*, the *cricoid cartilage*, and the *epiglottis*, together with the smaller paired *arytenoid, corniculate,* and *cuneiform cartilages* support the larynx. Several *membranes* connect these cartilages, and several *muscles* attach to them to change tension on the vocal cords during breathing and speaking or singing.

The thyroid cartilage consists of two plates or laminae of hyaline cartilage that are fused anteriorly to create a subcutaneous projection called the *laryngeal prominence* or *Adam's apple*. In the mature male the angle made by these laminae is smaller than that of the mature female, hence the sharper protrusion of the male larynx; the adult male larynx is also larger than that of

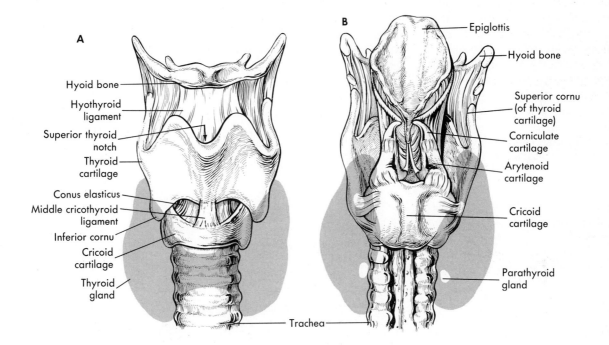

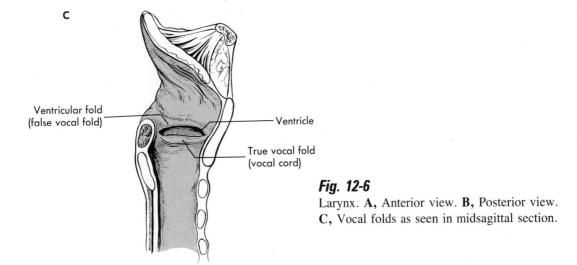

Fig. 12-6
Larynx. **A,** Anterior view. **B,** Posterior view. **C,** Vocal folds as seen in midsagittal section.

the female adult. The cartilage bears a *superior thyroidal notch* in the midline and carries *superior* and *inferior cornua* (horns). The superior cornua are connected to the hyoid bone by the *thyrohyoid ligaments,* while the inferior cornua attach to the cricoid cartilage. The *thyrohyoid membrane* connects the hyoid bone to the thyroid cartilage anteriorly.

The cricoid cartilage lies inferior to the thyroid cartilage and is a signet ring–shaped hyaline cartilage with a narrower anterior and wider posterior portion. A membrane called the **conus elasticus** connects the thyroid and cricoid cartilages, and an anterior **middle cricothyroid ligament** also connects the two cartilages.

The epiglottis is a leaf-shaped elastic cartilage. Its lower end is attached to the posterior aspect of the angle of the thyroid laminae just below the thyroid notch, and it projects upward and backward over the opening into the larynx. When a person is swallowing, this cartilage seals the laryngeal orifice.

The arytenoid cartilages are pyramid-shaped masses of hyaline cartilage that sit atop the posterior portions of the cricoid cartilage. Several muscles, to be described later, attach to these cartilages, as do the vocal cords. These are the main cartilages, since they are caused to move by the muscles and alter the tension of the vocal cords and thereby the pitch of the voice.

The pitch of sound produced when air is forced across the vocal cords is determined by their length and the tension applied to them by the laryngeal muscles. Longer cords, as in the male, produce a lower pitched sound. More tension on the cords produces a higher pitched sound. The tongue, lips, and cheeks are used to shape this sound into speech.

The corniculate cartilages lie on the apexes of the arytenoids. They are composed of elastic cartilage.

The cuneiform cartilages are small masses of elastic cartilage that lie on the lateral aspects of the arytenoid cartilages.

The cavity of the larynx contains two folds of tissue with a space between. The superior folds are the **false vocal folds,** the cavity is the **ventricle,** and the lower folds are the true **vocal folds** (*vocal cords*). The **glottis** is the space between the two laterally placed vocal cords.

Muscles

Extrinsic and intrinsic muscles are found within the larynx.

The extrinsic muscles are those that connect the larynx with some outside structure and include such muscles as the sternothyroideus, thyrohyoideus, and inferior pharyngis constrictor.

The intrinsic muscles (Fig. 12-7) connect the various cartilages and are responsible for altering the size of the glottis and changing the tension on the vocal cords. While it is beyond the scope of this text to fully describe these muscles, they are shown in Fig. 12-7, and their names suggest the cartilages they connect. The muscles are:

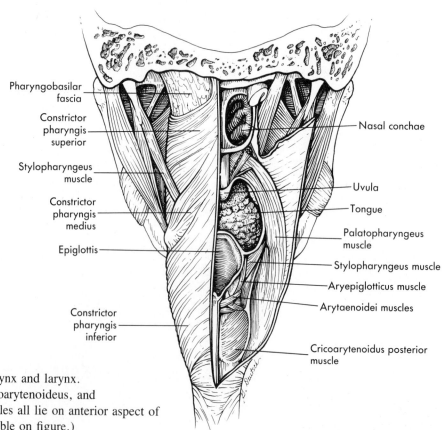

Fig. 12-7
Some muscles of pharynx and larynx. (Cricothyroideus, thyroarytenoideus, and thyroepiglotticus muscles all lie on anterior aspect of larynx and are not visible on figure.)

Cricothyroideus
Cricoarytenoideus, posterior and lateralis
Arytenoideus
Thyroarytenoideus
Aryepiglotticus
Thyroepiglotticus

Microscopically the mucous membrane of the larynx presents features similar to the membrane of the pharynx. Stratified squamous epithelium covers the anterior and upper posterior surfaces of the epiglottis and the vocal folds. Pseudostratified ciliated epithelium covers all other areas. Mucus-secreting glands lie in the lamina except over the vocal folds, and the laryngeal membranes or cartilages form the third layer of the organ wall.

Trachea and bronchi

The *trachea (windpipe)* attaches to the inferior aspect of the larynx and extends for about 11 cm into the thorax, where it terminates by branching into two **primary bronchi** *(bronchi)* (Fig. 12-8). Its diameter is 2 to 2.5 cm, and it is always larger in the adult male than in the adult female. The portion of the trachea that lies in the neck is associated with the thyroid gland (on anterior and lateral surfaces) and with several neck muscles (see Fig. 8-15).

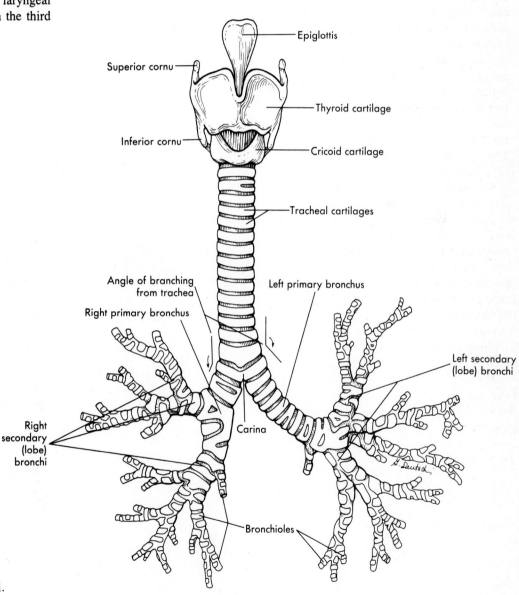

Fig. 12-8
Larynx, trachea, and bronchi.

The thoracic portion lies within the **mediastinum,** or central of the three cavities of the thorax (Fig. 12-9). It receives its blood supply from the inferior thyroid artery.

Laryngitis refers to inflammation of the laryngeal mucosa, most commonly by a virus. The mucosa may swell to the point at which the glottis is drastically narrowed, with difficulty in breathing developing. The vocal cords may become swollen, leading to hoarseness and an altered quality of the voice.

Microscopically (Fig. 12-10) the trachea possesses a pseudostratified ciliated epithelium containing goblet cells, underlaid by a connective tissue lamina. The latter contains many seromucous glands. Mucus, produced by the goblet cells and the laminal glands, together with the ciliated epithelium, further cleanses and removes particulate matter from the inhaled air, moving the mucus toward the pharynx. Six-

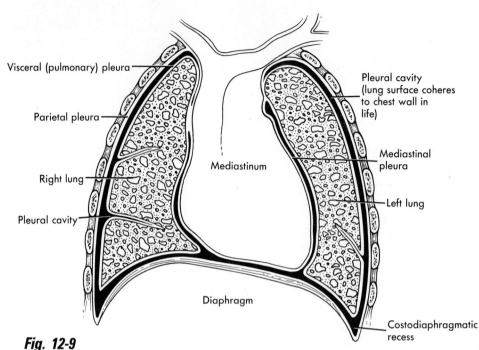

Fig. 12-9
Mediastinum (heart, its great vessels, trachea, and bronchi have been removed).

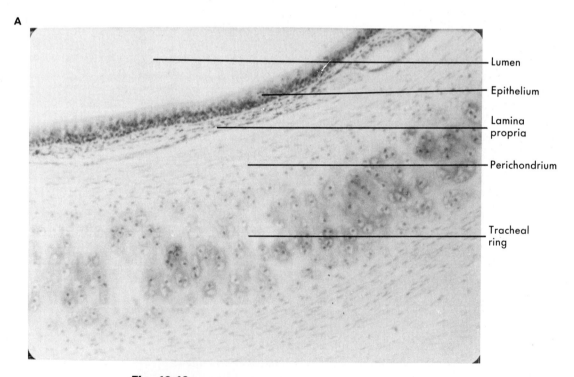

Fig. 12-10
Histology of lower respiratory system. **A,** Trachea. **B,** Bronchiole. **C,** Terminal bronchiole. (Terminal bronchiole appears collapsed because of smooth muscle contraction on tissue fixation.)

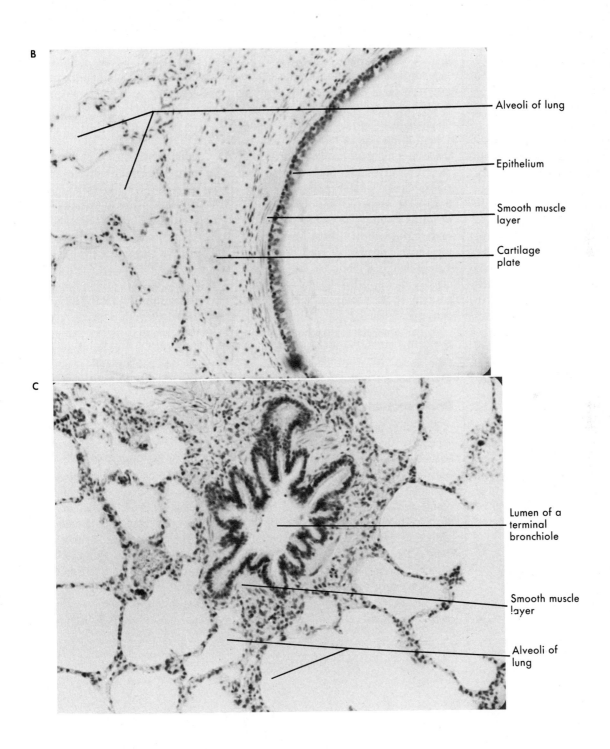

teen to twenty C- or Y-shaped rings of hyaline cartilage maintain the trachea in an open state during breathing. The open portions of these rings are directed posteriorly and are closed by smooth muscle *(trachealis muscle)* and by the previously mentioned tissue layers. The last tracheal cartilage is known as the *carina* and has a lower hooklike process that encloses the bronchial branches and their bifurcation from the trachea.

A *tracheotomy* opens the trachea anteriorly about 1 cm below the cricoid cartilage between the second and third tracheal rings.

The **primary bronchi** are two in number and are designated as the *right* and *left bronchi*. The right bronchus is wider and shorter (2.5 cm as opposed to 5 cm) and branches from the trachea at a greater angle than does the left bronchus (see Fig. 12-2). Their microscopic structure is the same as that of the trachea.

Inspirated objects most commonly end up in the right bronchus because it is more nearly in a direct line with the trachea.

Secondary or **lobe bronchi** arise from the primary bronchi. There are three on the right side, corresponding to the three lobes of the right lung and two on the left, again corresponding to the two lobes of the left lung. Further branching of the bronchial tree will be considered in the discussion of the lungs.

Lungs

Relationships of the lungs to the thoracic cavity

The paired **lungs** (Fig. 12-11) lie within the two **pleural cavities** of the thorax. Membranes known as the **pleurae** line the cavities and are reflected onto the lung surface at the **root** of the lung (the point where the bronchi enter). The name *parietal pleura* is given to the membrane lining the thoracic and diaphragmatic surfaces. Further subdivision of this part of the pleura may be made according to location; for example, cervical (in the neck), costal (on rib surfaces), and diaphragmatic portions. The term *visceral* (pulmonary) *pleura* is given to the membrane on the lung surface itself. These membranes are serous membranes and produce a watery fluid that lubricates the lung and chest surfaces, allowing them to slide easily over one another as the lungs expand and deflate.

In the adult the right lung averages 625 g, the left 562 g. In men the lungs form about $\frac{1}{37}$ of the body weight, in women about $\frac{1}{43}$ of the body weight.

Gross anatomy

The lungs are conical with a narrow superiorly directed **apex** that extends about 1 cm above the level of the clavicles and a broad inferiorly directed **base** that rests on the diaphragm. The **costal surfaces** are rounded to match the curvatures of the ribs. The medial surface is indented to form a hilum *(hilus),* the point of entry and exit of bronchi, pulmonary arteries and veins, and pulmonary nerves. The medial aspect of the left lung bears the **cardiac impression,** a concavity that houses the heart. The right lung has three **lobes,** *superior, middle,* and *inferior,* while the left lung has two lobes, *superior* and *inferior.* On the right lung the *horizontal fissure* separates superior and middle lobes, and an *oblique fissure* separates middle and inferior lobes. On the left lung an *oblique fissure* separates the two lobes.

Bronchioles

Within the substance of the lungs the secondary bronchi divide dichotomously (one tube gives rise to two) about 21 to 24 times, creating increasing numbers of smaller tubes that form the **bronchioles** and respiratory portions of the respiratory tree.

A bronchiole varies from about 0.5 cm to about 1 mm in diameter and has the usual three-layered wall structure we have come to expect. A pseudostratified, ciliated, goblet cell–containing epithelium is present; there is a thin lamina; and cartilaginous plates replace the rings of the bronchi. As the tubes become smaller, cartilage decreases at the same time that the amount of circularly disposed smooth muscle increases. At the **terminal bronchiole,** the last of the organs of the conducting division, no cartilage is present, the epithelium is a simple or pseudostratified columnar ciliated epithelium lacking goblet cells, and the lamina is very thin. The major tissue in the wall of the terminal bronchiole comprises several layers of smooth muscle. Since there is no rigid supporting tissue in these tubes, spasm of the muscle (as in asthma) may severely restrict air flow.

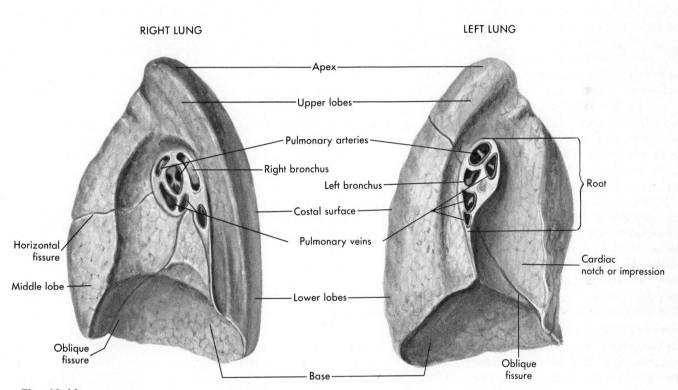

Fig. 12-11
Lungs. Each lung is viewed from medial aspect.

Respiratory portion of the system
(Fig. 12-12)

Gas exchange between the air in the respiratory system and the blood vessels of the lungs can occur only when there are *alveoli,* small simple squamous-lined cavities, in the tubes.

Respiratory bronchioles are branches of the terminal bronchioles that have a few scattered alveoli along their courses. Epithelial lining is simple cuboidal except where there are alveoli, and laminal tissue consists of only a few fibers of mainly elastic connective tissue.

Alveolar ducts rise from the respiratory bronchioles and consist of simple squamous-lined alveoli along their entire course. A few strands of elastic connective tissue lie around the ducts, and masses of smooth muscle adorn the septa between alveoli.

Alveolar sacs attach to the ducts. Each sac consists of several alveoli surrounded by elastic fibers.

Table 12-1 presents some facts about the respiratory tree. Within the alveoli lie two other types of cells in addition to the simple squamous epithelium that allows gas exchange. Rounded cells called **pneumonocytes** secrete a phospholipid material called *surfactant.* This substance lowers the surface tension of the alveoli and is an important factor in preventing alveolar collapse on deflation of the lungs. *Alveolar macrophages* engulf any particulate matter that may have reached the alveoli. The cells then move into the ciliated tubes and are eliminated from the system.

Table 12-1. Some characteristics of the respiratory system

Organ	Generation of branching	Number of organs (estimated)	Diameter (mm)	Total cross-sectional area (cm^2)
Trachea	0	1	18-25	2.5
Bronchus	1	2	12	2.3
Lobe bronchi	2	4-5	8	2.1
Small bronchi	5-10	1000	1.3	13.4
Terminal bronchioles	14-15	32,750	0.7	113.0
Respiratory bronchioles	16-18	260,000	0.5	534
Alveolar ducts	19-22	4.2 million	0.4	5880
Alveoli	23-24	300 million	0.2	(50-70 m^2)

Adapted from McClintic, J.R.: Basic anatomy and physiology of the human body, ed. 2. Copyright © 1980. Reprinted by permission of John Wiley & Sons, Inc.

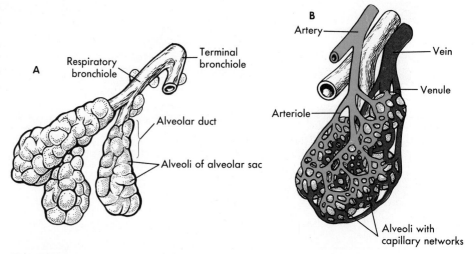

Fig. 12-12
A, Respiratory division of respiratory system, and, **B,** its blood supply.

Bronchopulmonary segments and lobules

Bronchopulmonary segments (Fig. 12-13) are portions of the lung lobes and are defined as that portion of a lobe supplied by a direct branch of a secondary bronchus. The segments are as definite as the lobes, being set off from one another by connective tissue septa. Surgery performed on the lung often uses these segments as logical areas for removal without damaging the rest of the lung.

Within the segments lie the **lobules** of the lung, again separated by thin connective tissue septa. The term *secondary lobule* is given to a group of terminal bronchioles and their branches. A **primary lobule** is often considered to be the basic respiratory unit of the lung and consists of one respiratory bronchiole and all of its branches.

Blood supply of the lungs

The two functional divisions of the respiratory system below the trachea receive their blood supply from two different sources. *Bronchial arteries* rise from the upper aorta and the upper intercostal arteries and supply the bronchi, bronchioles, and terminal bronchioles. The arteries follow alongside the tubes and after forming capillary beds in the walls of these vessels form bronchial veins that also course along the tubes to leave the lungs and empty into the *azygos veins* of the thorax. The *pulmonary arteries* form capillary beds around the respiratory division of the system and not only bring blood to the lungs for oxygenation and carbon diox-

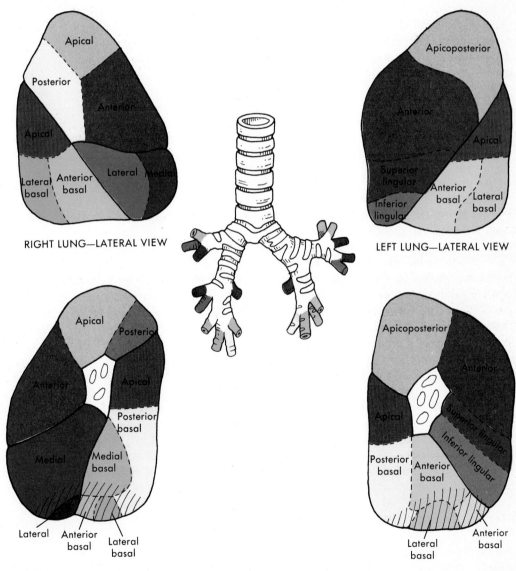

Fig. 12-13 Bronchopulmonary segments.

ide elimination but also nourish the respiratory division. The pulmonary veins formed from these capillary beds also receive the greater part of the blood brought to the lungs by the bronchial arteries.

Nerves of the lungs

Nerves that reach the lungs are derived from both the sympathetic and parasympathetic divisions of the autonomic nervous system. Sympathetic nerves are chiefly vasomotor in nature, supplying the muscular blood vessels of the lungs. Thus some degree of control over blood flow is provided. Parasympathetic nerves are primarily branches of the vagus nerve that serve a variety of respiratory reflexes concerned with rate and depth of breathing and coughing.

Breathing

Ventilation of the lungs involves two phases: *inspiration,* or drawing of air into the lungs, and *expiration,* or driving air from the lungs.

The inspiratory phase is always active, requiring muscular activity. Contraction of the diaphragm increases the vertical dimension of the thorax, and the contraction of the external intercostales muscles and several neck muscles causes the ribs to be elevated and to swing up and out. These combined actions increase the volume of the thorax (which is closed to outside communication) and thereby decrease the pressure within it. Pressure in the thorax *(intrathoracic pressure)* is always less than atmospheric pressure by 3 to 5 mm Hg and becomes even more so (7 to 8 mm Hg) on inspiration. The cavities of the lungs contain air *at* atmospheric pressure when no ventilation is occurring. The parietal and visceral pleurae cohere by virtue of the fluid films between them. As thoracic size increases, the lungs also enlarge, "following" the movements of the chest and diaphragm. As they enlarge, the pressure within the lungs *(intrapulmonic pressure),* previously at atmospheric level, drops 3 to 5 mm *below* atmospheric pressure. The pressure is now greater outside the lungs, and air moves into the lungs, eventually equalizing the pressure at atmospheric level at the end of inspiration. Expansion of the lungs also stretches the elastic tissue around the respiratory portions of the system.

Expiration is normally passive, that is, *not* involving muscular activity. Relaxation of the diaphragm and other muscles allows a return to normal thoracic volume and pressure. The elastic tissue of the lungs, stretched during inspiration, recoils and provides a rise above atmospheric pressure of 3 to 5 mm Hg. Air is driven out of the lungs, the pressure again equalizing at the end of expiration. Thus we see that intrapulmonic pressure cycles below and above atmospheric pressure, whereas intrathoracic pressure always stays less than atmospheric pressure.

The amount of air exchanged with each breath is called **tidal volume** and amounts to about 500 ml when a person is breathing at rest. After a normal inspiration, an additional volume of air, **reserve inspiratory volume,** can be drawn into the lungs. It amounts to about 3000 ml. The **reserve expiratory volume** can be emptied from the lungs after a normal expiration. It amounts to about 1100 ml. The *sum* of these three volumes is called the vital capacity; it represents the total exchangeable air of the lungs and totals 4600 ml in the "average" adult.

Air always remains in the lungs even after the most forcible expiration because the alveoli do not collapse. This air is called **residual volume** and amounts to about 1200 ml. Residual volume *plus* reserve inspiratory volume constitutes the **functional residual capacity.** This air oxygenates the blood during expiration; that is, breathing is intermittent but oxygenation is continuous.

The amount of effort required to ventilate the lungs depends on several factors.

Airway resistance refers to the ease with which air flows through the tubular portions of the system and depends on the lengths and diameters of those tubes. The longer or smaller the tube, the greater the effort required to move air through it.

Compliance refers to ease of expansion of lung tissue. The lungs are usually easy to inflate because of their content of elastic tissue. Replacement of this tissue by nonelastic fibers makes it harder to inflate them.

Elastance is a measure of recoil of the elastic fibers of the lungs and is the major force in expiration of air from the lungs. Again, loss of elastic tissue means that muscular activity may have to be employed to force air from the lungs.

Pulmonary disorders account for a major portion of disorders (10% to 15%) in people over 40 years of age. Several are described below with reference to what parameter of pulmonary function has been altered.

"*Black lung*" is a layman's term for *pneumoconiosis* in coal workers. Inhalation of airborne coal dust results in accumulation of particles in alveolar macrophages and in the lung tissue itself, restricting gas diffusion through alveoli. Carbon dioxide retention can lead to acidosis (fall of blood pH) as the gas reacts with water to form carbonic acid. This reaction emphasizes another function of the respiratory system, that of acid-base balance through elimination of CO_2. Small bronchioles often collapse, increasing airway resistance, and loss of elastic tissue decreases both compliance and elastance. Similar changes occur in **asbestosis,** caused by the inhalation of asbestos fibers. Inflammation is an additional complication, since the needlelike asbestos fibers penetrate and irritate the lungs.

Asthma is a bronchiospastic disorder in which the smooth muscle of small bronchioles suddenly contracts and greatly narrows those tubes, making it extremely difficult to get air into *or* out of the lungs. Causes of the spasm may be chemical, thermal, or psychological, and treatment is aimed at relaxing the muscle to permit normal airflow.

Emphysema is classed as a chronic obstructive pulmonary disease. Interalveolar septa are lost, reducing diffusing surface, small bronchioles collapse or kink, there is loss of elastic tissue in the lungs, and breathing is difficult even at rest. Little can be done except to prevent the condition.

Smoking is associated with emphysema-like symptoms, plus an increased risk of developing lung cancer, one of the deadliest forms of cancer. The tars of cigarette smoke are apparently the carcinogenic agents, and they and the nicotine in the smoke may cause epithelial changes leading to cancer, to loss of ciliated cells, and to production of a thick, hard-to-move mucus.

Hyaline membrane disease is referred to lack of surfactant with failure of alveoli to expand at birth or their later collapse. Autopsy reveals a "film" on the walls of alveoli and bronchioles that has a clear (hyaline) appearance. Life may be prolonged in infants with the disease by providing oxygen-rich air under intermittent positive pressure (to 14 mm Hg) to keep alveoli expanded and functional.

Controlling breathing and adjusting it to demands of activity depend on *central* and *peripheral influences*.

Central influences depend on *respiratory centers* located in the medulla and pons of the brainstem. *Inspiratory centers* in the medulla respond to CO_2 or H^+ levels of the bloodstream by initiating nerve impulses that pass to the muscles of respiration for inspiration of air. Interruption of this basically continuous activity (since CO_2 and H^+ are always present) is provided by expiratory and pneumotaxic centers and vagal nuclei.

Peripheral influences on breathing are provided by baroreceptors in the lungs and the chemoreceptors of carotid and aortic bodies. The baroreceptors set a limit to lung stretch and provide impulses to the expiratory and inspiratory centers that cause expiration. Chemoreceptor response is triggered by rise of blood CO_2 or fall in O_2 and causes an increase in rate *and* depth of breathing to alleviate either condition.

Apnea refers to a temporary cessation of breathing and may reflect a reduced response of the respiratory centers to their normal stimuli. Sudden infant death syndrome *(SIDS)* may be due to such changes in respiratory center sensitivity.

Lest the reader believe that breathing an oxygen-rich atmosphere is all to the good, let it be indicated that long-term breathing of high-level oxygen atmospheres can cause ***oxygen toxicity***. Lung linings are irritated; the nervous system malfunctions; and in infants, retinal detachment may occur. A limit of 40% oxygen for 12 or more hours of exposure is recommended.

Summary

1. The respiratory system provides a means of bringing atmospheric air into contact with the circulatory system for purposes of oxygen and carbon dioxide exchange and for removal of particulate matter from inhaled air; it also plays a role in phonation.
2. Nose and nasal cavities develop at $3\frac{1}{2}$ to 4 weeks from the olfactory placode and pit. The rest of the system begins development at about 4 weeks from a laryngotracheal groove that forms bronchopulmonary buds. The buds repeatedly branch to form the respiratory tree. Alveoli develop at about 24 weeks.
3. Organs of the system may be divided in several ways:
   a. Conducting division, those organs too thick walled to allow gas exchange: nose, nasal cavities, pharynx, larynx, trachea, bronchi, bronchioles, and terminal bronchioles.
   b. Respiratory division, permitting gas exchange: respiratory bronchioles and alveolar ducts and sacs.
   c. Upper respiratory system: nose, nasal cavities, and pharynx.
   d. Lower respiratory system: larynx to alveolar sacs.
4. The nose is a cartilage- and bone-supported structure. Its supports include nasal bones, maxillary bones, and septal and alar cartilages.
   a. The nostrils form the entry of air into the nose.
   b. The vestibule is the first 0.5 cm of the nose inside the nostrils.
5. The nasal cavities lie within the nose and skull.
   a. They have a respiratory and olfactory portion lined with different types of epithelium.
   b. Pseudostratified ciliated epithelium with goblet cells and an underlying lamina provide warming, moistening, and cleansing functions for the respiratory portion of the cavities.
   c. The paranasal sinuses (frontal, ethmoidal, maxillary, and sphenoidal) connect to the nasal cavities, as do the nasolacrimal ducts from the eyes.
6. The pharynx opens by choanae from the nasal cavities, has nasal, oral, and laryngeal portions, and is surrounded mainly by the three pharyngeal constrictors.

7. The larynx consists of six cartilages and contains the vocal cords.
   a. The single thyroid and cricoid cartilages and the epiglottis are the major cartilages of the larynx.
   b. Arytenoid, corniculate, and cuneiform cartilages (paired) support the vocal cords.
   c. Extrinsic muscles attach to the larynx and aid in swallowing.
   d. Intrinsic muscles connect the cartilages and enable alteration of vocal cord tension.
8. The trachea extends from the larynx into the chest, and from it rises the bronchi.
   a. The trachea has cervical and thoracic portions, measures 2 to 2.5 cm in diameter by about 11 cm in length, and has a pseudostratified ciliated epithelium and incomplete cartilaginous rings for support.
   b. Primary bronchi are the first branches of the trachea. There are right and left bronchi, and wall structure is similar to that of the trachea.
   c. Secondary (lobe) bronchi rise from the primary bronchi and number three on the right and two on the left.
9. The paired lungs lie within the two pleural cavities of the thorax, are covered by a visceral pleura that lies against the parietal pleura, and contain all parts of the respiratory system beyond the secondary bronchi.
   a. Grossly each lung has an apex, base, and costal and diaphragmatic surfaces. The right lung has three lobes and the left has two.
   b. Microscopically the lungs are composed of bronchioles, terminal bronchioles, alveolar ducts, and alveolar sacs.
   c. Wall structure of the lungs shows changes toward simple squamous epithelium, loss of cartilage, increase in muscle that virtually disappears in the respiratory division, and increase in capillary beds for gas exchange with the bloodstream.
10. Bronchopulmonary segments are areas of a lung supplied by the first branches of a secondary bronchus. Secondary lobules consist of a group of terminal bronchioles and their branches; primary lobules are a respiratory bronchiole and its branches.
11. Blood supply of the lungs is via bronchial vessels for the conducting division and pulmonary vessels for the respiratory division.
12. Nerves to the lungs serve blood vessels and respiratory reflexes.
13. Breathing is the process that ventilates the lungs, and it has inspiratory and expiratory phases.
    a. Inspiration requires muscular effort that increases the volume and decreases the pressures (intrathoracic and intrapulmonic) in both the thoracic cavity and lungs.
    b. Expiration is normally passive, not requiring muscular effort, and is brought about mainly by elastic recoil of lung tissue that is stretched during inspiration.
    c. During breathing certain volumes of air are exchanged or remain in the lungs: tidal volume cycles in and out during breathing—its volume is about 500 ml; reserve inspiratory volume is air that can be drawn *in* above tidal volume—its volume is about 3000 ml; reserve expiratory volume can be expelled *beyond* tidal volume—its volume is about 1100 ml; residual volume remains in the lungs after a forcible expiration—its volume is about 1200 ml.
    d. Vital capacity is the sum of tidal, reserve inspiratory, and reserve expiratory volumes—it is the total exchangeable air of the lungs. Functional residual capacity is the sum of reserve expiratory and residual volumes—it oxygenates the blood during expiration.
14. Several disorders of lung function are described, including black lung, asthma, emphysema, and hyaline membrane disease.
15. Control of breathing is supplied by central and peripheral influences.
    a. The central influence depends on several respiratory centers, including inspiratory, expiratory, pneumotaxic, and vagal groups.
    b. Peripheral influences are provided by baroreceptors and chemoreceptors in the lungs and blood vessels of the body.

Questions

1. What purposes does a respiratory system have?
2. List the organs of the respiratory system and relate question 1 to each organ that you list.
3. Considering the epithelium of the respiratory system from nose to alveolar sacs, what changes occur? What are the functional correlations of these changes?
4. What are the cartilages of the nose?
5. What are the paranasal sinuses and how are they related to the nasal cavities?
6. How does the pharynx differ structurally and functionally from the rest of the respiratory system?
7. Describe the structure of the larynx, giving names and relationships of its cartilages and muscles.
8. Compare trachea, bronchi, and bronchioles as to size, location, and wall structures.
9. Describe the lungs as to external anatomy, location in the thorax, and relationship to the pleural membranes.
10. Describe:
    a. A bronchopulmonary segment.
    b. A secondary lobule.
    c. A primary lobule.
11. What parts or organs of the respiratory system would be included in 10a to 10c?
12. What differences exist in blood supply to the conducting and respiratory divisions of the system?
13. How is air drawn into and expelled from the lungs?
14. What are the several volumes and capacities associated with air exchange? What is the adult volume for each? What is the value or importance of each volume or capacity?
15. Describe some disorders of lung function and explain how each affects compliance, elastance, and airway resistance.

Readings

Gil, J.: Organization of microcirculation in the lung, Annu. Rev. Physiol. **42**:177, 1980.

Fraser, R.G., and Pare, J.A.P.: Structure and function of the lung, Philadelphia, 1971, W.B. Saunders Co.

Strang, L.B.: Growth and development of the lung: fetal and postnatal, Annu. Rev. Physiol. **39**:253, 1977.

13

Digestive system

Objectives

Development of the system
   Formation of the mouth and gut
   Disorders of formation of the gut

Organs of the system

Mouth and oral cavity
   Lips and cheeks
   Teeth and gums
   Tongue
   Salivary glands

Fauces

Pharynx

General plan of tissue layers in the alimentary tract

Esophagus

Abdomen
   Boundaries
   Peritoneum
   Mesenteries and omenta

Stomach
   Gross anatomy
   Microscopic anatomy

Small intestine

Large intestine

Rectum and anal canal

Liver
   Gross anatomy
   Blood supply
   Microscopic anatomy

Gallbladder and bile ducts

Pancreas
   Gross anatomy
   Microscopic anatomy

Digestion in the small intestine

Blood vessels of the system

Selected disorders of the system

Summary

Questions

Readings

OBJECTIVES After studying this chapter, the reader should be able to:

List and briefly describe the functions a digestive system serves.

Discuss briefly the development of the digestive system and use this to account for some common congenital anomalies of the system.

Describe gross anatomy, relevant microscopic anatomy, and functional significance, including secretions produced and control of release of the secretion for each region or organ of the system (include mouth, tongue, teeth, salivary glands, esophagus, stomach, small and large intestines, liver, and pancreas).

Describe the general tissue plan of the walls of the alimentary tract.

Discuss the relationships of the abdomen to the organs it contains and to the peritoneal and mesenteric membranes of the cavity.

Briefly outline the blood supply to the various organs of the system.

Describe the relationships of the gallbladder and bile ducts to the liver and duodenum.

Comment on the nature of the more common disorders of the system.

The digestive system carries out the following four major activities:

Ingestion. The intake of food and drink is the initial step in the acquisition of energy-containing substances required to sustain cellular activity.

Digestion. Many organic compounds that we ingest are composed of "building blocks" that represent the form of the material that cells actually use. By a combination of mechanical subdivision and enzymatic breakdown, large molecules are reduced by the process of digestion to a form available to cells.

Absorption. Even though digestion of foodstuffs has occurred, the products are not available for body use until they have passed through the walls of the digestive organs and have entered blood or lymph streams for distribution to cells. This passage constitutes absorption.

Egestion. Included in foodstuffs we eat are indigestible materials. Excess secretions of digestive glands are not always taken back into the body. Egestion (elimination) rids the body of unusable or excess materials.

Incidental to these functions, the digestive organs may offer routes of synthesis or excretion of a variety of materials and protection against potentially harmful substances. Such functions will be discussed in connection with the particular organ involved.

Development of the system

(Fig. 13-1 and Table 13-1)

Formation of the mouth and gut

The mouth is formed as a depression of the surface ectoderm anterior to the first branchial arch of the head at about 22 days of development (Fig. 13-1). The depression is known as the *stomodeum*, and it is separated from the foregut or primitive pharynx by a two-layered *oropharyngeal membrane*. The stomodeum and the external layer of the membrane are ectodermally lined; the foregut and the internal surface of the membrane are endodermally lined.

The oropharyngeal membrane ruptures at about 24 days of development and provides a communication of the primitive gut with the amniotic cavity by way of the mouth.

During the fourth week of development a *primitive gut* is formed as the cavity of the yolk sac is enclosed by folds of the embryo's lateral body walls, which are endodermally lined. Three portions of the primitive gut are recognized; they are named, from anterior to posterior, the *foregut, midgut,* and *hindgut*. Each section is supplied by a major branch of the aorta that will form its blood supply throughout life. The *celiac artery* supplies the foregut, the *superior mesenteric artery* supplies the midgut, and the *inferior mesenteric artery* supplies the hindgut. Each portion will differentiate into particular portions of the digestive system, as indicated in Table 13-1.

The *anus* develops as an indentation at the posterior end of the embryo called the *proctodeum*. It, like the mouth, is separated from the hindgut by a *cloacal membrane* that normally ruptures at about 7 weeks of development.

The muscular and connective tissue layers of the tract are mesodermally derived and begin formation at about 12 weeks.

Table 13-1. Derivatives of the gut tube

Tube part	Derivatives	Development
Foregut	Esophagus	Recognizable at about 4 wk.
	Stomach	Recognizable at about 4 wk; becomes baglike at about 8-10 wk; and assumes typical form at about 12 wk.
	Duodenum to entrance of bile duct	Recognizable at about 4 wk.
	Liver and pancreas	Develop as outgrowths of gut at about 4 wk; liver lobes form by 6 wk; pancreas complete by 10 wk.
	Bile ducts and gallbladder	Outgrowth of duodenum; connection retained to duodenum forms ducts; bladder is outgrowth of duct.
Midgut	Rest of small intestine	Recognizable at about 4 wk; elongates and coils at 5 wk; villi at 8 wk.
	Cecum, appendix, one half of large intestine	Separated by 5 wk; completed by 8 wk.
Hindgut	Remainder of large intestine, rectum, anal canal	Formed by 4 wk; completed by 7 wk.

Adapted from McClintic, J.N.: Basic anatomy and physiology of the human body, ed. 2. Copyright © 1980. Reprinted by permission of John Wiley & Sons, Inc.

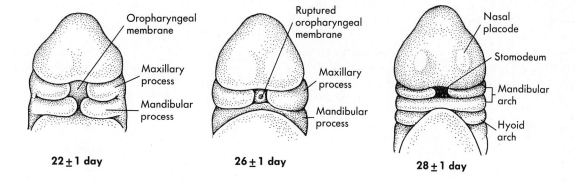

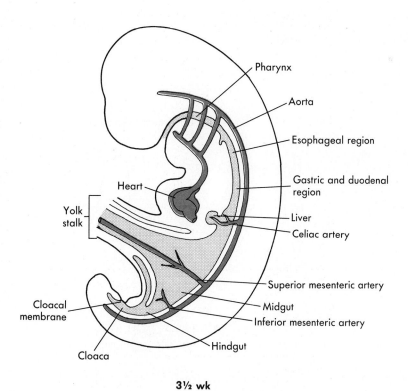

Fig. 13-1
Above, Development of oral cavity. *Below*, Development of gut.
After Moore.

Disorders of formation of the gut

As the changes described previously are occurring, several other important events take place.

A tracheoesophageal septum separates the esophagus from the trachea (fourth week).

The endoderm of the esophagus proliferates, and the lumen (cavity) of the esophagus is obliterated (7 weeks) to be recanalized at about 9 to 10 weeks.

The small intestine elongates and "herniates" into the umbilical cord through the as yet unclosed opening from the midgut to the cord (5 weeks); then, as the embryo's body grows and provides more room in the abdominal cavity, the intestine returns to the body (about 10 weeks).

The cloaca (the terminal portion of the hindgut) is partitioned into an upper rectum and anal canal and a lower urogenital sinus, thereby effecting a separation of the digestive and urogenital systems (about 7 weeks).

If such processes as these do not occur normally or development is arrested at some stage, malformations of the tract may occur. Several of the more common congenital disorders of the system are presented in Fig. 13-2 and Table 13-2.

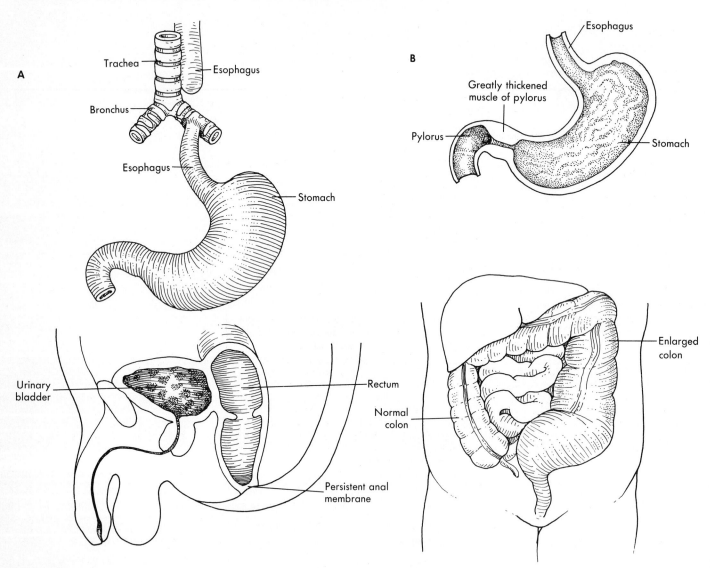

Fig. 13-2
Some common congenital disorders of digestive system. **A,** Esophageal atresia with connection to bronchus or trachea (85% to 95% of atresias have connection with respiratory system), *top,* and imperforate anus (failure of anal membrane to rupture), *bottom.* **B,** Pyloric stenosis (blockage or narrowing of pylorus as result of thickening of muscle), *top,* and megacolon (lack of nervous elements leads to fecal retention and colon enlargement), *bottom.*

Table 13-2. Some common congenital disorders of the digestive system

Disorder	Frequency (no./births)	Cause of disorder	Comments
Esophageal atresia	1/2500-1/3000	No recanalization of esophagus; improper separation from respiratory system; no passage to stomach	Infant shows excess of saliva and poor nutrition since foods cannot reach stomach; reflux of foods is common.
Pyloric stenosis	1/200 male 1/1000 female	Excessive development of muscle fibers of distal end of stomach that prevents passage of foods to intestine	Blockage of stomach exit causes vomiting of feedings, weight loss, and dehydration.
Imperforate anus	1/5000	Failure of anal membrane to rupture; no anus formed	Surgery is necessary to perforate the membrane.
Megacolon	1/25,000	Failure of nerve cells to innervate a section of the colon	Part without nerves does not move contents onward, accumulates feces, and dilates; removal of noninnervated section and anastomosis to healthy colon usually cures the condition.

Adapted from McClintic, J.R.: Basic anatomy and physiology of the human body, ed. 2. Copyright © 1980. Reprinted by permission of John Wiley & Sons, Inc.

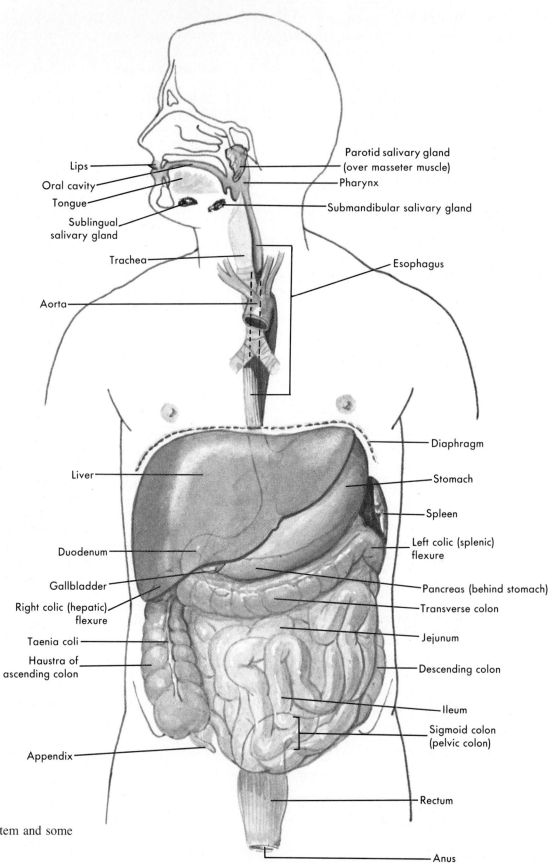

Fig. 13-3
Organs of digestive system and some associated structures.

Organs of the system

The many organs of the digestive system (Fig. 13-3) may be divided into two groups.

The *alimentary tract* (*digestive tube*) consists of the tubular organs serving as the areas for reception of food, chemical digestion, absorption, and elimination. The specific organs include the mouth, pharynx, esophagus, stomach, small and large intestines, and the anal canal and anus.

The *accessory organs* of digestion consist of those organs that develop from the wall of the digestive tube and contribute to the process of digestion. Included here are the teeth, tongue, salivary glands, liver, and pancreas.

Mouth and oral cavity

The *mouth* (Fig. 13-4) serves as the normal avenue for intake of food. The mouth has as its limits the lips anteriorly, the cheeks laterally, the hard palate superiorly, the tongue inferiorly, and the soft palate posteriorly. Between the lips and cheeks and the teeth and gums is a slitlike *vestibule*. The *oral cavity* is that part of the mouth enclosed by the teeth and is thus a smaller area than the mouth itself.

Lips and cheeks (see Fig. 13-4)

The lips are the two fleshy red folds that surround the orifice of the mouth. They are covered externally by thin skin and internally by a mucous membrane consisting of an uncornified ("soft") stratified squamous epithelium on a rather thick lamina of connective tissue. The central portion of the lips contains the fibers of the orbicularis oris muscle. Numerous small labial glands empty into the vestibule, producing mucus and serous fluids that lubricate and moisten the mucous membrane.

The cheeks are continuous anteriorly with the lips and form the sides of the face. Their structure is similar to that of the lips with the buccinator being the muscle involved. Buccal fat is abundant in the skin of the cheek, especially in infants. Buccal seromucous glands are found in the mucous membrane of the cheek. Several of these glands are larger than the rest, empty into the vestibule opposite the last molar teeth, and are called the *molar glands*.

Teeth and gums

The teeth normally number 52 and are the primary agents for subdivision of food.

Four types of teeth occur in the human mouth (see Fig. 13-4). *Incisors* present a broad but narrow profile, which gives them a chisellike shape. There are four incisors in each jaw that meet to give a shearing or scissorslike action useful in biting. *Canines* (*eyeteeth*) lie next in order in each jaw. A total of two canines are present in each jaw. Canines are conical teeth, form the "fangs" in certain animal groups, and are used by the human primarily for holding or shredding foods. *Premolars* (*bicuspids*) are two in number behind each canine, for a total of eight in the mouth. Their surfaces bear small elevations (cusps), and they usually have two roots. Premolars are adapted to grind foods, achieving the finest subdivision.

So far the teeth listed total 20. These teeth are replaced during our lifetime. We receive 20 *deciduous teeth* replaced by 20 *permanent teeth* for a total of 40. The remaining *molars* erupt only once. These teeth usually have three roots and are also grinding teeth;

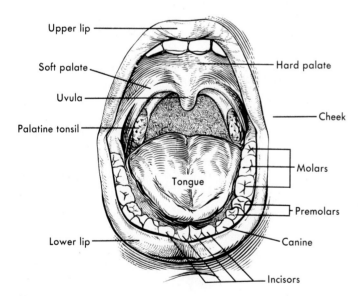

Fig. 13-4
Mouth and oral cavity.

they number six in each jaw for a total of 12. Thus 40 + 12 = 52 teeth. If we were to look at the number of each type of tooth in *one half of each jaw,* we would arrive at what is called the human dental formula of 2 1 2 3 (incisors, canines, premolars, molars, from anterior to posterior). Time for eruption of the teeth is shown in Table 13-3.

A vertical section through a tooth in the jaw (Fig. 13-5) reveals its structure and the relationships to the socket in which it resides.

The **crown** of the tooth projects into the mouth cavity and is covered with the **enamel.** The term *anatomical crown* is often used to refer to that part of the tooth covered with enamel; the term *clinical crown* is used to refer to the part of the crown projecting above the gum line in those in which the gums have not receded. The **root** of the tooth lies within the jawbone and is covered with **cementum,** a bonelike material. Between crown and root lies the **cervix** *(neck)* of the tooth, a narrow region where enamel and cementum are at their thinnest. This is an area particularly liable to attack by decay-producing substances ("gumline cavities"). The bulk of the tooth is formed of **dentin,** a yellowish soft material surrounding the **pulp cavity** of the tooth. The latter communicates to the outside via the *root canal* and is filled with **dental pulp.** Pulp consists of a loose type of connective tissue containing blood vessels, nerves, and lymphatic vessels that nourish the tooth. On the walls of the pulp cavity are **odontoblasts,** cells that produce dentin. Thus the cavity becomes smaller with age as new dentin is laid down on the *inner* surface of a tooth.

The tooth is held in its **alveolus** *(socket)* by a **periodontal membrane** that acts also as the source of cementum and as a periosteum for the jawbone. The membrane holds the tooth firmly but not rigidly in its socket.

The **gums** are composed of dense fibrous tissue (continuous with the periodontal membrane) covered with the oral mucous membrane. They also aid in holding the tooth in place and prevent food from lodging between gum and tooth (particularly if we brush as we should—from "root to crown").

In referring to the location of cavities in teeth, dentists often use the following terminology:

Labial or *buccal*—the surface of a tooth facing the lips or cheeks
Lingual—the surface of a tooth facing the tongue
Surface of contact—sides of a tooth facing other teeth (Contact surfaces may be further specified by anterior, posterior, medial, or lateral.)

Table 13-3. Time of eruption of teeth

Deciduous teeth	Months	Permanent teeth	Years
Lower central incisors	6-8	First molars	6
Upper central incisors	9-12	Central incisors	7
Upper lateral incisors	12-14	Lateral incisors	8
Lower lateral incisors	14-15	First premolars	9-10
First molars	15-16	Second premolars	10
Canines	20-24	Canines	11
Second molars	30-32	Second molars	12
		Third molars	17-18

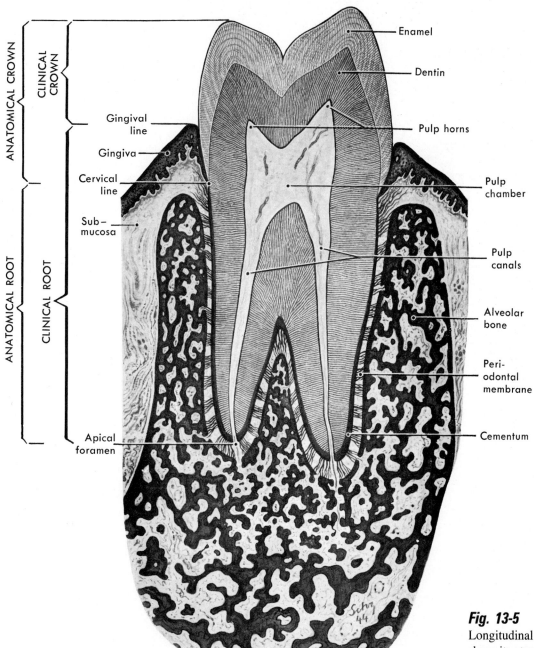

Fig. 13-5
Longitudinal section of molar tooth to show its structure and relationships to its bony socket.

Blood vessels and nerves of the teeth (Fig. 13-6) provide sensation and nourishment for the organs. Half of the lower jaw may be anesthetized by injecting an anesthetic in the area of the mandibular foramen, whereas the upper teeth require injection in the area of a particular tooth because nerve supply does not "come together" at one particular "reachable" point.

Tongue

The tongue (Fig. 13-7) is the main organ of the sense of taste, is an important organ in word formation, and aids in guiding food between the teeth for chewing and in swallowing. The tongue has a *root* or *base,* the posterior part that attaches to the hyoid bone by the hyoglossus and genioglossus muscles. Its *apex* or *tip* is the anterior limit. The inferior surface is connected to the floor of the mouth by the membranous *frenulum*. A frenulum that is too short may restrict tongue movements ("tongue-tied"). The upper surface of the tongue bears a longitudinal *median sulcus* that terminates in the *foramen cecum*. The latter marks the point where the thyroid gland developed. A variety of

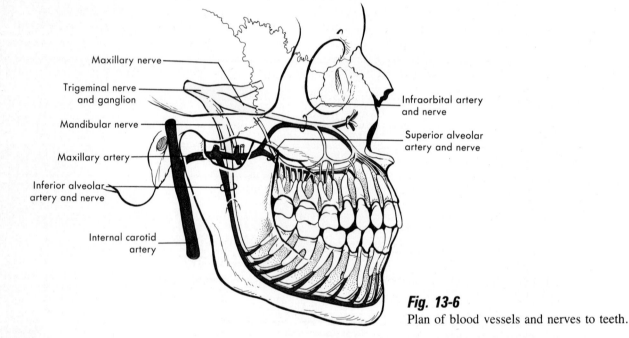

Fig. 13-6
Plan of blood vessels and nerves to teeth.

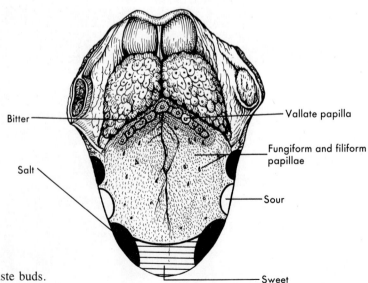

Fig. 13-7
Tongue: structure and location of taste buds.

papillae are found on the dorsum and sides of the tongue.

- *vallate (circumvallate) papillae* number 8 to 12 and form a V on the posterior base of the tongue. The main part of the papilla is buried in the tongue and is surrounded by a moat or trench. Taste buds are found in the walls of the papilla facing the trench. The buds sense bitter taste.
- *fungiform papillae* are found on the dorsum, sides, and apex of the tongue and, as their name suggests, resemble mushrooms projecting above the general surface of the tongue. Those on the sides and apex carry taste buds sensitive to sour, sweet, and salt; those on the dorsum usually bear no buds and serve to roughen the tongue for food guiding and for swallowing.
- *filiform papillae* are conical, cover the anterior two thirds of the tongue, and also serve to roughen the tongue.

The muscles of the tongue have been described elsewhere (Chapter 8). There are intrinsic muscles that lie entirely within the tongue (vertical, transverse, and longitudinal lingual muscles) and extrinsic muscles whose origins lie outside the tongue (genioglossus, hyoglossus, styloglossus, palatoglossus). The fibers of both sets of muscles form a three-way arrangement within the tongue. Between muscle fibers are **von Ebner's glands,** which secrete a watery fluid that aids in moistening food and the mouth lining. Additional structures found in the tongue include the lingual mucous glands, the sublingual salivary gland (to be described later), and the lingual tonsil in its base.

Salivary glands

Three pairs of salivary glands (Fig. 13-8) empty their secretions into the mouth. These are the parotid, submandibular (mandibular), and sublingual glands.

The **parotid glands,** largest of the three pairs, weigh 14 to 28 g and lie over the masseter muscle. Their ducts (Stensen's ducts) empty into the vestibule opposite the second upper molar. Histologically the parotid gland is seen to be composed only of **serous cells,** small granular and dark-staining cells. Such cells produce a watery fluid rich in an enzyme called salivary amylase (ptyalin) that initiates the digestion of starches.

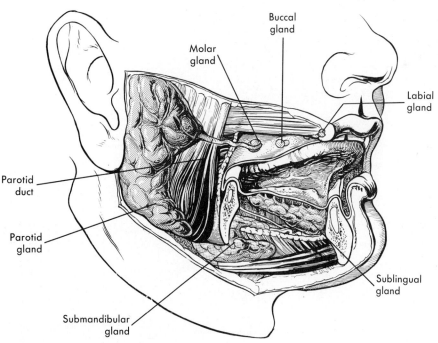

Fig. 13-8
Locations of salivary and oral glands.

The *submandibular glands* are about the size of walnuts and lie medial to the mandibular angles. Their ducts (*Wharton's* ducts) open into the mouth on either side of the frenulum of the tongue. Histologically (Fig. 13-9), the glands consist mostly of serous cells with a few mucous cells. Where both types of cells occur in the same secretory units, the serous cells form **demilunes** *(of Giannuzzi)* around the mucous cells.

The *sublingual glands* are almond sized and lie medial to the mandibular body near the symphysis of the mandible. Their ducts *(of Rivinus)* join those of the submandibular glands to empty either side of the frenulum. Histologically the sublingual gland contains mostly mucous cells with a few serous cells. Demilunes are present.

The collective secretion of the salivary glands is the **saliva**. It consists of 99.5% water and 0.5% solids. It has a pH of 5.8 to 7.1 (slightly acidic to about neutral) that aids in keeping microorganism growth diminished in the mouth. The water of the saliva *dissolves* soluble materials in foods, *softens* the mass, *cleanses* the teeth and mouth linings, and *moistens* the mucous membranes. Its content of mucus *lubricates* food for swallowing and aids in holding a mass of food into a *bolus* for swallowing. The major solids of the saliva include sodium, chloride, and potassium ions; urea, a product of amino acid metabolism; and the enzyme *salivary amylase* (ptyalin). The amylase digests cooked starches to dextrins (units of several to several hundred glucose molecules) and, if the food could remain in the mouth long enough, to disaccharides (units containing two glucose molecules). Foods remain for such a short time in the mouth that the dextrins are the major products (95% to 97% of the products versus 3% to 5% disaccharides).

Control of salivary secretion is entirely by nervous reflex with the taste buds as receptors, the seventh and ninth cranial nerves as afferent pathways, the salivary centers of the medulla as the central connections, the seventh and ninth cranial nerves as the efferent paths, and the glands themselves as the effectors. Although saliva production is continuous, the greatest amount is secreted when food is in the mouth.

Little absorption of digestion products occurs in the mouth because they are too large as yet. Some medicines may be absorbed through the mouth lining (lozenges held under the tongue).

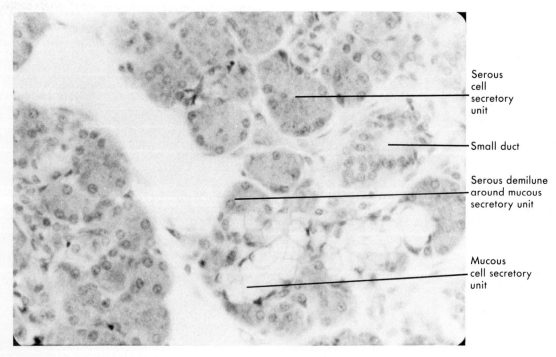

Fig. 13-9
Section of submandibular salivary gland.

Fauces

The opening of the mouth posteriorly into the pharynx is termed the *fauces* (see Fig. 13-4). The soft palate forms the superior border of the tissues guarding the fauces, whereas the *palatoglossal arch* forms the lateral and anterior margins. The *palatopharyngeal arch* lies posterior to the previous arch. Between the two arches on the lateral sides lie the *palatine tonsils*. The tonsils have been described in connection with the lymphatic system.

Pharynx

The structure of the pharynx has been described on p. 408 and should be reviewed now.

General plan of tissue layers in the alimentary tract

The organs of the alimentary tract, particularly those from the esophagus to the anal canal, have the same basic arrangement of tissue layers in their walls. Each organ does have special features of its own that enable histological identification, and these will be indicated in the discussions of each organ. Our purpose here is to become familiar with the terminology applied to these tissue layers.

Four basic layers of tissue (Fig. 13-10) are described:

mucosa (tunica mucosa), the innermost layer, consists of a lining *epithelium, an underlying layer called the lamina* (tunica) *propria*, and a thin layer of smooth muscle, the *muscularis mucosae*. The mucosa is typically glandular.

submucosa (tunica submucosa) consists of vascular connective tissue, may contain glands, and is the site of location of the *submucosal plexus (Meissner's plexus)*. The latter is a nerve plexus involved in the control of activity of the smooth muscle of the wall.

muscularis (tunica muscularis or *muscularis externa)* is the muscular coat typically composed of two layers of muscle that are skeletal in the upper esophagus and smooth from the lower third of the esophagus to the anal canal. The inner layer of muscle is in the form of a tight spiral *(circular layer)*, and the outer layer is in the form of a very loose spiral *(longitudinal layer)*. Between the two layers may be found the nerve cells and fibers of the *myenteric plexus (Auerbach's plexus)*.

serosa or *adventitia* is the outermost coat of the tract and is composed of a thin layer of connective tissue and a mesothelium on those organs that lie in the abdominal cavity. The term *adventitia* refers to a layer of connective tissue that blends without demarcation into the surrounding tissue. Such a layer occurs around the pharynx and the thoracic portion of the esophagus.

The reader should become quite familiar with these terms, since histological description of organs to follow will assume knowledge of their meaning and location.

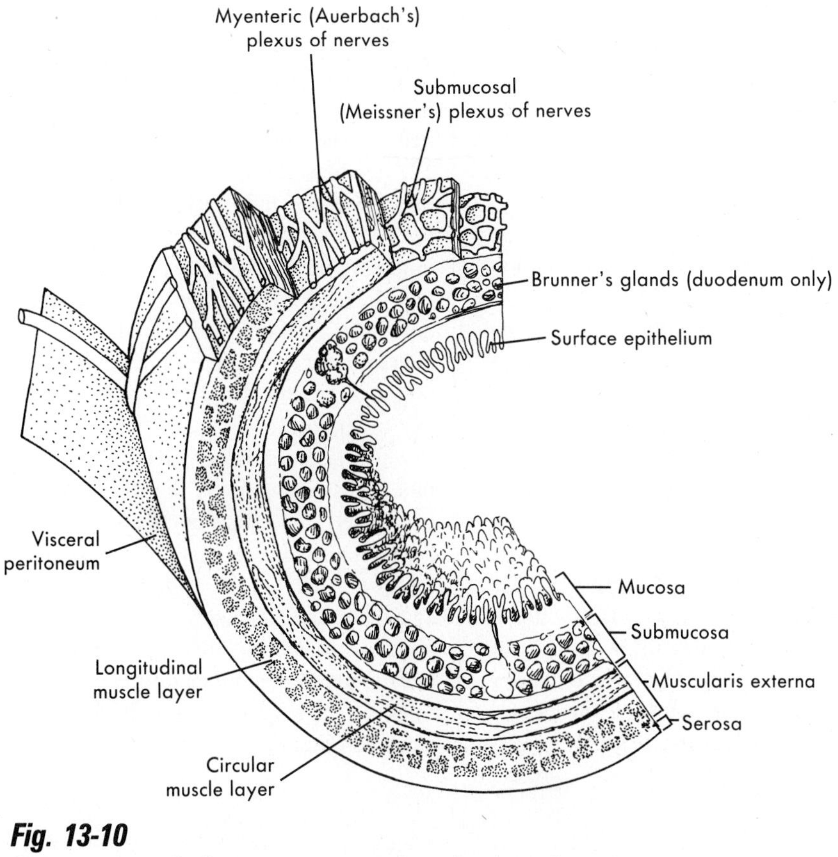

Fig. 13-10
Cross-section of alimentary tract to show its tissue layers.

Esophagus

The esophagus is a muscular tube extending from the pharynx about 24 cm to the stomach. It passes through the esophageal hiatus in the diaphragm (see Fig. 8-16, *A*) to make its connection to the stomach. It serves the function of *transporting* swallowed food and drink and has no digestive function.

The structure of the esophagus shows it to have a **mucosa** composed of a thick (500 to 800 μm) stratified squamous epithelium (for protection against mechanical attrition), a lamina propria that contains mucous glands known as *esophageal cardiac glands* (near the junction with the stomach), and a thick layer of smooth muscle, the muscularis mucosae. The **submucosa** is thick and vascular, has nerve cells of Meissner's plexus in it, and contains *esophageal glands* of mucous nature throughout its length. The **muscularis externa** is two layered, consisting of skeletal muscle in its upper third, mixed skeletal and smooth muscle in its middle third, and only smooth muscle in its lower third. Arrangement of muscle is in the previously described inner circular and outer longitudinal layers. Auerbach's plexus lies between the two layers. In the cervical and thoracic portions of the esophagus, the outermost coat is an ***adventitia***, whereas on the abdominal portion a typical *serosa* is present.

The esophagus receives blood vessels from the inferior thyroid artery, aorta, bronchial arteries, and left gastric artery. The main nerve supply is derived from the vagus nerve for control of muscular activity related to swallowing.

Abdomen

Boundaries

The abdomen and its related cavity, that is, the abdominopelvic cavity, contain the remaining organs of digestion, as well as the organs of the urinary and reproductive systems and several endocrine organs.

Superiorly the abdominal cavity is limited by the dome-shaped diaphragm that rises to the level of the fifth rib. The floor is formed by the pelvic diaphragm. The remaining limits are set by the vertebral column posteriorly and by the abdominal muscles laterally and anteriorly. Apertures in the wall of the abdomen include the aortic and esophageal hiatuses superiorly, as well as the opening for the vena cava, all occurring in the diaphragm.

Inferior are openings for the femoral vessels and in the male the opening for the spermatic cord. The abdomen is divided into quadrants and nine smaller areas (see Fig. 1-7), and the abdominal viscera may be projected on the abdominal surface (see Fig. 1-8).

Peritoneum

The lining of the abdominopelvic cavity and its organs consists of a membrane called the **peritoneum**. It has the structure of a serosa (serous membrane), that is, a simple squamous mesothelium underlaid by connective tissue. More specifically, the name *parietal peritoneum* is applied to the lining of the cavity walls itself, and the term **visceral peritoneum** is given to the membranes that are reflected over the abdominal organs. The potential space between these two layers is the *peritoneal cavity*. The cavity is divided into a greater and a lesser portion. The latter is related to the stomach's dorsal surface and is sometimes called the *omental bursa*. The two portions communicate with one another by the *epiploic foramen (foramen of Winslow)* that lies between the liver and duodenum.

Mesenteries and omenta

Certain of the abdominal viscera are completely surrounded by peritoneum that forms their serosal coverings. Such organs are suspended from the posterior abdominal wall (Fig. 13-11) by double folds of peritoneum that are called **mesenteries**. The term *mesentery*, if used alone, refers to the membrane suspending the small intestine; membranes suspending other organs receive more specific names that reflect what organ is being suspended. Examples include mesocolon, mesoappendix, mesovarium (ovary), and mesosalpinx (uterine tube). Other organs are not suspended by such membranes but have some part of their surface covered by peritoneum. Such organs are said to be **retroperitoneal**, that is, lying behind the peritoneum. The kidneys, ureters, and pancreas are examples of retroperitoneal organs.

There are two *omenta* (sing., *omentum*); they are double folds of peritoneum. The *lesser omentum* (Fig. 13-12) lies between the lesser curvature of the stomach and duodenum and the liver. The cavity of this omentum is the omental bursa. The *greater omentum* (see Fig. 13-11) attaches to the greater curvature of the stomach and the duodenum and lies as a curtain over the abdominal viscera. No cavity is found in this omentum.

The mesenteries provide a route for blood vessels to reach the organ being suspended, and the connective tissue component contains many macrophages that can act as a protective device against microorganisms that might enter the mesentery. Also, particularly in the greater omentum, significant quantities of fat may be stored in the connective tissue of the membranes.

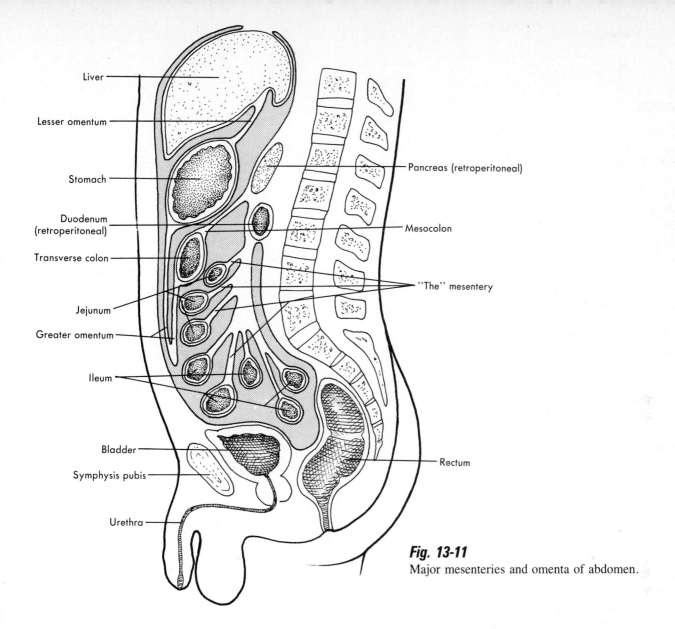

Fig. 13-11
Major mesenteries and omenta of abdomen.

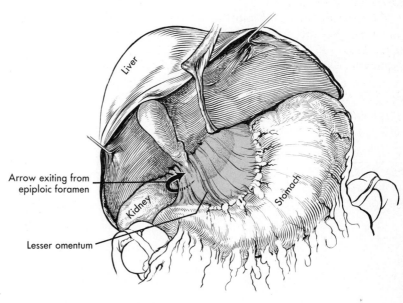

Fig. 13-12
Lesser omentum.

Stomach

The stomach (Fig. 13-13) lies primarily in the left upper quadrant of the abdomen. Its shape varies with fullness, tone of the gastric muscle, and the degree of "encroachment" of the adjacent intestines. In most cases it is described as being "J" shaped, with the lowest limit being usually at or near the umbilicus.

Gross anatomy

The esophagus connects to the stomach at the *cardia*. At this junction the muscularis externa of the esophagus is thickened to form a functional *cardiac sphincter*. This is not a true sphincter, that is, a more or less definitive circular layer of muscle that can close off one organ from another. The inefficiency of this sphincter is sometimes evidenced by the passage of acid from stomach into lower esophagus to give "heartburn." The left and upward directed blind pouch of the stomach is the *fundus*. The *body* of the stomach extends to a more or less obvious indentation on the stomach called the **angular incisura**. Beyond this notch is the *pylorus*, divided into a proximal *pyloric vestibule* and a distal *pyloric antrum* by

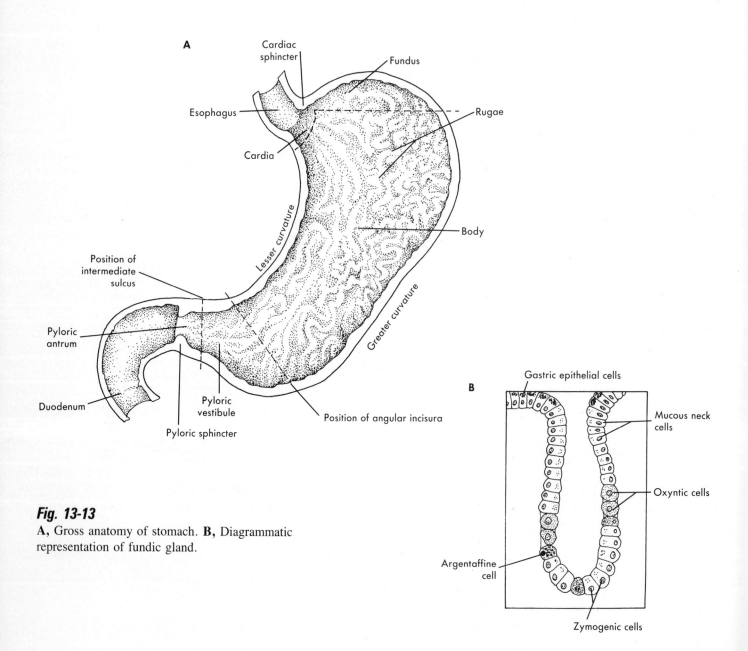

Fig. 13-13
A, Gross anatomy of stomach. **B,** Diagrammatic representation of fundic gland.

an *intermediate sulcus*. The latter feature may be quite obvious or not. The stomach opens into the duodenum, a junction that *is* guarded by a true sphincter, the **pyloric sphincter**. A **greater curvature** is directed to the left and a **lesser curvature** to the right.

A longitudinal and frontal section of the stomach shows a number of **rugae** or folds that run more or less longitudinally within the stomach. The rugae tend to flatten but not disappear as the organ fills. Close inspection of the wall with the naked eye will disclose innumerable pinpoint-sized holes in the internal stomach wall. These are the **gastric pits** *(foveolae)* that open into an estimated 35 million gastric glands.

Microscopic anatomy

The stomach wall has the four basic layers of tissue described earlier. The **mucosa** has an *epithelium* of simple columnar type, in which the outer half or so of the cells are filled with mucus that is constantly produced by these cells. The continuous mucus layer thus formed protects the stomach wall from autodigestion or damage by the enzymes and acid within the stomach. The *lamina propria* consists of a few strands of connective tissue between the simple tubular **gastric glands** (see Fig. 13-13) that fill the laminal layer. Within the glands are found four types of cells.

mucous neck cells *(neck chief cells)* lie in the upper or neck portion of the glands. They are cuboidal in shape, have a compressed or flattened nucleus, and contain mucus.

oxyntic *(parietal) cells* are spherical cells with a central nucleus and finely granular cytoplasm, and they take an acidophilic stain. These cells lie primarily in the central or body portion of the gland and produce the hydrochloric acid (HCl) of the gastric juice.

zymogenic *(chief) cells* lie mainly in the end or fundus of the gastric glands. They are cuboidal, granular, and basophilic-staining cells with basal nuclei. They produce *pepsinogen*, a precursor of *pepsin*, which is the main digestive enzyme of the stomach. Pepsin initiates protein digestion.

argentaffine cells require special stains to demonstrate them. They lie scattered between the other cells in the glands and have eosinophilic basally placed granules and apical nuclei. They may be the source of intrinsic factor of the stomach or of serotonin, a vasoconstrictive chemical.

The mucosa terminates with a **muscularis mucosae** of smooth muscle. Some fibers extend vertically between the gastric glands and may aid emptying of the glands by compressing the entire mucosa.

The **submucosa** is as described in the typical tissue plan, is vascular, and extends into the rugae.

The **muscularis externa** may be resolved with difficulty into three layers: an *inner oblique*, a *middle circular*, and an *outer longitudinal* layer. The muscularis tends to act as a unit, and the directions of the different layers have no particular significance. A typical *serosa* forms the outer coat.

Glands in the cardiac and pyloric regions of the stomach contain almost entirely the mucous neck cells; fundic and body glands have all four types of cells present.

The secretion of the gastric glands is called the **gastric juice**. Foods enter the stomach, where they are "stored" for 3 to 4 hours while the juice works on certain components of the foods. Adult gastric juice has a pH of 1.9 to 2.6 (strongly acid) that kills many microorganisms that enter the stomach from the mouth and sets the proper pH environment for the action of the major enzyme of gastric juice, *pepsin*. Gastric juice is thus mainly (99.4%) a watery solution of HCl. The 0.6% solids include K, Na, Cl, Ca, and Mg ions; much mucus (to 15 g/L); and pepsin.

Pepsin commences the digestion of proteins, reducing them to proteoses and peptones, units containing four to twelve amino acids. Other enzymes are found in gastric juice: *pepsin B*[5] acts more effectively in a pH of 3 and is more important in the infant stomach, where full acid production has not yet been achieved; *gastricsin* is an additional protein-digesting enzyme also operating best at pH 3; *gastric lipase* is able to break down short-chain triglycerides and is of considerable importance in the infant stomach for digestion of butterfat in milk.

Control of gastric secretion is handled by both nervous and chemical (humoral) mechanisms.

The *neural* or *cephalic phase* uses the same receptor and afferent path as for salivary secretion. Branches pass to the vagal nuclei and nerves, causing gastric glands to produce 50 to 150 ml of gastric juice. This juice, sometimes called "ignition juice," commences protein digestion with release of proteoses and peptones.

The proteoses and peptones first produced trigger the second or *gastric phase* of secretion. In this phase the products of digestion cause a cell resembling the argentaffine cell to produce a chemical called **gastrin**. This material is absorbed into the venous drainage of the stomach, enters the general circulation, and is distributed to the entire body, including the stomach. All the gastric glands are caused to secrete, producing up to 750 ml of juice.

The *intestinal phase* is also humoral (chemically controlled). When foods enter the duodenum from the stomach, a factor ("intestinal gastrin") is produced that follows the same kind of path that gastrin did—into the veins and back to the stomach. An additional small quantity of gastric juice is thus produced.

Small intestine

The small intestine extends from the pylorus of the stomach about 7 m* to its termination at the large intestine. Most of it is supported in the central and inferior portions of the abdominal cavity by the mesentery and is divisible into three portions: duodenum, jejunum, and ileum.

The **duodenum** is the first 25 cm of the small intestine and is the shortest, widest, and most fixed portion of the organ. It has no mesentery and is only partially covered by the peritoneum. The *superior portion* is about 5 cm long and extends to the neck of the gallbladder. The *descending portion* is 7 to 10 cm in length, curves along the medial border of the right kidney, and ends at the fourth lumbar vertebra. The *horizontal portion* is 5 to 8 cm long and passes to a point anterior to the aorta. The *ascending portion* is about 2.5 cm long and ascends to the left to a level of the second lumbar vertebra, where it turns abruptly ventralward to become the jejunum.

The name **jejunum** is given to the next 2.5 m of the intestine, and the **ileum** is the remaining 3.75 m or so. There is no morphological point of change between these portions, the subdivision being an arbitrary one.

Microscopically the small intestine follows most exactly the typical tissue plan described earlier.

The **mucosa** is composed of a simple columnar epithelium with goblet cells, the latter increasing in number through the length of the organ. The lamina propria is folded to form the cores of the **villi** (sing., *villus*), small fingerlike projections that increase the surface area of the intestine for absorptive purposes (Fig. 13-14). The villi contain a centrally placed lymph capillary called a **lacteal** and a network of blood capillaries. Extending into the lamina between villi are the **crypts** (*of Lieberkühn*), or intestinal glands. These glands are lined with the same type of epithelium as covers the villi. A narrow muscularis mucosae completes the mucosa.

The **submucosa** is thrown into circularly disposed **plicae** that are covered with mucosa. Further surface area increase is afforded by the plicae. This layer is, as in other parts of the gut, a vascular layer. It contains mucous glands (*Brunner's glands*) in the duodenum.

The **muscularis externa** consists of the usual inner circular and outer longitudinal layers of smooth muscle.

A **serosa** is found on all parts except the first part of the duodenum.

Peyer's patches are aggregated groups of lymph nodules that are most numerous in the ileum. They measure about 1 cm × 3 to 5 cm and appear as whitish ovals to the naked eye.

The duodenum receives the ducts from the liver and pancreas. The functions of their fluids will be considered later, as will the digestive action of the intestine.

---

*This length is most commonly a postmortem measurement. The intestine is about 2 m shorter because of tone of the muscle in the living subject.

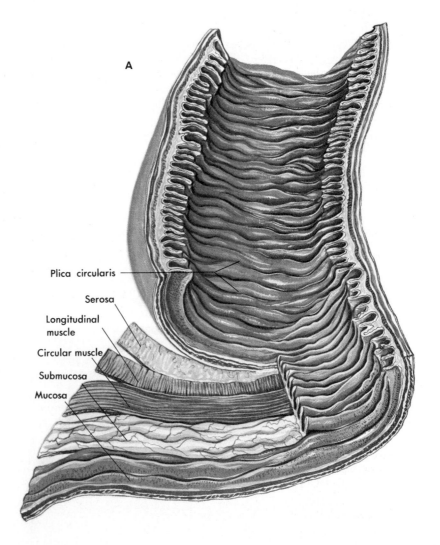

Fig. 13-14
A, Plicae and tissue layers of small intestine. **B,** Structure of villus, its blood and lymphatic supply, and relations to wall of organ.

Large intestine

The *large intestine* extends from the junction with the ileum some 1.5 m to the anal canal and is easily distinguished from the small intestine by its larger diameter and its saclike **haustrations** *(haustra)*. The ileum joins the large intestine above a blind pouch called the **cecum,** from which extends the **appendix** (vermiform appendix). This junction is marked by the **ileocecal sphincter.** The term *colon* refers to that portion of the large intestine between the cecum and the rectum and receives the following names:

ascending colon passes upward along the right side of the abdominal cavity to the level of the liver, where it bends medially at the *hepatic* (right colic) *flexure*.
transverse colon passes from right to left across the abdominal cavity and terminates at the *splenic* (left colic) *flexure* by the spleen.
descending colon passes downward along the left side of the abdominal cavity to the level of the midsacrum.
sigmoid colon is an S-shaped bend that returns the colon to the midline; it terminates in the rectum.

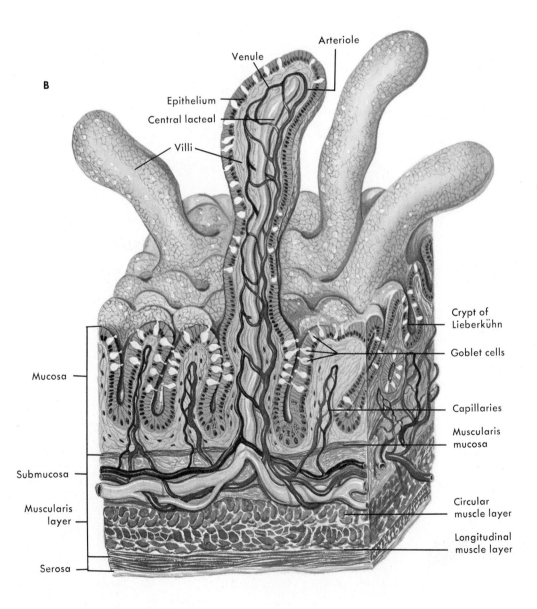

Rectum and anal canal

The *rectum* (Fig. 13-15) lacks haustrations, is about 12 cm long, and has the *rectal columns* that contain hemorrhoidal blood vessels.

All parts of the colon are supported by the mesocolon. The rectum has no mesentery.

The *anal canal* is 2.5 to 4 cm long and passes through the pelvic floor to the exterior body surface through the *anus*. This portion is guarded by a smooth muscle *internal anal sphincter* and the *external anal sphincter* formed by skeletal fibers of the pelvic diaphragm.

Microscopically the four typical layers are present:

- The *mucosa* lacks villi, possesses *crypts*, and its epithelium is predominantly goblet cells. The lamina propria lies between the crypts and may contain solitary lymph nodules. A thin muscularis mucosae is present.
- The *submucosa* is vascular loose connective tissue.
- The *muscularis externa* consists of a complete inner circular layer of smooth muscle; the outer coat is in the form of three longitudinal bands of smooth muscle placed equidistantly around the circumference of the organ. These strips are the *taenia coli.*
- A typical *serosa* is present on all parts except the rectum.

The appendix has the usual four coats, being smallest in diameter (5 to 10 mm) of any part of the tract. It has much lymphatic tissue in the mucosa and may serve as part of the defense mechanism of the body by virtue of its lymphatic tissue.

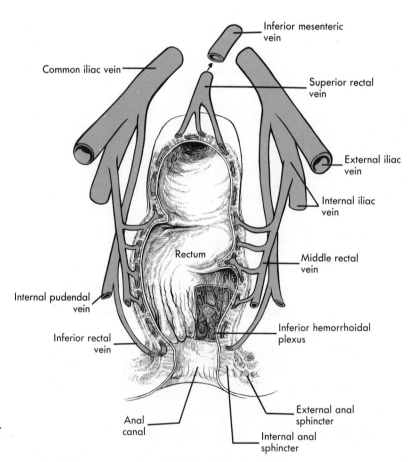

Fig. 13-15
Rectum and anal canal and their venous drainage.

Liver

The liver is the largest gland of the body, serving a wide variety of synthetic, storage, and excretory functions. It occupies most of the right hypochondriac region, weighs 1.4 to 1.6 kg in the adult male, and weighs 1.2 to 1.4 kg in the adult female.

Gross anatomy (Fig. 13-16)

The liver is roughly hemispherical in shape, the upper or *diaphragmatic surface* following the curvature of the diaphragm. It has four lobes: a large *right lobe*, a smaller *left lobe*, and much smaller *caudate* and *quadrate lobes*.

The liver is connected to the undersurface of the diaphragm and anterior abdominal wall by five ligaments:

falciform ligament lies between the right and left lobes and attaches to the diaphragm.
coronary ligament extends over the medial surface of the right lobe and attaches to the diaphragm.
right and *left triangular ligaments* lie on the lateral aspects of the right and left lobes and pass to the diaphragm.
round ligament is contained partially within the falciform ligament and extends to the umbilicus. It is the degeneration product of the umbilical vein.

Blood supply

The liver receives blood from two sources. Oxygen-rich arterial blood reaches the organ by way of the *hepatic artery*, a vessel derived from the celiac artery of the abdomen. Nutrient-rich blood from the stomach and intestines passes to the organ via the *portal vein*. After forming interlobar and

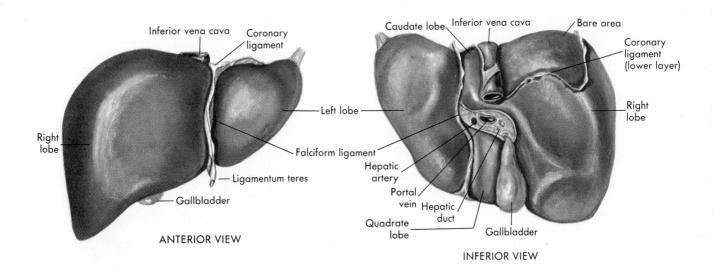

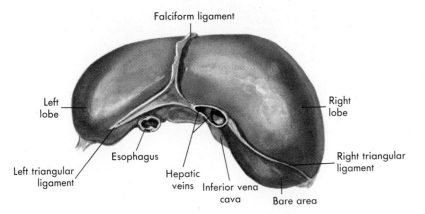

Fig. 13-16
Gross anatomy of liver.

smaller branches within the liver, the two afferent supplies come together at the sinusoid (discussed later). Outflowing vessels come together to form several **hepatic veins** that empty into the inferior vena cava just below the diaphragm.

Microscopic anatomy

The structural unit of the liver is the *lobule* (Fig. 13-17). These are 5- to 7-sided more or less cylindrical units about 2 mm in diameter. They are incompletely invested by connective tissue that carries the branches of the hepatic artery and portal vein into the liver. The name given to the "channels" of connective tissue that penetrate into the organ, carrying these vessels, is the **portal canal,** and it is *inter-lobular* in placement. The portal canal carries not only branches of the artery and vein but also a bile duct and sometimes lymphatic vessels. The artery, vein, and bile duct are of constant occurrence in the canals and are sometimes designated as the "portal trinity."

The lobule has as its center a blood vessel called the **central vein** (see Fig. 13-17). From the vein, **plates** or **cords of hepatic cells** are arranged in more or less radial fashion and are connected with one another crosswise. Blood spaces lie between the cords of cells; these are the **sinusoids** into which the hepatic artery and portal vein blood passes. The sinusoids are lined partly with typical endothelial cells (squamous epithelium) and partly with macrophages known as **Kupffer cells.** The latter phagocytose aged erythrocytes and microorganisms that may have entered the body through the intestines. Each hepatic cell faces, on at least one of its surfaces, a sinusoid; removes from that blood its nutrients; and also passes into that blood its products.

Each hepatic cell has a network of **bile canaliculi** around it. The bile, or product of erythrocyte breakdown and hepatic cell metabolism, is placed into these canaliculi. The bile moves toward the periphery of the lobule (opposite to the blood flow) and is collected in the interlobular bile ducts of the portal canal.

To be a bit more specific about the functions of the liver I may describe the following.

The *production of bile* is continu-

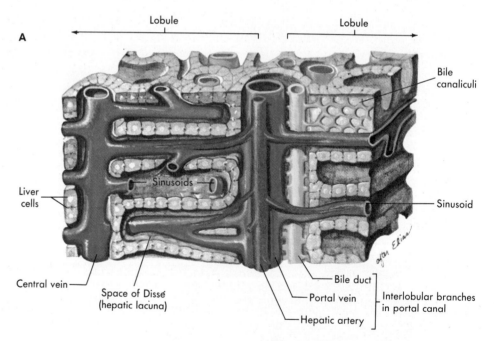

Fig. 13-17
Microscopic structure of liver. **A,** Relationship of blood vessels and bile ducts to liver cells. (After Elias.) **B,** Low-power micrograph of liver lobule. **C,** Portal canal area between lobules, showing vessels and ducts.

ous, with the fluid being stored in the gallbladder until required in the digestive process. The bile has two main jobs: to *emulsify* fats in the intestine, that is, to coat fat droplets and prevent their coalescence into a mass that would be extremely difficult to digest; and to exert a *hydrotropic action*, in which bile salts combine with lipids to form more water-soluble compounds that are more rapidly absorbed.

The liver can *"store"* up to 1 L of blood. This volume is transferred to the systemic circulation and can materially increase the circulating blood volume.

Production of blood cells is a *fetal* function of the liver. It may occur in the adult in certain pathological states because the stem cells are still retained within the organ.

Detoxification or rendering harmless materials that could damage the body is a very important function.

The liver **stores** iron, vitamins, and glycogen ("animal starch").

It **synthesizes** a host of plasma proteins.

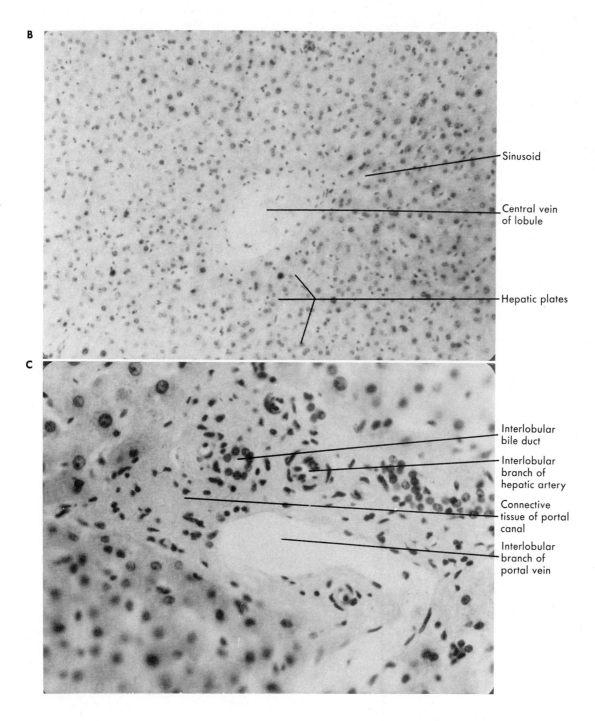

Gallbladder and bile ducts

The gallbladder (see Fig. 13-16) roughly resembles an elongated pear and lies on the inferior surface of the right lobe of the liver. It measures 7 to 10 cm in length by about 2.5 cm in width at its blind end and can contain 1 to 2 ml of bile/kg of body weight. It not only "stores" bile but also concentrates it by water absorption. It is attached to the liver by connective tissue and also by the peritoneum that forms the covering of both the gallbladder and liver.

Microscopically (Fig. 13-18) the gallbladder shows four major layers of tissue as follows:

mucosa layer consists of a lining of simple columnar *epithelium* in which the cells bear microvilli. The epithelium is underlaid by a *lamina propria* in which are found many thin-walled blood vessels, chiefly venous.

muscularis layer is loosely arranged and lies external to the mucosa. The smooth muscle bundles are rather widely separated by collagenous and elastic fibers.

perimuscular connective tissue layer surrounds the muscularis. This layer may contain significant numbers of adipose cells, blood vessels, and an occasional group of embryonic remnants of bile ducts *(Luschka's ducts)*. The latter connect with no other structure but remain in the layer.

serosa layer consists of an outer mesothelial layer underlaid by a submesothelial connective tissue layer and forms the outermost coat of the organ.

From each of the major lobes (right and left) of the liver, a **hepatic duct** collects bile from the liver. These two ducts fuse to form the **common hepatic duct,** which has a length of about 4 cm. From the common hepatic duct the **cystic duct** branches at an acute angle to connect to the gallbladder. At the junction with the gallbladder, the duct is surrounded by a mass of circular smooth muscle that forms the *spiral valve of Heister*. This sphincter controls entry and exit of bile into and out of the gallbladder. From the cystic duct the **common bile duct** carries the bile toward the small intestine and is joined by the *pancreatic duct* near the duodenum. The common bile duct is about 7.5 cm long and is about the diameter of a soda straw. Both ducts empty into the duodenum about 8 cm from the pylorus through an elevation in the wall of the duodenum known as the *papilla of Vater*. The relationships of these ducts are shown in Fig. 13-19. The entry of the ducts into the duodenum is surrounded by *Oddi's sphincter*. Thus control of entry of both bile and pancreatic secretions into the duodenum is

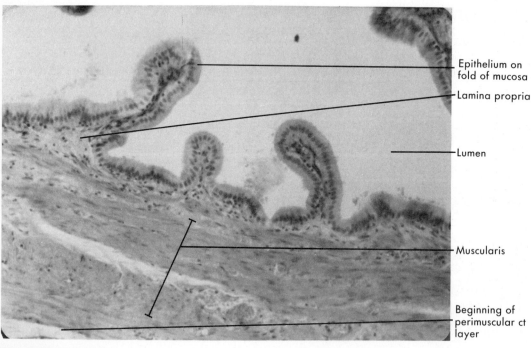

Fig. 13-18
Histology of gallbladder.

afforded. The structure of the common bile duct, which may be taken as typical of all the large extrahepatic ducts, is shown in Fig. 13-20.

A substance designated *cholecystokinin* (CCK) is produced by the duodenal mucosa when fats and acids enter the duodenum from the stomach. It enters the venous drainage of the duodenum and is ultimately distributed to the gallbladder, pancreas, and sphincters of the bile ducts. CCK causes contraction of the gallbladder muscle, relaxes the duct sphincters, and causes the release of enzyme-rich pancreatic secretions. Thus a pathway for bile and pancreatic juice is opened into the duodenum.

Gallstones form when withdrawal of water from the bile causes the solution to become saturated with the bile salts. Precipitation occurs and a stone is formed. Small stones may plug the bile duct, causing obstructive jaundice to occur as bile accumulates in the bloodstream instead of passing to the intestine.

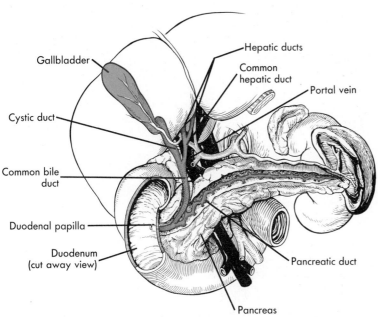

Fig. 13-19
Bile ducts and gallbladder.

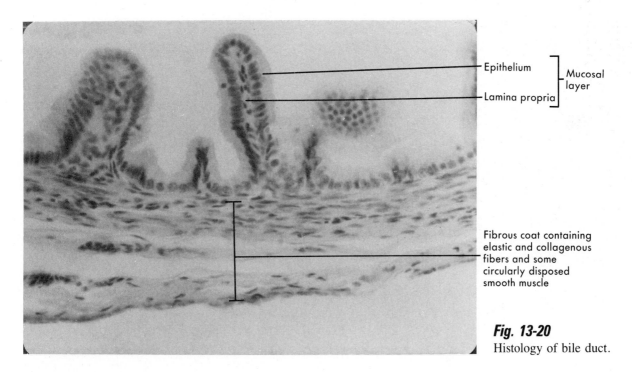

Fig. 13-20
Histology of bile duct.

Pancreas

Gross anatomy

The *pancreas* (Fig. 13-21) is a carrot-shaped organ that lies retroperitoneally between the greater curvature of the stomach and the duodenum. It measures 12.5 to 15 cm in length and weighs 85 to 100 g. A broad *head* faces the curve of the duodenum and a *body* extends beneath the stomach, tapering into a *tail* that points toward the spleen. The gland is divided into *lobes* and *lobules* by connective tissue.

The major duct of the pancreas *(pancreatic duct* or *duct of Wirsung)* collects from interlobar and interlobular ducts supplying the exocrine portion of the pancreas (the part that produces pancreatic juice) and joins with the common bile duct to empty into the duodenum. In most people an *accessory duct (of Santorini)* empties into the duodenum proximal to the pancreatic duct.

Microscopic anatomy (Fig. 13-22)

The exocrine portion of the pancreas consists of a compound alveolar (saccular) gland with grapelike secretory units, known as *acini,* connected to the duct system. The cells composing these acini are called *acinar cells,* and they produce the alkaline and enzyme-rich secretion of the pancreas. Concentrated in the distal body and tail of the pancreas are an estimated 500,000 to 1.5 million masses of endocrine tissue known as the **islets** *(of Langerhans).* The islets are composed mainly of alpha cells (20%) and beta cells (75%) that produce glucagon and insulin respectively. Glucagon stimulates the release of glucose from glycogen, thereby raising the blood glucose level; insulin promotes the formation of glycogen from blood glucose and stimulates cellular uptake of glucose, effects that lower the blood glucose level. These hormones are emptied into the blood vessels of the pancreas and work together to maintain normal blood glucose levels (80 to 120 mg/100 ml of blood).

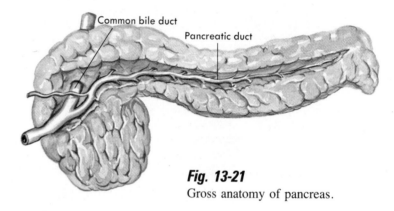

Fig. 13-21
Gross anatomy of pancreas.

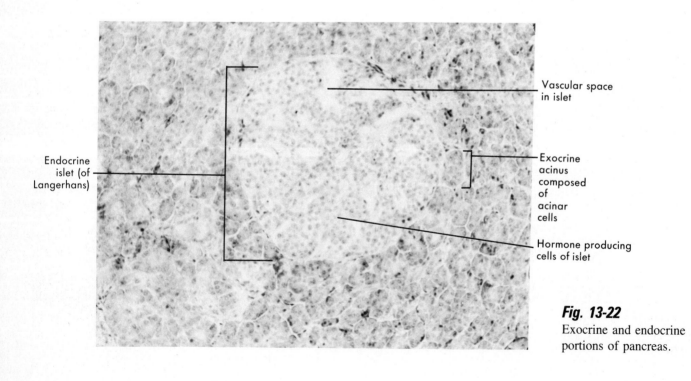

Fig. 13-22
Exocrine and endocrine portions of pancreas.

Digestion in the small intestine

Salivary amylase begins the digestion of starches, and gastric pepsin commences the digestion of proteins. Completion of digestion of foods occurs in the small intestine, and the products are absorbed mainly from this organ. Digestion in the small intestine involves the action of pancreatic juice and intestinal juice.

Pancreatic juice is an alkaline fluid (pH 7.5 to 8.8) that is the product of the exocrine cells of the pancreas. It is about 98% water; is rich in bicarbonate, sodium, and chloride ions; and contains about as much glucose as does the bloodstream. Enzymes present include ones acting on all three major foodstuff groups.

Protein-digesting enzymes are secreted as *proenzymes* or inactive materials that are then converted to active substances.

trypsinogen is converted to *trypsin* when acted on by *enterokinase,* an enzyme produced by duodenal epithelial cells.
chymotrypsinogen is converted to *chymotrypsin* when acted on by trypsin.
carboxypeptidase results when trypsin or enterokinase acts on *procarboxypeptidase.*

Each enzyme will attack specific peptide bonds within the protein, and the result of their combined activity is to create many *free amino acids* and *dipeptides,* units containing two amino acids.

Pancreatic amylase is the same enzyme as the one found in saliva and digests any remaining dextrins to disaccharides. Several of the disaccharides produced are sucrose, lactose, and maltose.

Pancreatic lipase splits fatty acids from their attachment to glycerol in a triglyceride. This activity is aided by the bile.

Nucleases split nucleic acids into their constituent nucleotides.

The control of pancreatic secretion is humoral and requires two substances for full secretion.

Secretin is released by cells of the upper intestine when acid or partially digested proteins enter from the stomach. It is absorbed into the veins of the intestine and is distributed to the pancreas, where it stimulates the release of a buffered watery secretion deficient in digestive enzymes.

Cholecystokinin is produced when partially digested proteins or fatty acids enter the intestine. It also goes into the circulation and eventually reaches the pancreas where it causes release of enzymes by the acinar cells. CCK is also involved in bile release from the gallbladder.

Intestinal juice produced by the cells of the crypts of the small intestine completes carbohydrate and protein digestion. The enzymes that do this job are located mainly in the cell membranes that cover the microvilli of the intestinal epithelial cells.

Final digestion of disaccharides to monosaccharides occurs by specific amylases that work on a particular disaccharide; for example, *lactase* works on lactose and releases glucose and galactose, *maltase* releases two glucose molecules from maltose, and *sucrase* acts on sucrose to release glucose and fructose.

Aminopeptidase and *dipeptidase* reduce any remaining dipeptides to their constituent amino acids.

Control of intestinal secretion is believed to be achieved humorally by a substance called *enterocrinin,* produced when digesting foods or acid reaches the intestinal wall. It follows the pathway described for other chemicals produced by the gut.

Absorption of foods occurs mainly from the small intestine. Monosaccharides, vitamins, salts, and water pass mainly into blood vessels in the first third of the intestine. Larger fats pass into lacteal vessels primarily in the middle third, whereas bile salts are absorbed mainly in the lower third of the organ.

The large intestine has no digestive function but does absorb water (300 to 400 ml/day) and inorganic substances (sodium, 60 mEq/day; bicarbonate, 30 mEq/day; chloride, 18 to 38 mEq/day; potassium, 36 to 66 mEq/day). Flora, or microorganisms, reside in the colon and produce nutritionally significant quantities of vitamins B_1, B_2, B_{12}, and K that are absorbed by the human host.

Blood vessels of the system

Structures of the alimentary tract in the region of the head and neck are supplied by branches of the external carotid artery, and the esophagus is supplied by branches from the aorta. The stomach, all of the small intestine, and the proximal two thirds of the large intestine are supplied by the superior mesenteric artery, and the remaining portion of the alimentary tract receives its blood from the inferior mesenteric artery. The liver is supplied by the hepatic artery, a branch of the celiac artery, and the portal vein. The pancreas is supplied from branches of the splenic artery, the latter derived from the celiac artery.

Selected disorders of the system

The digestive system seems to suffer a wider variety of disorders than any other group of body organs. It may be upset by dietary excesses or poor choice of foods, by psychological stress, and by aging. Some of the more common disorders are described in the following.

Caries refers to tooth decay, believed to be the result of bacterial action on foods in the mouth producing acids that destroy the tooth structure. A streptococcus appears to be a major culprit, and recent research indicates the possibility of effective vaccination against the microorganism.

Gastritis is inflammation of the gastric mucosa, most commonly as a result of consumption of irritant foods or alcohol.

Peptic ulcers result when pepsin erodes the wall of the tract, particularly in the stomach or duodenum. Production of gastric juice when there is no food for it to work on appears to be a major factor in creation of an ulcer. Stress acting through the hypothalamus may be a cause and route by which ulcers are produced.

Food poisoning is most often the result of contamination of food by microorganisms or their toxins. *Botulism* is most commonly the result of poor home-canning methods that fail to destroy the spores of *Clostridium botulinum* or the organism itself. The toxin causes paralysis of skeletal muscle and can cause suffocation by affecting the diaphragm.

Appendicitis is inflammation of the appendix, usually resulting when the lumen of the organ becomes blocked and it cannot clear itself of accumulated fecal material. If the organ ruptures, the entire abdominopelvic cavity may become infected as *peritonitis* occurs.

Tumors of the tract are among the more common cancers in the body. Rectal cancer has a high cure rate if discovered early.

Hepatitis appears to be reaching near-epidemic proportions. It is viral caused, and vaccines being developed offer some hope for its control.

Summary

1. The human digestive system provides for the following:
   a. Food intake
   b. Digestion of foodstuffs
   c. Absorption of end products of the digestive process
   d. Elimination of residues of the digestive process
   e. Protection against harmful materials
2. The digestive system is composed of two general groups of organs.
   a. The alimentary tract is the tubular part of the system beginning with the mouth and ending with the anal canal.
   b. Accessory organs lie inside or outside the alimentary tract and include the tongue, teeth, salivary glands, liver, and pancreas.
3. The mouth receives food and contains the oral cavity.
   a. The lips and cheeks are composed of epithelium, connective tissue, and skeletal muscle.
   b. There are four types of teeth: incisors, canines, premolars and molars. Each has a root, cervix, and crown; is composed of cementum, dentin, and enamel; has a pulp cavity filled with pulp; and is held in its socket by the periodontal membrane and gums.
   c. The tongue is composed of skeletal muscle and has filiform, fungiform, and vallate papillae on its dorsal surface.
   d. The three major salivary glands, parotid, submandibular, and sublingual, lie outside the mouth and ducts empty their secretions into the mouth. They produce saliva that moistens, cleanses, and lubricates the mouth and that initiates starch digestion by the enzyme amylase.
   e. The fauces are the posterior openings of the mouth into the throat.
4. There is a typical tissue plan forming the wall of the alimentary tract.
   a. An inner mucosa consists of an epithelium, lamina propria, and muscularis mucosae and is glandular in most organs of the tract.
   b. The submucosa is a vascular layer external to the mucosa.
   c. The muscularis lies external to the submucosa and is usually two layered.
   d. An adventitial or serosal coat covers the external portion of the tube.
5. The esophagus connects the throat to the stomach and merely conducts food.
6. The abdomen contains the remaining organs of the system.
   a. It is a primarily muscle-delimited cavity that contains apertures for the aorta, vena cava, and esophagus superiorly and for the femoral vessels and (in the male) spermatic cord.
   b. The membranous lining of the cavity is the peritoneum, divided into parietal, mesenteric or omental, and visceral portions.
   c. Mesenteries suspend certain organs in the cavity, distribute blood vessels to them, and form, between certain organs, the omenta.
7. The stomach has the following characteristics:
   a. It is a J-shaped organ.
   b. It has cardiac, fundic, body, and pyloric portions.
   c. It has internal folds called rugae.
   d. It contains gastric glands lined with mucous, zymogenic, oxyntic, and argentaffine cells that collectively produce gastric juice. The juice is mainly water and HCl, and the enzyme pepsin digests proteins to smaller units. Control of secretion is by both nerves and a chemical.
8. The small intestine is about 6 m long.
   a. Its three portions are the duodenum (first 25 to 30 cm), the jejunum (next 2.5 m or so), and the ileum (the rest).
   b. Its wall has villi, plicae, and a typical tissue plan.
9. The large intestine is about 1.5 m long.
   a. There are haustrations through most of its length.
   b. It has a cecum, from which arises the appendix and ascending, transverse, descending, and sigmoid colons.
   c. It terminates in the rectum and anal canal.
   d. There is an outer muscular layer in the form of three strips (the taenia coli).
10. The liver is the largest gland of the body and carries out many important metabolic activities.
    a. It has two major lobes (right and left) and two minor lobes (caudate and quadrate).

b. Several ligaments (falciform, coronary, triangular, round) attach it to the diaphragm and abdominal walls.
c. It receives blood from the hepatic artery and the portal vein.
d. Lobules are its structural and functional units. Lobules are composed of hepatic cells and blood spaces that contain macrophages.
e. It produces bile, stores blood, produces red cells fetally, detoxifies materials, stores nutrients, and synthesizes many materials.
11. The gallbladder is characterized by the following:
    a. It stores bile.
    b. It is associated with a system of hepatic, cystic, and bile ducts.
12. The pancreas has the following characteristics:
    a. It is an exocrine organ producing digestive enzymes. The acini form this part of the organ and are ducted to the duodenum.
    b. It is an endocrine organ that has islets producing blood sugar level–raising (glucagon) and lowering (insulin) hormones.
13. Digestion in the small intestine requires pancreatic and intestinal juices and the bile.
    a. Pancreatic juice contains several proteinases that carry proteins to dipeptide stage, an amylase that produces disaccharides, and a lipase to digest fats. Its production is controlled by two chemicals, secretin and CCK.
    b. Intestinal juice completes carbohydrate and protein digestion to simple sugars and amino acids using specific amylases and two proteinases.
14. Several disorders of the system are described, including caries, ulcers, and inflammatory disorders.

Questions

1. Why do we need a digestive system inasmuch as we eat food to supply our bodies' needs?
2. What are the functional reasons for having several types of teeth in our heads? Describe the structure of a tooth.
3. What nondigestive functions does the tongue have? What accounts for its versatility?
4. Describe the structure of the wall of the alimentary tract, and correlate it with production of digestive juices and the motility the tract exhibits.
5. How does musculature in the esophagus differ from that in the stomach and intestines? What functional correlations are evident here?
6. Describe the gross and microscopic anatomy of the stomach and explain the source of the enzyme and HCl the stomach produces.
7. Describe the boundaries of the abdomen and explain what parietal and visceral peritoneum are.
8. What is a mesentery? What are its functions?
9. What are some of the features possessed by the small intestine that increase its surface area?
10. Describe differences between the small and large intestines at both gross and microscopic levels.
11. How does the colon differ from the large intestine?
12. Describe gross and microscopic anatomy of the liver.
13. Describe the intra- and extrahepatic bile duct systems.
14. What is meant when it is said that the liver has a "double blood supply"?
15. Describe the structure and functions of the pancreas.
16. Compare the process of digestion in the several parts of the system, listing substrates, enzymes, and end products.

Reading

Davenport, H.W.: A digest of digestion, Chicago, 1978, Year Book Medical Publishers, Inc.

14

Urinary system

Objectives
Development of the system
Organs of the system
  Kidneys
  Physiology of the kidney
  Calyces, renal pelvis, and ureters
  Urinary bladder
  Urethra
Disorders of the system
Summary
Questions
Readings

OBJECTIVES After studying this chapter, the reader should be able to:

Describe the "three kidneys" of the human.
List and accurately locate the organs making up the urinary system.
Describe the size, relationships, and gross anatomy of the kidney, particularly the anatomy revealed by a frontal section of the organ.
Diagram and describe the structure of a nephron, the ultrastructure of the Bowman's capsule, and the nature of the cellular linings of the tubular portions of the nephron.
Relate the nephron to the blood supply of the kidney, naming the vessels of the organ.
Discuss how the kidney forms urine and regulates blood composition.
Describe the structure, gross and microscopic, of the renal calyces, pelvis, ureter, bladder, and urethra and indicate any differences in the structure between male and female.

Cellular metabolism produces a great variety of substances, some of which are toxic threats to the body economy. Intake of foods and fluids provides inorganic substances, water, and other substances in excess of body requirements. It is the primary task of the urinary system, particularly the kidneys, to regulate the blood concentrations of a wide variety of materials and to excrete the solute wastes of metabolism and thus preserve the homeostasis of the body fluids.

Development of the system

First evidence of formation of the urinary system is seen at about $3\frac{1}{2}$ weeks of embryonic development (Fig. 14-1). A **urogenital ridge** forms on either side of the vertebral column in the abdominal cavity. By 4 weeks a series of tubules and a duct have developed in the ridge. This first series of tubules is the **pronephros** or "forekidney." The tubules remain functionless as excretory organs and degenerate. The duct that the tubules connected to, the **pronephric duct,** does not degenerate and remains to serve a second set of tubules that develop posterior to the pronephros. This second or "midkidney" is called the **mesonephros,** and the duct to which *these* tubules connect is referred to as the **mesonephric duct.** The tubules of the mesonephros develop cuplike indentations on their ends that become associated with blood vessels in the urogenital ridge, establishing the connection that will enable the kidney to control the blood composition. Degeneration of mesonephric tubules occurs after about 5 weeks of development, and a **metanephros** or "hindkidney" forms posterior to the mesonephros. It forms in the same manner as the mesonephros and remains as the functional organ of the human. As the tubules of the metanephros are developing, an outgrowth of the mesonephric duct is proceeding toward the developing kidney. This outgrowth, the ureteric bud, will ultimately form the ureters, renal pelvis and calyces, and the terminal portions (collecting tubules) of the nephrons themselves. The bladder is formed as the cloaca is divided into a urogenital sinus and a rectal portion.

Congenital anomalies of the kidney are relatively common (Table 14-1).

Table 14-1. Some common congenital anomalies of the kidney

Anomaly	Frequency	Nature of disorder	Comments
Bilateral agenesis	1/5000 autopsies	Both kidneys fail to develop	Fetus makes it to term using placenta to excrete its wastes; may die of uremic poisoning after birth.
Unilateral agenesis	1/1200 autopsies	One kidney fails to develop	One kidney, if normal, will sustain life
Aberrant or extra blood vessels	1/105 autopsies	More than normal number of vessels to kidney	No threat to life
Horseshoe kidney	1/200 individuals	Lower poles of kidneys are fused across midline	Asymptomatic; no threat to life
Bifid pelvis	10% of individuals	Two pelves on a single kidney	Asymptomatic; no threat to life

Adapted from McClintic, J.R.: Basic anatomy and physiology of the human body, ed. 2. Copyright © 1980. Reprinted by permission of John Wiley & Sons, Inc.

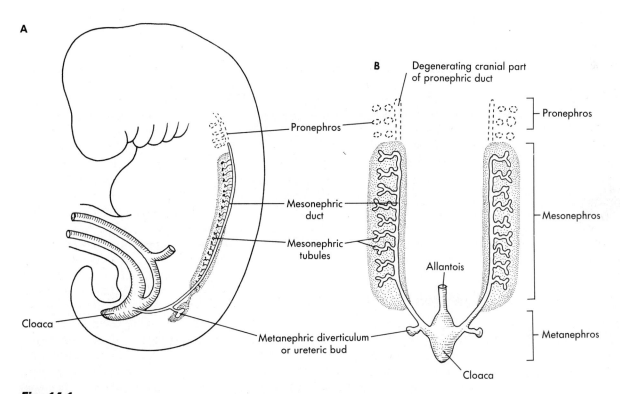

Fig. 14-1
Development of urinary system. **A,** Lateral view. **B,** Ventral view.

Organs of the system

The urinary system includes paired *kidneys* and *ureters*, a single *urinary bladder*, and the single *urethra* conducting urine to the exterior of the body.

Kidneys

Size, location, and attachments (Fig. 14-2). The living human kidneys are dark red bean-shaped organs that are located to either side of the twelfth thoracic to third lumbar vertebral bodies. They lack a mesentery and are thus said to be *retroperitoneal*, lying behind the parietal peritoneum. An *adipose capsule (perirenal fat)* surrounds each organ and aids in fixing them in place; the *renal fascia*, divided into anterior and posterior layers, surrounds the adipose capsule and is more instrumental in affixing the kidneys in place. Each kidney measures about 11 cm × 5 to 7.5 cm wide × 2.5 cm thick and weighs about 140 g, on the average.

Gross anatomy. Each kidney is surrounded by a fibrous *capsule* separate from the surrounding fat and fascia. It may be easily removed, but in so doing numerous small blood vessels are torn. A medially directed concavity is the *hilus;* the hilus leads to a hollow known as the *renal sinus* that pene-

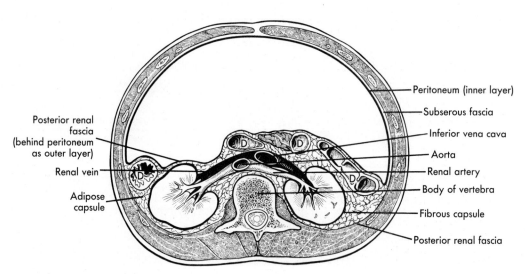

Fig. 14-2
Relationships of kidney to abdominal wall. *D,* Portions of duodenum that are retroperitoneal.

trates about halfway into the kidney. The sinus carries parts of the ducts of the kidney, blood vessels, and connective tissue and is not easily appreciated on a whole kidney. A frontal section of the organ (Fig. 14-3) reveals the upper funnel-shaped end of the ureter, the *renal pelvis,* occupying most of the renal sinus and its primary and secondary branches, the *major* and *minor calyces.* Each minor calyx terminates around an apex of the 8 to 18 *medullary pyramids* of each kidney. The pyramids form a discontinuous medulla or

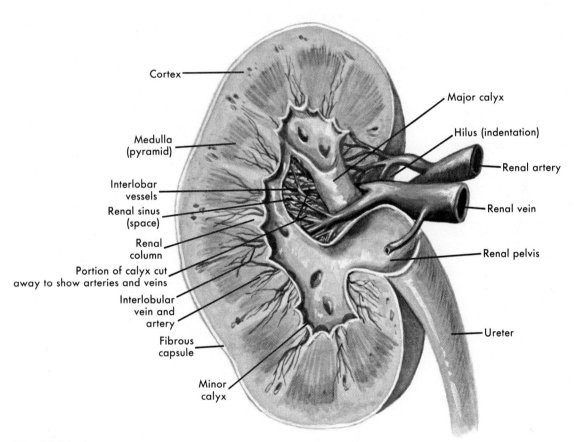

Fig. 14-3
Frontal section of kidney.

inner portion of the kidney. The *cortex* arches around the bases of the pyramids and lies between the pyramids as the *renal columns*. Cortical and medullary portions together form the *parenchyma* or cellular portion of the kidney. The parenchyma is in turn formed by the parts of 1 to 1.5 million nephrons or functional units in each kidney.

Microscopic anatomy. The *nephrons* (Fig. 14-4) are the microscopic and mostly tubular structural and functional units of the kidney. Two types of nephrons called *cortical* and *juxtamedullary* nephrons are recognized as occurring in the kidney. The former lie entirely within the cortex of the kidney and have short loops; the latter have long loops extending into the medullary portion of the kidney.

A nephron begins with a double-walled cup, the *glomerular (Bowman's) capsule,* which surrounds a capillary tuft, the *glomerulus.* The inner wall of this capsule lies in intimate contact with the capillary walls, is known as the *visceral layer* of the capsule, and is composed of unique cells known as *podocytes* (Fig. 14-5). The podocytes have fingerlike processes that interdigitate to form filtration slits *(slit pores)* around the capillaries. A space, *Bowman's space,* lies between the visceral and outer or *parietal layer* of the capsule. The latter is composed of a simple squamous epithelium. The capsule and its contained capillary network is known as a *renal corpuscle.* Renal corpuscles are 150 to 250 μm in diameter.

A *proximal convoluted tubule (PCT)* attaches to each glomerular capsule. These tubules are about 14 mm long, with a diameter of 60 μm, and they form the bulk of the renal cortex. The epithelium of the PCT is simple cuboidal; the cells appear thus in cross-sections of the tubule but are actually shaped like a shoe box with the long axis of the cell parallel to the length of the tubule. Nuclei are located in different parts of the cells, and thus in cross-section some nuclei may not be cut and some cells may appear to lack nuclei. The inner margins of the epithelial cells carry large microvilli forming a *brush border.* The interior of the PCT thus appears fuzzy, and in certain kidneys (for example, kitten) the microvilli are large enough to be seen in the light microscope.

The *loop of Henle* is the next segment of the nephron. It is shaped like a hairpin with the closed end of the pin directed toward the center of the kidney. The *descending limb* of the loop carries fluid centripetally, and the *ascending limb* carries fluid centrifugally.

The descending limb contains a narrowed portion about 15 μm in diameter lined with a simple squamous epithelium whose nuclei bulge into the tubular lumen; this part constitutes the *thin segment* of the loop, and it may continue into the ascending limb. The ascending limb becomes wider as the *thick segment* of the loop, assuming a diameter of about 35 μm. Sections of this part show tall squamous (short cuboidal?) cells with slightly compressed nuclei.

The *distal convoluted tubule (DCT)* lies in the area occupied by the PCTs, is about 5 mm long, and has a diameter of 20 to 50 μm. The cells of the tubule are all nucleated, form a simple cuboidal epithelium, and have no brush border.

Collecting tubules connect to the DCTs and join with similar tubules from other nephrons to form progressively larger tubes that course toward the apexes of the medullary pyramids. Collecting tubules are 20 to 22 mm in length × about 50 μm in diameter. Larger tubes are formed as collecting tubules come together in the apexes of the pyramids and are called *papillary ducts;* they measure 100 to 200 μm in diameter and are lined by a simple columnar epithelium.

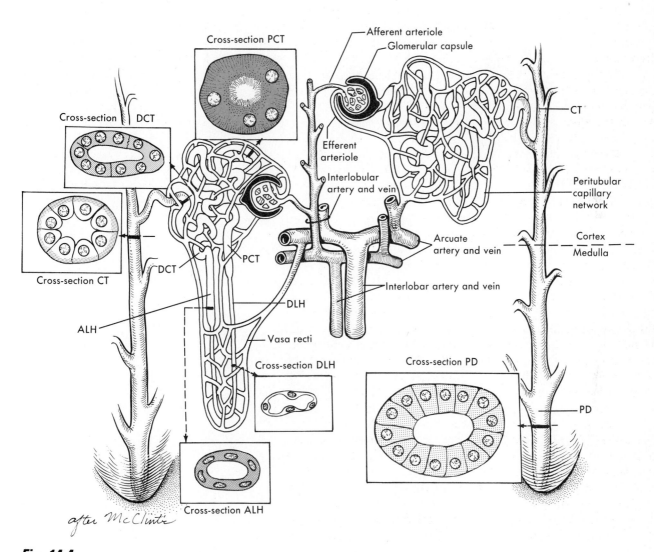

Fig. 14-4
Two nephrons of kidney and their associated blood vessels. Sections of different parts of nephrons depict their cellular structure. *ALH*, Ascending limb of Henle's loop; *CT*, collecting tubule; *DCT*, distal convoluted tubule; *DLH*, descending limb of Henle's loop; *PCT*, proximal convoluted tubule; *PD*, papillary duct.

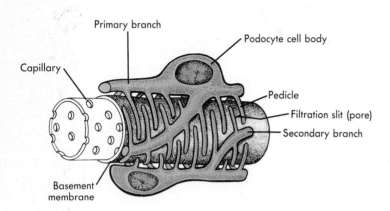

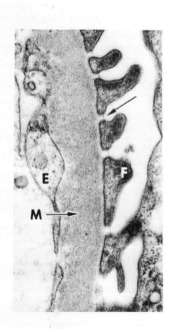
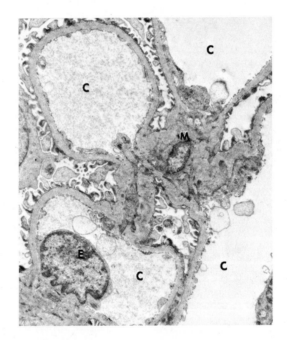

Fig. 14-5
Ultrastructure of renal corpuscle. *E*, Endothelial cell; *F*, foot processes of podocyte; *arrowed M*, basement membrane; *C*, capillaries; *M*, mesangial cell that "cleans" filtration slits (indicated by unlabeled arrow).

Blood supply. The usual blood supply (Fig. 14-6) consists of a single *renal artery* branching to each kidney from the aorta. At or just inside the renal sinus, the artery branches to form a dozen or so *interlobar arteries* that penetrate the parenchyma of the kidney by passing between the pyramids. On reaching the bases of the pyramids, the interlobar vessels branch to form *arcuate arteries* that curve around the pyramid bases. The arterial branches of a given interlobar artery *do not* connect with those of any other interlobar vessel; thus high blood pressure is ensured to the glomeruli. From the arcuate arteries, radially directed *interlobular arteries* arise. Many *afferent arterioles* arise as side branches from these arteries to form the capillary networks of the *glomerulus*. An *efferent arteriole* exits from each glomerulus and forms *pericapillary networks* around the nephron tubules, or a looped *vasa recti* that follows the long loops of Henle.

To carry the blood out of the kidney, a system of veins begins with *interlobular veins* that course along the arteries of the same names. *Arcuate veins*, which *do* connect with one another, receive the interlobular veins or directly receive the vasa recti veins. *Interlobar veins* gather the arcuate vessels and coalesce to form (usually) a single *renal vein* from each kidney.

Physiology of the kidney

The initial step in urine formation by the nephrons is *glomerular filtration*. Blood, with its wastes and "good" materials, arrives at the glomeruli at a pressure of 60 to 75 mm Hg. The glomerular capillaries and their associated podocytes act as a coarse sieve, permitting pressure-caused passage of materials from blood vessel to Bowman's space according to size or molecular weight. The slit pores formed between podocyte processes have a width of 20 to 30 μm and permit free passage of any substance having a molecular weight (MW) of 10,000 or less. Beyond 10,000 MW, substances pass with increasing difficulty, and an absolute upper limit appears to occur at about 200,000 MW. Nearly all materials in the plasma except formed elements are capable of being filtered through the glomerular wall. Glucose, amino acids, vitamins and minerals, water, and wastes are all small molecules and filter easily; about 30 g of protein also passes—chiefly albumins, the smallest (70,000 MW) of the plasma proteins.

The force causing filtration is the blood pressure, represented as P_B. The larger amount of plasma proteins do not filter and exert an osmotic pull (P_O) on filtered water that *opposes* filtration to the extent of 30 mm Hg. Also opposing filtration are two quantities that provide a resistance (P_R) to movement of fluid into tubes already filled with fluid. These two total some 20 mm Hg of pressure. Thus the original blood pressure is opposed by a total of 50 mm Hg. The *net pressure* actually bringing filtration about is called the **effective filtration pressure** (P_{eff}). It is calculated by:

$$P_{eff} = P_B - (P_O + P_R)$$
$$P_{eff} = (60 \text{ to } 75) - (50)$$
$$= 10 \text{ to } 25 \text{ mm Hg}$$

Filtration will not proceed adequately if a normal P_{eff} is not maintained, and toxic wastes will accumulate in the bloodstream.

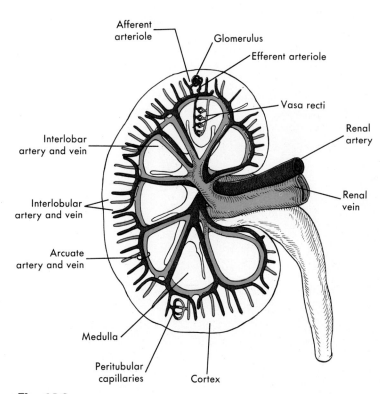

Fig. 14-6
Pattern of blood vessels of kidney.

After filtration has occurred, the fluid formed, called the *filtrate,* enters the proximal convoluted tubule. Here it is subjected to **tubular transport,** using *active transport systems. Reabsorption* involves the active movement of materials out of the filtrate into tubule cells and eventually into blood vessels that will return the substances to the circulation. *Secretion* implies active movement of materials *into* the filtrate, that is, in a direction opposite to reabsorption. Reabsorption of 80% to 90% of the filtered physiologically important solutes (for example, salts, vitamins, amino acids, and 100% of filtered glucose) occurs in the proximal tubule. As solutes are moved out, water follows osmotically, and it too achieves a reabsorption of 80% to 90%. Some substances secreted into the filtrate in the proximal tubule are organic acids, antibiotics, and organic bases. In general, wastes, because they have no active transport systems, are not reabsorbed (unless their concentration is high enough to result in diffusion through the tubule walls) and pass onward in the filtrate.

After leaving the proximal tubule, the filtrate enters the loop of Henle of the nephron and here, especially in the juxtaglomerular nephrons, the filtrate encounters the **countercurrent multiplier** and **exchanger.** The *purpose* of these mechanisms is to create and maintain a high salt concentration in the fluid around the nephron tubules to enable concentration of the filtrate later on (a water conservation measure). In general and simplified terms the multiplier operates in the following manner.

The cells of the ascending limb of the loop of Henle actively transport chloride and sodium ion out of the tubule, but *water does not follow.* This means that salt accumulates outside the tubule and is not diluted by water movement. Thus the fluid around the tubules is much more concentrated than the fluid inside.

The countercurrent exchanger is a vascular mechanism whereby blood in the vasa recti tends to pick up salt as it follows the descending limb of the loop of Henle and then to lose it as it follows the ascending limb; a recycling of salt thus *keeps* its concentration high in the interstitial fluid.

Fluid entering the distal convoluted tubule is hypotonic; that is, it contains little salt but much water in contrast to the fluid outside the tubules. In the distal tubule some additional reabsorption of nutrients occurs and a *secretion* of hydrogen ion occurs that **acidifies** the filtrate. This secretion is very important in regulating the acid-base balance of the body fluids.

The acidic, hypotonic filtrate next enters the collecting tubule, where it is surrounded by the salt-rich fluid "created" by the multiplier. There is a tendency for water to move by osmosis out of the tubule, but this does not occur unless *antidiuretic hormone* (a product of the hypothalamus) is present. The hormone permits water to pass through collecting tubule membranes, conserving it and causing the filtrate to become two to three times more concentrated than the blood.

In summary we see that the blood is filtered, essential nutrients are reabsorbed, acid-base balance is maintained, and the body is enabled to conserve its water supplies.

Calyces, renal pelvis, and ureters

The urine that is formed by the kidneys is emptied into the minor calyces through the apexes of the medullary pyramids. Fluid next passes into the major calyces and into the renal pelvis (see Fig. 14-3). The ureters drain the pelves. The ureter proper extends from the beginning of the funnel-shaped pelvis some 28 to 34 cm to the urinary bladder. Throughout their course the ureters are retroperitoneal in position.

Microscopically these portions of the system all show a similar structure (Fig. 14-7) as follows:

mucosa consists of a *transitional epithelium* underlaid by a *lamina propria.* In the calyces the number of layers of epithelial cells is small (two to three), and the numbers of layers increase to five or six in the pelvis and ureter. The lamina also becomes thicker through pelvis and ureter.

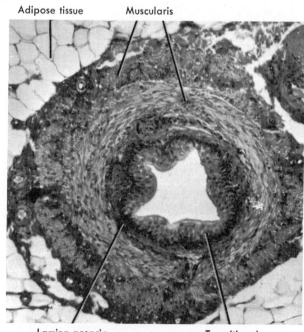

Fig. 14-7
Microscopic appearance of ureter.
From Bevelander, G., and Ramaley, J.A.: Essentials of histology, ed. 8, St. Louis, 1979, The C.V. Mosby Co.

muscularis composed of smooth muscle forms the middle coat. In calyces and pelvis an inner circular and outer longitudinal layer of muscle is present, each layer having a half dozen or so layers of smooth muscle in it. Thickness of these muscle layers increases onto the ureter, where the muscle bundles assume a looser arrangement with much fibrous tissue separating the muscle cells. In the lower third of the ureter, a third, inner, longitudinal layer of muscle is added.

fibrous coat, or *adventitia,* forms the outermost layer on the ureter. It blends with the fibrous tunic of the kidney at its upper end and with the fibrous coat of the bladder at its lower end.

Urinary bladder

The ureters enter the bladder obliquely on its posterolateral aspects. This arrangement acts to effectively close the ureters and prevent reflux of urine backward up the ureter. The muscular contractions of the ureter that move fluid toward the bladder are strong enough to open the junction.

The bladder is located posterior to and a bit higher than the level of the pubic symphysis. When empty, it lies in the true pelvis separated from the symphysis by a **prevesicular space**. In the male the bladder is separated posteriorly from the sigmoid colon by the **rectovesical fascia;** in the female a **uterovesical space** separates the bladder from the uterus. The bladder is held firmly to the pelvic floor by fibrous bands that attach to the base of the bladder only. This allows the upward expansion of the organ as it fills with urine.

Internally (Fig. 14-8), the bladder appears wrinkled when empty, except in the area of the trigone, where it is always smooth. The trigone is an area that would be outlined by an imaginary line connecting the ureteral and urethral orifices with one another.

Microscopically the bladder shows an arrangement of tissues similar to that of the ureter as follows:

mucosa, consisting of *transitional epithelium* and *lamina propria,* lines the internal surface of the organ. It is this layer that folds when the bladder is empty.

submucosa is described by some histologists as connecting the lamina to the muscular coat. Others regard the submucosa as lacking and consider that the deeper layers of the lamina are only looser in nature than the more superficial ones and do not deserve a designation as a separate layer.

muscularis is extremely thick on the bladder and is arranged as on the lower ureter; that is, it consists of inner longitudinal, middle circular, and outer longitudinal layers. Of these, the circular layer is heaviest and is called the *detrusor muscle.* It provides most of the force that empties the bladder.

serosa, composed of peritoneum, is on the superior surface of the bladder. Elsewhere an *adventitia* is present (no peritoneum involved).

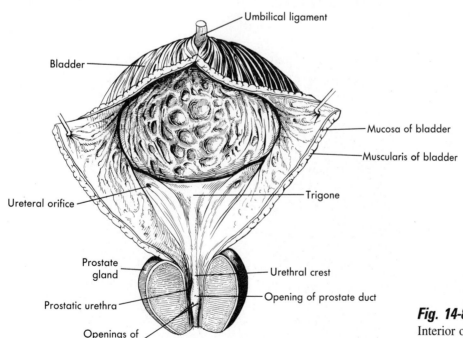

Fig. 14-8
Interior of urinary bladder and some associated structures.

Urethra

The *urethra* conveys urine from the bladder to the exterior. In the male it also serves to transport semen; in the female it is entirely separate from the reproductive organs.

Differences in male and female. In the male (Fig. 14-9) the urethra has a total length of 17.5 to 20 cm and is divided into three parts as follows:

prostatic urethra constitutes the first 3 cm or so below the bladder, the part surrounded by the prostate gland. Posteriorly this part shows a longitudinal fold, the *urethral crest,* perforated on both sides by openings of the ducts of the secretory portions of the prostate gland. The ejaculatory ducts and the *prostatic utricle* open from an elevation that lies about midway along the crest. The utricle is homologous to the uterus and vagina of the female.

membranous urethra passes through the floor of the pelvis (urogenital diaphragm) and is about 2 cm long.

cavernous urethra lies within the penis and is about 15 cm in length.

Microscopically the male urethra shows an inner *epithelium* that is usually transitional in the prostatic portion, pseudostratified in the membranous and most of the cavernous part, becoming stratified squamous in the last centimeter or so. A *lamina propria* affixes the epithelium to the tissues around the urethra (prostate gland, pelvic floor, or cavernous bodies).

The female urethra is a membranous canal about 4 cm long that opens anterior to the vagina. Its *epithelium* is transitional close to the bladder and becomes stratified squamous distally. A thin *lamina* connects the epithelium to spongy tissue and contains many veins that are sometimes designated as *erectile tissue*. A **muscularis,** continuous with the bladder musculature, forms a circular layer for the length of the urethra.

The bladder and urethra receive nerves from both the sympathetic and parasympathetic divisions of the autonomic nervous system. The ureters receive fibers from the sympathetic division only. The ureters are caused to exhibit peristaltic contractions that move urine in squirts to the bladder. The sympathetic fibers to the bladder inhibit contraction of the detrusor muscle and stimulate the sphincter at the bladder neck. Thus urine is allowed to accumulate and be retained within the bladder. The parasympathetic fibers stimulate the detrusor to contract and cause

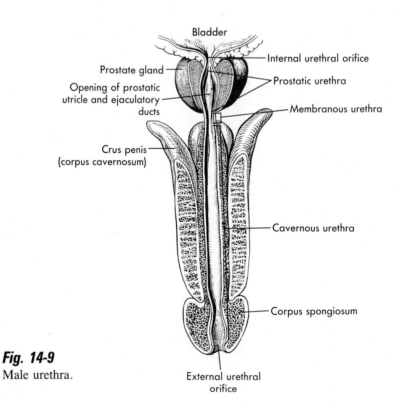

Fig. 14-9
Male urethra.

the sphincter to relax, allowing urine to be emptied from the bladder.

Desire to urinate usually occurs when 200 to 300 ml of urine accumulates in the bladder. There appear to be baroreceptors in the bladder wall that are triggered by bladder distention. Impulses pass over the pelvic nerves to the spinal cord, where motor impulses are sent back to the bladder over parasympathetic fibers. Urine is emptied from the bladder. Voluntary control is afforded by fibers from a brainstem center, so that some degree of control over urination results (after a year or two of age).

The artificial kidney relies on the principle that substances in the bloodstream will diffuse through a semipermeable membrane if a gradient for their diffusion is established. Many meters of dialysing tubing are surrounded by a solution that contains essential substances in the same concentration as in the blood; *zero* concentration of wastes is in the fluid. Net movement from blood to solution of wastes occurs. The fluid is constantly replaced to ensure continued removal of toxins. About 500 ml of blood is in the machine, a loss to which the patient easily adjusts. Blood is generally withdrawn and returned to the radial vessels. The blood is rendered incapable of clotting so that it does not clog the machine. A patient is typically "on the machine" for 8 to 12 hours two or three times a week.

Disorders of the system

Disorders associated with the urinary system include conditions affecting all organs.

Genetic disorders of the kidney usually result in failure to synthesize the enzymes or carriers required for reabsorption of substances. Loss of amino acids or water may result. **Wilms' tumor** is a cancer of embryonic origin that is the most common abdominal tumor of infancy. It usually occurs in one kidney—removal of the kidney leaves enough capacity in the remaining kidney to sustain life.

Inflammatory conditions may affect the entire system. **Pyelonephritis** arises from a blood-borne bacterium or one that has entered the urethra and has infected bladder, ureters, and kidney. **Glomerulonephritis** results from damage to the filtering membrane of the renal corpuscle by toxins from a streptococcus.

Nerve damage to the bladder or its sphincters can lead to the following:

Atonic bladder, the result of interruption of the sensory nerves from the bladder; the organ can become greatly distended with no urge to void.

Hypertonic bladder, the result of interruption of the voluntary pathways from brainstem to lower cord; small distentions result in an uncontrolled desire to urinate.

Automatic bladder, the result of cord damage above S_1; seen in quadriplegic or paraplegic patients, the bladder will fill and empty by reflex alone, and no voluntary control is possible.

Summary

1. The urinary system provides the major route for excretion of solid wastes of metabolism and excess water and inorganic materials.
2. Three separate excretory organs, called the pronephros, mesonephros, and metanephros develop in the human. The ducts of the first two are used by the third.
3. The organs composing the system include paired kidneys and ureters and single bladder and urethra.
4. The kidney has the following characteristics:
   a. It measures about $11 \times 5 \times 2.5$ cm, is retroperitoneal at the level of T12 to L3 vertebrae, and is held in position by adipose capsule and renal fascia.
   b. It has a hilus, renal sinus, pelvis, calyces, medullary pyramids, and cortex with renal columns.
5. The nephrons form the structural and functional units of the kidney. Each unit has the following:
   a. A capsule is composed of a visceral layer formed of podocytes, a space within the capsule, and a parietal layer of simple squamous epithelium.
   b. A proximal convoluted tubule is lined with cuboidal cells bearing microvilli.
   c. A loop of Henle is divided into descending and ascending limbs; each limb may contain simple squamous and simple cuboidal epithelium.
   d. A distal convoluted tubule is lined with simple cuboidal epithelium.
   e. A collecting tubule is lined with simple cuboidal epithelium.
   f. Papillary ducts that collect from many collecting tubules empty into the minor calyces.
6. The blood supply to the kidney is derived from the aorta and consists of the following vessels: renal artery, interlobar arteries, arcuate arteries, interlobular arteries, afferent arterioles, glomeruli, efferent arterioles, peritubular or recti vessels, interlobular veins, arcuate veins, interlobar veins, and a renal vein.

7. The kidney forms urine and regulates the blood composition.
   a. A filtration, depending on blood pressure, removes materials from the bloodstream in the glomerulus, forming the filtrate.
   b. Reabsorption of important solutes and secretion of materials occurs in the proximal tubule.
   c. The loop of Henle actively removes salt but no water from the filtrate.
   d. The distal tubule reabsorbs additional materials and secretes hydrogen ion to acidify the filtrate.
   e. Water passage is permitted from collecting tubules in the presence of antidiuretic hormone to concentrate the filtrate and conserve body water.
8. The calyces, pelvis, and ureter conduct urine to the urinary bladder. They have mucosal, muscular, and fibrous coats, details of which are given in the text.
9. The urinary bladder lies in the pelvic cavity, shows a trigone internally, and also has three coats of tissue in its wall.
10. The urethra carries urine to the exterior and has a different structure in the two sexes.
    a. The male urethra has prostatic, membranous, and cavernous portions and also serves the reproductive system.
    b. The female urethra is undifferentiated and separate from the reproductive system.
11. Innervation of the bladder and urethra is from sympathetic and parasympathetic divisions of the autonomic nervous system. Ureters have only a sympathetic innervation.
    a. Sympathetic fibers stimulate ureter contraction, inhibit bladder muscle, and stimulate bladder sphincters. Urine is retained.
    b. Parasympathetic fibers stimulate bladder muscle and inhibit sphincters. Urine is voided.
12. Disorders of the system include genetic, inflammatory, and nerve damage.
    a. Wilms' tumor is common in infants. It is inherited.
    b. Bacterial infection can cause pyelonephritis and glomerulonephritis.
    c. Nerve damage creates malfunctions in the urine-voiding reflexes.

Questions

1. Describe the size, location, and devices that secure the position of the kidneys.
2. What structures enter the kidney at the hilus?
3. What are the renal pelvis and calyces and what are the medulla, cortex, and renal columns?
4. Sketch a nephron, labeling its parts, and explain what occurs in each section as the kidney forms urine and regulates blood composition.
5. What are podocytes and what is their function in the glomerular capsule?
6. Compare the structure and innervation of the renal pelvis, ureter, and urinary bladder.
7. What are some differences between the location and structure of the male and female urethras?
8. What are some disorders associated with the system?

Readings

Beeuwkes, R., III.: The vascular organization of the kidney, Annu. Rev. Physiol. **42:**531, 1980.

Lassiter, W.E.: Kidney, Annu. Rev. Physiol. **37:**371, 1975.

Vander, A.J.: Renal physiology, ed. 2, New York, 1980, McGraw-Hill Book Co.

15

Reproductive systems

Objectives

Development of the systems

Female reproductive system
 Ovaries
 Uterine tubes
 Uterus
 Vagina
 External genitalia and pelvic floor
 Mammary glands
 Endocrine relationships
 Disorders of the female reproductive system

Male reproductive system
 Scrotum and testes
 Ducts of the testes
 Accessory glands
 Penis and pelvic floor
 Endocrine relationships

Summary

Questions

Readings

OBJECTIVES After studying this chapter, the reader should be able to:

Briefly outline the development of the reproductive systems, with special attention to how the two sexes differ in their formation.

List the organs making up each system.

Describe the structure of the ovary, the stages in development of an ovarian follicle, and the fate of the tissues remaining in the ovary after ovulation.

Describe the extent, structure, and functions of the uterine tube and uterus.

Outline the stages in the menstrual cycle, with special attention to the nature of the endometrium in each stage.

Discuss the structure and functions of the vagina, external female genitalia, and mammary glands.

Describe the structure of the scrotum and testes.

Discuss the stages of spermatogenesis and the function of the support cells of the testis.

Describe the location, characteristics, and histology of the ducts of the male reproductive system.

List and describe the structure of the accessory glands of the male reproductive system.

Describe the structure of the penis.

Show similarities and differences in the endocrine relationships between the pituitary gland and the gonads.

The term *reproduction* implies to most persons the act of mating between the two sexes with consequent production of offspring. The reproductive systems of male and female human beings provide the basis for production of sperm and ova and for delivering these sex cells to appropriate areas for fertilization, development, and nourishment of a new individual.

Development of the systems

The early development of the reproductive systems (Fig. 15-1) is associated with the development of the urinary system. The mesodermal masses developing in the posterior portion of the coelom are known as the urogenital ridges, and parts of the developing kidneys serve as portions of the reproductive systems, particularly in the male.

During early development the reproductive systems pass through similar stages. The gonad (sex gland; ovary or testis) develops at about 5 weeks. At this time a *germinal epithelium* develops on the medial side of the urogenital ridge, which sends cords of epithelial cells, the *primary sex cords,* into the underlying mesenchyme. An outer cor-

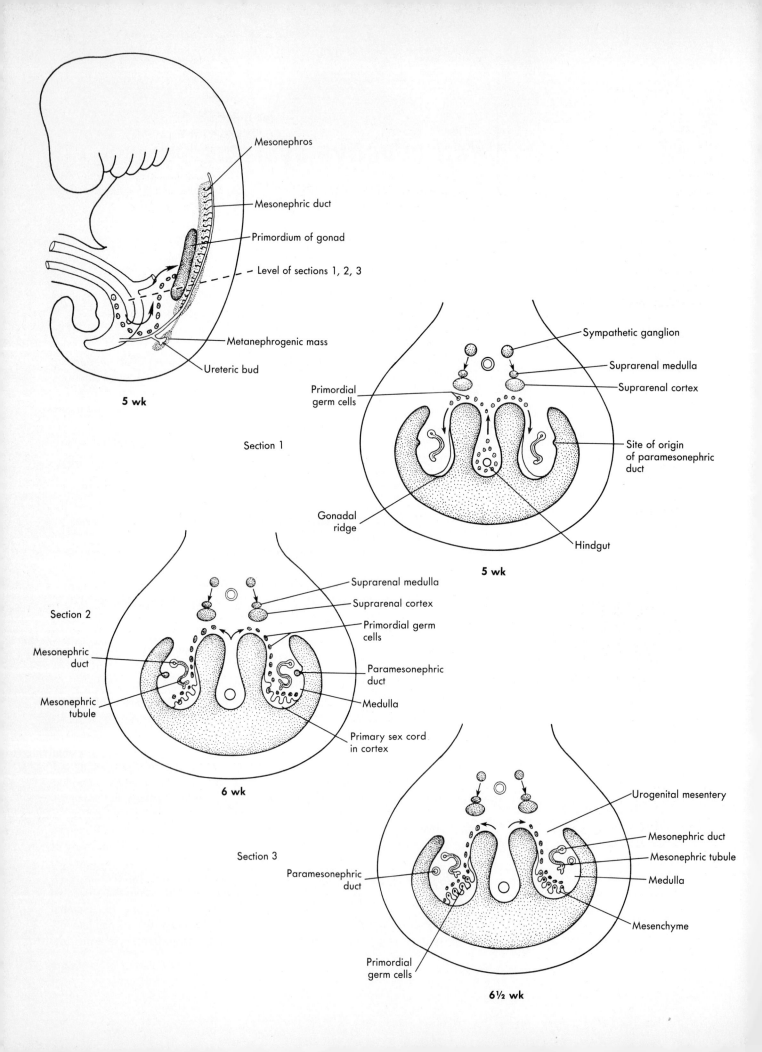

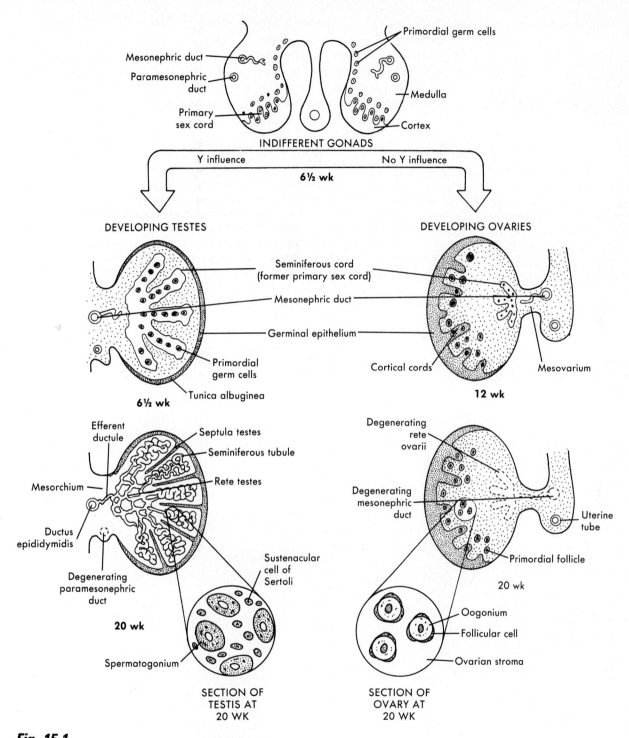

Fig. 15-1
Early development of gonads and their ducts.
After Moore.

tex and inner medulla may be distinguished at this time, and the gonad is said to be in an *indifferent stage,* capable of developing into either a testis or an ovary. In the female, XX chromosome constitution ensures development of the *cortex* into an ovary, whereas in the male, with an XY chromosome constitution, the *medulla* differentiates into the testicular tissue, and the cortex undergoes regression.

At about 7 weeks of fetal development, the primary sex cords of the male branch and form the seminiferous tubules and the straight and rete tubules of the duct system. Mesenchymal cells between the cords give rise to the interstitial (Leydig's) cells that synthesize the testicular hormone (testosterone). The epididymis, efferent ductules, and vas deferens are formed from the mesonephric tubules and duct that remain from the development of the mesonephros. Accessory glands (prostate gland, seminal vesicles, and bulbourethral glands) form as outgrowths of the vas deferens and urethra. Testes descend into the scrotum at about 8 months of fetal development.

Ovaries develop more slowly and can be recognized as such at about 10 weeks. In the female some of the cells of the sex cords undergo mitosis to form some 400,000 primordial follicles before birth. These then undergo meiosis as far as the first meiotic prophase (see Chapter 2). There is no postfetal formation of follicles. The primordial follicles (the most primitive stage) consist of the oögonium (primitive egg or ovum) and a surrounding layer of flattened follicular cells derived from the connective tissue of the ovary. The uterine tubes, uterus, and vagina develop from the paramesonephric *(müllerian)* ducts, tubes that form from the coelomic epithelium on the lateral sides of each mesonephros. Originally paired, these ducts normally fuse in their distal ends to form the single uterus and vagina.

Fig. 15-2
Development of external genitalia. Homologous organs are shown in pattern of hatching.

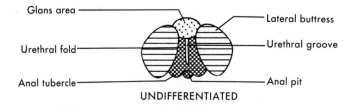
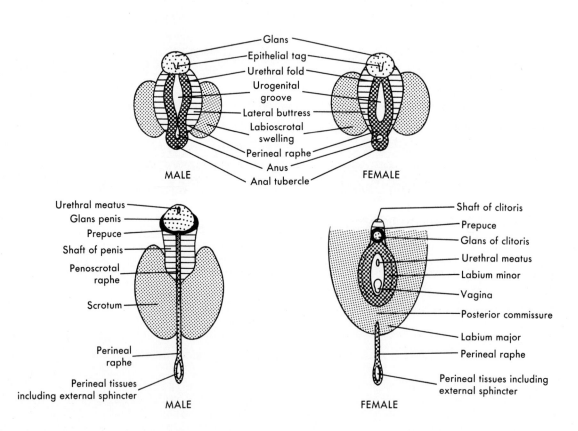

External genitalia (Fig. 15-2) also pass through an indifferent stage at about 4 or 5 weeks. A *genital tubercle* presages development of the scrotum or labia majora; a medial *phallus* will form the penis or clitoris. At about 8 weeks of fetal development, the two sexes can be differentiated by the characteristic appearance of the external genitalia. Table 15-1 summarizes some of the homologies in the male and female reproductive systems.

Table 15-1. Some homologies in the reproductive systems

Name of structure in indifferent stage	Derivative of structure in male	Derivative of structure in female
Gonad	Testis	Ovary
Cranial part of mesonephric duct	Efferent ductules and epididymal head	Epoophoron (a vestigial structure)
Remainder of mesonephric duct	Epididymis, vas deferens	Duct of epoophoron and Gartner's duct (both vestigial structures)
Upper part of müllerian duct	No derivative	Uterus
Middle part of müllerian duct	No derivative	Uterus
Lower part of müllerian duct	No derivative	Vagina
Urogenital sinus	Prostatic urethra	Vestibule
Phallus	Penis	Clitoris
Lips of urogenital groove (urethral fold)	Penoscrotal raphe	Labia minora
Genital tubercles	Scrotum	Labia majora

Congenital malformations of the genital systems (Table 15-2) are based primarily on incomplete development of a given system by sex or simultaneous development of organs of both systems. In either case, both internal and external reproductive organs may be ambiguous or characteristic of neither system.

Table 15-2. Some malformations of the reproductive systems

Condition	Sex chromosome constitution	Frequency	Comments
Gonadal agenesis (Turner's syndrome)	XO*	Found in 15% to 50% of females who do not menstruate	Ovaries fail to develop. There are immature genitalia and skin and skeletal defects. Sterility is present.
Hemaphroditism	Normal for sex (XX or XY)	Very rare	Functioning ovarian *and* testicular tissue are present in one person. Sex organs are abnormal, since both male and female hormones are produced.
Klinefelter's syndrome	XXY	Not known	Atrophic testes are present.
Pseudohermaphroditism	Normal for sex (XX or XY)	1/61,000 live births	Genitalia may resemble those of opposite sex because of failure of sex hormone production.
Cryptorchidism (undescended testes)	XY	4% of live male births show one or both testes not in scrotum	By 1 yr, 80% will show descended testes; rest may require surgery.
Bicornate uterus	XX	Not known	Müllerian ducts do not fuse; a "two-horned" uterus.

*O signifies a missing chromosome.

Female reproductive system

The organs of the female reproductive system (Fig. 15-3) may be grouped under two headings:

Internal organs include the *ovaries*, paired structures that produce ova and two important hormones; the *uterine tubes*, which convey ova to the uterus; the *uterus*, which houses a developing individual; and the *vagina*, which receives the penis and serves as the birth canal.

External organs include the *clitoris*, a homologue of the penis; the *labia majora* and *minora*, constituting, with the clitoris, part of the external genitalia; and the *mammary glands*, modified skin glands that are considered here because of their close relationship to the hormones of the reproductive system and to nourishment of offspring.

Ovaries

The *ovaries* are paired ovoid bodies measuring in the adult about 3 cm long × 2 cm wide × 1 cm deep. They lie lateral to the uterus and are attached to the broad ligament of the uterus by the *mesovarium,* to the uterus by the *ovarian ligament,* and to the posterolateral pelvic wall by the *suspensory ligament.*

Each ovary is covered by a layer of

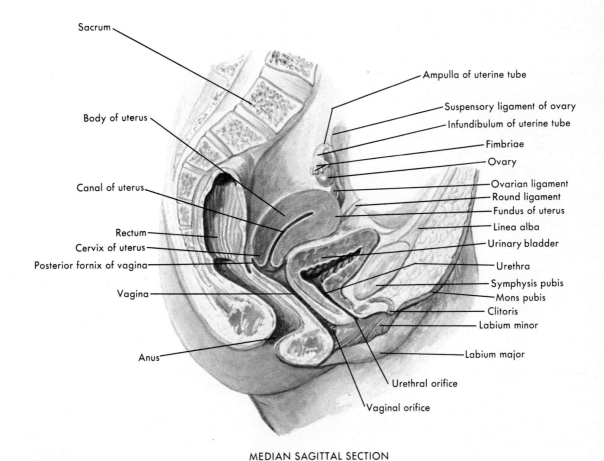

Fig. 15-3
Female reproductive organs and associated structures.

(typically) simple cuboidal epithelium that constitutes its **germinal epithelium.** After birth this epithelium forms no ova, acting only as an epithelial layer. Beneath the epithelium lies a layer of collagenous tissue forming a capsulelike covering for the ovary. This layer is designated the **tunica albuginea.** A connective tissue *stroma* forms the internal framework of the organ and is divided into a central **medulla** *(zona vasculosa)* and an outer **cortex.** *Ovarian follicles* are found in the cortex. These features are depicted in Fig. 15-4.

Ovarian follicles. Ovarian follicles (Fig. 15-5) are estimated to number

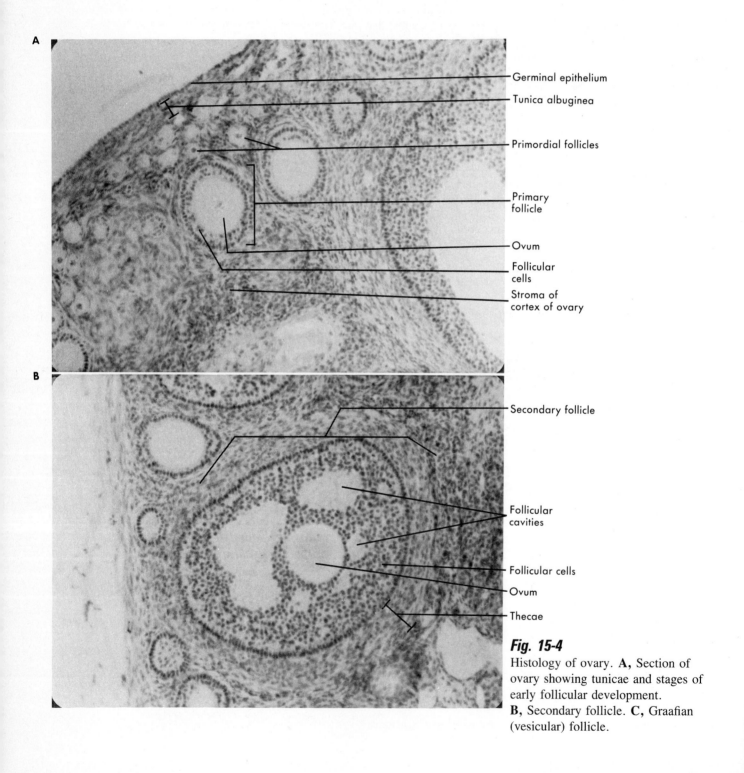

Fig. 15-4

Histology of ovary. **A,** Section of ovary showing tunicae and stages of early follicular development. **B,** Secondary follicle. **C,** Graafian (vesicular) follicle.

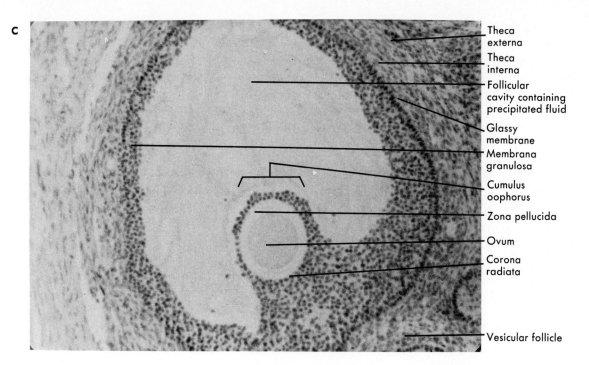

Fig. 15-4, cont'd
For legend see opposite page.

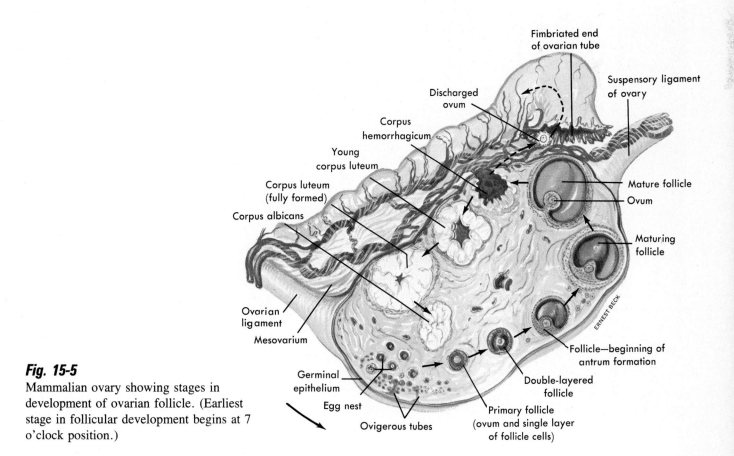

Fig. 15-5
Mammalian ovary showing stages in development of ovarian follicle. (Earliest stage in follicular development begins at 7 o'clock position.)

about 400,000 in both ovaries before birth. They develop to the stage of first meiotic prophase and remain "arrested" in this condition until puberty, when hormonal changes initiate further follicular development and the menstrual cycle. Over the reproductive life of the female, about 400 ova will complete their development and be capable on fertilization of becoming a new individual. The remaining ova will undergo a degenerative process known as *atresia*.

The most primitive stage in follicular development consists of *primordial follicles*. It is in this general stage that follicles remain until puberty. Such follicles are the most numerous at any time in the life of the female and are the only type of follicle present before puberty. They measure 40 to 50 μm in diameter and are found in the peripheral layers of the cortex in groups. They show an ovum with a single layer of flattened stromal cells surrounding the ovum (see Fig. 15-4). Further development of follicles consists of enlargement of the ovum and division of the stromal cells to form several layers of cells around the ovum. Simultaneously an obvious membrane, the *zona pellucida*, forms around the ovum. Follicles showing the features described are known as *primary follicles* and measure about 80 μm in diameter. At this stage the several layers of cells around the ovum are called *follicular cells*. Next, several small cavities develop within the layers of follic-

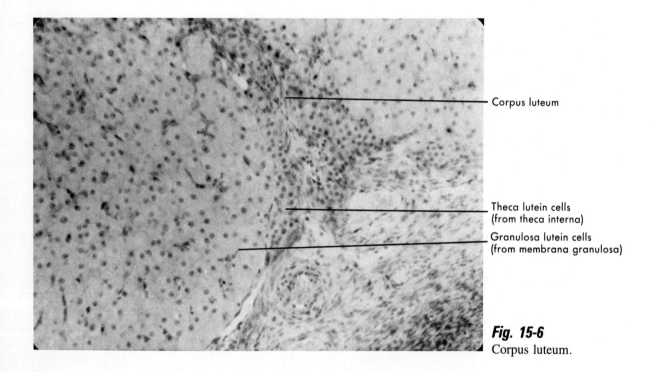

Fig. 15-6
Corpus luteum.

ular cells, the whole unit grows in size, and connective tissue *thecae* develop in two layers around the follicular cells. At this stage the follicle is known as a *secondary follicle.* This follicle continues to grow in size to 10 to 13 mm, the originally several separate cavities fuse to form a single cavity, and the spherical unit is designated as a *vesicular* (graafian) *follicle* (Fig. 15-4).

The vesicular follicle is the source of estradiol, a hormone involved in uterine, mammary, and secondary sex characteristic development.

Ovulation next occurs, releasing the ovum and a surrounding layer of cells (the corona radiata) from the follicle. The tissues remaining in the follicle collapse, bleeding occurs into the collapsed follicle, and a *corpus hemorrhagicum* is formed. Next, cells of the membrana granulosa and theca interna divide and ultimately fill what was formerly the cavity of the vesicular follicle. A structure called the *corpus luteum* develops (Fig. 15-6).

The corpus luteum produces the hormone progesterone that ensures uterine development for nourishment of a fertilized ovum and develops the mammary tissue further.

The luteum will remain functional in the ovary for 3 to 4 months of a pregnancy and will then degenerate. If fertilization and implantation of a zygote do *not* occur, the luteum degenerates in 2 to 3 weeks after ovulation. Replacement of the degenerating luteum by scar tissue results in the formation of a *corpus albicans* (Fig. 15-7).

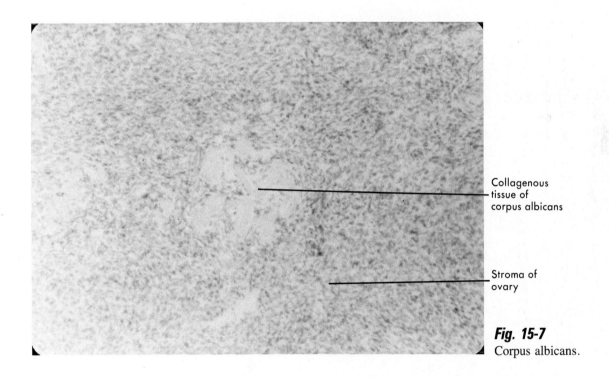

Collagenous tissue of corpus albicans

Stroma of ovary

Fig. 15-7
Corpus albicans.

Uterine tubes

The paired *uterine tubes* (*fallopian tubes, oviducts*) (see Fig. 15-3) extend from the vicinity of each ovary about 7.5 cm to the uterus. They are supported by a mesentery called the *mesosalpinx*. Each tube has an outer expanded end called the *infundibulum*, which in turn bears fingerlike *fimbriae*. The fimbriae terminate close to the ovary but do not attach to it. An *ampulla* extends from the infundibulum to the uterus and passes through the uterine wall as the *isthmus* of the tube. The tubes receive ovulated ova and convey them to the uterus.

Microscopically (Fig. 15-8) the uterine tubes show an inner *mucosa* composed of a simple columnar *epithelium* and an underlying *lamina propria*. The epithelium contains ciliated and secretory cells ("peg" cells) that aid in moving the ovum down the tube and providing it with nourishment. A *muscu-*

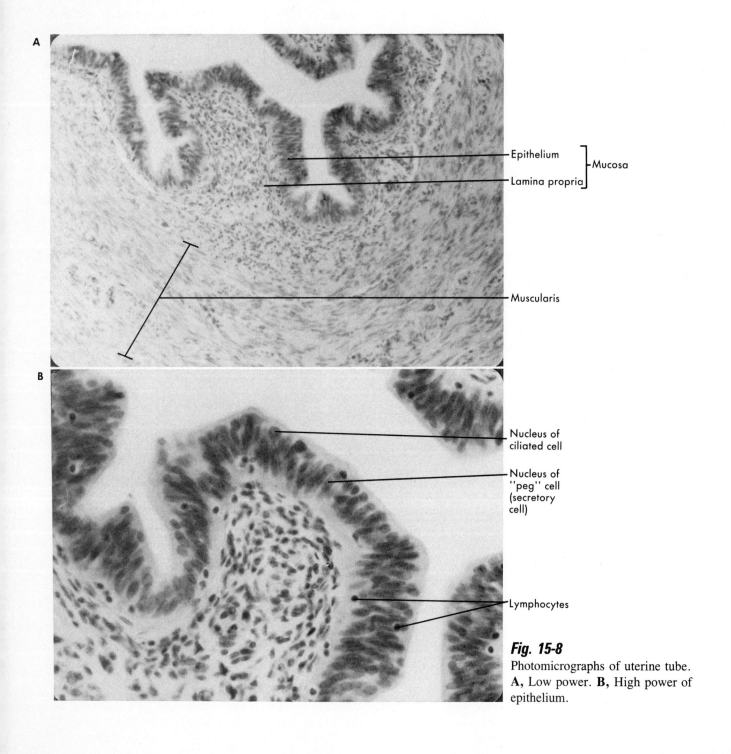

Fig. 15-8
Photomicrographs of uterine tube. **A**, Low power. **B**, High power of epithelium.

laris, consisting of a thick, inner circular layer of smooth muscle and a thin outer longitudinal layer of smooth muscle, forms the middle layer of the organ. An outer *serosa* of mesothelium completes the wall.

Ova remain capable of being fertilized only about 24 hours after ovulation. The journey through the uterine tube requires about 3 days. Thus if fertilization is to occur, it must occur in the distal one third or so of the tube. Fertilized eggs normally implant in the uterus; implantation elsewhere (tube, pelvic cavity) constitutes an *ectopic pregnancy.*

Uterus

The uterus (see Fig. 15-3) in a female who has never been pregnant may be described as a pear-shaped organ measuring about 7.5 cm in length × 5 cm in width × 2.5 cm in thickness. It lies between the urinary bladder anteriorly and the rectum and sigmoid colon posteriorly. It lies at about a 90-degree angle to the vagina. Several ligaments support the uterus:

paired broad ligaments run to the lateral walls of the pelvic cavity.
paired round ligaments pass to the anterior pelvic cavity wall.
uterosacral ligament is a band of fascia that curves along the posterolateral wall of the pelvic cavity to attach to the posterior aspect of the neck of the uterus.
cardinal ligament runs from the vagina-cervix junction to the lateral pelvic walls.

From below the uterus is supported by the pelvic diaphragm, particularly the levator ani muscle (Fig. 15-9).

The uterus is composed of a *fundus,* the portion of the organ lying superior to the entrance of the uterine tubes; the *body,* between the uterine tube entry and the pelvic floor; and the *cervix,* which projects into the superior end of the vagina.

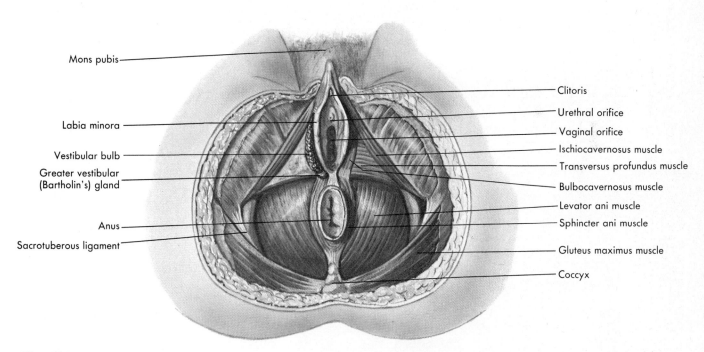

Fig. 15-9
Female perineum.

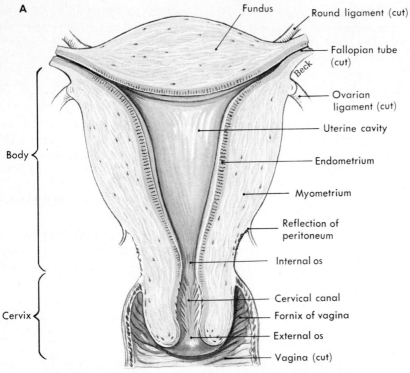

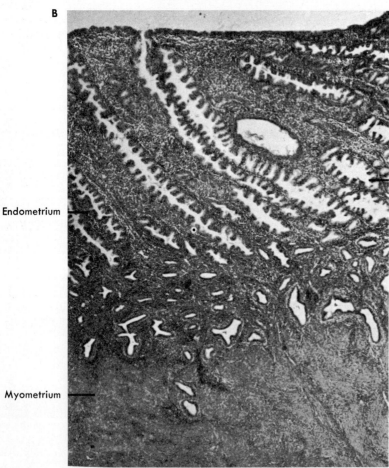

Microscopically (Fig. 15-10) the uterine wall may be seen to be composed of three layers of tissue as follows:

perimetrium is the outermost layer of tissue and is an outer serous coat derived from the peritoneum. It is absent on the cervix.
myometrium is the 15-mm thick muscularis of the organ. It consists of four ill-defined, circular, longitudinal and obliquely oriented smooth muscle layers.
endometrium is the mucosa of the organ and undergoes cyclical changes during the menstrual cycle. A deep-lying *basalis* is not shed during menses and serves as the tissue from which endometrial regrowth occurs. The superficial *functionalis* is shed during menses and consists of a vascular glandular layer of tissue lined with a simple columnar epithelium.

Menstrual cycle. The endometrium presents a different appearance during the different stages of the menstrual cycle because of the effects of

Fig. 15-10
A, Gross and, **B,** microscopic structure of uterus.

From Bevelander, G., and Ramaley, J.A.: Essentials of histology, ed. 8, St. Louis, 1979, The C.V. Mosby Co.

ovarian hormones on it. The cycle is usually divided into four stages.

menstrual stage constitutes the first 3 to 5 days of the cycle and is characterized by external show of blood and tissue. This stage results from deprivation of blood supply to the endometrium and subsequent degeneration and sloughing of the functionalis.

proliferative (follicular) stage occupies the next 7 to 15 days of the cycle and is characterized by a regrowth from the basalis of endometrial tissue. Connective tissue and glandular and vascular structures develop to a thickness of about 2 mm; this regrowth is occasioned by estradiol from the developing follicle.

secretory (luteal or progravid) stage is the next 14 to 16 days of the cycle. Progesterone from the corpus luteum increases the vascularity and glandularity of the endometrium in preparation for possible implantation of a fertilized ovum. A thickness of 4 to 5 mm is reached by the endometrium during this stage.

With no implantation, the luteum begins its degeneration, the vascular elements atrophy, and endometrial degeneration begins. This *premenstrual stage* may be defined only histologically and terminates with external show of blood.

Cycles show individual variation from 24 to 35 days. Ovulation does *not* always occur 14 days after the beginning of menstruation.

Vagina

The *vagina* (see Fig. 15-3) extends from the vestibule, the area of the external surface of the pelvic floor, to the uterus. It usually exists in a collapsed state, with its walls in contact. The uterine cervix projects into the upper end of the vagina, creating moatlike *fornices* (recesses) around the anterior, posterior, and lateral margins of the cervix.

Histologically the vagina shows a *mucosa* consisting of a stratified squamous epithelium and an underlying lamina. The cells of the epithelium are rich in glycogen, and their degeneration releases the carbohydrate to be acted on by the normal vaginal flora to produce an acid environment (about pH 5) in the organ. This pH effectively retards the growth of microorganisms in the vagina (population control). A *muscularis* consists of a thin inner circular and thicker outer longitudinal layer of smooth muscle. An *adventitial coat* forms the outer layer of the organ. It contains many veins and is erectile.

External genitalia and pelvic floor
(Fig. 15-11)

Collectively the external genitalia are designated as the *vulva* or *pudendum.* The individual members of the vulva are as follows:

mons pubis is the fatty eminence anterior to the pubic symphysis. A large portion of the *pubic hair* covers this area starting at puberty.

labia majora are paired, fat-filled, and, after puberty, hair-covered folds of skin. They extend backward from the mons pubis to about 2.5 cm from the anus.

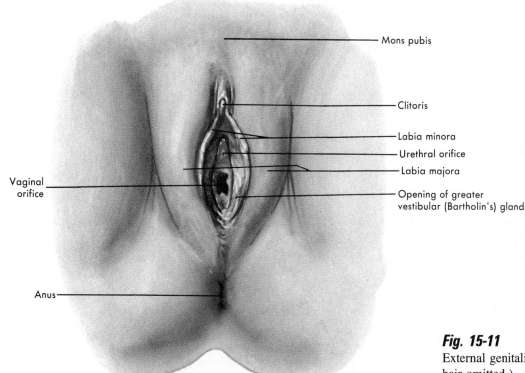

Fig. 15-11
External genitalia of human female (pubic hair omitted.)

labia minora are folds of mucous membrane that enclose the clitoris anteriorly and surround the urethral and vaginal openings. They bear no fat or hair.

clitoris is the female homologue of the male penis and is constructed of two small masses of erectile tissue (cavernous bodies). The clitoris bears a glans and is regarded as the chief source of excitement leading to the female orgasm.

The **vestibule** is the area between the labia minora. The **hymen** is a membranous structure that may nearly or not occlude the vaginal opening. The term **obstetrical perineum** refers to the area of the pelvic floor between the posterior termination of the major labia and the anal opening. It may be cut during childbirth to permit greater ease of passage of the fetal head; this is termed an *episiotomy*.

Emptying into the vestibule are the ducts of the **greater vestibular (Bartholin's) glands,** which provide most of the lubrication for insertion of the penis into the vagina during sexual intercourse.

Note the arrangement and names of the muscles forming the female pelvic floor (see Fig. 15-11).

Mammary glands

The mammary glands (Fig. 15-12) are modified apocrine sweat glands. In infants, children, and men they are present in undeveloped form. Chiefly through the influence of estradiol, they begin their development in the adolescent female. Each gland consists of a comma-shaped mass of fat and collagenous tissue. The "tail" of each gland extends to the anterior axillary fold, and the main mass lies over the pectoralis major muscle on the anterior chest wall. Each gland bears a **nipple** surrounded by an **areola.** The nipple usually lies at the level of the fifth intercostal space. Position of the gland is maintained primarily by the **suspensory (Cooper's) ligaments** of the breast. The ligaments are actually condensations of the connective tissue of the glands and are rather easily stretched, particularly in large-breasted females.

Internally each gland is seen to be composed of about 20 lobules of secretory tissue separated by connective tissue. Three to five **lactiferous ducts** collect the secretion of the glands and empty through the nipple.

Lymph drainage from the glands is mainly toward the axillary lymph nodes but also toward substernal and diaphragmatic nodes. Malignancy in the breast may thus be spread to a rather wide area of the body.

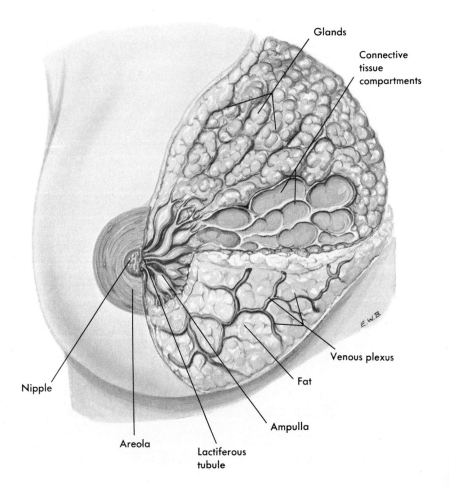

Fig. 15-12
Mammary gland.

Endocrine relationships

Maturation of ovarian follicles is determined by anterior pituitary secretion of *follicle-stimulating hormone (FSH)* beginning at puberty. The developing follicles then secrete estradiol, which initiates the menses and development of female secondary characteristics. *Luteinizing hormone (LH)* is also produced by the anterior pituitary gland and is required for ovulation, corpus luteum formation, and placenta formation in case of implantation. Progesterone secretion by the corpus luteum is determined by LH levels. Thus these *pituitary gonadotropins (FSH and LH)* are absolutely necessary for development of the female system and sexual characteristics. The relationships of these and the ovarian hormones to follicular development and uterine changes is depicted in Fig. 15-13. Attention is directed to the preovulatory "surges" of FSH and LH, which seem to be essential for ovulation to occur.

Disorders of the female reproductive system

From the preceding discussion, it should be apparent that normal ovarian function depends on a pituitary-ovary relationship that involves FSH, LH, estradiol, and progesterone. The same is true for establishment of normal menstrual cycles and development of secondary sex characteristics. Any deviations from a normal pattern of development require investigation not only of the function of the ovary but of that of the pituitary gland as well. The conditions to be described do not include all possible disorders, but they are the most commonly encountered or mentioned in the communications media.

If ovarian function exists at all, the secondary characteristics and genitalia will show some degree of development. Most commonly some disorder of menstrual function is the most obvious sign of a malfunction in the pituitary-ovary axis.

Disorders of ovarian dysfunction are generally grouped into six descriptive categories:

sexual precocity describes onset of menstrual cycles, growth of pubic hair, and breast development before the age of 8 years. Tumors of the ovary involving overproduction of estradiol or excessive FSH/LH secretion by the pituitary are common causes of this condition.

primary amenorrhea is defined as delay of onset of menstrual cycles beyond age 18. Investigation of pituitary gonadotropin (FSH and LH) secretion is warranted.

secondary amenorrhea refers to cessation of menstruation in a woman who has previously menstruated. In the absence of clear proof of pregnancy, a woman's missing three or four cycles indicates tests to determine pituitary and gonadal function.

oligomenorrhea refers to a decreased frequency of menstruation, usually to three or six cycles per year. The condition is common in women who have just commenced cycles and in premenopausal females. At any other time it may indicate anovulatory cycles.

hypomenorrhea is defined as a diminished quantity of menstrual flow. If it occurs only occasionally, it does not usually signify ovarian difficulty. A physical examination may be the only test required.

hypermenorrhea refers to an increase in quantity or duration of menstrual flow. It may indicate excessive hormone production.

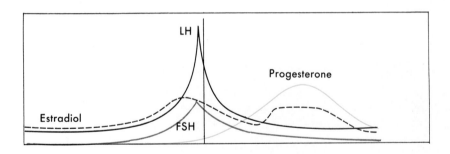

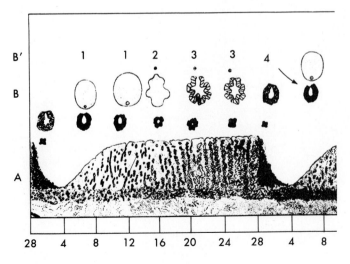

Fig. 15-13
Interrelationships of hormones and ovarian and uterine changes.

Other disorders include the following conditions:

***polycystic ovary syndrome** (Stein-Leventhal syndrome)* results when multiple follicular cysts are found in the ovary. Surgical removal of the cysts usually reverses the syndrome.

endometriosis refers to endometrial tissue that is located in any area outside the uterus. Surgical removal of the tissue is often required to prevent bleeding from the tissue as menstrual cycles occur.

cervical cancer may develop as a result of trauma, infection, or carcinogens that act on the cervix of the uterus. A "Pap test," in which smears of cervical cells are made and examined for abnormalities, has a 95% efficiency in diagnosing the condition.

endometrial cancer is most common in postmenopausal women and is usually treated by uterine removal (hysterectomy).

vaginal cancer has been shown to be more common in daughters of women who have used *d*iethyl*s*tilbesterol (DES), a synthetic hormone used to control menstrual disorders.

obstruction of the uterine tubes results in inability of an ovum to be fertilized and implant in the uterus. Removal of a mature ovum from the ovary, its external fertilization, and insertion into a receptive uterus forms the basis of the so-called test-tube baby. At least one successful pregnancy and the birth of a normal child have resulted from this technique.

cystic disease of the breast is the most common disease of the female breast. The cysts are benign and are usually surgically removed. It and ***fibroepithelial tumors*** are well encapsulated, noninvasive conditions.

breast cancer remains the most common malignant condition among females. The condition is usually not painful and may exist undetected for long periods of time. Metastasis (spread) from a tumor is common. Self-examination of the breast at regular intervals is a most important procedure that can aid in early diagnosis of the condition.

Male reproductive system

The organs of the male reproductive system (Fig. 15-14) may be grouped under three headings:

paired testes (sing., testis), the essential organs of the system, are responsible for the production of *male sex cells* (sperm) and of the male sex hormone *testosterone*. The testes are contained in the *scrotum*.

system of ducts conveys the sperm, and secretions contributed by glands to the exterior. The specific structures here include straight tubules, rete tubules, efferent ductules, epididymis, vas (ductus) deferens, ejaculatory ducts, and urethra.

accessory glands of the system include the paired seminal vesicles, bulbourethral (Cowper's) glands, and the prostate gland.

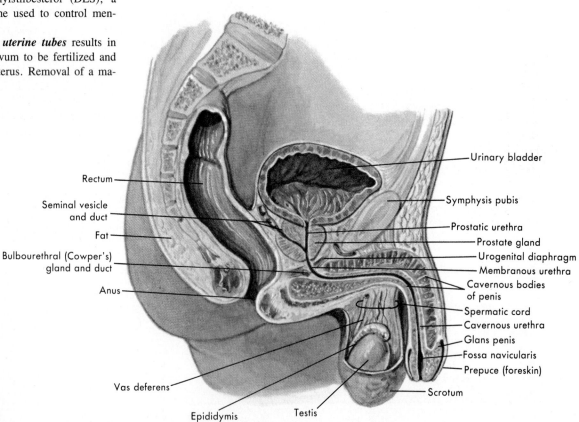

Fig. 15-14
Male reproductive organs and associated structures.

Median sagittal section

Scrotum and testes

The testes develop within the pelvic cavity and descend into a cutaneously derived pouch, the scrotum. The scrotum represents a herniation of the abdominal wall, anterior to the pubic symphysis, which progresses into the skin posterior to the penis. The initial outpouching from the abdomen creates a baglike *processus vaginalis* that enlarges the scrotal cavity itself. The canal through which the "herniation" occurs remains as the **inguinal canal** that conveys the spermatic cord from scrotum to the pelvic cavity. Descent of the testes begins at about 28 weeks of intrauterine life and is completed by about 32 weeks. The steps in this process are presented in Fig. 15-15.

As they pass from pelvis to scrotum, the testes acquire coverings contributed by both the abdominal wall and the scrotum itself.

The *skin* of the scrotum is thin and is provided with numerous sebaceous glands and kinky hairs whose roots are usually visible through the skin.

The **dartos tunic** consists of a thin layer of smooth muscle beneath the skin of the scrotum. It is associated with a fascial layer, and these two components divide the scrotum into lateral compartments, each housing a testis and its associated structures. Contraction of the dartos, as when it is exposed to cold, elevates and wrinkles the scrotum. Also elevating the scrotum and testes are the **cremaster muscles,** one to each testis. These muscles are derived from the obliquus internus abdominis muscles and are composed of skeletal muscle.

Tunics investing the testes are the following:

tunica vaginalis is a serous covering for each testis. It is actually the peritoneal lining of the processus vaginalis that becomes associated with the testis as it descends into the scrotum.

tunica albuginea is a dense layer of collagenous tissue that forms a capsule for the testis. It is greatly thickened on the medial side of the testis to form the ***mediastinum*** of the testis. **Septa** are derived

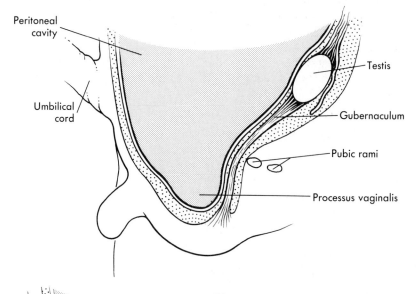

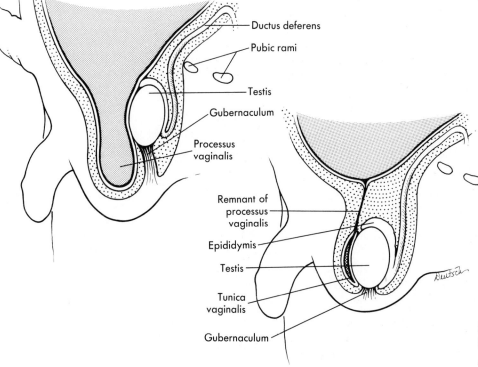

Fig. 15-15
Origin and descent of testis.
After Moore.

from the albuginea portion of the tunic, and these extend into the testis, dividing it into about 250 *lobules*. The latter house the seminiferous tubules and endocrine cells.

vascular tunic (*tunica erythroides*) lies internal to the albuginea and extends for short distances along the septae. The layer represents a *region,* rather than an actual coat, where small blood vessels are aggregated.

Features described are presented in Fig. 15-16.

Seminiferous tubules. Lying within each lobule of the testis are one to four highly convoluted and often joined *seminiferous tubules.* They measure 70 to 80 cm in length and 0.15 to 0.3 mm in diameter. They are lined with a *germinal epithelium* that contains two types of cells.

Cells that can or do undergo meiosis and differentiation into sperm, designated as *spermatogenic cells,* show five developmental stages if the tubule is engaged in sperm production (Fig. 15-17).

spermatogonia lie on the periphery of the epithelium against a basement membrane. They present rounded apexes and rather small dark-staining nuclei and are the only type of cell present in the prepuberal testis. Spermatogonia are diploid cells and undergo mitosis to produce the next stage.

primary spermatocytes are the largest of the sperm-producing cells. They lie internal to the spermatogonia.

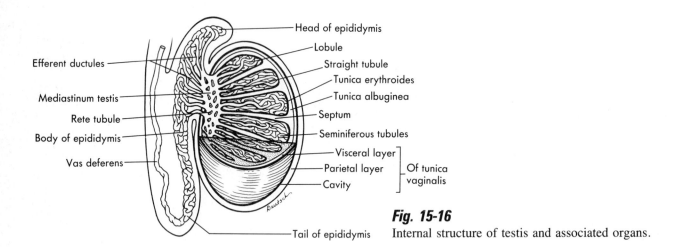

Fig. 15-16
Internal structure of testis and associated organs.

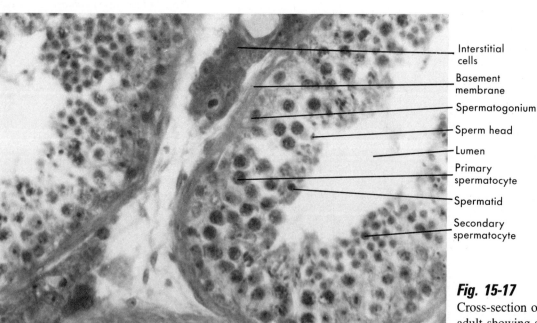

Fig. 15-17
Cross-section of seminiferous tubule in adult showing stages of sperm development.

secondary spermatocytes are haploid cells. The reduction division of meiosis occurs as these cells are formed. They resemble the primary spermatocytes in appearance but are smaller and more centrally placed in the epithelium.

spermatids are derived in the second meiotic division from the secondary spermatocytes. They are quite small and typically have dark-staining round nuclei.

Spermatids metamorphose into ***spermatozoa*** (sperm). Rearrangement of nuclear and cytoplasmic constituents creates a highly specialized cell capable of independent movement (Fig. 15-18).

The other variety of cells in the germinal epithelium is the **supporting cells** (Sertoli's or "sperm mother" cells). They reach from the basement membrane to the tubule lumen and are columnar in shape. Their nuclei are oval and typically show a large round nucleolus (see Fig. 15-17). These cells are secretory and nutritive in function and provide energy for the final metamorphosis of the spermatids.

Lying external to the basement membrane of the seminiferous tubules are a few smooth muscle cells. Contraction of these cells may aid in movement of the sperm out of the tubules. In the spaces between the tubules is connective tissue, and in that tissue are groups of **interstitial cells** (Fig. 15-19). These are the endocrine cells of the testis that produce the hormone **testosterone**. It aids in maturation of the male and

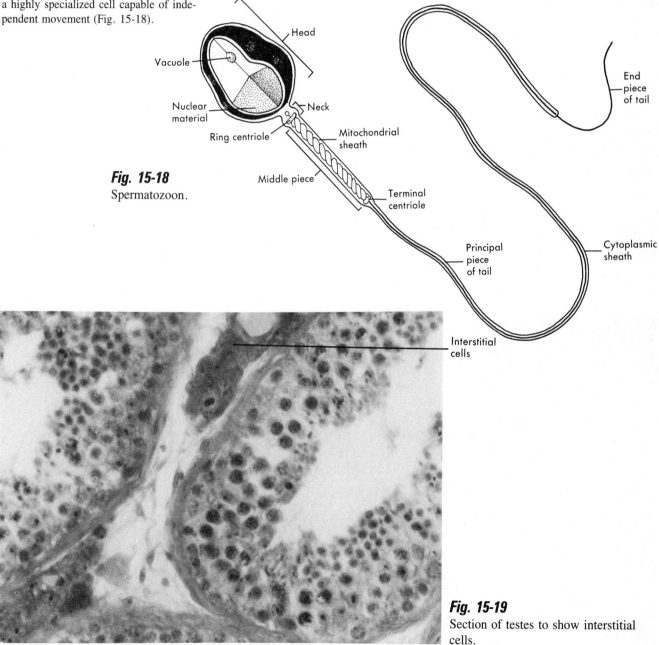

Fig. 15-18
Spermatozoon.

Fig. 15-19
Section of testes to show interstitial cells.

brings about the development of the male's secondary sex characteristics.

Ducts of the testes

The seminiferous tubules connect to *straight tubules* (Fig. 15-20; see Fig. 15-16) that proceed to the mediastinum of the testis. The straight tubules have a regular diameter and a low simple cuboidal lining. A few smooth muscle cells lie around this portion. At the junction with the seminiferous tubule only Sertoli's cells are present.

At the edge of the mediastinum the straight tubules connect to the *rete tubules* (Fig. 15-21; see Fig. 15-16), a series of branching and anastomosing irregular tubules that occupy the mediastinum. These have the same wall structure as do the straight tubules.

Twelve to sixteen *efferent ductules* (Fig. 15-22; see Fig. 15-16) carry sperm out of the testes. These tubules appear to be folded, but closer inspection will show them to be lined by alternating groups of tall and short cells. Several layers of smooth muscle surround the epithelium.

The *epididymis* (see Figs. 15-16 and 15-22) is a tube about 7 m long that lies coiled in the surface of the testis. Its superiorly directed *head (caput)* receives the efferent ductules, there is a central *body (corpus)*, and the vas deferens arises from the inferior *tail (cauda)*. Epithelium here is pseudostratified with the columnar cells bearing microvilli (here called stereocilia). Several layers of smooth muscle surround the tube. Because of its length, sperm take time to pass through the epididymis and while in it are considered to be "in storage."

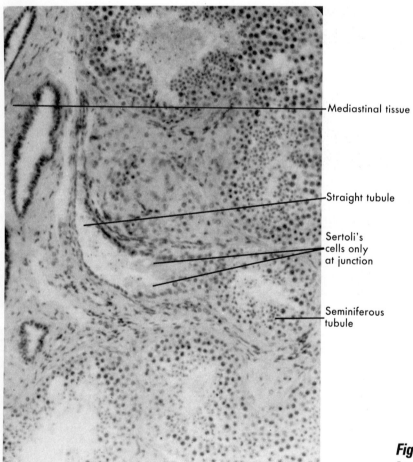

Fig. 15-20
Junction of seminiferous and straight tubule.

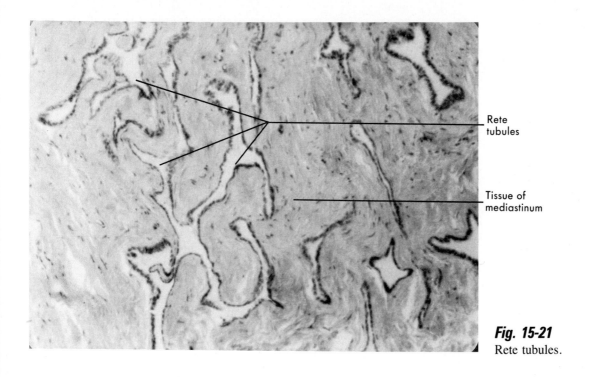

Fig. 15-21
Rete tubules.

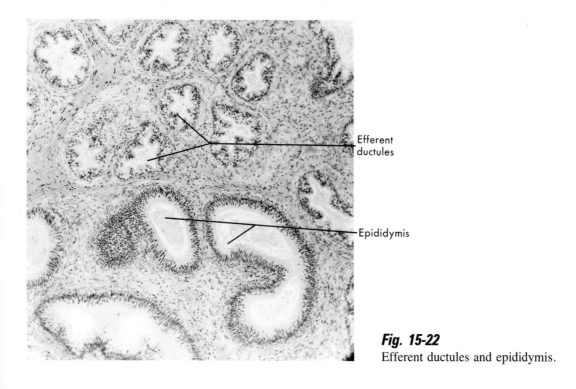

Fig. 15-22
Efferent ductules and epididymis.

The *vas deferens* (Fig. 15-23; see Fig. 15-16) is a muscular tube that passes from the tail of the epididymis through the inguinal canal, arches behind the urinary bladder, and joins the *ejaculatory duct* that empties into the prostatic urethra. A pseudostratified *epithelium* lies on a folded *lamina propria*. Inner longitudinal, thick middle, and outer longitudinal layers of *smooth muscle* propel sperm rapidly through the tube. An outer *adventitial coat* not only serves as a covering for the vas deferens but binds it, together with arteries, veins, nerves, lymphatic vessels, and the cremaster muscle, into the spermatic cord that passes through the inguinal canal.

A *vasectomy* is a procedure to sterilize a male. The vas deferens is usually severed within the scrotum and its ends folded back and tied. This prevents sperm from leaving the scrotum and becoming available for fertilization. Since some sperm may remain in the tract after surgery, two consecutive negative sperm counts done monthly after surgery are usually required as evidence of sterility. Side effects occur in some males, particularly because of products of degenerating spermatogenic cells or sperm that can act as antigens and call forth antibody production. Autoimmunization can occur. The operation can be reversed in about half the patients.

The structure of the **urethra** has been described in the previous chapter.

Accessory glands (Figs. 15-24 and 15-25)

The *seminal vesicles* are paired, hollow, tortuous sacs that lie on the posterior aspect of the urinary bladder. The duct from a vesicle joins with the vas deferens of that side to form the ejaculatory duct. The vesicles do not store sperm, although some may pass into the vesicles as ejaculation occurs. The vesicle secretion is rich in fructose and ascorbic acid and makes up about 60% of the volume of an ejaculate.

The *prostate gland* is about the size of a walnut and surrounds the first portion of the urethra. It consists of 30 to 50 separate secretory units that empty by 16 to 32 separate ducts into the urethra. The *secretory units* are bound to-

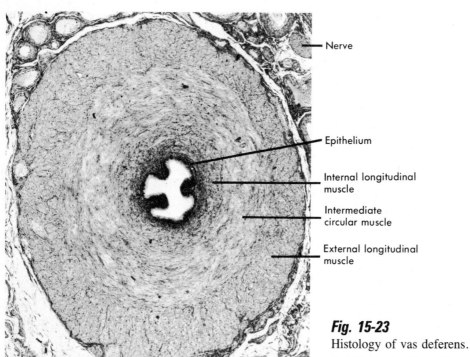

Fig. 15-23
Histology of vas deferens.

From Bevelander, G., and Ramaley, J.A.: Essentials of histology, ed. 8, St. Louis, 1979, The C.V. Mosby Co.

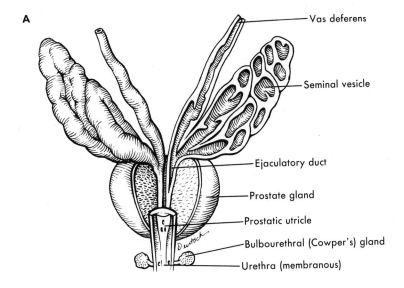

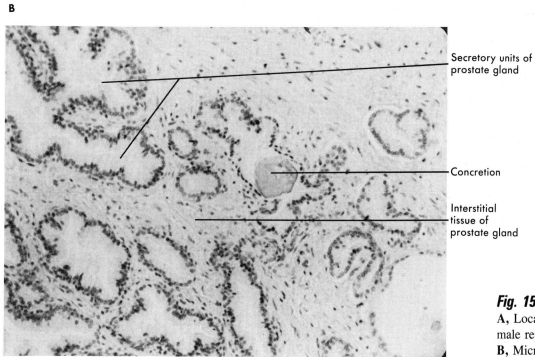

Fig. 15-24
A, Location of accessory glands of male reproductive system.
B, Micrograph of prostate histology.

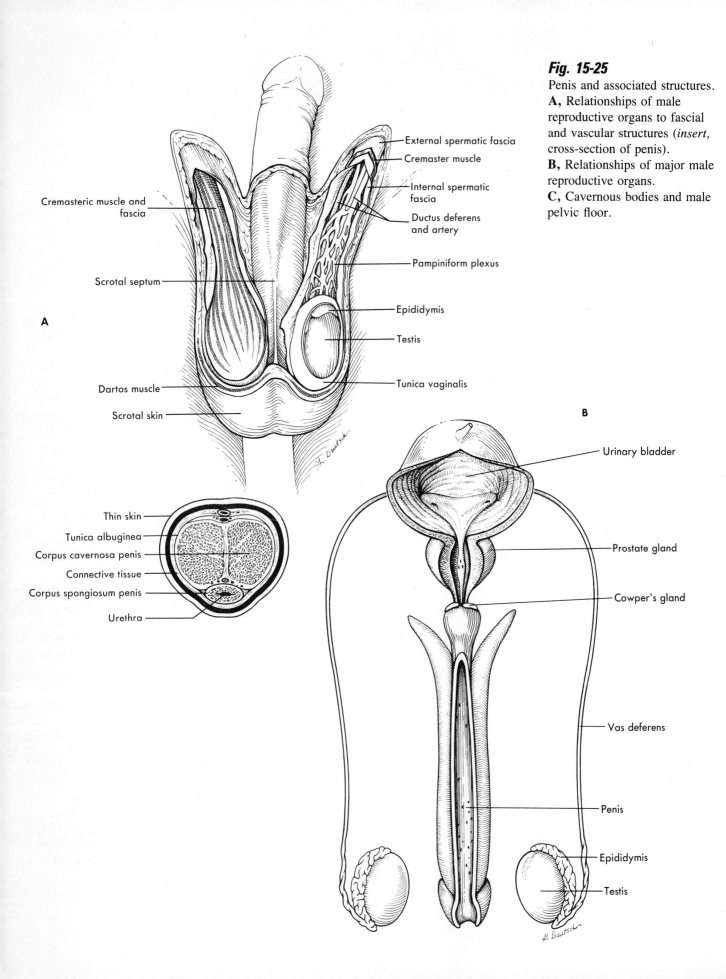

Fig. 15-25
Penis and associated structures. **A**, Relationships of male reproductive organs to fascial and vascular structures (*insert*, cross-section of penis). **B**, Relationships of major male reproductive organs. **C**, Cavernous bodies and male pelvic floor.

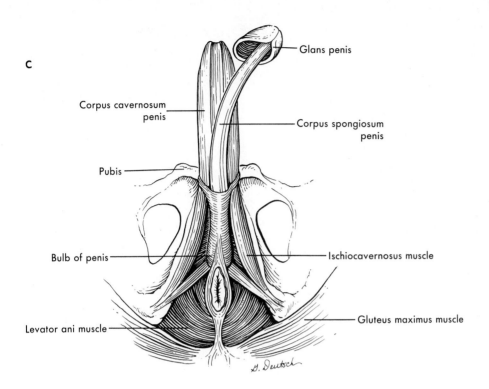

gether by *interstitial tissue* rich in fibrous tissue and smooth muscle, and they may contain lamellar *concretions*. The prostatic secretion is viscous, has a characteristic odor, and is thought to activate the heretofore immobile sperm. It accounts for about 37% of the volume of an ejaculate.

The term **semen** refers to the combined secretions of these glands *and* the sperm (300 to 400 million per normal ejaculate).

The **bulbourethral** *(Cowper's)* **glands** lie in the urogenital diaphragm and empty into the lower membranous or first part of the cavernous urethra. They are the size of a pea and secrete a few drops of alkaline mucus, which aids in neutralization of urine that may remain in the urethra.

Prostatic cancer is common in males over 65 years of age. Difficulty in urination or blood and pus in the urine may indicate enlargement of the gland that may be due to cancer.

Penis and pelvic floor

The ***penis*** (see Fig. 15-25), serves the reproductive system as the copulatory organ to introduce sperm into the vagina of the female. The basic structure of the organ is set by three cylindrical masses of tissue known as the ***cavernous bodies***. Each is surrounded by a *tunica albuginea,* and all are bound together by connective tissue and covered with thin hairless skin. The skin at the head or glans of the penis folds back on itself as the ***prepuce*** (foreskin).

Two of the cavernous bodies are larger and are dorsally placed in the organ. These are the ***corpora cavernosa penis***. The smaller ventrally placed body is the ***corpus spongiosum penis*** (also known as the *corpus cavernosum urethrae*). It carries the cavernous urethra. Its anterior portion is expanded to form the ***glans penis***.

Blood supply to the penis arrives mainly from the pudendal artery (a branch of the internal iliac artery). This vessel forms branches that course through the cavernous bodies and along the dorsal surfaces of the corpora cavernosa penis. These are all muscular arteries with heavy smooth muscle media. These vessels open into the endothelial-lined spaces within the cavernous bodies. Veins lie mostly superficially beneath the skin of the penis and form the external pudendal veins. Erection of the penis is accomplished by loss of tone in the arteries, allowing them to dilate, increasing the filling of the cavernous spaces; this filling also compresses the veins of the organ, retarding blood outflow. Constriction of the arteries allows greater outflow of the blood, and it returns to a flaccid state.

Compare the muscles of the male pelvic floor (see Fig. 15-25) with those of the female.

Endocrine relationships

Spermatogenesis is controlled by ***follicle-stimulating hormone*** *(FSH)* from the anterior pituitary gland. Some evidence exists to suggest that Sertoli's cells produce an as yet unnamed hormone as an additional response to FSH. ***Luteinizing hormone*** *(LH),* also from the anterior pituitary gland, controls the interstitial cell secretion of testosterone.

Summary

1. The reproductive systems first show development at about 5 weeks.
   a. Germinal epithelium and sex cords appear, and the gonad passes through an indifferent stage, where it is neither male nor female in character; it possesses a cortex and medulla.
   b. An XX hormone constitution for the female and an XY constitution for the male cause cortical development and ova production and medullary development and formation of seminiferous tubules.
   c. External genitalia also pass through an indifferent stage at 5 to 8 weeks. After 8 weeks the organs of the two sexes can be differentiated. One sex has organs homologous to the other, although some are suppressed during development.
   d. Congenital malformations are presented and may often be attributed to chromosome disorders.
2. The female reproductive system includes the following:
   a. Internal organs are ovaries, uterine tubes, uterus, and vagina.
   b. External organs are clitoris, two pairs of labia, and, because of its relationship to reproductive organs, the mammary glands.
3. The ovaries produce ova and two important hormones.
   a. They are located in the pelvic cavity and are held in position by the mesovarium, ovarian ligaments, and suspensory ligaments.
   b. Each ovary is covered by a germinal epithelium and a capsulelike tunica albuginea and has a stroma, divided into an inner medulla and outer cortex. Follicles are found in the cortex.
   c. Stages in follicular development include primordial follicle, primary follicle, secondary follicle, and vesicular (graafian) follicle. Characteristics of each are presented.
   d. After ovulation the ovarian follicle becomes a corpus hemorrhagicum, a corpus luteum, and a corpus albicans.
   e. The vesicular follicle produces estradiol as its major hormone; the corpus luteum produces progesterone.
4. The uterine tubes extend from the ovary to the uterus.
   a. The tubes have an infundibulum bearing fimbriae, an ampulla, and an isthmus.
   b. The wall is composed of a mucosa with ciliated and secretory cells, two layers of smooth muscle, and a serosa.
5. The uterus is the organ wherein new individuals undergo development.
   a. It is pear shaped and lies posterior to the urinary bladder.
   b. It is supported by broad, round, uterosacral, and cardinal ligaments.
   c. It has a fundus, body, and cervix.
   d. Its wall is composed of (from the outside inward) a perimetrium, myometrium, and endometrium.
   e. The endometrium undergoes cyclical changes during a menstrual cycle.
   f. The menstrual cycle has four stages: menstrual (shedding), proliferative (regrowth), secretory (increase of vascularity and glandularity), and premenstrual (involution of vessels and degeneration).
6. The vagina is characterized by the following:
   a. It acts as the birth canal and receptacle for the penis.
   b. Fornices lie between the upper end of the vagina and the uterine cervix.
   c. It is three layered with a mucosa, muscularis, and fibrous coat.
7. The vulva (pudendum) is the name given to the female external genitalia.
   a. The mons pubis lies anterior to the pubis.
   b. The labia majora extend from the mons to behind the vaginal opening and lie outside the minora.
   c. The labia minora surround the vaginal orifice.
   d. The clitoris is enclosed by the anterior part of the labia minora and is composed of erectile tissue.
   e. Greater vestibular (Bartholin's) glands open into the vestibule.
8. The mammary glands are modified apocrine sweat glands that produce milk for nourishment of a newborn.
   a. A nipple and areola characterize the external surface.
   b. Internally, lobules of secretory tissue, ducts, and connective tissue compose the gland.
   c. Cooper's ligaments (suspensory ligaments) support the glands.
9. FSH and LH from the anterior pituitary gland control development of ova and secondary characteristics, acting through the follicle and corpus luteum. Several disorders in this system are presented.
10. The male reproductive system includes the following:
    a. The testes are paired organs that produce sperm and the hormone testosterone.
    b. A system of ducts conveying or storing sperm includes straight tubules, rete tubules, efferent ductules, epididymis, vas deferens, ejaculatory ducts, and urethra.
    c. Accessory glands contributing to the semen are seminal vesicles, prostate gland, and bulbourethral (Cowper's) glands.
    d. Testes and parts of the duct system are contained within the scrotum.
11. The scrotum is an evagination of the pelvic cavity covered by skin. The testes descend from the pelvis through the inguinal canal into the scrotum between 28 and 32 weeks. It has two compartments, each housing a testis and its associated structures.
12. The testis has several tunics investing it.
    a. The tunica vaginalis is a mesothelial external coat.
    b. The tunica albuginea is a collagenous capsule for the organ.
    c. The vascular tunic is a layer of blood vessels.
    d. Septa extend from the albuginea, forming lobules; the mediastinum is a thickened medial part of the capsule.
13. Seminiferous tubules lie in the testicular lobules.
    a. Each is lined with germinal epithelium.
    b. In the germinal epithelium are spermatogenic cells and support (Sertoli's) cells.
    c. Spermatogenic cells show five stages of development after puberty: spermatogonia, primary spermatocytes, secondary spermatocytes, spermatids, and sperm. Sertoli's cells are secretory.
14. Between the coils of the seminiferous tubules lie interstitial cells, endocrine

cells producing testosterone, the male sex hormone.
15. The duct system includes the following:
   a. The straight and rete tubules lie within the testes; all other ducts are extratesticular.
   b. The epididymis "stores" sperm and the vas deferens propels it during ejaculation. The rest of the ducts convey sperm to the exterior.
16. Accessory glands include the following:
   a. Paired seminal vesicles lie behind the urinary bladder.
   b. A single prostate gland is around the prostatic urethra.
   c. Paired bulbourethral glands are in the urogenital diaphragm.
17. The penis is composed of three cavernous bodies.
   a. Corpora cavernosa penis refers to paired dorsally placed bodies.
   b. Corpus spongiosum penis is a single ventrally placed body that contains the urethra and forms the glans penis.
   c. Erection of the organ depends on blood filling the vascular spaces of the cavernous bodies.
18. FSH controls spermatogenesis within the testis; LH controls interstitial cell secretion.

Questions

1. What ensures the development of male or female gonads and external genitalia?
2. Describe the structure of the ovary.
3. What are the stages in the development of an ovarian follicle, and what characterizes each stage?
4. Describe the gross and microscopic structure of the uterine tubes, and speculate on the functions of ciliated and secretory epithelial cells in the lining of the tubes.
5. Describe the scrotum as to origin and structure.
6. What are the tunics of the testis? Make a simple diagram to illustrate them and the septa, lobules, and mediastinum.
7. Describe the stages in the formation of a sperm.
8. What are Sertoli's cells? Interstitial cells?
9. What structures support the ovary? The uterus?
10. Trace a sperm from its origin through the duct system, describing briefly the structure and location of each duct it would pass through.
11. Describe the structure of the accessory glands of the system.
12. Describe the structure of the penis.
13. Compare the endocrine relationships of ovary and testis.

Readings

Kolodny, R.C., Masters, W.H., and Johnson, V.E.: Textbook of sexual medicine, Boston, 1979, Little, Brown & Co.

Shearman, R.P., editor: Human reproductive physiology, ed. 2, Oxford, 1979, Blackwell Scientific Publications.

Shiu, R.P.C., and Friesen, H.B.: Mechanism of action of prolactin in the control of mammary gland function, Annu. Rev. Physiol. **42**:83, 1980.

Thomas, J.A., and Singhal, R.L.: editors: Advances in sex hormone research, Baltimore, 1980, Urban & Schwarzenberg, Inc.

16

Endocrine structures

Objectives
Development of the endocrine structures
Criteria for determining endocrine status
Pituitary gland (hypophysis)
   Location, size, and structure
   Blood supply
   Cells of the pituitary
Thyroid gland
   Location, size, and structure
   Blood supply
   Thyroid follicles
Parathyroid glands
   Location, size, and structure
   Cells of the parathyroid glands
Adrenal glands
   Location, size, and structure
   Blood supply
   Adrenal cortex
   Adrenal medulla
Pineal gland
Review of endocrine structures described in connection with other systems
   Pancreas
   Testes
   Ovaries
   Thymus
Gastrointestinal "hormones"
Prostaglandins
Disorders of the endocrine glands
   Disorders of the pituitary gland
   Disorders of the thyroid gland
   Disorders of the parathyroid glands
   Disorders of the pancreas
   Disorders of the adrenal glands
   Disorders of the gonads
Summary
Questions
Readings

OBJECTIVES After studying this chapter, the reader should be able to:

Give a general outline of the development of the body's endocrine organs.
Define the criteria used to evaluate endocrine status.
Describe, for the pituitary, thyroid, parathyroid, adrenal, and pineal glands, islets of the pancreas, testes, ovaries, and thymus, the following: origin, gross and microscopic structure, hormone(s) produced, and the general effect(s) of each of those hormones.
Describe some examples of other chemical substances that may have specific body effects but are not generally considered true hormones.

The endocrine structures do not form a system in the usual sense of that word. Systems heretofore discussed consisted of several organs working together to accomplish some process such as circulation or digestion. In the endocrine structures there are diverse organs producing a great variety of chemicals (hormones) that are used for controlling such things as growth, metabolism, and several aspects of homeostasis. Thus the influence of the endocrine structures and their chemical messengers is more widespread than the effects of other body systems and is in many cases vital to the survival of the organism.

Development of the endocrine structures

Endocrine structures first begin their development at about $3^{1}/_{2}$ weeks of intrauterine life. The three basic germ layers, ectoderm, endoderm, and mesoderm, are all involved in formation of endocrine glands. The anterior and posterior lobes of the pituitary gland, pineal gland, and adrenal medulla are ectodermal derivatives of the oral cavity and CNS. The thyroid and parathyroid glands are derivatives of the pharynx and are thus of endodermal origin. The pancreas endocrine tissue develops from the ducts of the exocrine pancreas, itself a derivative of the endoderm of the gut. The ovarian, testicular, and adrenal cortical endocrine structures are of mesodermal origin. According to germ layer of origin, there are rather clearly defined differences in the chemical nature of the hormones produced. For example, ectodermally derived endocrine structures usually produce small molecular weight amines; endodermally derived endo-

Table 16-1. Development of some of the endocrine structures

Endocrine structure	Age of first appearance (weeks)	Age of typical structure (weeks)	Area(s) formed in or from	Germ layer or origin
Adrenal cortex	5	7	Dorsal coelom	Mesoderm
Adrenal medulla	5	7	Neural crest by spinal cord	Ectoderm
Pancreas islets	12	14	Foregut	Endoderm
Parathyroid glands	7	9	Pharynx	Endoderm
Pineal gland	7	11	Brainstem	Ectoderm
Pituitary gland	4	16	Roof of mouth and brain	Ectoderm
Testes and ovaries	6	8	Pelvic cavity	Mesoderm
Thyroid gland	3½	12	Pharynx floor	Endoderm

Adapted from McClintic, J.R.: Basic anatomy and physiology of the human body, ed. 2. Copyright © 1980. Reprinted by permission of John Wiley & Sons, Inc.

crine structures produce large polypeptides or compound protein hormones; mesodermally derived endocrine structures produce steroid (fatty) hormones. Fig. 16-1 and Table 16-1 present some additional information related to the origins and times of development of the major endocrine structures.

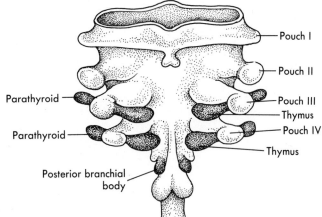

Fig. 16-1
Origins of some of the endocrine organs from branchial pouches.

Criteria for determining endocrine status

Although of diverse origins and widespread in body location, there are certain things that all endocrine structures have in common. These "things" serve as the criteria by which a particular structure may be judged to be an endocrine in the broadest sense.

Endocrine structures are *ductless;* that is to say, they do not pass their secretions through a duct to some epithelial-lined surface. Rather, they *place their products directly into blood vessels* within the endocrine gland itself. It is only by this device, the circulation, that the hormone produced can reach all body cells for possible effect.

Endocrine structures are *very vascular.* They synthesize their products from raw materials brought by their blood supply and use the bloodstream as the distributing route for their products. Endocrine structures are among the most vascular in the body on a milliliter of blood per gram of tissue basis.

Endocrine structures usually consist

of cells that are obviously different structurally and functionally from other body cells. That is, they are *specific, usually circumscribed groups of cells.*

They produce specific chemicals that *have specific effects* on body function. A physiologically active material can usually be isolated from a suspected endocrine structure. If this material is injected, it will have an effect on the body, and removal of the suspected endocrine structure will take the isolated material out of the body. Either procedure results in clearly defined and characteristic alterations in body function.

Using such criteria as these, the following are usually considered endocrine structures: the hypophysis or pituitary gland, the thyroid gland, the parathyroid glands, the adrenal glands, the pineal gland, the islets of the pancreas, the ovaries, the testes, and the thymus (Fig. 16-2). As research has expanded in the area of chemical production by body cells, many other substances produced by the body have been shown to have specific effects. Some of these substances have been called hormones, although it is not always possible to pinpoint a specific cell of origin. Thus the "hormones" *of* the gut that control secretory activity *by* the gut, hypothalamic regulating factors, and the prostaglandins will be discussed in this chapter as examples of other chemicals that are important to body function.

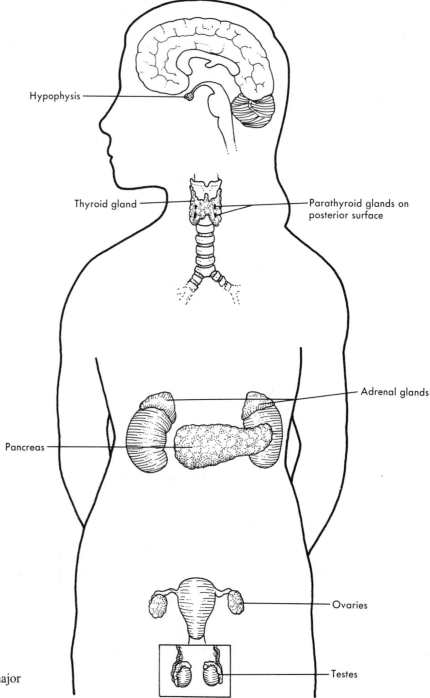

Fig. 16-2
Locations and names of major endocrine organs.

Pituitary gland (hypophysis)

Location, size, and structure

The *pituitary gland* lies in the sella turcica just posterior to the optic chiasma and below the inferior surface of the hypothalamus. It measures about 1 cm long × 1 to 1.5 cm wide × 0.5 cm deep. It weighs about 0.5 g in adult males, slightly more in adult females, and may reach 1 g in weight in a lactating female. Despite its small size the pituitary gland is one of the most important body organs, producing or releasing at least nine different hormones and maintaining intimate connections with the CNS and with other endocrine structures.

The organ has a dual origin, from the roof of the oral cavity and from the floor of the diencephalon. The portion derived from the mouth is termed the **adenohypophysis** and forms the *pars distalis* (anterior lobe), *pars intermedia* (intermediate lobe), and *pars tuberalis*. The portion derived from the diencephalon is called the **neurohypophysis** and consists of the *pars nervosa* (posterior lobe or infundibular process), *infundibulum* (or stalk), and *median eminence*. Fig. 16-3 shows the relationships of these portions of the gland.

Blood supply

Blood arrives at the organ through the ***superior*** and ***inferior hypophyseal arteries***. The superior artery supplies the stalk and the neighboring parts of the distalis; the inferior artery supplies the posterior lobe of the gland (Fig. 16-4). The pars distalis receives most of its supply as venous blood derived from capillaries in the median eminence that pass to the pars distalis as the ***pituitary portal system***. The portal system also provides a route for hypothalamically produced neurohumors (hormones?) that control activity of pars distalis cells. Venous blood drains from the en-

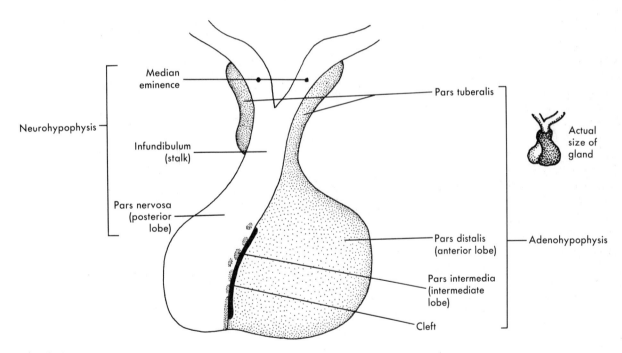

Fig. 16-3
Divisions of pituitary gland.

tire organ by several veins into the cavernous sinus that lies on either side of the optic chiasm.

Cells of the pituitary gland

Pars distalis. Cells of the pars distalis (Fig. 16-5) fall into three categories according to how they stain with common stains such as hematoxylin and eosin. **Acidophils** stain pink or red, **basophils** stain blue or purple, and **chromophobes** reject both stains. Further subdivision of acidophils and basophils is made on the basis of special stains, and on this basis, two subdivisions of acidophils and four of basophils may be made. All this gives six types of pars distalis cells, and six well-defined hormones are known to be produced by the pars distalis. One hormone may thus be assigned to each cell type. Table 16-2 shows the nomenclature assigned to these cells, their appearance, and the hormones each cell is believed to secrete.

It may be noted that four of the pars distalis hormones are **trophic** (tropic) hormones that influence the activity of other endocrine structures (adrenocorticotrophic hormone [ACTH], thyroid-

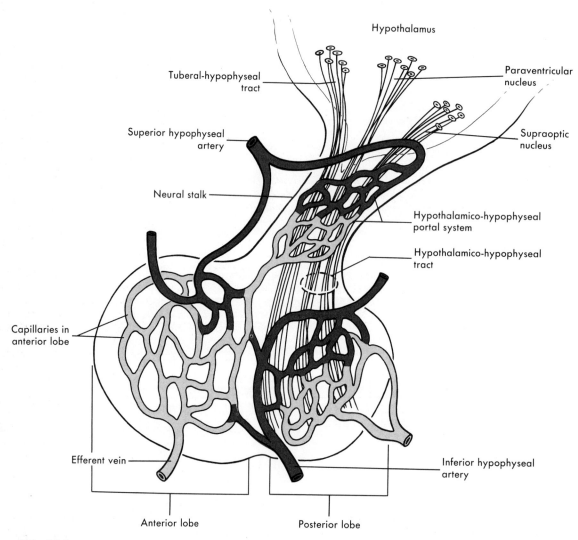

Fig. 16-4
Hypothalamic-pituitary connections.

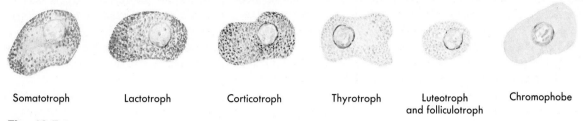

Fig. 16-5
Cells of pars distalis.

Table 16-2. Cells, hormones, and hormone effects of the pars distalis

Name of cell (synonyms included)		Staining preference of cell under special stain	Description of cell	Hormone produced (synonyms included)	General effects of hormone
Ordinary stains*	**Special stains**				
Acidophil (red, pink, or orange cytoplasmic granules)	α₁, Orangophil, somatotroph	Orange G	Round/oval small acidophilic granules	Growth hormone, human growth hormone, somatotropin	Controls growth of hard and soft tissues of body
	α₂, Carminophil, lactotroph	Azocarmine	Round/oval large acidophilic granules	Prolactin, lactogenic hormone, luteotropic hormone (LTH)	Causes milk secretion from mammary glands
Basophil (blue/purple cytoplasmic granules)	β₁, Corticotroph	PAS†	Irregular shape, eccentric nucleus	Adrenocorticotropic hormone (ACTH), corticotropin	Controls all activities of inner two zones of adrenal cortex
	β₂, Thyrotroph	PAS and AT‡	Irregular shape, eccentric nucleus	Thyroid-stimulating hormone (TSH), thyrotropin	Controls all activities of thyroid gland
	Δ₁, Luteotroph	PAS and AT	Round shape, central nucleus	Luteinizing hormone (LH), luteotropin (also known as ICSH—interstitial cell–stimulating hormone—for its action on the interstitial cells of the testes)	Female: necessary for ovulation, implantation, placenta formation, corpus luteum formation Male: controls activity of interstitial cells of testes
	Δ₂, Folliculotroph	PAS	Round shape, central nucleus	Follicle-stimulating hormone (FSH), folliculotropin	Female: causes maturation of ovarian follicles Male: controls spermatogenesis
Chromophobe (little/no stain taken by cytoplasm)	Chromophobe	None	Irregular shape, eccentric nucleus	None known; may be degranulated chromophils or reserve cells	—

*Such as hematoxylin and eosin. These stain cytoplasmic granules acidophilically or basophilically only.
†Periodic-acid Schiff. Stains glycoproteins.
‡Aldehyde thionine. Stains glycoproteins and protein polysaccharides.

stimulating hormone [TSH], follicle-stimulating hormone [FSH], and luteinizing hormone [LH]). For this reason the pars distalis is often called the "master gland of the endocrine system." It in turn is under the control of *hypothalamic regulating factors* (Table 16-3), so the CNS and pituitary gland are chemically tied together.

Pars intermedia. The pars intermedia is in the human represented by a series of cell-lined spaces (Rathke's cysts) that lie between the pars distalis and pars nervosa. The cells themselves are basophilic in stain and are thought to be producers of *melanocyte-stimulating* hormone (MSH). The hormone causes dispersion of pigment granules in pigment cells of the skin of lower animals. It has no known physiological role in the human.

Pars tuberalis. Cells of the pars tuberalis are small and both acidophilic and basophilic in stain. They contain much glycogen but produce no known hormone.

Neurohypophysis. All parts of the neurohypophysis show a similar structure (Fig. 16-6). Cells called *pituicytes* are found in the three parts of the neurohypophysis. They are interspersed among about 100,000 unmyelinated nerve fibers that course between the hypothalamus and the pars nervosa. The fibers constitute the *hypothalamico-hypophyseal tract (HHT)* that conveys *hormones produced in the hypothalamus* to the pars nervosa where they are stored in and released from the pars nervosa. The pituicytes are related to these nerve fibers as are glia elsewhere in the CNS, but their physiological role is uncertain.

Oxytocin and *vasopressin-ADH* (ADH, antidiuretic hormone) are the hormones released from the pars nervosa. Oxytocin is often called the "hormone of labor" because of its role in stimulating uterine contractions during childbirth. Vasopressin-ADH stimulates contraction of vascular smooth muscle and raises blood pressure (vasopressin activity) and increases water reabsorption by the collecting tubule of the nephron (ADH activity).

Table 16-3. Hypothalamic "neurohumors*" controlling the pituitary anterior lobe

Factor*/synonyms	Controls secretion of (hormone)	Effect on secretion	Comments
Growth hormone releasing factor (GHRF)	GH	Stimulates	GHRF released by low blood sugar, increased amino acid intake, stress, sleep
Growth hormone inhibiting factor (GIF)	GH	Inhibits	
Prolactin releasing factor (PRF)	Prolactin	Stimulates	Produced after childbirth; allows lactation to occur
Prolactin inhibiting factor (PIF)	Prolactin	Inhibits	Present continually in nonpregnant and antenatal pregnant females
Corticotropin releasing factor (CRF)	ACTH	Stimulates	Factor released by direct effect of hormones of target glands or via hypothalamus
Thyrotropin releasing factor (TRF)	TSH	Stimulates	
Luteinizing hormone releasing factor (LRF)		Stimulates	
Follicle stimulating hormone releasing factor (FSHRF)	FSH	Stimulates	

*These factors were formerly called *releasing factors* (RF). Since it is now known that these factors may both stimulate or inhibit secretion, it is appropriate to call them *regulatory factors* when speaking collectively of the chemicals.
†ICSH and LH are identical materials. ICSH denotes interstitial cell–stimulating hormone, a name sometimes used to denote its effects in the male.
Adapted from McClintic, J.R.: Basic anatomy and physiology of the human body, ed. 2. Copyright © 1980. Reprinted by permission of John Wiley & Sons, Inc.

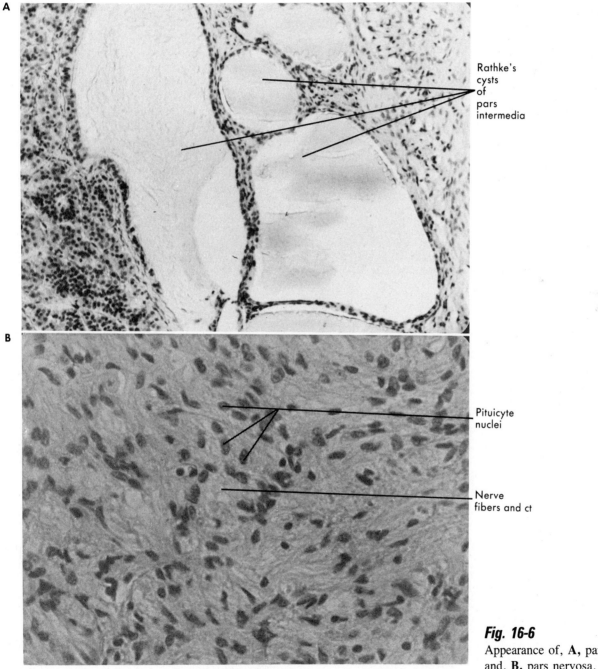

Fig. 16-6
Appearance of, **A,** pars intermedia and, **B,** pars nervosa.

Thyroid gland

Location, size, and structure

The *thyroid gland* (Fig. 16-7) lies on the anterior and lateral aspects of the lower larynx and upper trachea. Two *lateral lobes* are connected across the anterior midline by an *isthmus,* from which, in some people, a midline *pyramidal lobe* extends superiorly. The entire gland weighs about 20 g in the adult and receives 80 to 120 ml of blood/min (more than the kidney on a milliliter per gram basis).

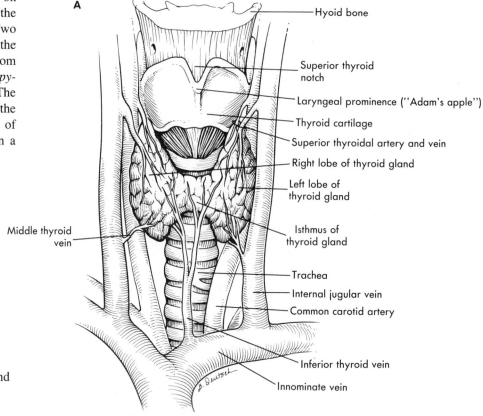

Fig. 16-7
Thyroid gland. **A,** Gross anatomy and location. **B,** Microscopic anatomy.

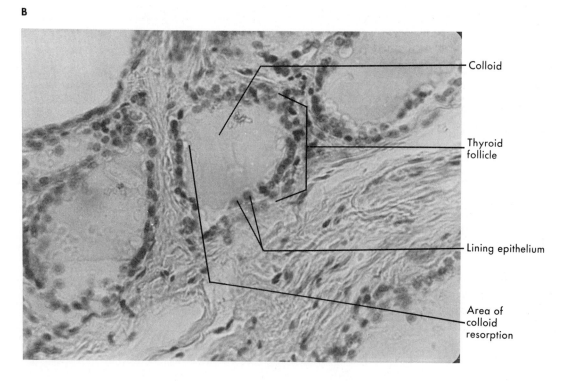

Blood supply (see Fig. 16-7)

Two pairs of vessels, the *superior* and *inferior thyroid arteries*, enter the lobes from above and below. Venous drainage occurs through the *anterior thyroidal veins*.

Thyroid follicles

The functional units of the thyroid gland are its **follicles,** spherical hollow structures measuring 0.2 to 0.9 mm in diameter (see Fig. 16-7). They are lined with a simple epithelium (usually cuboidal, but height depends on how much TSH is reaching the gland) and are filled with acidophilic-staining **colloid.** The colloid is a stored product and may show areas of resorption. The follicles are surrounded by dense capillary networks resembling those of the lungs.

Cells of the follicles. Two types of cells occur in the follicular epithelial lining (Fig. 16-8):

follicular (principal) cells are cuboidal cells with central nuclei that are believed to produce *thyroxin,* the metabolic-stimulating, iodine-containing hormone of the gland.

parafollicular (C) cells are oval, larger than the principal cells, and produce *calcitonin,* a hormone that is necessary for calcification of bones, cartilage (in bone formation), and teeth.

Thyroxin has no particular target organ. Any cell that contains systems of enzymes that degrade materials (catabolism) responds to the hormone. Thyroxin accelerates catabolic chemical reactions with a resulting increase in heat production that can be measured as an increase in the basal metabolic rate (BMR). This effect of thyroxin is called its **calorigenic action.** Thyroxin is essential in **growth,** particularly of the brain. Deficiency during brain development results in fewer and smaller neurons and mental retardation. Thyroxin *excess interferes with adenosine triphosphate (ATP) synthesis,* leading to muscular weakness.

Control of synthesis and release of the hormone is basically under the control of pituitary TSH. Other factors can alter the basic activity of the gland. Some of these are:

goitrogens, compounds that cause enlargement of the gland (a goiter) by blocking one or more steps in the synthesis of thyroxin, are usually thioureas (sulfur-containing substances), thiocyanates (sulfur and cyanide-containing compounds), or phenols.

pregnancy increases all aspects of thyroid activity in the mother, since fetal competition for thyroxin building blocks occurs.

stress, particularly that of a cold environment, stimulates release of thyroxin. Heat production is increased.

Calcitonin is a hormone whose production and release are accelerated by elevated blood calcium levels. Acting through osteoblast cells, the hormone increases mineralization of bones and teeth, a particularly important effect during skeletal growth. It also increases kidney reabsorption of calcium and phosphate.

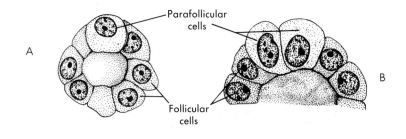
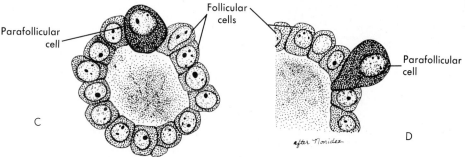

Fig. 16-8
Cells of thyroid follicles. *A* and *B,* Cat, hematoxylin stain. *C* and *D,* Dog, silver nitrate stain.

Parathyroid glands

Location, size, and structure

There are usually four *parathyroid glands* (Fig. 16-9) located on the posterior aspect of the thyroid lobes. They measure 3 to 8 mm long × 2 to 5 mm wide × 0.5 to 2 mm thick. They may be buried in the thyroid substance or may lie on the surface of the lobes, but in either case they have their own connective tissue capsules. They consist of closely packed masses or cords of cells.

Cells of the parathyroid glands

Two types of cells form the parathyroid glands.

principal (chief) cells are 7 to 10 μm in diameter with a larger pale-staining nucleus

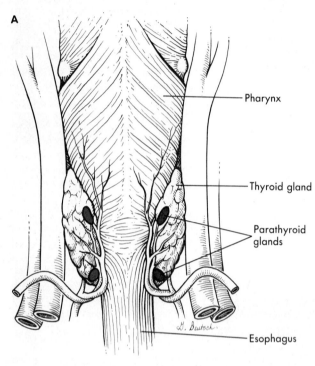

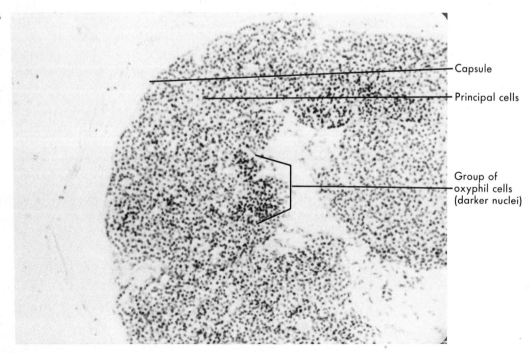

Fig. 16-9
A, Location of parathyroid glands.
B, Microscopic appearance of parathyroid gland.

and a lightly-staining slightly granular cytoplasm. These cells constitute the bulk of cells in the glands and are said to be the source of **parathyroid hormone (PTH)**.

oxyphil cells are larger than the principal cells, have a small dark-staining nucleus, and possess an obviously granular cytoplasm. They occur in groups among the principal cells. They may be reserve cells and produce no hormone.

PTH is involved in the metabolism of calcium, phosphate, magnesium, and citrate ions. Calcium and phosphate are easily recognized as the major constituents of bone. Magnesium is essential in ATP synthesis and protein synthesis. Citrate is a major source of energy for ATP synthesis.

Absorption, excretion, and maintenance of exchange between cells and body fluids of these substances is controlled by PTH. Release of hormone is increased by a low blood calcium level.

PTH increases absorption of all ions from the gut.

PTH, acting through osteoclasts, increases the destruction and demineralization of hard body tissues.

PTH increases kidney reabsorption of calcium and magnesium but increases phosphate excretion.

All these effects tend to maintain adequate blood levels, particularly of calcium ion.

Adrenal glands

Location, size, and structure

The *adrenal (suprarenal) glands* (Fig. 16-10) are paired organs that lie atop the superior poles of the kidneys. They measure 5 cm long × 3 cm wide × 1 cm deep and together weigh about 15 g. An outer **cortex** is of mesodermal origin, and an inner **medulla** is of ectodermal origin.

Blood supply

Blood supply is via the *inferior phrenic artery, adrenal branches of the aorta,* and *branches from the renal arteries.* Outflow is provided by the *adrenal vein* that joins the inferior vena cava or renal vein.

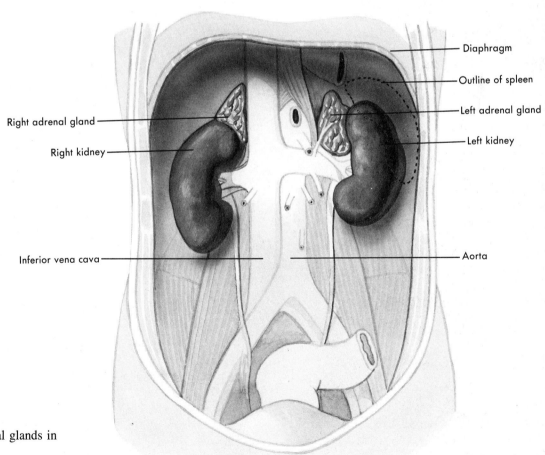

Fig. 16-10
Location of adrenal glands in abdominal cavity.

Adrenal cortex

The cortex (Fig. 16-11) constitutes the bulk of the adrenal gland and has three clearly defined zones in the adult. The zones are distinguished by *arrangement* of cells, rather than by cell type.

zona glomerulosa, the outermost zone, has groups of cells arranged in rounded or oval masses. It constitutes in the human about 7% of the total width of the cortex. These cells produce a category of hormones called **mineralocorticoids,** of which the most potent is *aldosterone.* This steroid hormone is concerned with kidney reabsorption of sodium ion.

zona fasciculata, the middle zone of the cortex, has more or less radially arranged cords of cells. It constitutes about 78% of the cortical width and the class of hormones produced here are called **glucocorticoids,** the best known of which is *cortisol.* Glucocorticoids are anabolic, concerned with protein synthesis, blood formation, and glucose metabolism.

zona reticularis is the inner zone of the cortex, in which cords of cells are arranged in a netlike fashion. **Sex hormones,** both androgenic and estrogenic, are produced here. The zone constitutes about 15% of the cortical width.

Adrenal medulla

The adrenal *medulla* (see Fig. 16-11) is composed of groups or cords of what are called **chromaffin cells** (from the brown granules they show when stained with chromium salts). Eighty percent of the secretion of the medulla consists of the hormone *epinephrine* (adrenalin), whereas 20% is composed of *norepinephrine* (noradrenalin). Both hormones are sympathomimetic; that is, their effects simulate the effects of stimulation of the sympathetic nervous system. The hormones fit the body to resist acute stress (the "fight-or-flight reaction").

Fig. 16-11

Microscopic structure of adrenal gland.

From Bevelander, G., and Ramaley, J.A.: Essentials of histology, ed. 8, St. Louis, 1979, The C.V. Mosby Co.

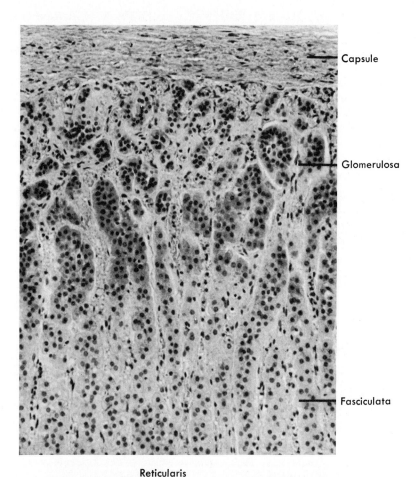
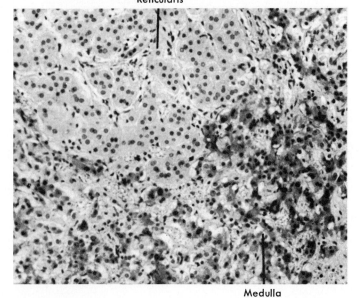

Pineal gland

The *pineal gland* (Fig. 16-12) is a flattened conical structure about 6 mm in length and diameter that is derived from the diencephalon. It is said to contain several types of epitheloid cells called **pinealocytes** *(chief cells)*. They resemble and are oriented like glia in the CNS. The gland commonly contains lamellar concretions called **corpora arenacea** *(acervuli, "pineal sand")*.

The only chemical produced by the pineal gland *solely* is **melatonin**, which seems to exert an antigonadotropic effect and may be involved in entraining cyclical fluctuations in many body functions *(circadian rhythms)*.

Review of endocrine structures described in connection with other systems

Pancreas

The pancreas contains islets of endocrine tissue among the acini that produce digestive enzymes. The islets contain **alpha cells** that produce **glucagon**, a hormone that brings about increased breakdown of liver glycogen into blood-borne glucose. It is thus a *hyperglycemic factor*. **Beta cells** of the islets produce **insulin**, which causes certain body cells to increase their uptake of blood glucose and also increases the liver and muscle formation of glycogen from glucose in the bloodstream. These effects reduce blood glucose levels, which is called a *hypoglycemia* effect. The two hormones are thus instrumental in keeping blood glucose levels within the normal range of 80 to 120 mg/100 ml.

The alpha cells constitute about one fourth of the islet cell population, the beta cells about three fourths. Both types of cells are arranged in cordlike structures between large capillary spaces in the islets. The hormones enter the circulation in response to changes in blood glucose levels, thus maintaining the normal values.

Testes

The endocrine structures of the testes are the **interstitial cells** that lie between the seminiferous tubules in the testicular lobules. These cells produce the hormone **testosterone**, which is responsible for developing the male reproductive organs and for developing the secondary sex characteristics of the male. The latter include hair pattern, muscular development, and voice changes that occur at puberty.

Ovaries

The ovary contains two endocrine structures. The **vesicular *(graafian) follicle***, specifically its *membrana granulosa cells*, is the site of the production of **estradiol**, the major follicular estrogen. This hormone is responsible for the development of female reproductive organs and the mammary gland and for establishing menstrual cycles. After ovulation the theca interna cells and granulosa cells form the **corpus luteum**, which secretes **progesterone**. Progesterone acts mainly on the uterine endometrium to prepare it to receive a fertilized egg and on the mammary gland to develop its secretory tissue.

Thymus

The *thymus* produces a hormone called **thymosin** that appears to be concerned with developing the body's im-

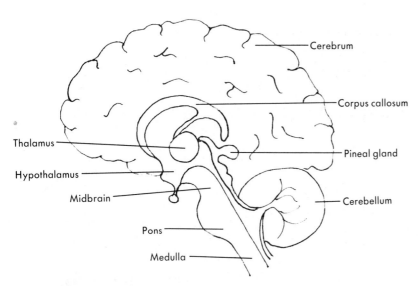

Fig. 16-12
Location of pineal gland, as shown on midsagittal section of brain.

mune system. Resistance to various types of infectious agents depends on the ability to produce antibodies, chemicals that can attack infectious agents and destroy them. The two-lobed thymus lies behind the sternum and is essential in the development of cells that can produce antibodies.

Gastrointestinal "hormones"

The stomach and small intestine produce a variety of substances that stimulate secretion of digestive juices by the stomach and pancreas and cause emptying of bile from the gallbladder. These substances are regarded by many as hormones, although their production cannot be attributed to specific cells or groups of cells in the gut. Some of the materials are discussed in the next three paragraphs.

Gastrin is produced by the stomach when partially digested proteins contact the mucosa of the stomach. It enters the bloodstream and causes secretion of large volumes of gastric juice.

Secretin is produced by duodenal cells when acid enters the duodenum from the stomach. It causes the pancreas to secrete an alkaline fluid that aids in neutralizing stomach acid.

Cholecystokinin (CCK) is produced when fats enter the duodenum from the stomach. It causes the pancreas to secrete an enzyme-rich fluid that includes *lipase,* an enzyme that digests fats, and it also causes expulsion of bile from the gallbladder. Bile aids fat digestion and absorption.

Prostaglandins

Prostaglandins are derivatives of a fatty acid and are produced by many body organs, including the lungs, liver, and reproductive organs. They appear to act as "intracellular switches" that can stimulate smooth muscle contraction (especially in the uterus— "cramps"), increase sodium excretion by the kidney, aid in keratinization of the skin, and exert other effects. They are under continued investigation as to their roles in body function.

Disorders of the endocrine glands

Disorders associated with endocrine glands are many and varied. Disorders are usually associated with excessive or deficient production of a given hormone. Although the following does not cover all known endocrine disorders, it does describe the more interesting or common ones.

Disorders of the pituitary gland

Giantism is the result of excessive production of growth hormone while the epiphyseal cartilages are still present in the long bones. Continued growth occurs, and the patient can reach great stature and weight (to 2.7 m and 150 or so kg). If excessive production of growth hormone occurs *after* epiphyseal closure, a condition known as *acromegaly* results. Intramembranous formation of bone can still occur on the surfaces of the body's bones, and the bones become thick and often misshapen. A tumor involving acidophil cells is the usual cause.

Pituitary infantilism results when there is deficient production of growth hormone. The body is well proportioned but short in stature and usually not sexually developed for the chronological age. If human growth hormone is available, its administration can cause dramatic increase in growth and development.

Disorders of the thyroid gland

Overproduction of thyroxin may be due to a tumor of the gland or to overstimulation by pituitary production of TSH. There is elevation of the basal metabolic rate, muscular weakness, elevated heart rate and blood pressure, and a bulging of the eyes called exophthalmos. There may be goiter associated with hyperfunction.

Deficiency of thyroxin production creates a low BMR, mental sluggishness, goiter, and myxedema, a drying of the skin with deposition of a pasty material in the dermis. Oral administration of the hormone is effective in controlling the disorder.

Disorders of the parathyroid glands

Hyperparathyroidism refers to excessive production of PTH. The most obvious effect of this condition is weak, easily fractured bones that present a moth-eaten appearance on x-ray examination. Remembering that PTH demineralizes bony tissue, we can easily understand the x-ray picture and the resulting hypercalcemia (high blood calcium levels). Kidney stones are common, and cardiac irregularities often occur because of high urine and blood calcium levels.

Tetany, or *hypoparathyroidism,* is the result of low blood calcium levels, which are usually due to deficient PTH production. Muscles undergo spasms and paralysis that can make respiration difficult. Raising the blood calcium level is the primary consideration in treatment.

Disorders of the pancreas

Insulin deficiency results in the development of **diabetes mellitus,** or sugar diabetes. There are high blood glucose levels (hyperglycemia), sugar appears in the urine since more filters than the tubule cells can reabsorb (glucosuria), there is an increased urine volume and thirst, and eventually there are atherosclerotic changes in blood vessels. According to severity and the presence of beta cells still capable of producing insulin, the disorder may require insulin injection, dietary measures, or oral chemicals to control it.

Some individuals seem to have hypersensitive beta cells. The normal response to ingested sugar is insulin release that removes it from the blood. Excessive insulin production after ingestion of carbohydrates is one cause of *hypoglycemia*. The blood glucose level goes way down, and the brain is affected immediately as its fuel source disappears. Coma and death may result if blood glucose levels are not raised.

Disorders of the adrenal glands

There is no disorder of deficient medullary production of epinephrine. The sympathetic nervous system can still maintain normal body function. In short, the medulla is not essential for life. A tumor of the medulla, called a pheochromocytoma, can pour such quantities of epinephrine into the bloodstream as to raise heart rate and blood pressure to dangerous levels.

The adrenal cortex may produce insufficient quantities of cortisol, causing **Addison's disease.** Anemia, muscular weakness, fatigue, and a peculiar bronzing of the skin, lips, and gums occurs. Administration of cortisol controls the disorder.

Cushing's syndrome is the result of excessive production of cortisol. There is trunk adiposity, abdominal striae, and a "moon-faced" appearance. A cortical tumor is a common cause and calls for surgical removal.

Disorders of the gonads

Overproduction of either male or female sex hormones is associated with accentuation of sex characteristics or their development at tender ages in precocious puberty. Deficiency of the hormones slows sexual development and appearance of secondary characteristics with failure to mature sex cells. Both aspects of production may be due to defective pituitary function.

Summary

1. The endocrine glands form a functional system that is oriented toward governing growth, metabolism, reproduction, and other basic body activities.
2. Development of endocrine structures occurs between $3\frac{1}{2}$ and 12 weeks. All three germ layers form endocrine tissue.
3. Certain criteria are generally used to define endocrine status.
    a. There is a lack of ducts; products are placed directly into the bloodstream.
    b. Endocrine structures are very vascular.
    c. Endocrine structures consist of specific cells that produce specific chemical substances (hormones).
    d. Hormones have specific and well-defined effects on the body. Removal of a suspected endocrine structure causes clear-cut effects (removes hormones); injection of principal(s) of suspected endocrine structure causes clear-cut effects—creates an excess of hormone(s).
4. The pituitary gland (hypophysis) lies in the sella turcica and is composed of two major parts.
    a. The adenohypophysis originates from the mouth and consists of the pars distalis (anterior lobe), pars intermedia (intermediate lobe), and pars tuberalis.
    b. The neurohypophysis originates from the diencephalon and consists of the median eminence, the infundibulum (stalk), and the pars nervosa (posterior lobe).
5. Blood supply to most parts of the gland except the pars distalis is provided by hypophyseal arteries. The pars distalis receives venous blood via a portal system that lies between the median eminence and the cells of the distalis.
6. The pars distalis contains seven types of cells and produces six hormones (see Table 16-2).
7. The pars intermedia consists of cell-lined spaces and produces a hormone controlling pigment cells.
8. The pars tuberalis produces no hormones.
9. All parts of the neurohypophysis have a similar structure consisting of glialike pituicytes and nerve fibers of the tract connecting the hypothalamus and pars nervosa. Hormones (oxytocin, vasopressin-ADH) are produced in the hypothalamus, pass down neuron axons, and are stored in and released from the pars nervosa.
10. The thyroid gland lies on the trachea and has two lobes and an isthmus.
    a. The functional units are follicles that are lined with principal and parafollicular cells and that contain colloid.
    b. Thyroxin (from principal cells) controls overall body metabolic rate; calcitonin (from parafollicular cells) is essential for formation of body hard tissue.
11. There are usually four parathyroid glands located on the posterior side of the thyroid lobes.
    a. Principal cells produce parathyroid hormone, which raises blood calcium levels by several means.
    b. Oxyphil cells may be reserve cells; they produce no hormone.
12. The adrenal glands lie above the kidneys and are of dual origin (cortex, mesoderm; medulla, ectoderm).
    a. The cortex has three zones: glomerulosa, which produces mineralocorticoids; fasciculata, which produces glucocorticoids; and reticularis, which produces sex hormones.
    b. The medulla consists of chromaffin cells and produces sympathomimetic epinephrine and norepinephrine.
13. The pineal gland lies above the midbrain and produces melatonin.
14. A review of endocrine structures discussed in other chapters is provided.
    a. The pancreatic endocrine tissue is its islets that produce glucagon (hyperglycemic) and insulin (hypoglycemic).
    b. Interstitial cells of the testes produce testosterone.
    c. The vesicular follicle of the ovary produces estradiol, and the corpus luteum produces progesterone.
    d. The thymus produces thymosin.
15. Other substances produced by body cells are often called hormones, although their specific origins may not be known.
    a. Gastrin, secretin, and CCK are examples of gastrointestinal hormones.

b. Prostaglandins are fatty acid derivatives that mediate intracellular reactions.
16. Some disorders of the endocrine structures are discussed.

Questions

1. What features do endocrine structures have in common?
2. From where are the parts of the pituitary gland derived? What are those "parts"?
3. Discuss the structure of the anterior and posterior lobes of the pituitary gland, and note what cells or areas produce what hormones.
4. Discuss the location, structure (gross and microscopic), and hormones produced by each of the following:
   a. Parathyroid glands
   b. Adrenal cortex
   c. Thyroid gland
   d. Pineal gland
   e. Pancreatic islets
   f. Graafian follicle
   g. Corpus luteum
   h. Interstitial cells of the testes
5. Give some examples of chemicals having specific effects on the body that are not "true hormones." Why do we not consider these as endocrine products?

Readings

Ezrin, C. Godden, J.O., and Volpe, R.: Systematic endocrinology, New York, 1979, Harper & Row, Publishers, Inc.

Kappers, A., and Pevet, P., editors: The pineal gland of vertebrates including man, New York, 1979, Elsevier North-Holland, Inc.

Labrie, F., et al.: Mechanisms of action of hypothalmic hormones in the adenohypophysis, Annu. Rev. Physiol. **41**:555, 1979.

Lands, W.E.M.: The biosynthesis and metabolism of prostaglandins, Annu. Rev. Physiol. **41**:633, 1979.

Notkins, A.L.: The causes of diabetes, Sci. Am. **241**:62, Nov. 1979.

Readings from Scientific American: Hormones and reproductive behavior, San Francisco, 1979, W.H. Freeman & Co. Publishers.

General references

Allen, G.: Life sciences in the twentieth century, New York, 1975, John Wiley & Sons, Inc.

Arey, L.B.: Developmental anatomy, ed. 7, Philadelphia, 1974, W.B. Saunders Co.

Barr, M.L.: The human nervous system, ed. 3, New York, 1979, Harper & Row, Publishers, Inc.

Beeson, P.B., and McDermott, W., editors: Textbook of medicine, ed. 15, Philadelphia, 1979, W.B. Saunders Co.

Bergersen, B.S., and Goth, A.: Pharmacology in nursing, ed. 14, St. Louis, 1979, The C.V. Mosby Co.

Bloom, W., and Fawcett, D.W.: A textbook of histology, ed. 10, Philadelphia, 1976, W.B. Saunders Co.

Cheek, D.B.: Fetal and postnatal cellular growth, New York, 1975, John Wiley & Sons, Inc.

Children are different: relation of age to physiologic functions, Columbus, Ohio, Oct. 1972, Ross Laboratories.

Diem, K., and Lentner, C., editors: Scientific tables, ed. 7, Basel, Switzerland, 1970, J.R. Geigy.

Goss, C.M., editor: Gray's anatomy of the human body, ed. 29, Philadelphia, 1973, Lea & Febiger.

Guyton, A.C.: Textbook of medical physiology, ed. 5, Philadelphia, 1976, W.B. Saunders Co.

Lockhart, R.D.: Living anatomy, London, 1970, Faber & Faber.

Moore, K.L.: Before we are born: basic embryology and birth defects, Philadelphia, 1974, W.B. Saunders Co.

Mountcastle, V.B., editor: Medical physiology, ed. 14, 2 vols., St. Louis, 1980, The C.V. Mosby Co.

Netter, F.H.: Ciba collection of medical illustrations, 5 vols., Newark, N.J., Ciba Pharmaceutical Products.

Nilsson, L.: Behold man, Boston, 1973, Little, Brown & Co.

Normal reference laboratory values, N. Engl. J. Med. **298**:34, 1978.

Rubin, A., editor: Handbook of congenital malformations, Philadelphia, 1967, W.B. Saunders Co.

Ruch, T.C., and Patton, H.D.: Physiology and Biophysics, 3 vols., Philadelphia, 1973-1977, W.B. Saunders Co.

Science, vol. 200, May 26, 1978 (entire issue on medicine and public health).

Scientific American Readings: Human physiology and the environment in health and disease, San Francisco, 1976, W.H. Freeman & Co. Publishers.

Smith, C.A., and Nelson, N.M.: The physiology of the newborn infant, ed. 4, Springfield, Ill., 1976, Charles C Thomas, Publisher.

Thompson, J.S., and Thompson, M.W.: Genetics in medicine, ed. 3, Philadelphia, 1980, W.B. Saunders Co.

Timiras, P.S.: Developmental physiology and aging, New York, 1972, Macmillan, Inc.

Vaughan, V.C. and McKay, R.J.: Nelson textbook of pediatrics, ed. 11, Philadelphia, 1979, W.B. Saunders Co.

Glossary

Abbreviations

abbr.	abbreviation
A.S.	Anglo-Saxon
Colloq.	Colloquial
e.g.	for example
esp.	especially
Fr.	French
Gk.	Greek
Ger.	German
i.e.	that is
L.	Latin
M.E.	Middle English
pert.	pertaining
pl.	plural
sing.	singular

Pronunciation

Pronunciations may be only approximately indicated unless all the markings in Webster's *New International Dictionary* are used, which is not practical in a glossary.

Accent. Indicates the stress on certain syllables.

Diacritics. Marks over the vowels to indicate the pronunciations. Only two diacritics are used: The *macron*, showing the long or name sound of the vowels, and the *breve*, showing the short vowel sounds, as

a in rāte	a in căt
e in ēat	e in ĕver
ee in seed	
i in īsle	i in ĭt
o in ōver	o in nŏt
oo in moon	oo in bŏok
u in ūnit	u in cŭt

a-, an- [G. *alpha*]. Prefix: without, away from, not.

ab- [L.]. Prefix: from, away from, negative, absent.

abdomen (ab-dō′men) [L.]. The part of the trunk between the chest and the pelvis.

absorb [L. *absorbere*, to suck in]. To take in.

Adapted from McClintic, J.R.: Basic anatomy and physiology of the human body, ed. 2. Copyright © 1980. Reprinted by permission of John Wiley & Sons, Inc.

absorption (ab-sorp′shun). Taking into body fluids or tissues, usually across a membrane.

abducens (ab-dū′senz) [L. drawing away from]. The sixth cranial nerve; innervates the lateral rectus muscle of the eye, which moves the eyeball outward or away from the median line of the body.

abduction (ab-dukt′shun) [L. *abducere*, to lead away]. Lateral movement away from the median plane of the body.

acceleration (ak-sel′e-rā-shun) [L. *acceleratus*, to hasten]. Increasing the speed of pulse, respiration, rate, or motion.

accept′or [L. *accipere*, to accept]. Compounds taking up or chemically binding or other substances, e.g., hydrogen acceptor.

accom′modate [L. *accomodare*, to suit]. To adjust or adapt; a decrease in nerve impulse discharge with continued stimulation; the adjustment of the eye for various distances by changing the curvature of the lens.

acetabulum (as-e-tab′ū-lum) [L. a little saucer for vinegar]. The rounded cavity of the os coxae that receives the head of the femur.

acetic acid (a-sē′tik a′cid) [L. *acetum*, vinegar, + *acidus*, sour]. The acid of vinegar. CH_3COOH.

acetone (as′e-tōn). A chemical [$(CH_3)_2CO$] produced when fats are not properly oxidized in the body. It is one of several substances called *ketone bodies* or *acetone bodies*.

acetylcholine (ă-sĕt-ĭl-kō′lēn). An ester of choline released from certain nerve endings; thought to play a role in transmission of nerve impulses at synapses and myoneural junctions. Easily destroyed by the enzyme, *cholinesterase*.

Achilles tendon (ăh-kil′ēz). Heel tendon of the gastrocnemius and soleus muscles of the leg.

acid [L. *acidus*, sour]. Any substance releasing H ion and neutralizing basic substances; a pH less than 7.00.

acid-base balance. State of equilibrium between acids and alkalies so that the H^+ concentration of the arterial blood is maintained between pH 7.35 and 7.45.

acid′ification [L. *acidum*, acid, + *factus*, to make]. Process of making acidic; becoming sour.

acidosis (as-ĭ-dō′sĭs) [L. *acidum* + Gk. *-osis*, condition]. Decrease of alkali in proportion to acid, or lowering of pH of extracellular fluid, caused by excess acid or loss of base.

acinus (as′ĭ-nus) (pl. *acini*) [L. grape]. Smallest division of a gland; a saclike group of secretory cells surrounding a cavity.

acne (ak′nē) [Gk. *acme*, point]. Chronic inflammatory disease of the sebaceous glands and hair follicles of the skin.

acoustic (a-koos′tik) [Gk. *akoustikōs*, hearing]. Pert. to sound or to the sense of hearing. The cochlear portion of the eighth cranial nerve.

acromegaly (ak-ro-meg′ă-lĭ) [Gk. *akros*, extreme, + *megas*, big]. Enlargement of the extremities (hands, feet) and jaw caused by hypersecretion of the somatotrophic or growth hormone from the anterior pituitary, after full growth has been achieved.

actin (ak′tĭn). A muscle protein responsible for the shortening of the muscle when it contracts. Forms a combination with the protein *myosin* when muscle contracts.

ac′tion potential (pō-tĕn′shăl) [L. *actio*, from *agere*, to do + *potentia*, power]. The measurable electrical changes associated with conduction of a nerve impulse or contraction of a muscle.

ac′tivator. Substances converting inactive materials to active ones, or causing activity.

ac′tive transport′ [L. *trans*, across, + *porta*, to carry]. Using carriers, energy (ATP), and enzymes of cell to cause a substance to cross a membrane.

acu′ity [L. *acuere*, to sharpen]. Clearness or sharpness, usually in reference to vision or hearing.

509

adap'tation [L. *adaptare*, to adjust]. The ability of an organism to adjust to environmental change; the ability of the eye to adjust to various intensities of light by change in size of the pupil.

adduction (ăd-dŭk'shŭn) [L. *ad*, toward, + *ducere*, to bring]. Movement toward the median plane of the body.

adenine (ad'ĕ-nēn). A nitrogenous base found in DNA and RNA (nucleic acids).

adenohypophysis (ad-ē-no-hi-poph'i-sis) [Gk. *adēn*, gland + *ypo*, under, + *physis*, growth]. The glandular lobes of the hypophysis or pituitary gland. Derived from the mouth, it includes the anterior lobe *(pars distalis)* and intermediate lobe *(pars intermedia)*.

adenoid (ad'ĕ-noid) [Gk. *adenoeides*, glandular]. Lymphoid tissue; the pharyngeal tonsil.

adenosine triphosphate (ad'ĕ-nō-sēn). The major source of cellular energy; found in all cells. Composed of a nitrogenous base *(adenine)*, ribose sugar, and three phosphoric acid radicals. Abbr: **ATP.**

ad'ipose [L. *adeps*, fat]. Refers to a type of connective tissue composed of fat-containing cells; fatty.

adolescence (ad-ō-les'ens) [L. *adolescere*, to grow up]. The period from the beginning of puberty until maturity; variable among individuals.

adrenal (ăd-rē'năl) [L. *ad*, to, + *ren*, kidney]. An internally secreting gland located above the kidney. Outer *cortex* produces hormones essential to life; inner *medulla* not essential to life.

adrenalin (ă-dren'ă-lin) [L. *ad*, + *ren*, + *in*, within]. Proprietary name for *epinephrine*, the main hormone of the adrenal medulla.

adrenergic (ad-ren-er'jik) [L. *ad* + *ren*, + Gk. *ergon*, work]. Term used to describe nerve fibers that release *norepinepherine* (noradrenalin or sympathin) at their endings when stimulated.

adsorption [L. *ad*, + *sorbere*, to suck in]. "Sticking together" of a gas, liquid, or dissolved substance on a surface. No bonding.

adventitia (ad'ven-tish'ē-ă) [L. *adventitius*, coming from abroad]. The outermost covering of an organ or structure.

af'ferent [L. *ad*, to, + *ferre*, to bear]. Carrying toward a center or toward some specific reference point, e.g., afferent glomerular vessels, or afferent nerves to the brain.

ag'onist [Gk. *agōn*, a contest]. The contracting muscle; a muscle whose contraction is permitted by relaxation of a muscle (antagonist) having the opposite action to the agonist.

albumin (al-bū'min) [L. *albus*, white; *albumen*, coagulated egg white]. One of a group of simple proteins in both animal and vegetable cells or fluids.

aldosterone (al-dos'tēr'ōn). The most active mineralocorticoid secreted by the cortex of the adrenal gland; regulates metabolism of sodium, potassium, and chloride.

-algia (al-je-a) [Gk. *algeis*, sense of pain]. Suffix: denoting pain.

alimen'tary [L. *alimentum*, nourishment]. Pert. to nutrition.

alkali [Arabic *al-qili*, ashes of salt wort]. A substance that can neutralize acids; combines with acid to form soaps; basic in reaction (pH).

alkali(ne) reserve. The amount of base in the blood, mainly bicarbonates, available to neutralize fixed acids.

alkalo'sis [Arabic *al-qili* + Gk. *-osis*, condition of]. Excessive amount of alkali, or rise of pH of extracellular fluid caused by excess base or loss of acid.

allograft (al'ō-graft) [Gk. *allos*, other, + L. *graphium*, grafting knife]. Grafts or transplants between genetically dissimilar members of the same species.

alveolus (al-vē'ō-lus) (pl. *alve'oli*) [L. small hollow or cavity]. A little hollow; an air sac of the lungs; a socket of a tooth.

ame'boid [Gk. *amoibē*, change, + *eidos*, resemblance]. Like an ameba; ameboid movements imply cellular motion by use of *pseudopods* (false feet or cytoplasmic outflows).

amenorrhea (a-men-ō-rē-ă) [Gk. *a-*, without, + *mēn*, month, + *rein*, to flow]. Absence or suppression of menstruation.

amine (ă-mēn'). A nitrogen-containing organic compound with one or more hydrogens of ammonia (NH_3) replaced by organic radicals.

amino acid (ă-mē'no). Unit of structure of proteins; a compound containing both an amine group ($-NH_2$), and a carboxyl group (-COOH). Two groups: *essential*, not synthesized in the body or not produced in amounts necessary for normal function; *nonessential*, synthesized in the body.

am'nion (am'nĭ-on) [Gk. little lamb]. The inner fetal membrane that holds the fetus suspended in amniotic fluid. By 8 weeks it fuses with the *chorion* to form the *bag of waters* or *caul*.

amphiarthrosis (am-fi-ar-thrō'sis) [Gk. *amphi*, on both sides, + *arthrosis*, joint]. A form of articulation in which the mobility in all directions is slight.

ampulla (ăm-pŭl'a) [L. little jar]. A saclike dilation of a duct or canal.

anabolism (an-ab'ō-lizm) [Gk. *anabolē*, a building up]. Synthetic or constructive chemical reactions.

anaphase (an'a-fāz) [Gk. *ana*, up, + *phainein*, to appear]. A stage in mitosis characterized by separation and movement of chromosomes toward poles of the dividing cell.

anastomosis (a-nas-to-mō'sis) [Gk. opening]. An opening of one vessel into another, or the joining of parts to form a passage between any two spaces or organs.

androgen (ăn'drō-jĕn) [Gk. *anēr, andros*, man, + *gen*, to produce]. A substance stimulating or producing male characteristics, as testosterone, the male sex hormone.

anemia (an-e'-mĭ-a) [Gk. *an*, without, + *aima*, blood]. A condition in which there is reduction of circulating red blood cells or of hemoglobin.

anesthesia (an-es-the'-zī-ă) [Gk. *an-*, without, + *aisthesis*, sensation]. Partial or complete loss of sensation with or without loss of consciousness.

aneurysm (an'ū-rism) [Gk. *aneurysma*, a widening]. Abnormal dilation of a blood vessel caused by weakness of the vessel wall or a congenital defect.

angina (ănjī'na) [L. *angere*, to choke]. Any disease characterized by suffocation or choking. *Angina pectoris* occurs when the heart muscle is deprived of blood flow and includes severe pain referred to the chest and left arm.

Angström unit (ong'strum) [Anders J. Angström, Swedish physicist, 1814-1874]. An international unit of wavelength measurement, $1/10$ millionth of a millimeter. Abbr: **Å** or **A**.

anion (an'ī-on) [Gk. *ana*, up, + *ion*, going]. A negatively charged ion or radical. In electrolysis, anions go toward the anode.

ann'ulus [L]. A fibrous ring surrounding an opening.

antag'onist [Gk. *antagōnizesthai*, to struggle against]. That which counteracts or has the opposite action of something else (*agonist*); esp. muscular contraction.

antenatal (ăn-tē-nā′tal) [L. *ante*, before, + *natus*, birth]. Occurring before birth.

ante′rior [L.] Before, or in front of, or the ventral (belly) portion.

antibiotic (ăn-tĭ-bī-ŏt′ik) [Gk. *anti*, against, + *bios*, life]. A substance produced by bacteria, molds, and other fungi, that has power to inhibit growth of or destroy other organisms, esp. bacteria.

antibody (an′tĭ-bod-ē). A protein (usually a globulin) developed against a specific substance (antigen), whether the antigen is foreign or not.

antigen (an′tĭ-jĕn) [Gk. *anti* + *gennan*, to produce]. A substance causing production of antibodies.

an′trum (pl. *antra*) [Gk. *antrom*, cavity]. Any nearly closed chamber or cavity, e.g., the antrum of the stomach (*pylorus*).

anus (ā′nŭs) [L.]. The outlet of the rectum lying in the fold between the buttocks.

aorta [Gk. *aorte*, aorta. The main trunk of the arterial system of the body arising from the left ventricle of the heart. It has a thoracic portion (ascending, arch, and descending subdivisions) and an abdominal portion (part of descending aorta). Branches from these supply the body with arterial (oxygenated) blood.

aperture (ap′er-tūr) [L. *apertura*, opening]. An opening or orifice.

apex (ā′peks) [L. apex, tip]. Summit or peak; the narrow portion or tip of an organ shaped in a pyramidal form.

aphasia (ă-fā′zĭ-ă) [Gk. *a-*, not, + *phasis*, speaking]. Loss of verbal comprehension, or inability to express oneself properly through speech.

apnea (ap′nē-a) [Gk. *a-* + *pnoē*, breath]. Temporary stoppage or cessation of breathing.

apocrine (ap′ō-krĕn, -krĭn) [Gk. *apo*, from, + *krinein*, to separate]. A method of production of a secretion by an externally secreting gland in which the cells lose part of their cytoplasm in forming the secretion, e.g., mammary gland, apocrine sweat glands.

aponeurosis (ăp-ō-nū-rō′sis) [Gk. *apo*, from, + *neuron*, sinew]. A white, flat, fibrous sheet of connective tissue that attaches muscle to bone or other tissues at their origin or insertion.

appendage (ă-pĕn′dĭj) [L. *appendere*, hang to]. Any part attached or appended to a larger part, as a limb or the appendix.

apposition (ap-ō-zĭ-shun) [L. *ad*, to, + *ponere*, to place]. A fitting together or contact of two substances, usually at a surface, e.g., appositional growth of cartilage by addition of new cartilage on the surface.

aqueduct (ak′we-dukt) [L. *aqua*, water, + *ductus*, duct]. A passage or canal to convey fluids; cerebrospinal fluid flows from the third to fourth ventricle through the cerebral aqueduct (of Sylvius).

aqueous (ā-kwē-ŭs) [L. *aqua*, water]. Watery; having the nature of water. Aqueous humor.

  a. humor [L. fluid]. Watery fluid in the anterior and posterior chambers of the eye.

arachnoid (ă-rak′noid) [Gk. *arachne*, spider, + *eidos*, form]. The central membrane covering the brain and spinal cord, having the appearance of a spider web.

arborization (ar-bor-ĭ-zā′shun) [L. *arbor*, tree]. A treelike structure; terminations of nerve fibers and arterioles.

ar′bor vi′tae [L. *arbor* + *vita*, life]. In the cerebellum the treelike outline of the internal white matter; a series of branching ridges on the uterine cervix.

areola (ă-rē′ō-lă) (pl. *areolae*) [L. a small space]. A small cavity in a tissue; around the nipple, the ringlike pigmented or red area; the part of the iris enclosing the pupil.

arrhythmia (ă-rith′mĭ-ă) [Gk. *a-*, not, + *rhythmos*, rhythm]. Lack of rhythm; irregularity.

arteriole (ar-tē′rĭ-ōl) [L. *arteriola*, small artery]. A small or tiny artery; its distal end leads into a capillary or capillary network.

arterioscloro′sis [L. *arteriola* + Gk. *sklērōsis*, hardening]. General term referring to many conditions in which there is thickening, hardening, and loss of elasticity of the walls of blood vessels.

artery (ar′ter-ĭ) (pl. *arteries*) [Gk. *arteria*, windpipe]. Any vessel carrying blood away from the heart to the tissues. Usually refers to the larger vessels.

arthritis (ar-thrī′tis) (pl. *arthritides*) [Gk. *arthron*, joint, + *itis*, inflammation]. Inflammation of a joint, usually accompanied by pain and frequently structural changes.

arthrosis [Gk. *arthron* + *osis*, increased]. A degenerative condition of a joint; a joint.

artic′ulate [L. *articulatus*, jointed]. To join together as in a joint; to speak clearly.

artic′ulation [L.]. A connection between bones; a joint.

-ase. Suffix used in forming the name of an enzyme; added to the part of or name of substance on which it acts.

associa′tion. Uniting or joining together; coordination with another structure, idea; a neuron lying between and joining an afferent and efferent neuron.

asthenia (as-thē′nĭ-ă) [Gk. *a-* + *sthenos*, strength]. Loss or lack of strength; weakness in muscular or cerebellar disease.

athero- (ath-er-ō) [Gk. *athērē*, porridge]. Prefix: referring to lipid (fatty) substances.

atherosclerosis. Arteriosclerosis characterized by lipid deposits in the inner layer of blood vessel walls (chiefly arteries).

asthma (az′mă) [Gk. panting]. Intermittent severe difficulty by breathing accompanied by wheezing and cough; caused by spasms or swelling of the bronchioles.

astrocyte (as′trō-sīt) [Gk. *astron*, star, + *kytos*, cell]. A star-shaped neuroglial cell possessing many branching processes. Found only in the CNS.

ataxia (ă-taks′ĭ-ă) [Gk. lack of order]. Muscular incoordination when voluntary muscular movements are attempted.

atlas. The first cervical vertebra by which the head articulates with the occipital bone. Named for Atlas, who was supposed to have supported the world on his shoulders.

atom [Gk. *atomos*, indivisible]. The smallest particle of an element that can take part in a chemical change, keeping its identity, and which cannot be further divided without changing its structure; composed of protons, neutrons, electrons.

atre′sia [Gk. *a-* + *trēsis*, a perforation]. Congenital absence of pathological closure of a normal opening; degeneration of ovarian follicles.

ATP. Abbr. for *adenosine triphosphate*.

atrium (ā′trĭ-um) (pl. *atria*) [L. corridor]. A cavity; one of the two upper chambers of the heart; tympanic cavity of the ear; a space that opens into the air sacs of the lungs.

at′rophy [Gk. *a-* + *trophe*, nourishment]. Wasting or decrease in size of a part resulting from lack of nutrition, loss of nerve supply, or failure to use the part.

aud′itory [L. *auditiō*, hearing]. Pert. to the sense of hearing.

auditory nerve. Eighth cranial nerve, sensory. Also, *vestibulocochlear* n.

auricle (aw′rĭ-kl) [L. *auricula*, the ear]. The outer ear flap on the side of the head; a small conical portion of the upper chambers of the heart, *not* synonymous with atrium.

auscultation (aws-kul-tā′shun) [L. *auscultāre*, listen to]. Listening for sounds in body cavities, esp. chest and abdomen, to detect or judge some abnormal condition.

autograft [Gk. *autos*, self, + L. *graphium*, grafting knife]. A graft or transplant taken from one part of the body to fill in another part on the same person.

autoimmune [Gk. *autos* + L. *immunis*, safe]. A condition in which antibodies are produced against a person's own tissue.

autonomic (aw-tō-nom′ik) [Gk. *autos* + *nomus*, law]. Self-controlling; refers to the involuntary or self-regulating portion of the PNS.

autosome (aw′to-sōm) [Gk. *autos* + *soma*, body]. Any of the paired chromosomes other than the sex (X and Y) chromosomes.

axial (aks′ĭ-al) [L. *axis*, lever]. Pert. to or situated in an axis.

axil′la [L. little pivot]. The armpit.

ax′is (pl. *axes*). A line running through the center of a body, or about which a part revolves. The second cervical vertebrae that has the dens *(odontoid process)* about which the atlas rotates.

ax′on [Gk. *axis*]. A neuron process that conducts impulses away from *(efferent)* the cell body.

balance. State of equilibrium; state in which the intake and output or concentrations of substances are approximately equal (homeostasis).

baroreceptors (baro-rē-sĕp′tor) [Gk. *baros*, weight, + L. a receiver]. Receptor organs sensitive to pressure or stretching.

bar′rier. An obstacle or impediment.

bā′sal [Gk. *basis*, base]. The base of something; of primary importance; the lowest level to maintain function.

ba′sal metab′olism [Gk. *basis* + *metabole*, change]. The amount of energy needed for maintenance of life when the subject is at physical, emotional, and digestive rest.

ba′sal metab′olic rate [Gk.]. The metabolic rate as measured under so-called basal conditions, expressed in large Calories/M^2 of body surface/hr. Abbr: **BMR**.

bāse [Gk. *basis*, base]. The lower or broader part of anything; an alkali, any substance that will neutralize an acid.

benign (bē-nīn′) [L. *benignus*, mild]. Mild; not malignant.

beta- (bā′tă) [Gk. β, Second letter of alphabet]. Prefix: to note isomeric position or variety in compounds of substituted groups.

biceps (bī′sĕps) [L. *bis*, two, + *caput*, head]. A muscle with two heads.

bifid (bī′fid) [L. *bis* + *findere*, to cleave]. Cleft or split into two parts.

bilirubin (bil-ĭ′ru-bin) [L. *bilis*, bile, + *ruber*, red]. The orange or yellow pigment in bile; derived from breakdown of heme ($C_{33}H_{36}O_6N_4$).

bio- [Gk. *bios*, life]. Prefix: life

-blast [Gk. *blastos*, germ or bud]. Suffix: immature or primitive.

blind spot [A.S. *blind*, unable to see]. Area in the retina in which there are no light receptor cells (no rods nor cones); point where optic nerve enters the eye *(optic disk).*

blood-brain bar′rier [A.S. *blōd*, blood, + *braegen*, cranial portion CNS, + *barrier*, an obstacle or impediment]. Special mechanism that prevents the passage of certain substances, such as dyes or toxins, from the blood to the CSF.

blood pressure. The pressures measured on a peripheral artery during contraction and relaxation of the heart.

  b.p., diastolic. The pressure remaining in the arteries during relaxation of the left ventricle.

  b.p., systolic. The peak pressure created by contraction of the left ventricle of the heart and the resistance to flow offered by the blood vessels.

bo′lŭs [Gk. a mass]. A mass of food ready for swallowing.

bone marrow [A.S. *bōn*, bone, + *mearh*, marrow]. Soft tissue within the hollow of certain bones of the skeleton; responsible for production of certain blood cells or for storage of fat.

-borne (born, [*bore*, to carry]. Refers to types of methods of transmission of communicable diseases, e.g., airborne, waterborne, foodborne.

Bowman's capsule [Sir Wm. Bōwman, Eng. phys., 1816-1892, + L. *capsula*, small box]. The capsule containing the glomeruli of the kidney; the cuplike depressions on the expanded ends of the renal tubules or nephrons that surround the capillary tufts *(glomeruli).*

brachium (brā′kĭ-um) [Gk. *brachiōn*, arm]. The upper arm from shoulder to elbow.

breathe (brēth) [A.S. *braeth*, odor]. To take in and release air from the lungs and respiratory system; respire.

bronchi (bron′kī) (sing. *bronchus*) [Gk. *bronchos*, windpipe]. The first two divisions of the trachea that penetrate the lungs and terminate in the bronchioles; air tubes to and from the lungs.

bronchiectasis (bron-kĭ-ek′tă-sis) [Gk. *bronchos* + *ektasis*, dilation]. Condition characterized by dilation of the bronchus(i) with secretion of large amounts of foul substance.

bronchiole (bron′kĭ-ōl) [L. *bronchiolus*, air passage]. A smaller division of the bronchi that leads to the alveolar ducts to the alveoli.

bucca (buk′a) [L. mouth, cheek]. The mouth; hollow part of the cheek.

buffer (bŭf′ĕr) [Fr. *buffe*, blow]. A substance preserving pH on addition of acids or bases; reduces the change in H^+ concentration

  b. pair, b. system. The combination of a weak acid and its conjugate base, which work together to maintain pH nearly constant.

bul′b [Gk. *bulbos*, bulb-shaped root]. An expansion of an organ, canal, or vessel.

bulk [M.E. *bulke*, a heap]. Size, mass, or volume; the main body of something.

bun′dle of His [M.E. *bundel*, bind, + W. His, Gr. anat., 1863-1934]. A bundle of fibers of the impulse-conducting system that initiates and controls contraction of the heart muscle. It extends in the A-V node and becomes continuous with Purkinje's fibers of the ventricles.

bur′sa [Gk. a leather sac]. A sac or cavity filled with a fluid that reduces friction between tendon and ligament or bone, or other structures where friction may occur.

buttocks (but′uks) [M.E. *butte*, thick end]. The gluteal muscle prominence, the ''rump'' or ''seat.''

-calcemia (kal-sē′mĭ-ă) [L. *calcarius*, pert. to calcium or lime, + Gk. *amia*, blood]. The calcium level of the blood.

calcification (kal-sif′ĭk-a shun) [L. *calx*, lime, + *ferre*, to make]. Deposit of lime salts in bones and other tissues.

calcitonin (kal-sĭ-tōn′ĭn) [Gk. *calci*, pert. to calcium + *tonos*, tone]. A hormone secreted by the thyroid gland that aids in regulating calcium-phosphorus metabolism.

calculus (kal′kū-lus) (pl. *calculi*) [L. pebble]. A "stone" formed within the body, usually composed of mineral salts.

cal′lus [L. hard]. Circumscribed area of the horny layer of the skin that has thickened; osseous material formed between the ends of a fractured bone.

calyx (kā′lix) (pl. *calyces*) [Gk. *kalix*, cup]. Cuplike divisions of the kidney pelvis.

canal (ka-nal) [L. *canā′lis*, canal]. A narrow channel, tube, duct, or groove.

canaliculus (kan-ă-lĭk′ū-lus). A tiny canal or channel.

can′cellous [L. *cancellus*, a grating]. Having a latticework or reticular structure, as the spongy tissue of bone.

cancer (kan′ser) [L. a crab; ulcer]. A malignant tumor or neoplasm (new growth); a carcinoma or sarcoma.

capillary (kap′ĭ-lā-rē) (pl. *capillaries*) [L. *capillaris*, hairlike]. Minute blood vessels that connect the arterioles with the venules to carry blood; minute lymphatic ducts. Capillary *networks* allow passage of oxygen and nutrients from the blood to the tissues, and of wastes from the tissues into the blood.

capitulum (ka-pĭt′ū-lum) [L. *caput*, head]. A small rounded end of a bone that articulates with another bone.

ca′put [L.]. The head; upper part of an organ.

carbohyd′rates (kar-bō-hĭd′ratz) [L. *carbo*, carbon, + G. *ydōr′*, water]. A class of organic compounds composed of C, H, and O, with H and O usually in the ratio of 2:1; sugars, starches and cellulose; the *monosaccharides, disaccharides,* and *polysaccharides.*

car′bon [L. *carbo*, carbon or coal]. Constituent of organic compounds, found in all living things; combines with H, O, N to make life possible.

carbon dioxide. A colorless gas, heavier than air; produced by complete combustion, fermentation, or decomposition of carbon (compounds) found in air, and exhaled by all animals. CO_2.

carbonic anhydrase. An enzyme catalyzing the reaction of CO_2 and water to form *carbonic acid* (H_2CO_3) and vice versa; present primarily in red blood cells.

carcinogens (kar′sĭ-nō-jens′) [Gk. *karkinos*, cancer, + *genere*, to produce]. Substances known to cause cancer or neoplasms.

carcinoma (kar-sĭ-nō′mă) [Gk. *karkinos* + *oma*, tumor]. A malignant cancer or neoplasm of epithelial tissues.

cardiac output. Amount of blood ejected from one ventricle per minute.

caries (ka′rēz) [L. rottenness]. Tooth or bone decay.

carotene (car′ō-ten) [Gk. *karōton*, carrot]. A yellow crystalline pigment obtained from yellow vegetables; stored and converted to *vitamin A* in the liver.

carpus (kar′pus) [Gk. *karpos*, wrist]. The eight bones of a wrist. Also, *carpals*.

carrier [Fr. *carier*, to bear]. A large molecule transporting substances across cell membranes in *active transport.*

carotid (kă-rot′ĭd) [Gk. *karōtides*, from *karos*, heavy with sleep]. Arteries providing the main blood supply to the head and neck; pressure on them may produce unconsciousness.

cartilage (kar′til-ăj) [L. *cartilagō*, gristle]. A form of connective tissue usually with no blood or nerve supply of its own; it is firm, elastic, and a semiopaque bluish white or gray; cells lie in cavities called *lacunae.*

catabolism (kă-tab′ō-lism) [Gk. *katabolē*, a casting down]. The destructive phase of metabolism; breaking down of complex chemical compounds into simpler ones, usually with the release of energy.

catalyst (kat′ă-list) [Gk. *katalysis*, dissolution]. A substance that alters or speeds up the rates of chemical reactions without being itself altered in the process.

cat′aract [Gk. *katarraktēs*, a rushing down]. Opacity or loss of transparency of the lens of eye, its capsule, or both.

cation (kat′ĭ-on) [Gk. *katiōn*, descending]. A positively charged ion or radical.

cauda equina (kaw′dă e-kwin′a) [L. *cauda*, tail, + *equus*, horse]. The terminal portion of the roots of the spinal nerves and spinal cord below the first lumbar nerve: resembles a horse's tail.

cavernous body (kăv′ĕr-nŭs) [L. *caverna*, a hollow, + *corpus*, body]. One of several columns of tissue containing blood spaces in the dorsum of the clitoris or penis that aids in the erection of these organs.

cavity (kav′ĭ-tĭ) [L. *cavitas*, hollow]. A hollow space, such as in a body organ or a decayed tooth.

cecum (sē′kŭm) [L. *caecum*]. *The blind pouch that forms the first portion of the large intestine.*

-cele [Gk. *hernia*, tumor]. Suffix: a swelling.

celiac (sē′lĭ-ak) [Gk. *koilia*, belly]. Pert. to the abdominal region.

coelom (sē′lom) [Gk. *koilōma*, a hollow]. The embryonic cavity between the layers of mesoderm that develops into the peritoneal, pleural, and pericardial cavities. Also **celom.**

center (sen′ter) [Gk. *kentron*, middle]. Midpoint of a body; usually refers to a group of nerve cells in the brain or spinal cord; controls some specific activity.

centimeter [L. *centum*, a hundred, + *metron*, measure]. Abbr. **cm.** 2.5 cm equals 1 inch; 1 cm equals $1/100$ meter, or about $2/5$ inch (0.3937 inch).

centriole (sĕn′trĭ-ōl). A cell organelle that is involved in cell division; forms a spindle.

centromere (sen′trō-mēr). The structure at the junction of the two arms (chromatids) of a chromosome.

centrosome (sen′trō-sōm) [Gk. *kentron*, center, + *soma*, body]. An area of the cell cytoplasm lying near the nucleus that contains one or two *centrioles;* active during cell division.

cephalic (sef-ăl-ic) [Gk. *kephalē*, head]. Cranial, pert. to the head; superior in position.

cerebellum (ser-ĕ-bel′um) [L.]. A dorsal portion of the brain that is involved in co-ordinating activity of skeletal muscles, and in the coordinating voluntary muscular movements.

cerebrum (ser′ĕ-brum) [L.]. The upper largest part of the brain; contains the motor, and sensory areas, and the association areas that are concerned with the higher mental faculties.

cerumen (se-rū′men) [L. *cera*, wax]. The waxlike secretion in the external ear canal.

cervical (ser′vĭ-kal) [L. *cervicalis*, pert. to the neck]. The neck region, or neck of an organ, e.g., *cervix* or neck of the uterus. The first *8* spinal nerves; the first *4* cervical spinal nerves form the *c. plexus.* The first *7* vertebrae.

chamber (cham′bèr) [Gk. *kamara*, vault]. A compartment or closed space, as the chambers of the heart.

chem′oreceptor. A sense organ or sensory nerve ending that is stimulated by a chemical substance or change, as in the taste buds.

chemotactic (kem-o-tak′-tĭk) [Gk. *chēmeia*, chemistry, + *taxikos*, arranging]. Responding to a chemical stimulus by being attracted or repelled.

chiasm, chiasma (ki'azm) [Gk. *chiazen*, to mark with letter X]. A crossing or decussation; optic chiasm; the point of crossing of the optic nerve fibers.

chemotherapy (kem-o-ther'-a-pi) [Gk. *chēmeia* + *therapeia*, treatment]. Treatment of disease by use of chemical agents that have a specific effect on the causative microorganism or agent, usually without producing toxic effects on the person.

choana (kō'ă-nă) (pl. *choanae*) [Gk. *choane*, funnel]. A funnel-shaped opening; the communicating passageways between the nasal cavities and the pharynx; the posterior nares or nostrils.

cholecystokinin (ko-le-sĭs-tō-kī'nin) [Gk. *cholē*, bile, + *kystis*, cyst, + *kinein*, to move]. A hormonelike chemical secreted by the duodenal mucosa that stimulates contraction of the gallbladder. Abbr. **CCK**.

cholesterol (ko-les'ter'ol) [*cholē* + *stereos*, solid]. A sterol (monohydric alcohol, $C_{27}H_{45}OH$) forming the basis of many lipid hormones in the body, e.g., sex hormones, adrenal corticoids; synthesized in the liver; a constituent in the bile.

cholinergic (kō'len-er-gik). Term applied to nerve endings that liberate acetylcholine.

chondroblast (kŏn'drō-blăst) [Gk. *chondros*, cartilage, + *blastos*, germ]. A primitive cell that forms cartilage.

chondrocyte (kŏn'drō-sīt) [Gk.]. A cartilage cell.

cholinesterase (kō-lĭn-ĕs'ter'ās). Any enzyme that catalyzes the hydrolysis of choline esters (organic acid + an alcohol) e.g., *acetylcholinesterase* catalyzes the breakdown of *acetycholine*.

chorion (kō'rĭ-ŏn) [Gk.]. Outer embryonic membrane of the blastocyst from which *villi* connect with the endometrium to give rise to the placenta.

chordae tendinae (kor'da tĕn'dĭn-ē) [Gk. *chorde*, cord, + L. *tendinōsus*, like a tendon]. Small tendinous cords that connect the edge of an atrioventricular valve to a papillary muscle and prevent valve reversal.

chorea (ko-re'ă) [Gk. *choreia*, dance]. Rapid, irregular, and wormlike movements of body parts; a nervous affliction.

chorioid (kō'rĭ-oid) [Gk. *chorioeidēs*, skinlike]. The middle vascular layer of the eyeball between the sclera and retina; a part of the *uvea*. Also, **choroid**.

choroid plexus (kō'roid plĕk'sŭs) [Gk. *chorioeidēs*, skinlike, + L. *plexus*, interwoven]. Vascular structures in the roofs of the four ventricules of the brain that produce *cerebrospinal fluid*.

chromatin (krō'ma-tin) [Gk. *chrōma*, color, + *in*, within]. Darkly staining substance within the nucleus of most cells; consists mainly of *DNA* and is considered to be the physical basis of *heredity;* forms *chromosomes* during mitosis and meiosis.

chromosome (kro'mō-sōm) [Gk. *chrōma*, color, + *soma*, body]. A microscopic J- or V-shaped body developing from chromatin during cell division. Contains the gene or hereditary determinants of body characteristics. In human, total **46** (23 pairs).

chronic [Gk. *chronos*, duration]. Long and drawn out with slow progress; a disease that is not acute.

chyle (kīl) [Gk. *chylos*, milklike]. The contents of intestinal lymph vessels. Consists mainly of absorbed products of fat digestion.

chyme (kīm) [Gk. *chymos*, juice]. The mixture of partially digested foods and digestive juices found in the stomach and small intestine.

cilia (sil'ĭ-ă) [L. *ciliaris*, pert. to eyelashes]. Motile hairlike projections from epithelial cells; eyelashes.

circumduction (sir-kum-duk'shun) [L. *circum*, around, + *ducere*, to lead]. Movement of a part in a circular direction; involves flexion, extension, abduction, adduction, and rotation.

cirrhosis (sĭ-rō'sĭs) [Gk. *kirros*, yellow, + *osis*, infection]. A chronic liver disease characterized by degenerative changes in the liver structure.

cister'na [L. a vessel]. A cavity or enlargement in a tube or vessel.

cleavage (kle'vej) [A.S. *cleofan*, to split]. Mitotic cell division of a fertilized egg; splitting a complex molecule into two or more simpler ones.

cleft [M.E. *clyft*, crevice]. A fissure; divided or split.

cleido (kli'do) [Gk. *kleis*, clavicle]. Prefix: pert. to the clavicle (collarbone).

clitoris (klī'to-rĭs, klit'ō-ris) [Gk. *kleitoris*, clitoris]. Female erectile organ that is homologous to the penis of the male.

clone (klōn) [Gk. *klon*, a cutting used for propagation]. A group of cells descended from a single cell; something reproduced exactly like its predecessor.

clot (klŏt) [A.S. *clott*]. The semisolid mass of fibrin threads plus trapped blood cells, which forms when the blood coagulates; to coagulate.

coagulation [L. *coagulatiō*]. The process of clotting by which liquid blood is changed into a gel to aid in preventing blood loss through injured vessels.

coccygeal (kok-sij'ē-al) [Gk. *kokkyx*, coccyx]. Pert. to the coccyx, tailbone, last four fused spinal bones.

cochlea (kok'lē-ă) [Gk. *kochilias*, a spiral]. A cone-shaped winding tube that forms the portion of the inner ear that contains the organ of Corti, the receptor for hearing.

coitus (ko'i-tus) [L. a uniting]. Sexual intercourse between man and woman. Also, *coition*, *copulation*.

collagen (kol'ă-jen) [Gk. *kolla*, glue, + *gennan*, to produce]. A fibrous protein forming the main organic structures of connective tissues.

collateral (ko-lat'er-al) [L. *con*, together, + *lateralis*, pert. to a side]. Side branch of a nerve axon or blood vessel.

colon [Gk. *kōlon*]. The portion of the large intestine from the end of the ileum to the rectum. It consists of the ascending, transverse, descending, and sigmoid portions.

co'ma [Gk. *kōma*, a deep sleep]. An abnormal deep stupor from which the person cannot be aroused by external stimuli.

compact' bone [L. *compactus*, joined together]. The hard or dense part of bones containing subunits of structure called osteons or haversian systems.

compensate [L. *cum*, with, + *pensdāre*, to weigh]. To make up for a defect or deficiency.

compliance. In the lung, ease of expanding.

component. A constituent part.

compound [L. *componere*, to place together]. A substance composed of two or more parts and having properties different from its parts.

commissure (kŏm'ĭ-shūr) [L. *commissura*, a joining together]. Band of nerve fibers passing transversely over the midline in the CNS.

concave (kon'kāv) [L. *con*, with, + *cavus*, hollow]. Having a rounded hollow or depressed surface.

concentration (kon-sen-tra'shun) [L. *contratiō*, in the center]. Amount of a solute per volume of solvent; increase in strength of a fluid by evaporation or removal of water by other means.

concep'tion [L. *conceptiō*, a conceiving]. The time when a sperm unites with an ovum to initiate formation of a new individual; *fertilization*.

concha (kong'kă) [Gk. *kogchē*, shell]. A shell-shaped structure, as the nasal turbinates.

conduc'tion [L. *conducere*, to lead]. The transmission of a nerve impulse by exciting progressive segments of a nerve fiber; the transfer of electrons, ions, heat, or sound waves through a conducting media.

congenital (kon-jen'ĭt-al) [L. *congenitus*, born together]. Present at birth.

cone (kōn) [Gk. *kōnos*, cone]. A receptor cell in the retina concerned with color vision.

conjunctiva (con-junk-tī'vă) [L. *con*, with, + *jungere*, to join]. Mucous membrane lining of the eyelid that is reflected onto the eyeball.

conscious (kon'shus) [L. *conscius*, aware]. Awake, being aware and having perception.

constric'tion [L. *con* + *stingere*, to draw]. The narrowing or becoming smaller in diameter as of a pupil of the eye or blood vessel.

continuity (kon-tĭ-nū'ĭ-tī) [L. *continuus*, continued]. The state of being intimately united, continuous, or held together.

contrac'tion [L. *contractio*, a drawing up]. Shortening or tightening as of a muscle, or a reduction in size.

convergent (kon-ver'jent) [L. *con* + *vergere*, to incline]. Tending toward a common point; coming together.

conver'sion [L. *convertere*, to turn around]. Change from one state to another.

con'vex [L. *convexus*, vaulted, arched]. Having an outward curvature or bulge.

convul'sion [L. *convulsio*, a pulling together]. Paroxysms or involuntary muscular contractions and relaxations.

coordination (ko-or-din-a'shun) [L. *co*, together, + *ordināre*, to arrange]. The working together of different body systems in a given process; the working together of muscles to produce a certain movement.

cope (kōp). The ability to effectively deal with and handle the stresses to which a person is subjected.

cor'nea [L. *corneus*, horny]. The transparent anterior portion of the fibrous coat of the eye, continuous, with the sclera.

corium (ko'rĭ-um) [Gk. *chorion*, skin]. The dermis or true skin that lies immediately under the epidermis; contains nerve endings, capillaries, and lymphatic vessels, and is composed of connective tissue.

coronary (kor'o-na-rĭ) [L. *coronarius*, pert. to a circle or crown]. Blood vessels of the heart that supply blood to its walls.

cor'pus (pl. *corpora*) [L. *body*]. Any mass or body; the principal part of any organ.

cor'puscle [L. *corpusculum*, little body]. An encapsulated sensory nerve ending; any small rounded body; old term for blood cell, erythrocyte or leukocyte.

cor'pus luteum (lū'tē-um) [L. *corpus* + *luteus*, yellow]. Yellow body of cells that fills the ovarian follicle in the stage following ovulation.

coro'na [Gk. *korōnē*, crown]. Any structure resembling a crown; *coronal* plane or section, same as *frontal*.

cor'tex (pl. *cortices*) [L. rind]. An outer portion or layer of an organ or structure.

cor'ticoid. One of many adrenal cortical steroid hormones. Also, *corticosteroid*.

corticospi'nal [L. *cortex* + *spīna*, thorn]. Pert. to cerebral cortex and spinal cord.

cos'tal [L. *costa*, rib]. Pert. to a rib.

countercurrent (kown'ter-ker'rent) [L. *contra*, against, + *currere*, to run]. Flow of gases or fluids in opposite directions in two closely spaced limbs of a bent tube.

crā'nium [L. from Gk. *kranion*, skull]. That portion of the skull that encloses the brain; consists of eight bones.

creatine (krē'ă-tĭn) [Gk. *kreas*, flesh]. A crystalline substance found in organs and body fluids; combines with phosphate and serves as a source of high-energy phosphate released in the anaerobic phase of muscle contraction.

crest [L. *crista*, tuft]. A ridge or long prominence, as on bone.

crista galli [L. *crista* + *galea*, helmet]. A crest or shelf on the ethmoid bone that attaches the *falx cerebri*.

criterion (pl. *criteria*) [Gk. *krites*, judge]. A standard or means of judging a condition or establishing a diagnosis.

crura (kroo'ra) (sing. *crus*) [L. legs]. The legs; a pair of long bands or masses resembling legs.

cross-linkage. Chemical bonds linking fibrous molecules.

crypt (kript) [Gk. *kryptein*, to hide]. A small cavity or sac extending into an epithelial surface; a tubular gland.

crys'talloid [Gk. *krystallos*, clear ice, + *eidos*, form]. A substance that is capable of forming crystals and that in solution diffuses rapidly through membranes; less than 1 μ in diameter; smaller than a *colloid*.

cuta'neous [L. *cutis*, skin]. Pert. to the skin.

cyanosis (sī-an-o'sis) [Gk. *kyanos*, dark blue, + *-osis*, state of]. A condition of blue color being given to skin and mucous membranes by excessive amounts of reduced hemoglobin in the bloodstream.

cycle (sī'kl) [Gk. *kyklos*, circle]. A series of events; something that has a predictable period of repetition; a sequence.

-cyte (sīt) [Gk. *kytos*, cell]. Suffix denoting cell.

cyto- [Gk.]. Prefix indicating the cell.

cytokine'sis [Gk. *kytos* + *kinēsis*, movement]. Division of cytoplasm in latter stages of mitosis.

cytoplasm (sī'to-plazm) [Gk. *kytos* + *plasma*, matter]. The cellular substance between the membrane and nucleus of a cell.

damping. Steady decrease in amplitude of successive vibrations, as in the cochlea.

de-. Prefix: down; from.

death (deth) [A.S. *dēath*]. Permanent cessation of all vital functions. Absence of life. A flat EEG for 48 hours is taken as a sign of loss of spontaneous brain activity and death.

decompression [*de* + *compressio*, a squeezing together]. Release or removal of pressure from an organ or organism.

decussate (de-kus'āt) [L. *decussāre*, to cross, as an X]. To undergo crossing, or crossed.

defecation (def-ĕ-kā'shun) [L. *defaecāre*, to remove the dregs]. Evacuation or emptying fecal matter from the large bowel.

de'fect. A flaw or imperfection.

deficit (def'e-sit) [L. *deficere*, to be lacking]. Lack of or lacking in; to be less than.

degeneration (di-jen'er-ā-shun) [L. *degenerāre*, to degenerate]. To "go bad" or deteriorate; to lose quality.

degrade (di-grāde) [L. *degradare*, to reduce in rank]. To tear down or break down, as to degrade a chemical compound.

dehydration (de-hī-drā'shun) [*de* + Gk. *ydōr*, water]. Removal of water from something; the result of water removal.

den'drite [Gk. *dendrités*, pert. to a tree]. A branched process of a neuron that conducts impulses to the cell body; form synapses.

dener′vate [L. *dē*, from, + Gk. *neuron*, nerve]. To block, remove, or cut the nerve supply to a structure.

de novo (dē-nō′vō) [L.]. Anew; again; once more.

dense [L. *densus*, compact]. Packed tightly together; in a bone, the outer layers of bony tissue that do not contain osteons.

deoxyribonucleic acid (dē-ok-sī-ri-bō-nu′klē-ik). Composed of nitrogenous bases, deoxyribose sugar, and a phosphate group. It is believed to be the site of determination of the body's hereditary characteristics. **DNA**.

depolarization (dē-pō-lar-ī-zā′shun) [*de-* + *polus*, pole]. Loss of polarity or the polarized state.

depressed (de-prest′). To push down; flattened or hollow; lowered in intensity; low in spirits.

depression (de-presh′un) [L. *depressiō*, a pressing down]. A lowered or hollowed region; lowering of a vital function; a mental state of dejection, lack of hope or absence of cheerfulness.

derivative (dĕ-riv′a-tiv) [L. *derivare*, to turn a stream from its channel]. A substance or structure originating from another substance or structure and having different properties than its source.

-derm- (durm) [L. *derma*, skin or covering]. A word used as a prefix or suffix referring to the skin or a covering.

derma (dermis). Pert. to the skin; the connective tissue layer of the skin under the epidermis or covering layer.

desquamate (des-kwam-āte) [*de-* + L. *squamāre*, to scale off]. Shedding of the epidermal surface cells.

detoxify (de-toks′ī-fī) [*de-* + Gk. *toxikon*, poison, + L. *facere*, to make]. To remove the toxic or poisonous quality of a substance.

development (Fr. *de′velopper*, to unwrap). Progress from embryonic to adult status; change from a less specialized to a more specialized state.

dextrose (deks′trōs) [L. *dexter*, right + *ose*, sugar]. Another name for glucose, a simple sugar (monosaccharide). Also, *grape sugar*.

di- [Gk.]. Prefix: twice or double.

dialysis (di-al′ĭ-sis) [Gk. *di-* + *lysis*, loosening]. Passage of a diffusible solute through a membrane that restricts passage of colloids; separation of solutes by their rates of diffusion.

diameter (di-am′ĕ-ter) [Gk. *di-* + *metron*, a measure]. A distance from any point on the periphery of a surface to the opposite point.

diapedesis (di-ă-ped-e′sis) [Gk. *di* + *pēdan*, to leap]. Movement of white blood cells through undamaged capillary walls.

diaphragm (di′ă-fram) [Gk. *di-* + *phragma*, wall]. Any thin membrane; a rubber or plastic cup placed over the cervix of the uterus for contraceptive purposes; the muscle and connective tissue partition between chest and abdomen (muscle of respiration).

diaphysis (di-af′i-sis) [Gk. *di-* + *plassein*, to grow]. The shaft or cylindrical part of a long bone.

diastole (di-as′tō-le) [Gk. *diastellein*, to expand]. The relaxation phase of cardiac activity, when fibers are elongating and filling of a chamber is rapid.

diencephalon (dī-en-sef′ă-lon) [Gk. *di-* + *enkephalos*, brain]. The second portion of the brain between telencephalon and mesencephalon; includes thalamus and hypothalamus.

differentiation (dif′er-en′shi-a′shun) [L. *dis*, apart, + *ferre*, to bear]. Acquiring a structure or function different from those originally present.

diffuse (dĭ-fūs′) [L. + *fundere*, to pour]. To spread, scatter, or move apart; usually caused by movement of molecules in a solution or suspension.

diffusion (dĭ-fu′zhun). Eventual even mixing of solutes and solvents as a result of motion of molecules.

digest [Gk. *di-* + L. *gerere*, to carry]. To break down foods using chemical or mechanical means; to convert foods to an absorbable state.

dilate (di′lāt) [L. *dilatāre*, to expand]. To become larger in size or diameter.

dilute (di-loot′) [L. *dilutus*, to wash away]. To thin down or weaken by addition of water or other liquids.

dimension (dĭ-men′shun) [Fr. *dimensio*, a measuring]. A measurable extent, such as length, width; extent or size; importance.

dip′loë [Gk. a fold]. The spongy bone between the inner and outer tables (layers) of dense bony tissue in the skull bones.

diploid (dĭp′loyd). Cell having twice the number of chromosomes present in the egg or sperm of a given species.

disaccharide (dī-sak′ĭ-rid) [Gk. *dis*, two, + *sakcharon*, sugar]. A sugar composed of two simple sugar molecules.

dissect (dis-sekt′) [L. *dissecare*, to cut up]. To separate tissues or organs of a cadaver for study.

dissociate (dis-so′si-āt) [L. *dissociare*, to separate from fellowship]. To separate into components or parts.

dissolve (dĭ-zolv) [L. *dossolvere*, to dissolve]. To make liquid, liquefy, or melt by placing in water or other liquids.

distal (dis′tal) [L. *distāre*, to be distant]. Farthest from the center of the body or point of attachment; opposite of proximal.

diuresis (di-u-re′sis) [Gk. *dia*, through, + *ourein*, to urinate]. Secretion and passage of abnormally large amounts of urine.

diuretic (di-u-ret′ik). A substance causing diuresis.

diverticulum (di-ver-tik′u-lum) [L. *diverticulāre*, to turn aside]. An outpocketing, sac, or pouch in the wall of a canal or hollow organ.

dominance (dom′i-nense) [L. *dominare*, to rule]. In inheritance of characteristics, the effects of one allele overshadow the effects of the other and the character determined by the dominant (stronger) gene prevails.

donor (dō′ner) [Fr. *donoer*, to give]. Someone who gives or donates (such as blood or organs).

dor′sal [L. *dorsum*, back]. Pert. to the upper or back portion of the body.

drug [Fr. *drogue*]. A medicinal substance used in the treatment of disorder or disease.

duct [L. *ducere*, to lead]. A tubular vessel or channel carrying secretions from a gland onto an epithelial surface.

ductus. A channel from one organ or part of an organ to another.

duplication (doo′plĭ-cā-shun) [L. *duplicare*, to double]. To double something; to form a copy.

dura (du′ră) [L. *durus*, hard]. Hard or tough; *dura mater*, the tough outer membrane around the brain and spinal cord.

dwarf (M.E. *dwerf*, little). An abnormally small, short, or disproportioned person.

dysmenorrhea (dis-men-o-re′a) [L. *dys*, bad, difficult, or painful + *men*, mouth, + *rein*, to flow]. Painful or difficult menstruation.

dyspnea (disp-ne′a) [L. *dys*, + Gk. *pnoē*, breathing]. Labored or difficult breathing.

dystrophy (dis′tro-fī) [L. *dys*, + Gk. *trephein*, to nourish]. Degeneration of an organ resulting from poor nutrition, abnormal development, infection, or unknown causes.

ectoderm (ek'tō-derm) [Gk. *ekto*, outside, + *derma*, skin]. The outer layer of cells in the embryo, from which develop the nervous system, special senses, and certain endocrines.

-ectomy (ek'tō-my; [Gk. *ektomē*, a cutting out]. Suffix pert. to surgical removal of an organ or gland.

ectopic (ek-top'ik) [Gk. *ek*, out, + *topia*, place]. In an abnormal place or position.

edema (e-de'ma) [Gk. *oidema*, swelling]. Swelling caused by increase of extracellular fluid volume; dropsy.

effect'or [L. *effectus*, accomplishing]. An organ that responds to stimulation; a gland or muscle cell.

ef'ferent [L. *ex*, out, + *ferre*, to carry]. Carrying away from a center or specific point of reference.

egest (ē-jest') [L. *egestus*, to discharge]. To rid the body of something. Also, eliminate.

ejaculation (ē-jak-u-la'shun) [L. *ejaculāri*, to throw out]. Ejection of semen from the male urethra during sexual excitement or the discharge of secretions from vaginal glands.

elastance (ē-last'ance) [Gk. *elasticos*, elastic]. The power of elastic recoil after being stretched. Also, *elasticity*.

electro- (ē-lec'trō) [Gk. *elektron*, amber]. Prefix pert. to electrical phenomena of one sort or another.

electrolyte (e-lek'trō-līt) [Gk. *elektron* + *lytos*, solution]. A substance that dissociates into electrically charged ions; a solution capable of conducting electricity because of the presence of ions.

elec'tron [Gk.]. A minute body or charge of negative electricity; a component of atoms.

el'ement [L. *elementum*, a rudiment]. A substance that cannot be separated into its components by conventional chemical means.

eliminate (e-lim'ĭ-nāte) [L. *ē*, out, + *limen*, threshold]. To rid the body of wastes by emptying of a hollow-organ; to expel.

em'bolus [Gk. plug]. A floating undissolved mass or bubble of gas brought into a blood vessel by the fluid flow.

embryo (em'brĭ-o) [Gk.]. The developing human from the *third* to *eighth weeks* after conception.

embryonic disc. The primitive two-layered structure formed about 2 weeks after fertilization, from which the embryo develops.

emeiocytosis (ēm-ē-ōsī-tō'sis) [Gk. *emein*, to vomit, + *cyte*, cell, + *osis*, state of]. Elimination of materials from a cellular vacuole by fusion of the vacuole with the cell membrane and release of contents to the outside of the cell.

emesis (em'ĕ-sis) [Gk. *emein*, to vomit). Vomiting.

-emia (ē'mi-a) [Gk. *aima*, blood]. Suffix: blood.

emmetropic (em-me-trō'pik) [Gk. *emmetros*, in due measure, + *opsis*, sight]. Normal in vision.

emphysema (em-fi-se'mă) [Gk. *emphysan*, to inflate]. Condition resulting from rupture or expansion of alveoli of the lungs.

en- [Gk. *en*, in]. Prefix: in.

encephalo- (en-sef'a-lō) [Gk. *egkephalos*, brain]. Prefix pert. to brain or cerebrum.

en'dō- [Gk. *endon*, within). Prefix: within; inner.

endocrine (ĕn'do-krīn,-krĭn) [Gk. *endon* + *krinein*, secrete]. Internally secreting into the bloodstream.

endoderm (en'do-derm) [Gk. *endon* + *derma*, skin]. Inner layer of cells of the embryo; gives rise to *linings* of gut and all outpocketings of the gut (e.g., lungs, pancreas, liver).

endometrium (en-do-me'trī-um) [Gk. *endon* + *mētra*, uterus]. The mucous membrane lining the inner surface of the uterus.

endoplasmic reticulum (en-dō-plas'mik re-tĭk'ū-lum). The series of tubular structures found in the cytoplasm of cells; transports substances through cells.

endothelium (en-dō-the'lī-um) [Gk. *endon* + *thēlē*, nipple]. The simple squamous epithelium forming the inner lining of blood and lymphatic vessels and the heart.

end product. The final product or waste of a digestive or metabolic process.

energy (en'er-ji) [Gk. *en*, in, + *ergon*, work]. Capacity to do work; heat or chemical bond capacity for work. Energy is seen as motion, heat, sound and so forth.

engorged (en-gorjd') [Fr. *engorger*, to obstruct or devour]. Swollen or distended as with blood.

engulf (in-gulf') [Fr. *engolfer*, to swallow up]. To surround, take in, or swallow.

en'tero- [Gk. *enteron*, intestine]. Prefix pert. to the intestines.

enteroceptive (en-ter-ō-sep'tĭv) [Gk. *enteron* + L. *capere*, to take]. Originating within body viscera.

environment (en-vī'rŏn-ment) [M.E. *envirounen*, about, + *ment*, from Fr. *en*, in, + *virer*, to turn]. The surroundings of a cell, organ, or organism.

enzyme (en-zīm) [Gk. *en*, in, + *zymē*, leaven]. An organic catalyst causing alterations in rates of chemical reactions without being consumed in the reaction.

eosin (ē'o-sin) [Gk. *ēōs*, dawn]. A rose-colored dye used for staining cells and tissues. Something exhibiting a preference for the dye is called *eosinophilic*.

epi- [Gk.]. Prefix: on, at, outside of, in addition to.

epididymis (ep-ĭ-did'ĭ-mis) [Gk. *epi-* + *didymos*, testes]. A long convoluted tubule or duct resting on the testis and conveying sperm to the vas deferens.

epiglottis (ep-ĭ-glot'is) [Gk. *epi-* + *glōttis*, glottis]. A leaf-shaped structure located over the superior end of the larynx. Aids in closing the larynx during swallowing.

epiphysis (ē-pif'ĭ-sis) [Gk. *epiphysis*, a growing on]. One of the two ends of a long bone, separated from the diaphysis or shaft by a growth line.

epithelial (ep-ĭ-thē'lī-al). Pert. to epithelia, the covering or lining tissues of the body.

equilibrium (e-kwil-ib'rī-um) [L. *aequus*, equal, + *libra*, balance]. A state of balance or rest, in which opposing forces are equal.

eruption (e-rup'shun) [L. *eruptio*, a breaking out]. Becoming visible, such as a lesion or rash appearing on the skin or a tooth appearing through the gums.

erythropoiesis (ē-rith-ro-poy-e'sis) [Gk. *erythema*, redness, + *poiesis*, making]. The formation of erythrocytes, or red blood cells.

eschar (es'kar) [Gk. *eschara*, scab]. A slough (dead matter or tissue); *debris* developing following a burn.

esthesia (es-the'zī-ā) [Gk. *aisthēsis*, sensation or feeling]. Perception, sensation, or feeling.

estrogen (es'trō-jen) [Gk. *oistros*, mad desire, + *gennan*, to produce]. A substance producing or stimulating female characteristics; the follicular hormone.

eu- (ū) [Gk. *eu*, well]. Prefix: normal, well, or good.

euphoria (u-fo′rĭ-ā) [Gk. *eu* + *pherein*, to bear]. An exaggerated feeling of well being.

evacuate (e-vak′ū-āte) [L. *evacuāre*, to empty]. To empty or discharge the contents of, especially the bowels.

evaporate (ē-va′por-āte) [L. *ē*, out, + *vaporāre*, to steam]. To change from liquid to gaseous form usually by addition of heat, as to evaporate water by boiling, or on the skin.

evoke (e-vōk) [Fr. *évoquer*, to call out]. To cause to happen, to call forth.

ex- [L.]. Prefix: out, away from, completely.

excitable (ĕk-sī′tă-bl) [L. *excitāre*, to rouse]. The property of being capable of responding to stimuli by alterations in ion separation.

excitatory [L.]. Something that causes changes in state of excitability; something that stimulates a function.

excrete (eks-krēt′) [L. *excernere*, to separate]. To separate and expel useless substances from the body.

exocrine (eks′ō-krēn) [Gk. *exō*, out of, + *krinein*, to separate]. External secretion by a gland, through ducts, on an epithelial surface.

exogenous (eks-oj′ĕ-nus) [Gk. *exō*, out of, + *gennan*, to produce]. Originating outside the cell or organism.

expiration (eks-pi-rā′shun) [L. *ex*, out, + *spirare*, to breathe]. Expulsion of air from the lungs. Death.

exponential (ek-spō-nen′shul) [L. *exponere*, to put forth]. Change in a function according to multiples of powers (e.g., square or cube), rather than numerically (e.g., one time, two times).

external (ek-sturn′al) [L. *externus*, outside]. Outside or toward the surface.

exteroceptive (eks-ter-o-sep′tiv) [L. *exter*, outside, + *ceptus*, to take]. Sensations or reflex acts originating from stimulation of receptors at or near a body surface.

extracellular [L. *extra*, outside of, + cell]. Fluids outside the cells. Abbr. **ECF**.

extraneous (eks-tra′ne-us) [L. *extraneus*, external]. Outside of or unrelated to something.

extrinsic (eks-trĭn′sik) [L. *extrinsecus*, outside]. From without or coming from without; separate or outside of an organ or cell.

facial (fā′shul) [L. *facies*, the face]. Pert. to the face.

facilitation (fah-sĭl-ĭ-tā′shun) [L. *ficilis*, to make easier]. An increased ease of passage or a nerve impulse across a synapse.

fascia (fash′ĭ-ă) [L. *fascia*, a band]. Fibrous membranes covering, supporting, and separating muscles. *Superficial f.* is the hypodermis; *deep f.* is around muscles.

fascicle (fas′ik-l) [L. *fasciculus*, a little bundle]. An arrangement resembling a bundle of rods. Applied to bundles or nerve and muscle fibers.

fatigue (fă-tēg) [L. *fatigare*, to tire]. A feeling of tiredness; the state of an organ, synapse, or tissue in which it no longer responds to stimulation.

feces (fē′sēz) [L. *faeces*, dregs]. The waste matter expelled from the bowel through the anus.

feedback. Detection of the nature of an output and using that to control the process producing the output. May be negative *(inhibitory)* or positive *(stimulating)*.

fe′male [L. *femella*, woman]. The designation given to the sex that produces ova and bears offspring; a girl-child or woman.

fertilization (fur-tĭl-ĭ-zā′shun) [L. *fertilis*, to bear]. The impregnation of an ovum by a sperm.

fetus (fē′tus) [L. *fetus*, progeny]. The name given to a developing human after it assumes *clearly human form;* the period from 6 to 8 weeks after conception to birth.

fiber (fī′ber) [L. *fibra*, thread]. A larger threadlike or ribbonlike structure; a muscle cell; usually composed of smaller units.

fibril (fī′brĭl) [L. *fibrilla*, a little fiber]. A very small, threadlike structure, often the component of a muscle or nerve cell, or a connective tissue fiber.

fibrillate (fī′bril-āte) [L. *fibrilla*, a little fiber]. Spontaneous uncoordinated quivering of a muscle fiber, as in cardiac muscle.

fibrin (fī-brĭn) [L. *fibra*, a fiber]. A white or yellowish insoluble fibrous protein formed when blood clots.

fibrinogen (fī-brĭn′ō-gen). A soluble plasma protein that is acted on to produce fibrin as blood or lymph clots.

fibrosis (fi-brō′sis). Abnormal deposition or increase in fibrous tissue in a body part, organ, or tissue.

fil′iform [L. *filum*, thread, + *forma*, form]. Having a threadlike or hairlike form.

fil′ter [L. *filtare*, to strain through]. To strain or separate on the basis of size; a device to strain liquids.

filtrate (fĭl′trāt). The name given to the fluid that has passed through a filter.

filtration (fĭl-trā′shun). The process of forming a filtrate by passing a fluid, under pressure through a filter or selective membrane.

flaccid (flak′sid) [L. *flaccidus*, flabby]. Relaxed, flabby, having no tone, as in a muscle.

flatus (fla′tus) [L. *flatulentiā*, a blowing]. The gas in the digestive tract.

flora (flō′ra) [L. *flos*, flower]. Plant life; the microorganisms of the bowels, adapted to live in that organ.

follicle (fŏl′ĭ-cul) [L. *folliculus*, a little bag]. A small hollow structure containing cells or secretion.

foodstuff (food′stŭf) [M.E. *fode*, to eat]. Any substance made into or used as a food; nutritive substances providing heat and energy to the body when metabolized.

fornix (for′nĭks) [L. *fornix*, arch]. Any structure having an arched or vaultlike shape; a band connecting lobes of a cerebral hemisphere; a cleftlike space of the vagina around the neck of the uterus.

frequency (frē′qwĕn-sĭ) [L. *frequens*, often or constant]. Number of repetitions of something in a given time period (e.g., cycles per second, *cps*).

fulcrum (fŭl′krŭm) [L. *fulcire*, to prop or support]. The point about which a lever turns or pivots.

function (fŭnk′shŭn) [L. *functia*, to perform]. The action performed by living matter (cell, tissue, organ, system, or organism); to carry on an action.

fundus (fŭn′dŭs) [L. *fundus*, sling or base]. The larger, usually blind end of an organ or gland.

fu′siform [L. *fusus*, spindle, + *forma*, shape]. Having a shape tapered at both ends.

fusion (fū′shun) [L. *fusio*, to meet]. Coming together; meeting or joining.

gait (gāt) [M.E. *gaite*, street]. The manner of walking.

galactose (gă-lak′tōs) [Gk. *gala*, milk, + *ose*, sugar]. Milk sugar; a simple hexose sugar found in milk. $C_6H_{12}O_6$.

gamete (gam′ēt) [Gk. *gamēte*, a wife or spouse]. A male or female reproductive cell, that is, a sperm or egg (ovum).

gametogenesis (gam-ē-tō-jen′ĕ-sis) [Gk. *gamēte* + L. *genesis*, birth or origin]. The formation of gametes; spermatogenesis or oögenesis.

gamma (găm'ŭh) [Gk. letter g of alphabet]. The third item of a series; a microgram (one millionth gram).

gamma globulins. A plasma protein carrying most of the blood-borne antibodies.

ganglion (gang'lĭ-ŏn) [Gk. *gagglion*, a tumor or swelling]. A mass of nerve cell bodies outside of the brain and spinal cord; a cystic tumor developing on a tendon.

gas'tric [Gk. *gaster*, stomach]. Pert. to the stomach.

gene (gēn) [Gk. *gennan*, to produce]. The unit of heredity responsible for transmission of a characteristic to the offspring; believed to be a part of a nucleic acid chain.

genetic (jen-ĕt'ik) [L. *genesis*, origin]. Pert. to genesis or origin of something.

genital (jen'ĭ-tal) [L. *genitalis*, genital]. Pert. to the organs of reproduction; *external genitalia*, the external organs of reproduction.

genotype (jĕn'ō-tīp) [Gk. *gennan*, to produce, + *typos*, type or kind]. The hereditary makeup of individuals as determined by their genes.

germinal (jer'mĭn-ăl) [L. *germen*, from *gignere*, to beget]. Pert. to a reproductive cell, or a "germ" (microorganism).

germ layer. One of the three basic tissue layers (ectoderm, endoderm, mesoderm) formed in the embryo that give rise to all body tissues and organs.

gerontology (jer-on-tol'ō-ji) [Gk. *geron*, old man, + *logos*, study of]. The study of the phenomena of old age.

gestation (jes-ta'shun) [L. *gestāre*, to bear]. The period of intrauterine development of a new organism; the time of pregnancy.

gingiva (jin'jĭ-vă) [L. *gingiva*, gum]. The gums, or tissues surrounding.

girdle (gĭr'dŭl) [A.S. *gyrdel*, to encircle]. A belt, zone, or a structure resembling a circular belt or band.

gland [L. *glans*, a kernel]. A secretory organ or structure.

glia (glē'ă) [Gk. *glia*, glue]. The nonnervous tissues of the nervous system; supporting and nutritive cells of various types.

globular (glob'ū-lar) [L. *globus*, a globe]. Having a rounded or spherical shape.

glomerulus (glō-mer'ū-lus) [L. a little skein, a tangle of thread or yarn]. A small round mass of cells or blood vessels.

glottis (glŏt'ĭs) [Gk. *glottis*, back of tongue]. The opening between the vocal cords in the larynx (voice box).

-glyc- (glīk) [Gk. *glykus*, sweet]. Pert. to glucose.

glyceride (glĭs'ĕr-īd) [Gk. *glykus*, sweet]. Glycerin (an alcohol) together with one or more fatty acids.

glycogen (glī'ko-jen) [Gk. *glykos*, sweet, + *genēs*, born]. A complex polysaccharide; "animal starch"; stored in liver and muscle.

goblet cell. A one-celled mucus-secreting gland found in epithelia of the respiratory and digestive systems.

goiter (goi'tĕr) [L. *guttur*, throat]. An enlargement of the thyroid gland, irrespective of cause.

gon'ads [Gk. *gonē*, a seed]. A general name for an organ producing sex cells; the name of the embryonic sex gland before differentiation into ovary or testis.

gradient (grā'dĭ-ent). A slope or grade; a curve representing increase or decrease of a function.

graft [L. *graphium*, grafting knife]. A tissue or organ inserted into a similar substance or area to correct loss or absence of that structure; a transplant; to carry out the procedure of grafting or transplantation.

-gram. Combining form pert. to the record drawn by a machine (e.g., cardiogram, telegram).

granulation (grăn-ū-lā'shun) [L. *granulum*, little grain]. To form granules; tissue appearing in the course of wound healing, characterized by its large numbers of blood vessels.

gran'ule [L.]. Any minute or tiny grain in a cell.

grav'ity [L. *gravitās*, weight]. The force of the earth's gravitation attraction that gives objects weight.

gray matter. The tissue of the CNS consisting mainly of nerve cell bodies.

gristle (grĭs'l) [A.S.]. Cartilage.

gross (grōs) [L. *grossus*, thick]. Large; anatomy seen with the naked eye.

ground substance. The fluid or semifluid material occupying the intercellular spaces in connective tissues.

growth [A.S. *grōwan*, to grown]. Progressive increase in size of a cell, tissue, or whole organism.

gut [A.S.]. The primitive or embryonic digestive tube (fore-, mid-, hindgut); Colloq. for intestine.

gyrus (jī'rus) (pl *gyri*) [Gk. *gyros*, circle]. An upfold of the cerebrum.

$[H^+]$. Symbol for hydrogen ion concentration.

hap'loid [Gk. *aploos*, simple]. Having one half the normal number of chromosomes characteristic of the species; sperm and ova are haploid cells.

haustra (haws'tra) [L. *haurīre*, to draw water]. The sacculations or pouches of the colon.

heat (hēet) [Gk. *heito*, fever]. High temperature; being hot; a form of energy; to make hot.

helix (hē'liks) [Gk. *helix*, a coil]. A coil or spiral.

hematocrit (he-mat'ō-krit) [Gk. *aima*, blood, + *krinein*, to separate]. The volume of red blood cells in a tube after centrifugation.

heme (hēm) [Gk.]. The iron-containing red pigment that, with globin, forms hemoglobin.

hemiplegia (hem-ĭ-plē'jĭ-ă) [Gk. *hemi*, half, + *plēgē*, a stroke]. Paralysis of one half of the body in a right-left direction.

hemoglobin (hē-mō-glō'bin) [Gk. *aima*, blood, + *globus*, globe]. The respiratory pigment of red blood cells that combines with O_2 and CO_2 and acts as a buffer.

hemorrhage (hem'o-rij) [Gk. *aima* + *rēgnunai*, to burst forth]. Abnormal loss of blood from the blood vessels, either internally or externally.

hemorrhoids (hem'o-royds) [Gk. *aimorris*, a vein liable to discharge blood]. Dilated veins in the rectal columns.

hemostasis (he-mō-stā'sis) [Gk. *aima* + *stasis*, stopping]. Stoppage of blood loss through a wound as a result of vascular constriction and coagulation.

he'par [Gk. *ēpar*, liver]. The liver.

hepatic (he-pat'ik) [Gk.]. Pert. to the liver.

hereditary (hĕ-red'ĭ-tar-ē) [L. *hereditas*, heir]. Passed or transmitted (as a genetic characteristic) from parent to offspring; handed down.

hernia (her'nĭ-ă) [Gk. *ernos*, a young shoot]. Protrusion or projection of an abdominal organ through the wall that normally contains it.

heterogeneous (het-er-ō-je'nē-us) [Gk. *eteros*, other, + *gennos*, type]. Composed of things different or contrasting in type of nature.

heterograft [Gk. *eteros* + L. *graphium*, grafting knife]. Grafting of tissues or organs between different species.

hex'ose [Gk. *hex*, six, + *ose*, sugar]. A six-carbon sugar (e.g., glucose, fructose).

hi′lus [L. *hilus*, trifle]. A depression or recess on the side of an organ; vessels and nerves usually enter and/or leave at this point.

histio-, histo- [Gk. *histos*, tissue]. Prefix: tissue.

holocrine (hŏl-o-krĭn) [Gk. *olos*, whole + *krinein*, to secrete]. A manner of production of a secretion by a gland in which a cell is the product or in which the whole cell enters the secretion.

homeostasis (hō-mē-ō-stā′sĭs) [Gk. *omoios*, like, + *stasis*, a standing]. The state of near constancy of body composition and function.

homogenous (hō-moj′ĕn-ŭs) [Gk. *homo*, likeness, + *genos*, kind]. Like or uniform in structure or composition.

homograft [Gk. *homo* + L. *graphium*]. Use of tissues or organs of the same species for grafting purposes. *Isograft.*

homologous (hō-mol′o-gus) [Gk. *homologos*, agreeing]. Similar in origin and structure.

hormone (hor′mōn) [Gk. *ormanein*, to excite]. The chemical produced by an endocrine gland.

humor (hū′mŭr) [L. *humor*, fluid, moisture]. Any fluid or semifluid substance in the body.

hyaline (hī′ă-lĭn) [Gk. *hyalinos*, glassy]. Clear, translucent, or glassy.

hyaluronidase (hi-ă-lur-on′ĭ-dās). An enzyme that liquefies the intercellular cement holding cells together.

hydrate (hī′drāt) [Gk. *ydōr*, water]. To cause water to combine with a compound; the compound after water has been added.

hydrogen (hī′drō-jen) [Gk. *ydōr* + *gennan*, to produce]. An inflammable gaseous element, the lightest of all known elements.

hydrostatic pressure [Gk. *ydōr* + *statikos*, standing]. Pressure exerted by liquids.

hyperemia (hī-per-ē′mĭ-ă) [Gk. *yper*, above, + *aima*, blood]. An extra amount of blood in an area.

hyperesthesia (hī-per-es-the′zē-ă) [Gk. *yper* + *aisthēsis*, sensation]. Excessive sensitivity to sensory stimulation.

hypermetropic (hi-per-mĕ-trō′pik) [Gk. *yper* + *metron*, measure + *ōps*, eye]. Pert. to farsightedness.

hyperphagia (hī-per-fā-jē-ă) [Gk. *yper*, + *phagein* to eat]. Excessive intake of food.

hyperplasia (hī-per-plā′zē-ă) [Gk. *yper* + *plassein*, to form]. Increase in size because of increased numbers of cells derived by division.

hyperpnea (hī-perp′nē-ă) [Gk. *yper* + *pnoe*, breath]. An increase in depth of breathing.

hypertension (hī-pĕr-tĕn′shun) [Gk. *yper* + L. *tensio*, tension]. High blood pressure.

hypertonic (hi-per-tŏn′ik) [Gk. *yper* + *tonos*, tension]. A solution containing a greater solute concentration and thus osmotic pressure than the blood or having an osmotic pressure greater than another reference solution.

hypertrophy (hi-per′trō-fē) [Gk. *yper* + *trophē*, nourishment]. Increase in size by adding substance and *not* by increasing numbers of cells.

hyperventila′tion [Gk. *yper* + L. *ventilatio*, ventilate]. Increased exchange of air by increasing both rate and depth of breathing.

hypnotic (hip-nŏ′tik) [Gk. *ypnos*, sleep]. An agent producing sleep or depression of the senses.

hypokalemia (hī-pō-kăl-ē′mē-ă) [Gk. *ypo*, under, + L. *kali*, potash, potassium, + *aima*, blood]. Low blood potassium levels.

hypoten′sion [Gk. *ypo* + L. *tensio*]. Low blood pressure.

hypothesis (hī-poth′ĕ-sis) [Gk. *ypo* + *thesis*, a placing]. An assumption made to explain something. It is unproved and is made to enable testing of its soundness.

hypotonic (hi-pō-tŏn′ik) [Gk. *ypo* + *tonos*]. A solution containing less solute, and thus having a lower osmotic pressure, than the blood, or a solution having a lower osmotic pressure than some reference solution.

hypoxemia (hī-poks-ē′mē-ă) [Gk. *ypo* + *oxys*, acid, + *aima*]. Lowered blood oxygen levels.

hypoxia (hī-poks′ē-ă) [Gk. *ypo* + *oxys*]. Low oxygen levels in inspired air or reduced tension in the tissues.

icteric (ik-tĕr′ik) [Gk. *ikteros*, jaundice]. Pert. to jaundice; excessive accumulation of bile pigments in the body tissues.

idiopathic (id-ĭ-ō-path′ik) [Gk. *idio*, own, + *pathos*, disease]. Any condition arising without a clear-cut cause; of spontaneous origin.

immune (ĭm-ūn′) [L. *immunis*, safe]. Protected from getting a given disease.

impermeable (im-pĕrm′ē-ă-bl) [L. *in*, not, + *permeāre*, to pass through]. Not allowing passage.

implantation (im-plan-tā′shun) [L. *in*, into, + *plantāre*, to plant]. Embedding of the blastocyst in the uterine lining; inserting something into a body organ.

impulse (im′puls) [L. *impellere*, to drive out]. A physicochemical or electrical change transmitted along nerve fibers or membranes.

inactivate (ĭn-ăk′tĭ-vāt) [L. *in*, not, + *activus*, acting]. To make inactive or inert.

inclusion (ĭn-klū′zhun) [L. *inclusus*, enclosed]. A lifeless, usually temporary constituent of a cell's cytoplasm.

independent irritability. Capable of reacting to external stimuli, or those delivered by other than the normal route.

indigestion (in-dī-jes′chŭn) [L. *in*, not, + *digerere*, separated]. Inability to digest food; incomplete digestion; dyspepsia. Usually accompanied by pain, nausea, gas (belching), and heartburn.

induce (in-dūs) [L. *inducere*, to lead in]. To cause an effect; induction may be used to express the way that a gene causes an effect.

infarct (in′farkt) [L. *infacire*, to stuff into]. An area of an organ that dies following blockage of blood supply.

infection (in-fek′shun) [L. *inficere*, to taint]. The state when the body has been invaded by a pathogenic (disease-producing) agent.

infectious (in-fek′shus) [L. *inficere*]. Capable of being transmitted with or without contact. Usually used in connection with microorganism-caused diseases.

inflame (in-flām) [L. *inflammare*, to set on fire]. To cause to become warm, swollen, red, sore, and feverish.

infundibulum (in-fun-dib′ū-lum) [L. *infundibulum*, funnel]. A funnel-shaped passage.

infusion (in-fū′zhun) [L. *infusiō*, from *infundere*, to pour in]. Introduction of liquid into a vein.

ingest (in-jest′) [L. *ingerere*, to carry in]. To take foods into the body.

inguinal (in′gwi-nal) [L. *inguinalis*, the groin]. Pert. to the groin.

inherit (in-her′ĭt) [L. *inhereditare*, to appoint as heir]. To receive from one's ancestors.

inhibit (in-hĭ′bĭt) [L. *inhibere*, to restrain]. To repress or slow down.

innervate (ĭn-nŭr′vāt) [L. *in*, in, + *nervus*, nerve]. To supply with nerves; to stimulate to action.

inorgan′ic [L. *in*, not, + Gk. *organon*, an organ]. A chemical compound not containing both hydrogen and carbon; not associated with living things.

input (in'pŏot). That which is put into something, as in the passage of nerve impulses to the brain for processing.

insertion (in-sŭr'shen) [L. *insertio*, to join into]. The addition of something by "setting it into" something else, as insertion of parts of a chromosome into another chromosome. The attachment of a muscle to a more movable bone.

in situ [L.]. In position.

inspiration (in-spĭ-rā'shun) [L. *inspiratio*, to breath in]. Taking air into the lungs.

integrate (in'tē-grāt) [L. *intergratus*, to make whole]. To blend, put together or unify.

integument (in-teg'ū-ment) [L. *integumentum*, a covering]. The outer covering of the body; skin.

intensity (in-ten'sĭ-tĭ) [L. *intensus*, tight]. Degree of strength, force, loudness, or activity.

internuncial (in-tĕr-nun'sē-al) [L. *inter*, between, + *nuncius*, messenger]. A connector; between two other items, as neurons.

interphase. The "resting state" of a cell, when it is not dividing.

interstitial (in-ter-stish'al) [L. *interstitium*, thing standing between]. Lying between, as, interstitial fluid lies between vessels and cells or between cells.

in'tima [L. innermost]. The inner coat of a blood vessel.

intracellular (in-tra-sel'ū-lar) [L. *intra*, within, + *cellula*, cell]. Within cells.

intravenous (in-tră-ve'nus) [L. *intra* + *vena*, vein]. Within or into a vein.

intrin'sic [L. *intrinsicus*, on the inside]. Located or originating within a structure.

in utero [L.]. Within the uterus.

inversion (in-ver'shun) [L. *in*, into, + *versiō*, a turning]. Turning, as upside down, or inside out; reversal.

in vitro [L.]. In glass, as in a test tube; *not* within the body.

in vivo [L.]. Within the body.

involuntary (in-vol'un-tĕr-i) [L. *involuntarius*, without act of will]. Occurring without an act of will.

involution (in-vō-lū'shun) [L. *in*, into, + *volvere*, to roll]. Change in a backward or diminishing direction.

i'on [G. *iōn*, going]. An atom or group of atoms (radical) carrying an electric charge.

ipsilateral (ip-sĭ-lăt'er-al) [L. *ipse*, same, + *latus*, side]. On the same side; opposite of contralateral (*contra*, opposite).

irritability (ĭ-rĭ-tă-bĭl'ĭ-tē) [L. *irritare*, to tease]. Excitability, or the ability to respond to a stimulus.

ischemia (ĭs-kē'mĭ-ă) [Gk. *ischein*, to hold back, + *aima*, blood]. Local temporary reduction of blood flow to a body organ or part.

i'so- [Gk. *isos*, equal]. Prefix: equal to.

i'sograft [Gk. *isos*, equal, + L. *graphium*, grafting knife]. A graft or transplant from another animal of the same species. Also, *homograft*.

isometric (ī-sō-mĕ'trik) [Gk. *isos*, equal, + *metron*, measure]. Refers to no change of length, as an isometric muscle contraction.

isoton'ic [Gk. *isos* + *tonos*, tension]. A solution having the same osmotic pressure as the blood or a reference solution; a muscular contraction in which shortening is allowed.

isthmus (ĭs-mus) [Gk. *isthmos*, narrow passage]. A narrow structure connecting two other structures.

-itis (ī'tĭs) [Gk.]. Suffix: inflammation of.

jaundice (jawn'dis) [M.E. *jaundis*, yellow]. A condition in which the skin is yellowed because of excessive bile pigment (bilirubin) in the blood.

joint [L. *junctura*, a joining]. An articulation or junction between bones or bone and cartilage.

jug'ular [L. *jugulum*, neck]. Large veins in the neck; pert. to the neck.

juice (jūs) [L. *jus*, broth]. Liquid secreted or expressed from an organism or any of its parts; esp., one of the digestive secretions.

ju'venile [L. *juvenilis*, young]. Pert. to youth or childhood; an immature white cell (metamyelocyte).

juxta- [L. near to]. Close or near to; in close proximity, as in juxtaglomerular.

karyoplasm (kar'ĭ-ōplăs-m) [Gk. *karyon*, nucleus, + *plasma*, a thing formed]. The nuclear protoplasm.

karyotype (kar'ĭ-ō-tīp) [Gk. *karyon* + *typos*, form]. A grouping or arrangement of the 46 human chromosomes based on the size of the individual chromosomes.

keratin (ker'ă-tĭn) [Gk. *keras*, horn]. A protein found in the superficial epidermal cells.

kil'ogram [Gk. *chilioi*, one thousand, + *gramma*, a weight]. A metric measurement of weight; equals 1000 g or 2.2 lb.

kinesthesia (kin-es-thē'zĭ-ă) [Gk. *kinesis*, motion, + *aisthēsis*, sensation]. The sensation concerned with appreciation of movement and body position.

kinetic (kĭ-net'ĭk) [Gk. *kinēsis*, motion]. Pert. to movement or work.

-kinin. Suffix denoting causing motion or action.

kyphosis (ki-fō'sis) [Gk. *kyphosis*, humpback]. An exaggeration of the normally anteriorly concave thoracic spinal curvature.

labium (lā'bĭ-um) [L. *labium*, lip]. A lip or liplike structure.

lacrimal (lak'rĭm-al) [L. *lacrima*, tears]. Pert. to the tears or any structure involved in producing or carrying tears.

lact- [L. *lac*, milk]. Prefix: milk.

lactation (lak-tā'shun) [L. *lactatio*, a suckling]. The function of secreting milk.

lactose (lak'tōse) [L. *lactatio* + *ose*, sugar]. Milk sugar, or a double sugar (disaccharide).

lacuna (lă-kū'nă) [L. *lacuna*, a pit]. A hollow space in the matrix of cartilage and bone in which lie characteristic cells.

lamella (lam-el'ă) [L. *lamella*, a little plate or leaf]. Circularly arranged layers, as in the elastic layers in a large artery; a ring of bony tissue around a haversian canal.

la'tent [L. *latēre*, to be hidden]. Not active, quiet.

leg [M.E.]. Specifically, the part of the lower limb between the knee and ankle; Colloq.: a lower appendage.

lemniscus (lĕm-nis'kŭs) [Gk. *lemniskōs*, a filet]. A bundle of sensory nerve fibers in the brainstem.

lesion (lē'zhun) [L. *laesio*, a wound]. A wound, injury, or infected area.

leuko-, leuco- (lu'ko) [Gk. *leukos*, white]. Prefix: white.

lever (lĕv'ĕr, lē'vĕr) [M.E. *lever*, to raise]. A rigid elongated structure used to change direction, force, or movement.

ligament (lĭg'ă-mĕnt) [L. *ligamentum*, a band]. A strong band or sheet of fibrous connective tissue commonly used to connect bones together; a structure formed by degeneration of a fetal blood vessel.

lim'bic system [L. *limbus*, border]. A group of nervous structures in the cerebrum and diencephalon serving emotional expression.

liminal (lĭm'ĭ-nal) [L. *lēmen*, threshold]. Threshold; just perceptible.

linear (lĭn'ē-ar) [L. *linea*, line]. Pert. to a line or lines; extended in a line.

linkage (link'ĭj) [M.E. *linke*, akin to]. In genetics, two or more genes on a chromosome tending to remain together during sex cell formation.

lipid (lĭp'id) [Gk. *lipos*, fat]. Fats or fatlike substances that are not soluble in water.

lipoprotein (lī-pō-prō'tēen) [Gk. *lipos + proteios*, major, of first importance]. A combination between a lipid and a simple protein.

liter (lē'tĕr) [Gk. *litra*, a weight]. Metric measurement of fluid volume. Equals 1000 ml or 1.06 quarts or 61 cubic inches.

-lith- [Gk. *lithos*, stone]. Stone; presence of stones, lithiasis.

lobe (lōb) [Gk. *lobos*, part or section]. A section or part of an organ separated by clear boundaries from other sections or parts.

localize (lō'kăl-īz) [L. *locus*, place]. To restrict to a small area.

locomotion (lō-kō-mō'shun) [L. *locus*, place, + *motus*, moving]. To move from one place to another; movement.

locus (lō'kŭs) [L. *locus*, place]. A place; in genetics the location or position of a gene in a chromosome; a place in the brain or heart where abnormal electric discharge may originate.

logarithmic (lŏg-ă-rĭth'mĭk) [Gk. *logos*, reason, + *arithmos*, number]. Progression of a function of powers (e.g., square or cube) and not by individual numbers.

loin (loyn) [Fr. *loigne*, long part]. The sides and back of the trunk between ribs and pelvis.

lozenge (loz-ĕnj) [Fr. *lozenge*, diamond shaped]. A solid medicine to be held in the mouth until dissolved.

lucid (lū'sĭd) [L. *lucidus*, clear]. Clear, as of thought, mind, or speech.

lumen (lū'mĕn) (pl. *lumina*) [L. *lumen*, light]. The cavity of any hollow organ; a unit of light.

luteo- (lū'tē-ō) [L. *luteus*, yellow]. Prefix: yellow.

lympho- (lĭm'fō) [L. *lympha*, lymph]. Prefix: lymph or lymphatic system.

-lysis (lī'sĭs) [Gk. *lysis*, dissolution]. To destroy or break down.

lysosome (lī'sō-som) [Gk. *lysis + soma*, body]. Cell organelle concerned with digestion of large molecules.

L-tubule. Longitudinally arranged tubules of the sarcoplasmic reticulum in muscle cells.

macro- (mak'rō) [Gk.]. Prefix: large or long.

mal- (mahl) [L. *malum*, an evil]. Prefix: ill, bad, or poor.

male (māl) [L. *masculus*, man]. The designation given to the sex that produces sperm, fertilizes ova, and begets offspring; a boy-child or man.

malignancy (mă-lĭg'năn-sĭ) [L. *malignus*, of bad kind]. A severe form of something; tending to grow worse.

malnutrition (mal-nū-trĭ'shun). Literally, "poor nutrition"; most often used to refer to absence of essential foodstuffs in the diet.

maltose (mawl'tōs) [A.S. *mealt*, grain]. Malt sugar, a double sugar (disaccharide) found in malt and seeds. $C_{12}H_{22}O_{11}$.

mammary (mam'ă-rĭ) [L. *mamma*, breast]. Pert. to breast.

mamillary (mam'ĭl-lar-ĭ) [L. *mammilla*, nipple]. Resembling a nipple.

mastication (măs-tĭ-kā'shun) [L. *masticāre*, to chew]. Chewing of food in the mouth.

matrix (mā'trĭks) [L. *matrix*, mother or womb]. Intercellular substance of cartilage; formative portion of a tooth or nail; the uterus.

mature (ma-tūr) [L. *maturus*, ripe]. Fully developed or ripened.

maximal (maks'ĭ-mal) [L. *maximum*, greatest]. Highest; greatest possible.

mean (mēn) [L. *medius*, in the middle]. The average (sum of numbers divided by the number of numbers).

meatus (mē-ā'tŭs) [L. *meatus*, opening]. A passage or opening.

mediastinum (mē-dĭ-ăs-tī'nŭm) [L. in the middle]. A septum or cavity between two parts of an organ or between two other cavities.

medulla (mĕ-dul'lă) [L. marrow]. Inner or central portion of an organ; the medulla oblongata of the brainstem.

medullated (med'ū-lāt-ĕd). [L. marrow]. Having a myelin sheath.

mega- [Gk.]. Large; 1 million.

meiosis (mī-ō'sĭs) [Gk. *meiōsis*, diminution]. A form of cell division that reduces chromosome number to haploid. Occurs in formation of sex cells.

melanophores (mel-ăn'ō-fōr) [Gk. *melas*, black, + *phoros*, a bearer]. A cell carrying dark pigment(s).

membrane (mĕm'brăn) [L. *membrana*, membrane]. A thin soft layer of tissue lining a tube or cavity, or covering or separating one part from another.

meninges (mĕn-ĭn'jēz) [Gk. *mēnigx*, membrane]. The membranes around the central nervous system; dura mater, arachnoid, pia mater.

meniscus (men-is'kus) [Gk. *mēniskos*, crescent]. The crescent-shaped cartilages of the knee joint.

menstrual (men'strū-al) [L. *menstruāre*, to discharge menses]. Pert. to menstruation or the sloughing of the uterine endometrium.

merocrine (mer'ō-krĭn) [Gk. *meros*, a part, + *krinein*, to secrete]. A method of production of a secretion in which no part of the secreting cell enters the secretion.

mesentery (mes'en-ter-ĭ) [Gk. *mesos*, middle, + *enteron*, intestine]. A double-layered fold or peritoneum suspending the small intestine from the posterior abdominal wall.

mesoderm (mĕs'ō-derm) [Gk. *mesos + derma*, skin]. The middle germ layer of the embryo; gives rise to connective tissue, muscle, blood, and the cellular part of many organs.

meta- Prefix: after or beyond, later or more developed.

metabolism (mĕ-tăb'ŏl-ĭzm) [Gk. *metabolē*, change, + *ismos*, state of]. The sum total of all chemical reactions in the body; any product of metabolism is a *metabolite*.

metaphase (mĕt'ă-fāz) [Gk. *meta + phasis*, a shining out]. A stage of mitosis in which chromosomes line up on the equator of the dividing cell.

metastasis (mĕ-tas'tă-sis) [Gk. *meta + stasis*, a standing]. Movement from one part of the body to another (e.g., cancer cells).

micro- [Gk.]. Small; one millionth.

micron (mī'kron) [Gk. *mikros*, small]. Metric unit of length; equals one one-thousandth of a millimeter. μ, Synonym for micrometer (μm).

microvillus (mī-kro-vĭl'ŭs) [Gk. *mikros + villus*, tuft of hair]. Very small fingerlike extensions of a cell surface.

micturition (mĭk-tū-rĭ'shŭn) [L. *micturīre*, to urinate]. Voiding of urine; urination.

milliequivalent (mil-e-e-kwĭv'a-lent) [L. *mille*, thousandth, + *aequis*, equal, + *valere*, to be worth]. One one-thousandth of an equivalent weight. (An *equivalent* weight is the quantity, by weight, of a substance that will combine with 1 g of H or 8 g of O.) Abbr: *mEq.*

millimeter (mil'ĭ-mēt-er) [L. *mille* + Gk.

metron, measure]. One one-thousandth of a meter. Abbr: **mm**. One **meter** equals 39.36 inches.

millimicron. One-thousandth of a micron. Abbr: **mµ**, Synonym for nanometer (nm).

milliosmole. One one-thousandth of an osmol. (An osmol is the molecular weight of a substance in grams divided by the number of particles each molecule releases in solution.) Abbr: **mosm**.

mimetic (mi-met′ĭk) [Gk. *mimētikos*, to imitate]. Imitating or causing the same effects as something else.

miotic (mī-ŏ′tĭk) [Gk. *meiōn*, less]. An agent causing pupillary contraction.

mitosis (mī-tō′sis) [Gk. *mitos*, thread]. A type of cell division that results in production of daughter cells like the parent.

mitral (mi′tral) [L. *mitra*, a miter, a bishop's hat]. Pert. to the bicuspid (mitral) valve between the left antrium and ventricle of the heart.

modality (mō-dal-ĭ-tē) [L. *modus*, mode]. The nature or type of a stimulus; a property of a stimulus distinguishing it from all other stimuli.

molecular weight. Weight of a molecule obtained by adding the weights of its constituent atoms. It carries no units. **mol wt, gram.** The molecular weight of something expressed in grams. Also, a **mol(e)**.

molecule (mŏl′ĕ-kūl) [L. *molecula*, little mass]. The smallest unit into which a substance may be reduced without loss of its characteristics.

monitor (mon′ĭ-tŭr) [L. *monere*, to warn]. To watch, check, or keep track of; one or something that watches.

monosaccharide (mŏn-ō-sak′ar-id) [Gk. *mono*, one, single, + *sakcharon*, sugar]. A simple sugar that cannot be further decomposed by hydrolysis.

morphology (mor-fol′ō-ji) [Gk. *morphē*, form, + *logos*, study]. The structure or form of something; study of the same.

motor (mō′tor) [L. *motus*, moving]. Refers to movement or those structures (e.g., nerves and muscles) that cause movement.

motor end plate. That part of a neuromuscular junction on a skeletal muscle fiber.

motor unit. One motor nerve fiber (axon) and the skeletal muscles fibers it supplies.

mucous (mū′kŭs) [L. *mucus*, mucus]. A mucus-secreting structure.

mucus (mū′kŭs) [L.]. The thick sticky fluid secreted by a mucous cell or gland.

multi- [L.]. Prefix: many.

mutation (mū-tā′shŭn) [L. *mutare*, to change]. A change or transformation of a gene; the evidence of a gene alteration.

mydriatic (mid-rĭ-at′ik) [Gk. *midriasis*, dilation]. An agent causing pupillary dilation.

myelin (mī′ĕ-lĭn) [Gk. *myelos*, marrow]. A fatty substance forming a sheath or covering around many nerve fibers. Speeds impulse conduction.

myelo- [Gk.]. Prefix: pert. to the spinal cord or to bone marrow.

myeloid (mī′el-oid) [Gk. *myelo-* + *eidos*, form]. Formed in bone marrow; resembling marrow.

myoglobin (mī-amo-glō′bĭn) [Gk. *myo*, muscle, + L. *globus*, globe]. A respiratory pigment of muscle that binds oxygen.

myopia (mī-ŏ′pē-ă) [Gk. *myein*, to shut, + *ōps*, eye]. Nearsightedness. Also, *hypometropia*.

myosin (mī′ō-syn) [Gk. *myo*, muscle]. A muscle protein acting as an enzyme to aid in initiating muscle contraction.

myotatic (mī-ō-tă′tik) [Gk. *myo* + L. *tactus*, touch]. Refers to muscle or kinesthetic sense (kinesthesia).

nasal (nā′zl) [L. *nasus*, nose]. Pert. to the nose.

narcotic (nar-kŏt′ik) [Gk. *narkōtikos*, benumbing]. A drug that depresses the CNS, relieving pain and producing sleep.

nares (nar′ēz) [L. *naris*, nostril]. Nostrils.

natremia (na-trē′mĭ-ă) [L. *natrium*, sodium, + Gk. *aima*, blood]. Blood sodium.

necrosis (něk-rō′sis) [Gk. *nekrōsis*, a killing]. Death of tissue or cells.

negative pressure. A pressure less than atmospheric pressure, that is, <760 mm Hg.

neoplasm (nē′ō-plăzm) [Gk. *neo-*, new, recent, + *plasma*, a thing formed]. A new and abnormal formation of tissue, as a cancer, tumor, or growth.

nephron (něf′ron) [Gk. *nephros*, kidney]. The functional unit of the kidney that forms urine and regulates blood composition.

nerve (nŭrv) [L. *nervus*, sinew]. A bundle of nerve fibers outside the CNS.

neurilemma (nŭr-ĭ-lěm′mă) [Gk. *neuron*, sinew, + *lemma*, rind]. A thin living membrane around some nerve fibers; aids in myelin formation and fiber regeneration.

neuroglial (nū-rŏg′lĭ-ăl) [Gk. *neuron* + *glia*, glue]. Pert. to the glial or nonnervous cells of the nervous system.

neuron (nū′rŏn) [Gk.]. The cell serving as the unit of structure and function of the nervous system; is excitable and conductile.

neurosis (nu-rō′sis [Gk. *neuron* + *osis*, disease]. A disorder of the mind in which contact with the real world is maintained.

neutralize (nū′tral-īz) [L. *neuter*, neither, from *ne-*, not, + *uter*, either]. To counteract, make inert, or destroy the properties of something.

newborn. A child under 6 weeks of age.

nigra (nī′gră) [L. *nigra*, black]. Black or blackness.

nitrogenous (nī-troj′ěn-ŭs) [Gk. *nitron*, soda, + *gennan*, to produce]. Pert. to or containing nitrogen.

node (nōd) [L. *nodus*, knot]. A constricted region; a knob, protuberance or swelling; a small rounded organ.

nostril [A.S. *nosu*, nose, + *thyrl*, a hole]. External opening of the nasal cavity.

nuchal (nū′kal) [L. *nucha*, the back of the neck]. Pert. to the back of the neck.

nuclear (nū′klē-ăr) [L. *nucleus*, a kernel]. Pert. to a nucleus (cell) or a central point.

nucleic acids (nū-klay′ik). Large molecules formed by nucleotides. They form the basis of heredity and protein synthesis. **DNA, RNA**.

nucleolus (nū-klē′ō-lŭs) [L. little kernel]. A spherical mass of nucleic acid within the nucleus. Stores and may synthesize **tRNA**.

nucleotide (nū′klē-ō-tīd) [L. *nucleus*]. A unit or compound formed of a nitrogenous base, a five-carbon sugar, and a phosphoric acid radical. Unit of structure of *DNA* and *RNA*.

nucleus (nū′klē-ŭs) [L.]. The controlling center of a cell; a group of nerve cells in the brain; the heavy central atomic region in which mass and positive charge are concentrated.

nutrient (nū′trĭ-ent) [L. *nutriens*, to nourish]. A food substance necessary for normal body functioning.

nutrition (nūtrī′shun) [L. *nutritiō*, a feeding]. The total of all processes involved in intake, processing, and utilization of foods.

nystagmus (nĭs-tag′mŭs) [Gk. *nystazein*, to nod]. Involuntary cyclical movements of the eyeball.

obese (ō-bēs′) [Gk. *obesus*, fat]. Excessively fat; overweight.

obligatory (ob-lĭg′ă-tō-rē) [L. *obligatorius*, to bind to]. Carries the idea of not having a choice; bound to or fixed in function.

oblique (ŏb-līk) [L. *oblique*, slanting]. A slanting direction or position.

obliterate (o-blīt′er-āte) [L. *obliterate*, to blot out]. To erase, leaving no traces; extinction; occlusion of a part by surgical, degenerative, or disease processes.

occlude (ŏ-klŭd) [L. *occludere*, to shut up]. To close, obstruct or block something, such as a blood vessel.

occlusion (ŏ-klū′zhŭn) [L. *occlusiō*, a closing up]. The state of being closed, blocked, or obstructed.

ocular (ok′ū-lăr) [L. *oculus*, eye]. Pert. to eye or vision.

-oid [Gk.]. Suffix: like, similar to, resembling.

olfactory (ŏl-fak′tō-rī) [L. *olfacere*, to smell]. Pert. to smell or the sense of smell.

oligo- (ŏl′ĭ-gō) [L. *oligos*, little]. Prefix: small, few, scanty or little.

oliguria (ol-ĭg-ū′rĭ-ă) [L. *oligo-* + *ouron*, urine]. Secretion of small amounts of urine.

-ology [Gk.]. Suffix: science of, study of, knowledge of.

omentum (ō-mĕn′tŭm) [L. *omentum*, a covering]. A four-layered fold of peritoneum lying between the stomach and other abdominal viscera; a site of fat storage.

oncotic (ŏng-kŏt′ĭk) [Gk. *ogkos*, tumor]. Pert. to the colloid osmotic pressure created by the plasma proteins.

opaque (ō-pāk′) [L. *opacus*, dark]. Not transparent; not allowing light to pass; dark.

ophthalmic (ŏf-thăl′mĭk) [Gk. *ophthalmos*, eye]. Pert. to the eye.

-opsin (ŏp′sĭn) [Gk. *opson*, food]. Suffix: pert. to the protein component of the visual pigments.

oral (ō′răl) [L. *os*, *or-*, mouth]. Pert. to the mouth or the mouth cavity.

orbicular (ŏr-bĭk′ū-lăr) [L. *orbiculus*, a small circle]. Circular in arrangement.

orbit (or′bĭt) [L. *orbita*, track]. The cavity holding the eyeball; to go around.

orchido- [Gk. *orchis*, testicle]. Combining form meaning testicle.

organ (or′găn) [Gk. *organon*, organ]. Two or more tissues organized to a particular job.

organelle (or-găn-ĕl′) [Gk. *organelle*, a small organ]. Submicroscopic formed structures within the cytoplasm that carry out particular functions.

organic (or-găn′ĭk) [Gk. *organon*, organ]. Pert. to organs; compounds containing carbon and hydrogen.

organic acids. An acid containing one or more carboxyl groups (-COOH).

organism (or′găn-ĭzm) [Gk. *organon*, organ, + *ismos*, condition]. A living thing.

orifice (or′ĭ-fĭs) [L. *orificium*, outlet]. The entrance or outlet to any chamber or hollow.

origin (or′ĭ-jin) [L. *origo*, beginning]. A source; the beginning a a nerve; the more fixed attachment of a muscle.

os (ŏs) [L.]. Bone; mouth; opening.

oscillate (ŏs′ĭll-āt) [L. *oscillāre*, to swing]. To move back and forth; to swing.

-ose (ōs). Suffix: pert. to sugar.

-osis [Gk.]. Suffix: caused by; state of; disease or intensive.

osmo- [Gk.]. Combining form, pert. to smell; pert. to osmosis.

osmol (os′mŏl). The molecular weight of a substance, in grams, divided by the number of particles it releases in solution.

osmolarity (os-mō-lar′ĭ-tĭ). The number of osmols per liter of solution.

osmoreceptor (ŏz-mō-rē-cĕp′tor). A receptor sensitive to osmotic pressure of a fluid.

osmosis (ŏz-mō′sis) [Gk. *osmōs*, a thrusting, + *-osis*, intensive]. The passage of solvent through a selective membrane.

osmotic pressure. The pressure developing when two solutions of different concentrations are separated by a membrane permeable to the solvent.

osseous (ŏs′ē-ŭs) [L. *osseus*, bony]. Bonelike, bony, or pert. to bones.

osteo- [Gk.]. Prefix: bone(s).

osteoid (ŏs′tē-oyd) [Gk. *osteo-* + *eidos*, resembling]. A substance resembling bone.

osteon (ŏs′tē-on) [Gk. *osteon*, bone]. The unit of structure of compact bone. Also, *haversian system*.

oto- [Gk.]. Combining form: the ear.

-otomy (ŏt′ō-mē) [Gk. *tomē*, incision]. Suffix: pert. to opening or repair of an organ without its removal.

output. The result or product of the operation of an organ or a machine.

ova (ō′vă) (sing. ovum) [L. *ovum*, egg]. Female reproductive cells; eggs.

ovale (ō′val′ē) [L. *ovum*, egg]. Egg-shaped or oval.

ovulate [L. *ŏvulum*, little egg]. Release of an egg from the ovary.

-oxia (ŏks′ĭ-ă). Suffix: Pert. to oxygen or oxygen concentration.

oxidation (ŏk′sĭ-dā′shŭn) [Gk. *oxys*, sharp]. The combining of something with oxygen; loss of electrons.

oxygen (ŏk′sĭ-jĕn) [Gk. *oxys*, sharp, + *gennan*, to produce]. A colorless, odorless, tasteless gas; forms more than 75% of organisms; symbol: **O**.

oxygen debt. A temporary shortage of oxygen necessary to combust products (lactic acid, pyruvic acid) of glucose catabolism.

oxyhemoglobin (ŏk-sĭ-hē-mō-glō′bĭn) [Gk. *oxys*, sharp, + *aima*, blood, + *globus*, globe]. The combination of oxygen and hemoglobin.

pacemaker (pās′mk̄-ĕr) [L. *passus*, a step, + A.S. *macian*, to make]. The sinoatrial (SA) node that sets the basic rate of heartbeat; artificial pacemaker is an electric device substituting for the SA node.

palate (păl′ăt) [L. *palatum*, palate]. The roof of the mouth; composed of hard and soft portions.

palpate (păl′pāt) [L. *palpāre*, to touch]. To examine by feeling or touching.

papilla (pă-pĭl′ă) [L. *papilla*, a nipple]. A small elevation or nipplelike protuberance.

para- [Gk.]. Combining form: near, past, beyond, the opposites, abnormal, or irregular.

paralysis (pă-ral′ĭ-sĭs) [Gk. *paralyein*, to disable at the side]. Temporary or permanent loss of the ability of voluntary movement; loss of sensation.

parasympathetic (păr-ă-sĭm-pă-thĕt′ĭk) [Gk. *para*, beside, + *sympathētikos*, suffering with]. Pert. to the portion of the autonomic nervous system that controls normal body functions; the craniosacral division.

paresthesia (păr-ĕs-thē′zĭ-ă) [Gk. *para* + *aisthēsis*, sensation]. Abnormal sensation without demonstrable cause.

parietal (pă-rī′ĕ-tăl) [L. *paries*, wall]. Pert. to outer lining or covering, or wall of a cavity; a cell of the stomach secreting HCl.

paroxysm (păr′ŏk-sĭzm) [Gk. *para*, beside, + *oxynein*, to sharpen]. A sudden, periodic attack or recurrence of symptoms of a disease.

pars (parz) [L. *pars*, a part]. Part or portion.

passive (păs'ĭv) [L. *passivus*, enduring]. Not active; in immunity, the acquisition of antibodies by administration from the outside rather than producing them oneself.

patent (păt'ĕnt, pā'tĕnt) [L. *patens*, from *patēre*, to be open]. Wide open; not closed.

pathological (păth-ō-lŏj'ĭk-l) [Gk. *pathos*, disease, + *logos*, study]. Diseased or abnormal.

peduncle (pĕdung'kl) [L. *pedunculus*, a little foot]. A band of nerve fibers connecting parts of the brain.

peptide (pĕp'tīd) [Gk. *peptein*, to digest]. A compound containing two or more amino acids formed by cleavage of proteins.

perforate (pŭr'fō-rāt) [L. *perforāre*, to pierce through]. To make a hole through; puncture.

perfuse (pŭr-fūz) [L. *perfundere*, to pour through]. To pass a fluid through something.

peri- [Gk.]. Prefix: around, or about.

peripheral (pĕr-ĭf'ĕr-ăl) [Gk. *peri-* + *pherein*, to bear]. Located away from the center; to the outside.

peristalsis (pĕr-ĭs-tăl'sis) [Gk. *peri-* + *stellein*, to place]. A progressive wavelike contraction of visceral muscle that propels liquids through hollow tubelike organs.

peritoneum (pĕr-ĭ-tō-nē'ŭm) [Gk. *peritonaion*]. The serous membrane lining the abdominal cavity and covering the abdominal organs.

permeable (per'me-ă-bl) [L. *per*, through, + *meare*, to pass]. Allowing passage of solutes and solvents in solutions.

pH. A symbol used to express the acidity or alkalinity of a solution. Strictly, the logarithm of the reciprocal of H ion concentration.

-phag- [Gk. *phagein*, to eat]. Combining form: an eater, or pert. to engulfing or ingestion.

phagocytosis (făg-ō-sī-tō'sĭs) [Gk. *phag-* + *kytos*, cell]. Engulfing of particles by cells.

-phil (fĭl). Combining form: love for; having an affinity for.

phospholipid (fŏs-fō-lĭp'ĭd) [Gk. *phōs*, light, + *pherein*, to carry, + *lipos*, fat]. A fatty substance combined with some form of phosphorus.

pia (pī'ă) [L. tender]. Tender or delicate; the inner meninx carrying blood vessels to cord and brain. Also, *pia mater*.

pigment (pĭg'mĕnt) [L. *pigmentum*, paint]. Any coloring substance.

pinocytosis (pi-nōsī-tō'sĭs) [Gk. *pinein*, to drink, + *kytos*, cell]. Intake of solution by cells through "sinking in" of the cell membrane.

pitch. That quality of a sound making it high or low in a scale; depends on frequency of vibration.

placenta (plă-sĕn'ta) [L. a flat cake]. The structure attached to the inner uterine wall through which the fetus gets its nourishment and excretes its wastes.

plane (plān) [L. *planus*, flat]. A flat surface formed by making a real or imaginary cut through the body or a portion of it.

-plasia (pla'zi-a) [Gk. *plasis*, a molding]. Combining form: Pert. to change or development. Also, *-plastic*.

-plasm (plăs'm) [Gk. *plasm*, a thing formed]. Combining form: Pert. to the fluid substance of a cell.

plasma (plăs'mă). The liquid portion of the blood.

plasma cell. One capable of producing antibodies in response to antigenic challenge. Also *plasmocyte*.

plasma membrane. The cell membrane.

pleural (plū'răl) [Gk. *pleura*, a side]. Pert. to the membrane(s) lining the two lateral cavities of the thorax, or covering the lung; or the cavities themselves that house the lungs.

plexus (plĕk'sŭs) [L. a braid]. A network of nerves or vessels (blood or lymphatic).

plica (plī'kă) [L. a fold]. A fold.

-pnea (nē'ă). Suffix: pert. to breathing. Also *pneo*.

pneumo- [Gk.]. Combining form: pert. to lungs or air.

pneumonia (nū-mō'nĭ-ă). Inflammation of the lungs.

-pod, -poda, -podo- [Gk.]. Combining forms referring to foot or feet.

poikilocytosis (poy-kĭl-ō-sī-tō'sĭs) [Gk. *poikilos*, various, + *kytos*, cell, + *osis*, intensive]. Variation in shape of red blood cells.

polarize (pō'lar-īz) [Fr. *polaire*, polar]. To create the state in which ions are in unequal concentrations on two sides of a membrane, with production of an electric potential across the membrane.

poly- [Gk.]. Prefix: many, much, or great.

polymer (pŏl'ĭ-mer) [Gk. *poly*, many, + *meros*, a part]. A substance formed by combining two or more molecules of the same substance.

pore (pōr) [Gk. *pōros*, a passage or hole]. An opening of small size.

positive pressure. A pressure greater than atmospheric, that is, >760 mm Hg.

post- [L.]. Prefix: after; in back of.

postganglionic (pōst-găng-lĭ-ŏn'ĭk). Beyond or past a ganglion or synapse.

postmortem (pōst-mor'tĕm) [L.]. After death.

potency (pō'tĕn-sĭ) [L. *potentia*, power]. Strength; power; force.

potential (pō-tĕn'shăl) [L. *potentia*]. An "electrical pressure." Implies a measurable electric current flow or state between two areas of different electrical strength.

pre- [L.]. Prefix: in front of; before.

precipitate (prē-sĭp'ĭt-āt) [L. *praecipitāre*, to cast down]. Something usually insoluble that forms in a solution; to cause precipitation or a casting down of an insoluble mass. To cause something to happen.

precocious (prē-kō'shŭs) [L. *praecoquere*, to mature before]. Matured or developed earlier than normal.

precursor (pri-kŭr'sŭr) [L. *praecurrere*, to run ahead]. Anything preceding something else or giving rise to.

preganglionic (prēgăng-lĭ-abon-ik). In front of or before a ganglion or synapse.

pregnancy (prĕg'năn-sĭ) [L. *praegnans*, with child]. The condition of carrying a child *in utero*.

premature (prē-mă-tūr'). Before term or full development.

primitive (prĭm'ĭ-tīv) [L. *primitivus*, first]. Original; early in development; embryonic.

primordial (prī-mor'dĭ-ăl) [L. *primordium*, the beginning]. Existing first or in undeveloped form.

pro- [L.; Gk.]. Prefix: for, from, in favor of.

process (prŏs'ĕs) [L. *processus*, a going before]. A method of action; a projection from something.

prognosis (prŏg-nō'sĭs) [Gk. foreknowledge]. Prediction of the course and outcome of a disease or abnormal process.

projection (prō-jĕk'shŭn) [L. *pro*, forward, + *jacere*, to throw]. Throwing forward; the efferent connection of a part of the brain.

proliferate (prō-lĭf'ĕr-āt) [L. *proles*, offspring, + *ferre*, to bear]. To increase by reproduction of similar forms or types.

propagation (prŏp-ă-gā'shŭn) [L. *propagāre*, to fasten forward]. Carrying forward; act of reproducing or giving birth.

prophase (prō′fāz) [Gk. *pro*, before, + *phasis*, an appearance]. A stage in mitosis characterized by nuclear disorganization and formation of visible chromosomes.

proprioceptive (prō-prĭ-ō-sĕp′tĭv) [L. *proprius*, one's own, + *cepius*, to take]. Pert. to awareness of posture, movements, equilibrium, and body position.

protein (prō′tē-in, prō′tēn) [Gk. *prōtos*, first]. A large molecule composed of many amino acids.

pseudo- (sū′dō) [Gk. *pseudēs*, false]. Prefix: false.

puberty (pū′bĕr-tĭ) [L. *pubertās*, puberty]. The time of life when both sexes become functionally capable of reproduction.

pulmo- [L.]. Combining form: lung.

pulse (pŭls) [L. *pulsāre*, to beat]. A throbbing caused in an artery by the shock wave resulting from ventricular contraction.

pupil (pū′pĭl) [L. *pupilla*, pupil]. The opening in the center of the iris of the eye.

pyelo- [Gk.]. Combining form: the pelvis.

pyramidal (pĭ-răm′ĭd-ăl) [Gk. *pyramis*, pyramid]. Having the shape of a pyramid.
  p. cell. Characteristic cell of cerebral cortex.
  p. tract. The bundle of nerve fibers from the primary motor area of the cerebrum to lower levels of the CNS.

pyrogen (pī′rō-jĕn) [Gk. *pyr*, fire, + *gennan*, to produce]. An agent causing a rise of body temperature.

quad- (kwŏd) [L. *quadri*, four]. Prefix: four.

quadriplegic (kwŏd-ră-plē′jik) [L. *quad-* + Gk. *plēgē*, a stroke]. Paralysis involving all four limbs.

quality (kwŏl′ĭ-tĭ) [L. *qualitās*, quality]. The nature or characteristic(s) of something.

quantity (kwŏn′tĭ-tĭ) [L. *quantitās*, quantity]. Amount.

radial (rā′dĭ-ăl) [L. *radius*, a spoke]. To pass outward from a specific center, like spokes of a wheel.

radiation (rā-dĭ-ā′shŭn) [L. *radiāre*, to emit rays]. Sending out heat or other forms of energy as electromagnetic waves; emission of rays from a center; to treat a disease by using a radioactive substance.

radical (răd′ĭ-kăl) [L. *radix*, root]. A group of several atoms that act as a unit in a chemical reaction.

ramus (rā′mŭs) (pl. rami) [L. *ramus*, a branch]. A branch of a vessel, nerve, or bone.

re- (rē) [L.]. Prefix: again; back.

receptor (rē-sĕp′tor) [L. a receiver]. A sense organ responding to a particular type of stimulus.

recessive (rĭ-sĕs′ĭv) [L. *recessus*, to go back]. To go back; in genetics, a characteristic that does not usually express itself because of suppression by a dominant gene (allele).

recipient (rē-sĭp′ĭ-ĕnt) [L. *recipiens*, receiving]. One who receives, as blood or a graft.

reciprocal (rē-sĭp′rō-kăl) [L. *reciprocus*, turning back and forth]. The opposite; interchangeable in nature.

reduction (rē-dŭk′shŭn) [L. *reductio*, a leading back]. Uptake of H by a compound, gain of electrons; restoring a broken bone to normal relationships.

referred (rē-ferd′) [L. *re*, back, + *ferre*, to bear]. Sent to another area or place, as, referred pain.

reflex (rēflĕks) [L. *reflexus*, bent back]. An involuntary, stereotyped response to a stimulus.

reflex arc. A series of nervous structures serving a reflex (receptor, afferent nerve, center, efferent nerve, effector).

refract (rē-frăkt′) [L. *refractus*, to break or bend]. To bend or deflect a light ray.

refractory (rē-frăk′tō-rĭ) [L. *refractus*]. Resistant to stimulation.

regeneration (rē-jĕn-ĕr-ā′shŭn) [L. *re-* + *generāre*, to beget]. Regrowth or repair of tissues or restoring a body part by regrowth.

regulate (rĕg′ū-lāt) [L. *regulatus*, from *regulare*, to direct or control]. To control or govern.

regurgitate (rē-gur′jĭ-tāt) [L. *re-* + *gurgitāre*, to flood]. To return to a place just passed through; backflow.

relative to (rĕl′e-tĭv) [L. *relativus*, to bring back]. Considered in comparison with something else.

remnant (rĕm′nent) [L. *re-* + *manēre*, to stay]. Something left over or behind; a fragment.

renal (rē′năl) [L. *renalis*, kidney]. Pert. to the kidney.

replicate (rĕp′lĭ-cāt) [L. *replicatio*, a folding back]. To duplicate.

repression (rē′prĕsh-ŭn) [L. *repressus*, to check]. To put down or prevent expression of something; in genetics, the way a gene shuts off its effect.

reproduction (rē-prō-dŭk′shŭn) [L. *re*, again, + *productio*, production]. Creation of a similar structure or individual.

respiration (rĕs-pĭr-ā′shŭn) [L. *respirāre*, to breathe]. Cellular metabolism; the act of exchanging gases between the body and its environment.

response (rĭ-spons′) [L. *respondum*, to answer]. A reaction to a stimulus.

resuscitation (rē-sŭs-ĭ-tā′shŭn) [L. *resuscitātus*, to revive]. Bringing back to consciousness.

reticular (rē-tĭk′ū-lar) [L. *reticula*, net]. In the form of a network; interlacing.

reticuloendothelial (re-tĭk′ū-lō-ĕn-dō-thē′lĭ-al) [L. *reticula*]. Pert. to the tissues of the reticuloendothelial system, that is, the relatively fixed phagocytic cells of the body and the plasma cells.

retina (rĕt′ĭ-nă) [L. *rete*, a net]. The third and innermost coat of the eye; contains visual receptors (rods and cones).

retro- [L.]. Prefix: behind or backward.

rhin-, rhino- [Gk.]. Combining form: nose.

rhodopsin (rō-dŏp′sĭn) [Gk. *rhodon*, a rose, + *opsis*, vision]. A visual pigment found in rod cells; "visual purple."

rhythm (rĭth′ŭm) [Gk. *rhythmos*, measured motion]. A regular recurrence of action or function.

ribo- (rī′bō). Combining form: pert. to ribose, a five-carbon sugar *(pentose)*.

rickets (rĭk′ĕts) [Gk. *rachitis*, spine]. A bone disease resulting from deficient or defective deposition of inorganic salts in new bone; primarily resulting from vitamin *D* deficiency.

RNA. Ribonucleic acid, composed of nitrogenous bases, ribose sugar, and phosphate. Three types: messenger-RNA, transfer-RNA, and ribosomal-RNA.

sac (săk) [Gk. *sakkos*, a bag]. Any bag or sacklike part of an organ.

saccharide (săk′ă-rīd) [Gk. *sakcharon*, sugar]. A sugar; a carbohydrate containing two or more simple sugar units.

sagittal (săj′ĭ-tăl) [L. *sagitta*, arrow]. Like an arrow or arrowhead; a suture in the skull; a plane dividing the body into right and left portions.

salt (sawlt) [A.S. *sealt*]. NaCl (sodium chloride); a chemical compound that results from replacing a hydrogen in the carboxyl group of an acid with a metal or cation.

saltatory (sal′tă-tō-rī) [L. *saltātio*, a leaping]. Dancing, skipping, or leaping.
  s. conduction. Conduction of a nerve im-

pulse down a myelinated nerve by leaping from node to node.

sarcolemma (sar-kō-lĕm-ă) [Gk. *sarco*, flesh, + *lemma*, rind]. Cell membrane of a skeletal muscle fiber.

sarcoma (sar-kō′mă) [Gk. *sarco* + *ōma*, tumor]. A neoplasm or cancer arising from muscle, bone, or connective tissue.

sarcomere (sar′kō-mēr) [Gk. *sarco* + *meros*, part]. The portion of a myofibril lying between two Z lines.

sarcoplasm (sar′kō′plăzm) [Gk. *sarco* + *plasma*, a thing formed]. Muscle protoplasm.

saturated (săt′ū-rā-tĕd) [L. *saturare*, saturate]. Holding all it can. A saturated fat has all the hydrogen it can hold on its chemical bonds.

scheme (skēm) [L. *schema*, a plan]. An orderly series of events or changes, as, a metabolic scheme.

sciatic (sī-ăt′ĭk) [Gk. *ischiadikos*, pert, to the ischium]. Pert. to the hip or ischium.
  s. nerve. Largest nerve in the body. Runs from pelvis to knee.

sclera (sklē′ră) [Gk. *sklēros*, hard]. The outer coat of the eyeball; the "white of the eye."

sclerosis (sklē-rō′sĭs) [Gk. *sklerosis*, a hardening]. Hardening or toughening of a tissue or organ, usually by increase in fibrous tissue.

sebaceous (sē-bā′shŭs) [L. *sebaceus*, fatty]. Pert. to sebum, a fatty secretion of the sebaceous (oil) glands.

secrete (sē-krēt′) [L. *secretus*, separated]. To separate; to make a product different from that originally presented as a starting material; active movement of substances into a hollow organ.

secretion. The product of secretory activity.

segmental (sĕg-mĕn′tăl) [L. *segmentum*, a portion]. Composed of segments, that is, individual parts or portions.

seizure (sē′zhŭr) [M.E. *seizen*, to take possession of]. A sudden attack of a disease, or pain.
  s., convulsive. Epilepsy.

selective (sē-lĕk′tĭv). Exhibiting choice, as a selective membrane passes some materials and not others.

semi- [L.]. Prefix: half.

seminiferous (sĕm-ĭn-ĭf′ĕr-ŭs) [L. *sēmen*, seed, + *ferre*, to produce]. Pert. to production and transport of sperm and/or semen.

senescence (sĕn-es′ĕns) [L. *senescere*, to grow old]. Growing old; the period of old age.

sensation (sĕn-sā′shŭn) [L. *sensatio*, a feeling]. An awareness of a change of condition(s) inside or outside the body because of stimulation of a receptor (requires the brain to interpret the change).

sensitivity (sĕn-sĭ-tĭv′ĭ-tē) [L. *sensitivus*, feeling]. The quality of being sensitive or able to receive and transmit sensory impressions; affected by external conditions.

sensitize (sĕn′sĭ-tīz). To make sensitive to an antigen by repeated exposure to that antigen.

sensory (sĕn′sō-rī) [L. *sensorius*]. Pert. to sensation or the afferent nerve fibers from the periphery to the CNS.

septum (sĕp′tŭm) [L. *saeptum*, a partition]. A membranous wall separating two cavities.

serological (sĕ-rō-lŏj′ĭk-ăl) [L. *serum*, whey, + Gk. *logos*, study]. Pert. to study of serum.

serous (sĕ-rŭs). Pert. to a cell producing a secretion having a watery nature; a watery product.

serum (sē′rum). [L. *serum*, whey]. The watery portion of the blood remaining after coagulation.

sesamoid (ses′am-oyd) [Gk. *sēsamon*, sesame, + *eidos*, form]. Like a sesame seed; specifically refers to bones that develop in tendons.

sex (sĕks) [L. *sexus*]. The quality that distinguishes between male and female.
  s. chromosomes. Chromosomes determining sex (XX, female; XY, male).
  s., nuclear. The genetic sex as determined by the presence or absence of sex chromatin *(Barr bodies)* in the nuclei of body cells.

shaft. The central portion of a long bone. Also *diaphysis*.

sheath (shēth) [A.S. *scēath*]. A covering surrounding something.

shock (shŏk) [M.E. *schokke*]. A state resulting from circulatory collapse (low blood pressure and weak heart action).

shunt (shŭnt) [M.E. *shunten*, to avoid]. A "shortcut"; passage between arteries and veins that bypasses the capillaries; a scheme for metabolism of a foodstuff that eliminates certain steps found in a scheme that metabolizes the same substance.

sickle cell. An erythrocyte that is crescent shaped because it contains an abnormal hemoglobin.

sinus (sī′nŭs) [L. a hollow or curve]. A cavity in a bone, or a large channel for venous blood to flow in.

sinusoid (sī′nŭs-oyd) [L. *sinus* + Gk. *eidos*, like]. A small irregular blood vessel found in the liver and spleen.

site (sīt). Position, place, or location.

solubility (sŏl-ū-bĭl′ĭ-tĭ) [L. *solubilis*, to dissolve]. Capable of being dissolved or going into solution.

solute (sŏl′ūt) [L. *solutus*, dissolved]. The dissolved, suspended, or solid component of a solution.

solution (sō-lū′shŭn) [L. *solutio*, a dissolving]. A liquid containing dissolved substance.

solvent (sŏl′vĕnt) [L. *solvens*, to dissolve]. A dissolving medium.

soma (sō′mă) [Gk. *sōma*, body]. Pert. to the body, as a cell body, or the body as a whole.

somatic (sō-măt′ĭk) [Gk. *soma*]. Pert. to the body exclusive of reproductive cells; pert. to skeletal muscles and/or skin.

somesthesia (som-es-thē′sĭ-ă) [Gk. *soma* + *aisthēsis*, sensation]. The awareness of body sensations.

somite (sō′mīt) [Gk. *soma*]. An embryonic blocklike segment of mesoderm formed alongside the neural tube.

spasm (spăzm) [Gk. *spasmos*, convulsion]. A sudden involuntary, often painful contraction of a muscle.

spastic (spăs′tĭk) [Gk. *spastikos*, convulsive]. Contracted or in a state of continuous contraction.

spatial (spā′shăl). Pert. to space.

species (spē′shēz) [L. a kind]. A specific type, kind or grouping of something having distinguishing and unique characteristics.

sperm (spŭrm) [Gk. *sperma*, seed]. The male sex cells produced in the testes. Also *spermatozoan; spermatozoa*.

sphincter (sfĭngk′tĕr) [Gk. *sphigktēr*, a binder]. A band of circularly arranged muscle that narrows an opening when it contracts.

spontaneous (spŏn-tā′nē-ŭs) [L. *spontaneus*, voluntary]. Occurring without apparent causes; activity seeming to occur "on its own."

squamous (skwā′mŭs) [L. *squama*, scale]. Flattened; scalelike.

stage (stāj) [L. *stare*, to stand]. A period, interval, or degree in a process of development, growth, or change.

stagnant (stăg′nănt) [L. *stagnare*, to stand]. Without motion, not flowing or moving.

stasis (stā′sĭs) job [Gk. *stasis*, halt]. Stoppage of fluid flow.
stenosis (stĕn-ō′sĭs) [Gk. *stenōsis*, a narrowing]. Narrowing of an opening or passage.
stereotyped (stĕr′ē-ō-tȳp′d) [Gk. *stereos*, solid, + *typos*, type]. Repeated, predictable response to stimulation.
steroid (stĕr′oyd). A lipid substance having as its chemical skeleton the phenanthrene nucleus.
stimulate (stĭm′ū-lāt) [L. *stimulare*, to goad on]. To increase an activity of an organ or structure.
stimulus (stĭm′ū-lŭs). An agent acting to cause a response by a living system.
strabismus (stră-bĭz′mŭs). Inability of the eyes to both focus at the same point. Crossed eyes or "squint."
stratified (străt′ĭ-fĭd) [L. *stratificare*, to arrange in layers]. Disposed in layers.
stratum (străt′ŭm) [L. *stratum*, layer]. A layer, as of the epidermis of the skin.
striated (strī′āt-ĕd) [L. *stria*, channel]. Striped or streaked.
stroke (strōk) [A.S. *strāk*, a going]. The total set of symptoms resulting from a cerebral vascular disorder. Also *apoplexy*.
sub- [L.]. Combining form: under, beneath, in small quantity, less than normal.
subconscious (sŭb-kŏn′shŭs) [L. *sub-* + *conscius*, aware]. Operating at a level of which one is not aware.
subcutaneous (sŭ-kū-tā′nēus) [L. *sub* + *cutis*, skin]. Beneath the skin, as an injection. Abbr. *Sq.*
subjective (sŭb-jĕk′tĭv). Arising as a result of activity by a person; a *personal* reaction or conclusion.
subliminal (sŭb-lĭm′ĭn-ăl) [L. *sub* + *limen*, threshold]. Less than that required to get a response; not perceptible.
substrate (sŭb′strāt) [L. *substratum*, a strewing under]. The substance an enzyme acts on.
sudoriferous (sū-dor-ĭf′er-ŭs) [L. *sudor*, sweat, + *ferre*, to bear]. Pert. to production or transport of sweat.
 s. gland. Sweat gland.
sulcus (sŭl′kĭs, sŭl′sus) [L. a groove]. A slight depression or groove, esp. in the brain or spinal cord.
super-, (*supra*) [L.]. Combining form: above, beyond, superior.
superficial (sū-pĕr-fĭsh′ăl) [L. *supra* + s *facies*, shape]. At or to a surface.
surface tension (alveolus). The phenomenon whereby liquid droplets tend to assume the smallest area for their volume. The surface acts as though it had a "skin" on it as a result of cohesion among the surface water molecules.
surfactant (sŭrf-ăk′tănt). A lipid-protein substance that reduces surface tension in the lung alveoli and thus reduces chances of alveolar collapse. Also, *surface active agent*.
suspension (sŭs-pĕn′shŭn) [L. *suspensio*, a hanging]. A state in which solute molecules are mixed but are not dissolved in a solvent and do not settle out.
sym-, *syn* [Gk.]. Combining form: with, along, together with, beside.
sympathetic (sĭm-pă-thĕt′ĭk) [Gk. *syn*, with, + *pathos*, suffering]. Pert. to the portion of the autonomic nervous system that controls response to stressful situations.
symphysis (sĭm′fĭs-ĭs) [Gk. *symphysis*, a growth together]. A type of joint in which the two bones are held together by fibrocartilage.
synapse (sĭn′ăps) [Gk. *synapsis*, to touch together]. A point of junction between two neurons (functional continuity).
synchronous (sĭn′krŏn-ŭs) [Gk. *syn* + *chronos*, time]. Occurring at the same time.
syncytium (sĭn-sĭsh-yŭm) [Gk. *syn* + *kytos*, cell]. Cells running together anatomically or functionally, and behaving as a single multinucleated mass.
syndrome (sĭn-drōm) [Gk. *syndromē*, a running together]. A complex or group of symptoms.
synergist (sĭn′ĕr-jĭst) [Gk. *syn* + *ergon*, work]. A muscle working with other muscles to "firm" an action.
synonym (sĭn′ō-nĭm) [Gk. *syn* + *onoma*, name]. Names used to refer to the same process, characteristic, or structure; an additional or substitute name.
synovial (sĭn-ō′vĭ-ăl) [Gk. *syn* + L. *ovum*, egg]. Pert. to the cavity or fluid in the space between bones of a freely movable joint.
synthesis (sĭn′thē-sĭs) [Gk. *syn* + *tithenai*, to place]. To form new or more complex substances from simple precursors.
syphilis (sĭf′ĭ-lĭs) [Origin uncertain; Gk. *syn*, with, + *philos*, love, or from *Syphilus*, shepherd who had the disease]. An infectious, chronic venereal disease transmitted by physical contact with an infected person.
system (sĭs′tĕm) [Gk. *systema*, an arrangement]. An organized group of related structures.

systemic (sĭs-tĕm′ĭk). Pert. to the whole or the greater part of the body.
systole (sĭs′tō-lē) [Gk. *systolē*, contraction]. Contraction of the muscle of a heart chamber.

tachy- [Gk.]. Combining form: rapid, fast.
tamponade (tăm-pōn-ād′) [Fr. *tampon*, plug]. To plug up.
 t., cardiac. A condition resulting from accumulation of excess fluid in the pericardial sac.
taxis (tăk′sĭs) [Gk. arrangement]. Response to an environmental change; a turning toward (positive) or away from (negative) the change.
telodendria (tĕl-ō-dĕn′drĭ-ă) [Gk. *telos*, end, + *dendron*, tree]. The treelike branching ends of an axon or its branches.
telophase (tĕl′ō-fāz) [Gk. *telos* + *phasis*, a phase]. The final stage of mitosis characterized by cytoplasmic division and reformation of nuclei.
temperature (tĕm′per-ă-tūr) [L. *temperatura*, proportion]. The degree(s) of heat of a living body.
temporal (tĕm′por-ăl) [L. *temporalis*, pert. to time]. Pert. to time.
tension (tĕn′shŭn) [L. *tensio*, a stretching]. The state of being stretched; a concentration of a gas in a fluid; the force developed by a muscle contraction as measured by a gauge.
terminal (ter′mĭn-al) [L. *terminus*, a boundary]. End or placed at the end; a disease that will end with death.
tetanus (tĕt′ă-nŭs) [Gk. *tetanos*, tetanus]. A sustained contraction of a muscle; a disease caused by a bacterium characterized by sustained contraction of jaw muscles ("lockjaw").
tetany. Muscular spasm caused by deficient parathyroid hormone. Also, *hypoparathyroidism*.
therapy (thĕr′ă-pī) [Gk. *therapein*, treatment]. The treatment of a condition or disease.
thermal (ther′măl) [Gk. *thermē*, heat]. Pert. to heat.
thorax (thō′răks) [Gk.]. The chest.
threshold (thrĕsh′ōld) [A.S. *therscwold*]. Just perceptible; the lowest strength stimulus that results in a detectable response or reaction. Also, *liminal*.
thrombus (thrŏm′bŭs) [Gk.]. A blood clot obstructing a vessel.
thrombosis. The formation of the clot, or the result of blocking the vessel.

tissue (tĭsh´ū) [L. *texere*, to weave]. A group of cells that are similar in structure and function (e.g., epithelial, connective, muscular, and nervous tissues).

-tome (tōm) [Gk.]. Combining form: a cutting, or a cutting instrument.

tone (tōn) [L. *tonus*, a stretching]. A state of slight constant tension or contraction exhibited by muscular tissue.

tonic (tŏn´ĭk) [Gk. *tonikos*, pert. to tone]. Having tone; continual.

tonicity (tō-nĭs´ĭ-tĭ) [Gk. *tonos*, tone]. The property of having tone; a reference to the osmotic pressure of a solution.

tortuous (tŏr´tū-ŭs) [L. *tortuosus*, twisted]. Twisted and turned; full of windings.

toxic (tŏks´ĭk) [Gk. *toxikon*, poison]. Poisonous.

toxin (tŏks´ĭn) [Gk. *toxikon*]. A poisonous substance derived from plant or animal sources.
  t., bacterial. Toxin produced by bacteria.

trace (trās) [L. *tractus*, a drawing]. A very small quantity.
  t. element. Metals or organic substances essential in minute amounts for normal body function.

tract (trăkt) [L. *tractus*, a track]. A bundle of nerve fibers in the CNS that carries particular kinds of motor or sensory impulses.

trans- (trăns) [L.]. Prefix: across, through, over, or beyond.

transcription (trăns-krĭp´shŭn). The process in which DNA gives rise to RNA.

transduce (trăns-dūs) [L. *trans*, across, + *ducere*, to lead]. To change one form of energy into another.

transient (trăn´shĕnt) [L. *trans*, over, + *ire*, to go]. Temporary, not permanent.

transition (trăns-ĭ´shŭn) [L. *transitio*, a going across]. Changing from one state or position to another.

translation (trăns-lā´shŭn) [L. *translatio*, to translate]. The process by which the code in messenger RNA is used to synthesize a specific protein.

transparent (trăns-păr´ĕnt) [L. *trans*, across, + *parēre*, to appear]. Allowing light through to permit a clear view through a substance.

transplantation (trăns-plăn-tā´shŭn) [L. *trans* + *plantāre*, to plant]. To take living tissue from one person or species and place it elsewhere on that person's body or in another person or species.

transport (trăns-pŏrt´) [L. *trans* + *portare*, to carry]. To transfer or carry across.

transverse (trăns-vĕrs) [L. *transversus*, turned across]. Crosswise.

trauma (traw´mă) [Gk. *trauma*, wound]. An injury or wound.

tremor (trĕm´or) [L. *tremor*, a shaking]. A quivering or involuntary rhythmical movement of a body part.

treppe (trĕp´eh) [Ger. *treppe*, staircase]. An increase in strength of muscular contraction when a muscle is stimulated maximally and repeatedly.

tri- [Gk.]. Combining form: three.

trophic (trō´fĭk) [Gk. *trophē*, nourishment]. Literally, nourishing. Used to refer to those hormones that control activity of other endocrine structures (e.g., ACTH, TSH, LH, FSH, ICSH).

-tropic (trō´pĭk) [Gk. *tropos*, turning]. Combining form: turning, changing, responding to stimulus.

T-tubule. Transversely arranged microscopic tubules that convey ECF to the interior of muscle cells.

tumor (tū´mor) [L. a swelling]. A swelling or enlargement. Also neoplasm.

tunic (tū´nĭk) [L. *tunica*, a sheath]. A coat, covering, or layer.

turbinate (tŭr´bĭn-āt) [L. *turbo*, a whirl]. Nasal concha.

turbulent (tŭr´bū-lĕnt) [L. *turbulentus*, to trouble]. Disturbed or not "smooth." Used to describe flow of blood in the circulatory system.

twitch (twĭch) [M.E. *twicchen*]. A single muscular contraction in response to a single stimulus.

ulcer (ŭl´ser) [L. *ulcus*, ulcer]. An open sore or lesion in the skin or a mucous membrane–lined organ (e.g., stomach, mouth, intestine).

umbilical (ŭm-bĭl´ĭ-kăl) [L. *umbilicus*, navel]. Pert. to the navel.
  u. cord. The structure connecting the fetus to the placenta.

un- [A.S.]. Prefix: not, back, reversal.

unconscious (ŭn-kŏn´shŭs) [A.S. *un*, not, + L. *conscius*, conscious]. Insensible, not aware of surroundings.

uni- [L.]. Combining form: one.

unit (ū´nĭt) [L. *unus*, one.]. One of anything; a single distinct object or part.

urea (ū-rē´ă) [Gk. *ouron*, urine]. A waste of metabolism formed in the liver from NH_3 and CO_2. $CO(NH_2)_2$.

uremia (ū-rē´mĭ-ă) [Gk. *ouron* + *aima*, blood]. A toxic disorder associated with elevated blood urea levels. A result of poor kidney function.

-uria (ūr´ĭ-ă) Suffix: pert. to urine.

uric acid (ū´rĭk) [Gk. *ouron*]. An end product of purine (a nitrogenous base) metabolism found in urine. $C_5H_4N_4O_3$.

urine (ū´rĭn) [L. *urina*, urine]. The fluid formed by the kidney and eliminated from the body via the urethra.

uro- [Gk.]. Combining form: pert. to urine.

vaccinate (văk´sĭn-āt) [L. *vaccinus*, pert. to a cow]. To inoculate with a vaccine to produce immunity against a disease.

vaccine (văk-sēn´) [L. *vacca*, a cow]. Killed or attenuated bacteria or viruses prepared in a suspension for inoculation.

vacuole (văk´ū-ōl) [L. *vacuolum*, a tiny empty space]. A fluid- or air-filled space in a cell.

vagus (vā´gŭs) [L. wandering]. The tenth cranial nerve.

valve (vălv) [L. *valva*, a fold]. A structure for temporarily closing an opening so as to achieve one-way fluid flow.

varicose (văr´ĭ-kōs) [L. *varix*, a twisted vein]. Pert. to distended, knotted veins.

vascular (văs´kū-lăr) [L. *vasculum*, a small vessel]. Pert. to or having blood vessels.

vasomotor (văs-ōmō´tor) [L. *vaso*, vessel, + *motor*, a mover]. Pert. to nerves controlling activity of smooth muscle in blood vessel walls. May cause vasodilation (enlargement) or vasoconstriction (narrowing) of the vessel.

vein (vān) [L. *vena*]. A vessel carrying blood toward the heart or away from the tissues.

venereal (vē-nē´rē-ăl) [L. *venerus*, from Venus, goddess of love]. Pert. to or resulting from sexual intercourse.

ventilation (vĕn-tĭl-ā´shŭn) [L. *ventilāre*, to air]. To circulate air, esp. into and out of the lungs.
  v., alveolar. Supplying air to alveoli.
  v., pulmonary. Supplying air to the conducting division of the respiratory system.

ventral (vĕn´trăl) [L. *venter*, the belly]. Pert. to belly or front side of the body.

ventricle (vĕn´trĭk-l) [L. *ventriculus*, a little belly]. One of the two lower chambers of the heart.

venule (vĕn´ul) [L. *venula*, a little vein]. A small vein; gathers blood from capillary beds.

vermis (vĕr´mĭs) [L. *vermis*, worm]. The middle lobe of the cerebellum. *Vermiform*, wormlike, as vermiform appendix.

vesicle (vĕs´ĭ-kl) [L. *vesicula*, a little bladder]. A small, fluid-filled sac.

vestibular (vĕs-tĭb′ū-lăr) [L. *vestibulum*, vestibule]. Used to refer to the equilibrium structures of the inner ear and their nerves; literally, pert. to a small space or cavity.

villus (vĭl′ŭs) [L. *villus*, tuft of hair]. Small fingerlike vascular projections from or of certain mucous membranes (e.g., intestine).

virus (vī′rŭs) [L. *virus*, poison]. An ultramicroscopic infectious organism that requires living tissue to survive and reproduce.

viscera (vĭs′ĕr-ă) [L. *viscus*, viscus]. The internal body organs, esp. those of the abdominal cavity.

visceral. Pert. to the viscera or an outer lining of an organ.

viscous (vĭs′kŭs) [L. *viscōsus*, sticky]. Sticky or of thick consistency.

visual (vĭzh′ū-ăl) [L. *visio*, a seeing]. Pert. to vision or sight.

vital (vī′tăl) [L. *vitalis*, pert. to life]. Essential to or contributing to life.

vitamin (vī′tă-mĭn) [L. *vita*, life, + *amine*[. An essential organic substance not serving as a source of body energy but working with enzymes to control body function.

vitreous (vĭt′rē-ŭs) [L. *vitreus*, glassy]. Glassy; the vitreous body (humor) of the eyeball found between the lens and retina.

voluntary (vŏl′ŭn-tā-rĭ) [L. *voluntas*, will]. Pert. to being under willful control.

wallerian degeneration [A.V. *Waller*, Eng. phys.-physiologist, 1816-1870]. Describes the degenerative changes occurring in an axon that has been severed from its cell body.

waste (wāst) [L. *vastāre*, to devastate]. Useless end products of body activity; to shrink or become smaller in size and/or strength.

white matter (hwīt măt′tĕr). That part of the CNS composed primarily of myelinated nerve fibers.

Willis, circle of [Thos. *Willis*, Eng. phys., 1621-1675]. An arterial circle on the base of the brain composed of anterior cerebral, anterior communicating, internal carotid, posterior communicating, and posterior cerebral arteries.

xeno (zē′nō) [Gk. *xeno*, foreign]. Combining form: strange, foreign.

xenograft (ze′nō-grăft) [Gk. *xeno* + L. *graphium*, grafting knife]. A graft of tissues or organs between different species.

XX. Complement of sex chromosomes giving rise to a female.

XY. Complement of sex chromosomes giving rise to a male.

yolk sac (yōk săk). A membranous sac surrounding the yolk in an embryo that contains yolk; in the human, a transitory structure incorporated into the gut.

Z line. An intermediate line in the striation of skeletal muscle; lies within the *I* line, and serves as the limits of a sarcomere.

zygote (zī′gōt) [Gk. *zygotōs*, yoked]. A cell produced by the union of an egg and sperm; a fertilized ovum.

Index

A

Abdomen, 434
 boundaries, 434
Abdominal regions, 6
 organs in, 7
Achilles tendon, 79, 80, *243*, 244
Achondroplasia, 50
Acromegaly, 506
Actin, *172*, 173, *173*
Adam's apple, 409
Addison's disease, 507
Adenine, 16
Adenohypophysis, 495, *495*
Adenosine triphosphate, 12
Adipose capsule, 452, *452*
Adipose tissue, *47*, 48
Adrenal glands, 503-504
 blood supply, 503
 cortex, 504, *504*
 disorders, 507
 hormones, 504
 location, 503, *503*
 medulla, 504, *504*
 size, 503
Adrenocorticotropic hormone, 497t
Airway resistance, 418
Albumin, 345
Aldosterone, 504
Allantois, 30
Alveolar duct, 416
Alveolar macrophage, 416
Alveolar sac, 416
Amenorrhea, secondary, 479
Aminopeptidase, 447
Amniocentesis, 30
Amnion, 28
 development, *29*
 functions of, 28
Amniotic cavity, 21, *21*
Amniotic fluid
 functions, 30
Ampulla (uterine tube), *469*, 474
Amygdaloid nucleus, 270, *270*, 280
Anal canal, 440, *440*
Anal sphincters, 440, *440*
Anatomical position, 2, *2*

Italicized entries indicate that a figure illustrates the point discussed. A *t* following an entry indicates a table that contains information pertinent to the subject. The index is organized mainly by organ and system.

Anatomy
 defined, 1
 developmental, 2
 gross, 1
 history, 1
 surface, 2
Anesthesia, 313
Aneurysm, 373, *373*
Ankle, *79*, 80
Anterior, 3
Anterior root, 300, *300*
Antidiuretic hormone, 458, 498
Aphasia, 337
Apnea, 419
Aponeurosis, 178
Appendicitis, 448
Appendix, *426*, 439
Aqueous humor, 320
Arachnoid, 286, *288*, *289*, *291*
Arachnoid granulation, 290, *291*
Arachnoid septa, *288*, 289
Arches of foot, *139*, 140
Arcuate nucleus, 277, *277*
Areola, 478, *478*
Arrector pili, 92
Arteries
 aorta, 374, *374*, 375t
 auditory, internal, *387*, 388
 axillary, 76, *379*, 379t, 380, *380*
  branches of, *380*
 basilar, *387*, 388
 brachial, 76, *379*, 379t, 380, *380*
 brachiocephalic, *374*, 375, 375t
 bronchial, *374*, 375, 375t
 carotid
  common, 74, *374*, 375, 375t, 378
   left, *374*, 375, 375t
  external, 378, *378*
  internal, 378, *378*, 386
 celiac, *374*, 375t, 376
  branches of, 376, *376*
 cerebellar, *387*, 388
  anterior inferior, *387*, 388
  posterior inferior, *387*, 388
  superior, *387*, 388
 cerebral, 386, *387*, 388
  anterior, 386, *387*
  middle, *387*, 388
  posterior, *387*, 388
 circle of Willis, *387*, 388
 circumflex, 360, *360*, 361
 communicating, 386, *387*, 388

531

Arteries—cont'd
    coronary, 360, *360*, *361*, 374, *374*
    costocervical, 378, 379t
    digital, *379*, 379t, 381, *381*
    dorsalis pedis, 80, 381
    esophageal, *374*, 375, 375t
    facial, 378, *378*
    femoral
        branches of, 381, *381*
    genicular, descending, 381, *381*
    gluteal, 376
    gonadal, *374*, 375t, 376
    iliac
        external, 381, *381*
        internal, *347*, 375t, 376, 381, *381*
            branches of, *381*
    intercostal, *374*, 375, 375t
    interventricular, right, 360, *361*
    labyrinthine, *387*, 388
    lingual, 378, *378*
    mammary, 378, *378*, *379*, 379t
    marginal, 360, *360*
    maxillary, 73, 378, *378*
    medial, 379t
    mesenteric
        inferior, *374*, 375t, 376
            branches of, 376, *377*
        superior, *374*, 375t, 376
            branches of, 376, *377*
    metatarsal, 381
    obturator, 376
    occipital, 378, *378*
    ophthalmic, 386
    pericardial, 375, 375t
    peroneal, 381, *381*
    pharyngeal, 378, *378*
    phrenic, *374*, 375, 375t
    plantar arch, 381, *381*
    pontine, *387*, 388
    popliteal, 80, 381, *381*
    pudendal, 376
    pulmonary, 390, *390*
    radial, 77, *379*, 379t, 380
        branches of, 379t, 380
    rectal, 376
    renal, *374*, 375t, 376
    sacral, middle (midsacral), *374*, 375t, 376
    spinal, anterior, *387*, 388
    subclavian
        left, *374*, 375, 375t
        right, *374*, 375, 375t, 378
    subscapular, *379*, 379t, 380, *380*
    suprarenal, *374*, 375t, 376
    temporal, superficial, 72, *72*, 378, *378*
    thoracic, internal, 378, *378*, *379*, 379t
    thyrocervical, 378, *378*, *379*, 379t
    thyroidal, superior, 378, *378*
    tibial, anterior, 381, *381*
    ulnar, *379*, 379t, 380
        branches of, 379t, 380
    umbilical, 391, *391*
    uterine, 376
    vaginal, 376
    vertebral, 378, *378*, 379t, *387*, 388

Arteries—cont'd
    vesical, 376
    volar arch, 380
Arteries of
    brain, 386, *387*
    head and neck, 378, *378*
    lower appendage, 381, *381*
    lungs, 390, *390*
    upper appendage, 378, *379*, 379t
Arteriosclerosis, 372
Asbestosis, 418
Asthma, 419
Astigmatism, 326
Astrocyte, 259, *259*
Atherosclerosis, 372, *372*
Athetoid movements, 270
Atom, 9
ATP; *see* Adenosinetriphosphate
Atresia, 472
Atrioventricular bundle, 359, *359*
    branches of, 359, *359*
Atrioventricular node, 358, *359*
Atrophy, 180
Auditory ossicles, 328, 333, *333; see also* specific bone names
    development, 328, *331*
Autonomic fibers, 252
Autonomic nervous system
    ganglions, 304, *305*
    parasympathetic division, 304, *305*
    plexuses, 304, *305*
    sympathetic division, 304, *305*
    visceral afferents, 303
    visceral efferents, 304
Axilla, 76
Axon, *254*, 255
    hillock, 255
    reaction, 262

B

Baroreceptor, 314, *314*
Bartholin's gland, 478
Basis pedunculi, 279, *279*
Basophil, *346*, 347, 348t
Bile canaliculi, 442, *442*
Bile duct(s), 444, *445*
Black lung, 418
Blastocoele, 20, *20*
Blastocyst, 20, *20*
Blastodermic vesicle, 20, *20*
Blastomeres, 20
Blood, 344-348
    development, 344, *345*
    formed elements, 345, *346*, 347, 348t
    functions, 345
    plasma, 345
Blood-brain barrier, 389
Blood vessel(s), 363-391
    development, 363, *363*, 364
    structure
        arteries, 366-368, *366*, 367t
        capillaries, 368, *369*
        veins, 370-372, *370*, *371*, 371t

Blood vessel(s)—cont'd
    types, 364
Body appearance, 67-71
    adolescent, 68, *68*
    adult, 68, *69*, 70
    aged, *71*
    child, 67
    newborn, 67
    premature infant, 67
Body cavities
    abdominopelvic cavity, 7, *8*
    cranial cavity, 7, *7*
    dorsal cavity, 7, *7*
    thoracic cavity, *7*, 8
    ventral cavity, *7*, 8
Body regions, 2, *3*
Bone, 50-52
    cancellous, 50, *51*
    cells, 52
    compact, 50, *51*
    dense, 52
    spongy, 50, *51*
    structure, 50
Bone(s)*
    atlas, 122, *122*
    axis, 122, *123*
    clavicle, 126, *127*
    coccyx, 123, *123*
    ear, 110, 328, 333, *333*
    ethmoid, 106, *106*
    femur, 137, *137*
    fibula, *138*, 139
    frontal, 104, *104*, 105, *105*
    humerus, 128, *128*, 129
    hyoid, 113, *113*
    ilium, 134, *134*, 135
    ischium, 134, *134*, 135
    lacrimal, 106, *106*
    mandible, 108, *108*
    maxilla, 106, *107*
    metacarpal, 130, *130*
    metatarsal, 139, *139*
    nasal, 104, *104*
    nasal concha, inferior, 106
    number of, in body, 101, 102
    occipital, *105*, 110, 111, *111*
    os coxae, 134, *134*, 135
    palatine, 106, *107*
    parietal, 108, *109*
    patella, 138, *138*
    phalanges
        foot, *139*, 140
        hand, 130, *130*
    pubis, 134, *134*, 135
    radius, 129, *129*
    ribs, 124, *124*, 125
    sacrum, 123, *123*
    scapula, 126, *127*
    sesamoid, 101
    shapes of, 101, *101*
    sphenoid, 110, 112, *112*

Figures cited present the best, but not necessarily the only, view of a given bone.

Bone(s)—cont'd
  sternum, 78, 125, *125*
  tarsal, 139, *139*
  temporal, 108, *109*
  tibia, 138, *138*
  ulna, 129, *129*
  vomer, 106, *107*
  wormian, 101
  wrist, 130, *130*
  zygomatic, 73, *73, 104, 109,* 110
Bony tissue
  blood vessels, 100
  formation, 97-100, *97, 98, 99*
  hormones and, 100
  vitamins and, 100
Botulism, 448
Bowman's capsule, 454, *455*
Brachial plexus, 300, *301*
  nerves rising from, *301,* 302
Brain
  defined, 263
  parts of, 263, *263*
Brainstem, 276-280
  external anatomy, *276*
  internal anatomy, 277-280, *277-280*
  subdivisions of, 276, *276*
Breathing, 418
  control of, 419
Bronchiole, *413,* 414
Brown-Séquard syndrome, 286
Bulbourethral gland, *480, 487, 488,* 489
Bypass, cardiac, 361

C

Calcitonin, 501
  actions of, 501
Calyces, 453, *453*
  wall structure, 458
Cancer, 18
  cells, 19
Capacity
  functional residual, 418
  vital, 418
Carbohydrate, 9
Carboxypeptidase, 447
Cardia, 436, *436*
Cardiac muscle
  disposition in heart, 357, *357*
  properties, 357
Cardiac sphincter, 436, *436*
Cardiogenic plate, 349
Caries, 447
Carina, *411,* 414
Carotene, 85
Cartilages
  elastic, *49,* 50
  fibrocartilage, *49,* 50
  growth, 50
  hyaline, *48,* 50
Cataract, 326
Cauda equina, *281,* 282
Caudate nucleus, 270, *270*
Cavernous bodies, *488, 489,* 489
CCK; see Cholecystokinin

Cecum, 439
Cell, 10
  structure, 10, *10*
Cementum, 428, *429*
Central nervous system, 250
  components, 250
Centrioles, 14, *14*
Centromere, 18, *18*
Centrosomes, 14, *14*
Centrosphere, 14
Cerebellum
  central white matter, 271
  deep nuclei, 272, *272*
  fissures, 271, *271*
  folia, 271, *273*
  functions, 272
  hemispheres, 271, *271*
  histology, 272, *273*
  peduncles, 271
Cerebral aqueduct, 290, *290*
Cerebrospinal fluid, 290, 291
  circulation, 290
  production, 290
Cerebrum, 263
  basal ganglions, 270, *270*
  cortex, 267
  fissures, 264, *264,* 265, *265*
  functional areas, 266, *266,* 267
  gyri, 265
  histology, 267, 268, *268*
  lobes, 264, *264*
  medullary body, 268, *269*
  sulci, *264, 265, 265,* 266
Cerumen, 332
Cervical cancer, 480
Cervical enlargement, 281, *281*
Cervical plexus, 300, *301*
Cheeks, 427
Chemoreceptor, 314, *314*
Cholecystokinin, 445, 447, 506
Choreiform movements, 270
Chorioid, *318,* 320, *320*
Chorion, 21, 28, *29*
  function, 28
  villi, 28
Choroid plexus, 290, *291*
Chromatin material, 16, *16*
Chymotrypsin, 447
Cilia, 15, *15,* 35
Ciliaris muscle, *318,* 320, *320*
Ciliary body, 320, *320*
Ciliary muscle, *318,* 320, *320*
Ciliary processes, 320, *320*
Ciliary zonule, *318,* 320, *320*
Circulatory system, 344-400
  subdivisions, 344
Cisterna chyli, *392,* 393
Cisternae, 290, *291*
Claustrum, 270, *270*
Cleavage, 20
Cleavage lines, 85
Clitoris, *469, 477,* 478
Cochlea, *332, 334,* 336
  structure, 336, *336,* 337

Cold receptor, 310, *311*
Collagenous connective tissue, *45,* 48
Collecting tubule, 454, *455*
Colon, *426,* 439
Color blindness, 326
Communicating ramus
  gray, 300, *300*
  white, 300, *300*
Compliance, 418
Conception, 20
Cone, 322, *322*
Conjunctivitis, 327
Connective tissue(s), 42-53
  adult, 42, 44
  bone, 50-52
  cartilages, 48-50
  classifications, *42*
  connective tissue proper, 44, *44-47,* 48
  embryonal, 42, 43, *43*
  formation, 50
  general characteristics, 42
  physiology, 50
  summary, 53t
  types, 42
Conus medullaris, *281,* 282
Cornea, 318, *318, 319*
  structure, *319,* 320
Corpus albicans, 473, *473*
Corpus callosum, *265, 266,* 269
Corpus luteum, *472,* 473
Corpuscle
  Krause's, 88, *89*
  Meissner's, 88, *88*
  pacinian, 88, *89*
  Ruffini's, 88, *89*
Cortisol, 504
Countercurrent exchanger, 458
Countercurrent multiplier, 458
Cowper's gland, *480, 487, 488,* 489
Cramps, 181
Cranial nerves, 252, 292-299, 293t
  abducent, *295,* 296
  accessory, 299, *299*
  facial, 296, *296*
  glossopharyngeal, 297, *297*
  hypoglossal, 299, *299*
  oculomotor, 294, *295*
  olfactory, 294, *294*
  optic, 294, *294*
  trigeminal, 295, *295*
  trochlear, 294, *295*
  vagus, *298,* 299
  vestibulocochlear, 296, *297*
Cremaster muscle, 481
Crest
  iliac, 80
  pubic, 80
Cricoid cartilage, 74, *409,* 409
Crista ampullaris; see Semicircular canal(s). crista
Cuneate nucleus, 277, *277*
Cuneate tract, 277, *284*
Cupula, 334, *335*

Cushing's syndrome, 507
Cuticle, 37, *37*
Cyanosis, 85
Cytokinesis, 18
Cytology, 2
Cytoplasm, 11
Cytoplasmic specialization at epithelial surface, 37
Cytosine, 16

D

Da Vinci, 1
Dartos tunic, 481
Deafness, 336
   nerve, 337
   transmission, 337
Defibrillation, 361
Dendrite, *254*, 255
Dense irregular connective tissue, 48
Dentate nucleus, 272, *272*
Denticulate ligament, *288*, 289, *289*
Deoxyribonucleic acid
   functions, 17
   structure, 16, *17*
Deoxyribose, 16
Desmosome, 38, *38*
Diabetes mellitus, 506
Diaphysis, 101, *101*
Diencephalon, 249, *249*, 274, *274*
   derivatives, 249, *249*
Differentiation, 20
Digestive system, 421-448
   congenital disorders, 424, *424*, 425t
   development, 422, 422t, *423*
   disorders, 447, 448
   functions, 421
   organs, *426*, 427
   tissue plan in wall, 433, *433*
Dipeptidase, 447
Diploid, 19
Directional terms, 3
Distal convoluted tubule, 454, *455*
Ductus arteriosus, 391, *391*
Ductus venosus, 391, *391*
Duodenum, 438
Dura mater
   cranial, 286, *287*, 291
   spinal, 288, *288*, 289
Dural sinuses, 286, *287*, 291
DNA; *see* Deoxyribonucleic acid
Dystrophies, 180

E

Ear, 328-337
   bones; *see* Auditory ossicles
   development, 328, *328-331*
   disorders, 336, 337
Eardrum, *330, 332*, 333
Ectoderm, 21, *21*, 248
   derivatives, 22t
Ectopic pregnancy, 475
Effective filtration pressure, 457

Efferent ductule, *482*, 484, *485*
Ejaculatory duct, 486, *487*
Elastance, 418
Elastic connective tissue, *46*, 48
Elements composing body, 9
Emboliform nucleus, 272, *272*
Embryo
   three-layered, 22, *23*
   two-layered, 20, 21, *21*
Embryonic disc, 20, 21, *21*, 248, *249*
Emmetropia, 325, *326*
Emphysema, 419
Enamel, 428, *429*
Endocardium, 354, *354*
Endocrine structures, 492-507
   criteria for classification, 493, 494
   development, 492, *493*, 493t
   disorders, 506, 507
   organs composing, 494, *494*
Endoderm, 21, *21*, 248
   derivatives, 22t
Endometrial cancer, 480
Endometriosis, 480
Endometrium, 476, *476*
Endomysium, 178, *178*
Endoneurium, 262, *262*
Endoplasmic reticulum, *11*
   functions, 11
   rough, 11
   smooth, 11
   structure, 11
Endoscopy, 2
Endosteum, 52
Endothelium, 31
Enterocrinin, 447
Eosinophil, *346, 347*, 348t
Ependyma, 259, *259*
Epicardium, 354, *354*
Epididymis, *480*, 484, *485*
   parts, 484
Epidural space, 288, *288*
Epiglottis, 409, *409, 410, 411*
Epimysium, 178, *178*
Epinephrine, 504
Epineurium, 262, *262*
Epiphysis, 101, *101*
Episiotomy, 478
Epithalamus, 274, *274*
Epithelial membranes, 40, 41
Epithelium, 31-41, 39t
   aberrant, 31, 34, *34*, 35, *35*
   classification, 31
   columnar, 31, *32*
   cuboidal, 31, *32, 33, 33*
   general characteristics, 31
   germinal, 35, *36*
   mesenchymal, 33
   neural, 35, *36*
   pseudostratified, 34, *34*
   simple, 31, *32*
   squamous, 31, *32*
   stratified, 31, *32*, 33
   surface modifications, 35-37
   syncytial, 35, *35*

Epithelium—cont'd
   transitional, 34, *35*
   types, 31
Erythrocyte(s), *346, 347*, 348t
Esophagus, *426*
   blood supply, 434
   function, 434
   structure, 434
Eustachian tube (pharyngotympanic tube), 328, *330, 332, 332*, 333
Excitability in neurons, 255
Expiration, 418
External auditory meatus, 72, *72, 330, 332, 332*
External ear, 328, 332, *332*
   development, 328, *331*
External genitalia, 467
   development, *466*
Eye, 317-327
   development, 317, *317*
   disorders, 325-327, *326*
   extrinsic muscles, 317, 324, *324*, 324t
   fundus, *323*
   structure, 318, *318*
   tunics, 318-322
Eyelids, 324, *325*

F

Fallopian tube; *see* Uterine tube
Falx cerebelli, 286
Falx cerebri, 286, *287*
Farsightedness, 325, *326*
Fasciae, 178
Fasciculation, 181
Fastigial nucleus, 272, *272*
Fauces, 433
Fertilization, 20
Fetal circulation, 350, *350*, 391, *391*
Fetal membranes, 28, *29*
Fetus, 25
Fibers
   association, 268, *269*
   commissural, 268, *269*
   projection, 268, *269*
Fibrillation, 181, 361
Fibrinogen, 345
Filum terminale, *281*, 282
Fimbrae, *469*, 474
Fissure
   calcarine, 264, *264*
   central, 264, *264*
   lateral, 264, *264*
   longitudinal, 264, *264*
   parietooccipital, 264, *264*
   transverse, 264
Flagella, 15, 35
Flexion crease, 85
Flexion line, 85
Flexor retinaculum, *186*
Follicle-stimulating hormone, 479, 479t
   relationship to ovarian and uterine changes, *479*
   relationship to testes, 489
Follicular cells of ovary, *470*, 472
Fontanels, *103*, 104, 104t

Food poisoning, 448
Foot, *79, 80*
Foramen (foramina)
  nutrient, 100
  optic, 105
  ovale, 391, *391*
  skull, 117t, 118t
Forebrain, 249, *249*
  derivatives, 249, *249*
Foreskin, 489
Fractures, 144, *145*

G

Gallbladder, *426, 441,* 444
  control of activity, 445
  functions, 444
  location, 444
  wall structure, 444, *444*
Gallstone, 445
Ganglion(s)
  collateral, 304
  geniculate, *296*
  lateral, 304
  semilunar, 295, *295*
  terminal, 304
Gastric gland, *436,* 437
Gastric juice, 437
Gastrin, 437, 506
Gastritis, 447
Gastrocnemius tendon, *79, 80, 243,* 244
Gastrointestinal hormones, 506
Germ layers, 22
  derivatives, 22t
Germinal epithelium, 463
  ovary, 470, *470*
  testis, 482, *482*
General references, 508
Geniculocalcarine tract, 322, *323*
Giantism, 506
Glabella, 104, *104*
Glands
  apocrine, 41, 42t
  cells, 41
  compound, 41, 41t
  endocrine, 41
  exocrine, 41
  holocrine, 41, 42t
  merocrine, 41, 42t
  multicellular, 41, 41t
  simple, 41, 41t
  structure, 41, 41t
  unicellular, 41, 41t
Glaucoma, 327
Glia
  functions, 259
  types, 259, *259*
Globose nucleus, 272, *272*
Globulin, 345
Globus pallidus, 270, *270*
Glomerular capsule, 454, *455*
Glomerular filtration, 457
Glomerulonephritis, 461
Glottis, 410

Glucagon, 505
Glucocorticoids, 504
Glycogenoses, 14
Golgi complex (body), *13*
  functions, 13
  location, 13
  structure, 13
Golgi tendon organ, 316, *316*
Gracile nucleus, 277, *277*
Gracile tract, 277
Greater omentum, 434, *435*
Greater vestibular gland, 478
Growth hormone, 497t
Guanine, 16
Gubernaculum, 481, *481*
Gums, 428, *429*
Gyrus (gyri)
  cingulate, *265,* 266
  cuneate, *265,* 266
  frontal, *265, 265*
  fusiform, *265,* 266
  lingual, *265,* 266
  orbital, *265,* 266
  parahippocampal, *265,* 266
  postcentral, *265,* 266
  precentral, *265, 265*
  precuneate, *265,* 266
  rectus, *265,* 266
  temporal, *265,* 266
  transverse temporal, 266

H

Habenular nucleus, 274, *274*
Hair(s)
  bud, *83,* 84
  development, 83, *83*
  lanugo, 84
  papilla, *83,* 84
  structure, 92, *92*
  terminal, 84
  vellus, 84
Haploid, 19
Haustra, 439
Haversian system, 50, *51*
Heart, 349-362
  blood vessels, 360, *360,* 361, *361*
  chambers and valves, 355, *355,* 356, *356*
  congenital anomalies, *351*
  cords, 349
  development, 349, *349,* 350
  location, 352, *352*
  nerves, 362, *362*
  pericardium, 353, *353*
  size, 352
  tissues composing, 357-359, *357, 358, 359*
  tubes, 349, *349*
  wall structure, 354, *354*
Hepatic portal system, 386, *386*
Hepatitis, 448
Hilus (kidney), 452, *453*
Hindbrain, 249, *249*
  derivatives, 249, *249*
Histology, 2
Homeostasis, 63

Human development, 20-28, 26t-29t
Hyaline membrane disease, 419
Hydrocephalus, 291
Hymen, 478
Hypermenorrhea, 479
Hypermetropia, 325, *326*
Hyperparathyroidism, 506
Hypoglycemia, 507
Hypophysis; *see* Pituitary gland
Hypotension, 368
Hypothalamic-regulating factors, 498, 498t
Hypothalamico-hypophyseal tract, *496,* 498
Hypothalamus
  functions, 275, *275*
  nuclei, 275, *275*
Hypothenar eminence, *77, 78*

I

Ileum, 438
Implantation, 21, *21*
Inclusions, *10*
  functions, 15
  types, 15
Incus, 333, *333*
Indifferent gonad, *465,* 466
Infarction, 361
Inferior colliculus, *276,* 279
Infundibulum, *469,* 474
Inguinal canal, *208, 209,* 481
Inguinal ligament, 209
Inner cell mass, 20, *20*
Inner ear, *332,* 334-336
  development, *329*
Inspiration, 418
Insula, 264
Insulin, 505
Intercellular surfaces, 38
Intercostal nerves, 302, *302*
Intermediate zone, *249,* 250
Internal arcuate fibers, 277, *277*
Internal capsule, 268
Internode, *254,* 255
Interstitial cells, 483, *483*
Interventricular foramen, 290, *290*
Intestinal juice, 447
  enzymes, 447
Intrapulmonic pressure, 418
Intrathoracic pressure, 418
Iris, *318, 320,* 320
Ischemia, 181, 361
Islets of Langerhans, *446*
  hormones produced by, 446
Isthmus (uterine tube), 474

J

Jaundice, 85
Jejunum, 438
Joint(s)
  acromioclavicular, *76, 76, 156, 156*
  ankle, *164, 165, 165*
  atlantoaxial, 153, *153, 154*
  atlantooccipital, 153, *153, 154*
  carpometacarpal, 158, *158*
  costovertebral, 154, *155*

Joint(s)—cont'd
   disorders, *162*, 166
   elbow, 157, *157*
   hip, 160, *160*
   interphalangeal, *158*, *159*, *164*, 165, *165*
   intertarsal, *164*, 165, *165*
   knee, *161*, 163
   metacarpophalangeal, 158
   metatarsophalangeal, *164*, 165, *165*
   sacroiliac, 159, *159*
   sensation, 315
   shoulder, 156, *156*
   sternoclavicular, 155, *156*
   sternocostal, 155, *155*
   tarsal-metatarsal, *164*, 165, *165*
   temporomandibular, 152, *152*
   tibiofibular, 164, *164*
   wrist, 158, *158*
Junctional complexes, *38*

K

Kidney
   blood supply, 457, *457*
   gross anatomy, 452, *453*
   microscopic anatomy, 454, *455*, *456*
   orientation in body, 452, *452*
   physiology, 457, 458
Kinesthesia, 315
   neural pathway for, 316
Krause's corpuscle, 310, *311*
Kupffer cell, 442
Kyphosis, 120, *120*

L

L-dopa, 271
Labia majora, *469*, 477, *477*
Labia minora, *469*, 477, 478
Lacrimal apparatus, 325, *325*
Lacrimal duct, 325, *325*
Lacrimal gland, 325, *325*
Lacrimal sac, 325, *325*
Lactase, 447
Lacteal, 438
Lactiferous duct (tubule), 478, *478*
Large intestine, *426*, 439
   functions, 447
Laryngitis, 412
Larynx
   cartilages, 409, *409*
   muscles, 410, *410*, 411
Lateral geniculate body, 322, *323*
Lateral lemniscus, 278, *278*
Law of adequate stimulus, 309
Law of specific nerve energies, 310
Leg, 80
Lens, *318*, 321
   placode, 317, *317*
   vesicle, 317, *317*
Lesser omentum, 434, *435*
Leukocyte(s)
   lymphoid, *346*, 347, 348t
   myeloid, *346*, 347, 348t
Levers, 183-185
   classes, 184, *184*

Levers—cont'd
   components, 183
Ligaments of spine, 121, *121*
Limb development, 25, *25*
Limbic system, 280
Lipid, 9
Lips, 427, *427*
Liver
   blood supply, 441
   functions, 442, 443
   gross anatomy, 441, *441*
   ligaments, 441, *441*
   lobes, 441, *441*
   lobules, 442, *442*, *443*
   microscopic anatomy, 442, *442*, *443*
Lobule (brain)
   inferior parietal, *265*, 266
   paracentral, *265*, 266
   superior parietal, *265*, 266
Loop of Henle, 454, *455*
Loose connective tissue, *44*
   cells, 44, 48
Lordosis, 120, *120*
Lumbar enlargement, 281, *281*
Lumbosacral plexus, 302, *303*
   nerves rising from, 302, 303, *303*
Lungs, 414-418
   blood supply, 417
   bronchopulmonary segments, 417, *417*
   gross anatomy, 414, *415*
   lobules, 417
   location, 414
   microscopic anatomy, *413*, 414
   nerves, 418
   weight, 414
Luteinizing hormone, 479, 497t
   relationship to ovarian and uterine changes, 479
   relationship to testis, 489
Lymph
   composition, 392, 393
   formation, 392
Lymph node(s)
   abdominal, *392*, 394
   axillary, *392*, 394
   cervical, *392*, 394
   distribution, *392*, 394
   functions, *392*, 394
   inguinal, *392*, 394
   mesenteric, *392*, 394
   structure, 394, *394*
   thoracic, *392*, 394
Lymph nodule, 393
Lymph vascular system
   fluid, 392
   organs, 393-398
Lymph vessel(s), 363, 364, *392*, 393
   areas of drainage, 393, *393*
   formation, 363
   functions, 393
   types, 393
Lymphocyte, *346*, 347, 348t
Lysosomes
   functions, 14

Lysosomes—cont'd
   structure, 14, *14*

M

Macula adherens, 38, *38*
Malleus, 333, *333*
Maltase, 447
Mammary gland, *84*, 85, 478, *478*
   cancer, 480
   cystic disease, 480
   suspensory ligament, 478
Mandibular sling, 193
Marginal zone, *249*, 250
Marrow cavity, 101
Mastoid process, 72, *72*
Medial lemniscus, 277, *277*
Median eminence, 495, *495*
Mediastinum, 412, *412*
Medulla (oblongata)
   external anatomy, 276
   internal anatomy, 277, *277*
Medullary pyramid, 453, *453*
Meiosis, 19
   results of, 19
   stages, 19
Meissner's corpuscle, 311, *311*
Melanin, 85
Melanocytes, 83, *83*
Melatonin, 505
Membrane
   epithelial, *40*, 41
   mucous, *40*, 41
   serous, 40, *40*
Membranous labyrinth, 328, *329*, 334, *334*
Meninges, 286-289
   blood vessels, 286
Menstrual cycle, 476
   length, 477
   stages, 477
Mesencephalon, 249, *249*
Mesenteries, 434, *435*
Mesoderm
   derivatives, 22t
   embryonic, 22, *23*
   extraembryonic, 21, *23*
   formation, 22, *23*
Mesonephric duct, 460, *461*
Mesonephros, 460, *461*
Mesosalpinx, 474
Mesothelium, 31
Metanephros, 460, *461*
Metencephalon, 249, *249*
   derivatives, 249, *249*
Microglia, 259, *259*
Microtubules
   functions, 15
   structure, 15
Microvilli, 36, *37*
Midbrain
   derivative, 249, *249*
   external anatomy, 276
   internal anatomy, 279, *279*
Middle ear
   components, 332

Middle ear—cont'd
   development, 328, *330*
   muscles, 333, 334
Mineralocorticoids, 504
Mitochondria, *12*
   cristae, 12
   functions, 12
   structure, 12
Mitosis
   duration, 18
   results, 18
   stages, 18, *18*
Mondino, 1
Monocyte, *346, 347*, 348t
Mons pubis, *469, 477, 477*
Morula, 20, *20*
Mouth, 427, *427*
Multiple sclerosis, 262, 286
Muscle
   actions, 185
   naming, 185
   spindle, 180, *180*, 315, *315*
Muscles
   abductor digiti minimi, *246*, 247, 247t
   abductor hallucis, *246*, 247, 247t
   abductor pollicis brevis, 228, *228*, 229t
   abductor pollicis longus, 77, *225, 226*, 228, *228*, 229t
   adductor brevis, *233*, 234t, 236
   adductor hallucis, 247, 247t
   adductor longus, *233*, 234t, 236
   adductor magnus, *187, 233*, 234t, *236*
   adductor pollicis, *228*, 229, 229t
   adductors of thigh, *79*, 80
   anconeus, *223*, 223t, *225, 226*
   aryepiglotticus, *410*, 411
   arytenoideus, *410*, 411
   auricularis, 190, *191*, 191t
   biceps brachii, *76*, 77, *186, 215, 219*, 222, *222*, 223t
   biceps femoris, *187*, 235t, *237*, 239
   brachialis, *76*, 77, *219, 222, 222*, 223t
   brachioradialis, *219, 222, 222*, 223t, *224*
   buccinator, *188*, 189, 189t, *191, 192*
   constrictor pharyngis inferior, *194*
   constrictor pharyngis medius, *194*
   constrictor pharyngis superior, *194*
   coracobrachialis, *215*, 219, 221t
   corrugator, *188*, 189, 189t, *191*
   cricoarytenoideus, *410*, 411
   cricothyroideus, 411
   deltoideus, 76, *76, 186, 187, 215, 219*, 220, 221t, *222*
   depressor anguli oris, *188*, 189, 189t, *191*
   depressor labii inferioris, *188*, 189, 189t, *191*
   diaphragm, *207, 212*, 213, 214t
   digastricus, *194, 196*, 197, 197t
   dilator pupillae, 321
   epicranius, 190, 191t
   erector spinae, 205, *206*
   extensor carpi radialis brevis, *226*, 227t
   extensor carpi radialis longus, *219*, 225, *226*, 227t
   extensor carpi ulnaris, *225, 226*, 227, 227t

Muscles—cont'd
   extensor digiti minimi, *226*, 227, 227t
   extensor digitorum communis, 226, *226*, 227t
   extensor digitorum longus, *186*, 240, *241, 242*, 245t
   extensor hallucis longus, *186*, 240, *241*, 245t
   extensor indicis, 226
   extensor pollicis brevis, *225, 226*, 228, *228*, 229t
   extensor pollicis longus, *225, 226*, 228, *228*, 229t
   external obliquus abdominus, 79, *186, 187, 206, 208*, 209, *209*, 211t, *215, 216*
   flexor carpi radialis, 77, 224, *225*, 227t
   flexor carpi ulnaris, 77, 224, *225*, 227t
   flexor digitorum accessorius, *246, 246*, 247t
   flexor digitorum brevis, *246*, 247, 247t
   flexor digitorum longus, *243*, 244, 245t
   flexor digitorum profundus, *224*, 226, 227t, *228, 231*
   flexor digitorum superficialis, *224*, 226, 227t, *231*
   flexor hallucis brevis, *246*, 247, 247t
   flexor hallucis longus, *243*, 244, 245t
   flexor pollicis brevis, 228, *228*, 229t
   flexor pollicis longus, 228, *228*, 229t
   frontalis, 190, 191t, *191*
   gastrocnemius, *79*, 80, *186, 187, 241*, 242, *242, 243*, 245t
   gemellus, 235t, *238*, 239
   genioglossus, 194, *194*, 195t
   geniohyoideus, *194, 196*, 197, 197t
   gluteus maximus, *187*, 235t, *237, 237*
   gluteus medius, *233*, 235t, *237*, 238, 239
   gluteus minimus, *233*, 235t, *238*, 239
   gracilis, *187, 233*, 234t, *236*, 237
   hamstrings, 80
   hyoglossus, 194, *194*, 195t
   iliacus, 210, 211t, 232, *233*, 234t
   iliocostalis, 205, 211t
   iliocostalis cervicis, 201, 205, *207*, 211t
   iliocostalis lumborum, 205, *207*, 211t
   iliocostalis thoracis, 205, *206, 207*, 211t
   iliopsoas, 210, 211t, 232, *233*, 234t
   infraspinatus, *187, 206, 216*, 220, 221t
   intercostalis externus, *207, 212*, 213, 214t, *215*
   intercostalis internus, *212*, 213, 214t, *215*
   internal obliquus abdominus, 79, *206, 208*, 209, 211t, *215, 216*
   interossei (hand), 230, *230*, 230t, *231*
   interossei (foot), *246, 246*, 247t
   interspinales, *201*, 203t, 205, *207*
   intertransversarii, 203t, 205, *207*
   latissimus dorsi, 76, *187, 206, 208, 215, 216*, 219, 221t, *222*
   levator anguli oris, *188*, 189, 190t, *191*
   levator costorum brevis, *207*, 214
   levator labii superioris, *188*, 189, 190t, *191*
   levator labii superioris alaeque nasi, 188, *188*, 190t
   levator scapulae, *196, 200, 201, 202, 206, 216*, 218, 218t
   longissimus, 205, 211t

Muscles—cont'd
   longissimus capitis, *201*, 205, 211t
   longissimus cervicis, *201*, 205, *207*, 211t
   longissimus thoracis, 205, *206, 207*, 211t
   longitudinal lingual, 194, *194*
   longus capitis, *199*, 203t, 204
   longus colli, *199*, 203t, 204
   lumbricales (foot), 246, *246*, 247t
   lumbricales (hand), 230, 230t, *231*
   masseter, 72, *72, 188, 191*, 193, 193t
   mentalis, *188*, 189, 189t, *191*
   multifidus, *201*, 205, *207*
   mylohyoideus, *196*, 197, 197t
   nasalis, 188, *188*, 189t
   obliquus externus (abdominis), 79, *186, 187, 206, 208*, 209, *209*, 211t, *215, 216*
   obliquus internus (abdominis), 79, *206, 208*, 209, 211t, *215, 216*
   obturator, 235t, *238*, 239
   occipitalis, *188*, 190, *191*, 191t
   omohyoideus, *191*, 196, 197t, 198, *198*
   opponens pollicis, *228*, 229, 229t
   orbicularis oculi, 188, *188*, 189t, *191*
   orbicularis oris, 188, *188*, 190t, *191*
   palatoglossus, 194, *194*, 195t
   palmaris longus, 224, *225*, 227t
   pectineus, *233*, 234t, *236*, 237
   pectoralis major, 76, *76, 186, 196, 208, 215*, 219, 221t, 222
   pectoralis minor, *215*, 218, 218t
   peroneus brevis, *186, 187, 241*, 242, *242, 243*, 245t
   peroneus longus, *186, 187, 241*, 242, *242*, 245t
   peroneus tertius, *241*, 242, 245t
   piriformis, 235t, *238*, 239
   plantaris, *187, 243*, 244, 245t
   platysma, *188*, 199, *199*, 202, 203t
   procerus, 188, *188*, 190t, *191*
   pronator quadratus, 223, 223t, *224*
   pronator teres, *222*, 223, 223t, *225*
   psoas major, 210, 211t, 232, *233*, 234t
   pterygoideus lateralis, *192*, 193, 193t
   pterygoideus medialis, *192*, 193, 193t
   pyramidalis, *208*, 209
   quadratus femoris, 235t, *238*, 239
   quadratus labii superioris, *188*, 190t
   quadratus lumborum, *207*, 210, 211t, *233*
   quadratus plantae, 246, *246*, 247t
   quadriceps, *79*, 80
   rectus abdominis, 79, *186, 208*, 209, *209*, 211t, *215*
   rectus capitis, *199, 201*, 203t, 204
   rectus femoris, 232, *233*, 234t
   rhomboideus, *187*, 218, 218t
   rhomboideus major, *200, 206, 216*, 218, 218t
   rhomboideus minor, *200, 206, 216*, 218, 218t
   risorius, *188*, 189, 190t, *191*
   sacrospinalis, 205, *206*
   sartorius, *79*, 80, *186*, 232, *233*, 234t
   scalenus, *199*, 202, 203t, 204, *207*
   semimembranosus, *187*, 235t, *238*, 239
   semispinalis capitis, *200, 201, 202*, 203t, 205
   semispinalis cervicis, *201, 202*, 203t, 205, *207*

Muscles—cont'd
  semispinalis thoracis, 203t, 205, *207*
  semitendinosus, *187*, 235t, *237*, 239
  serratus anterior, *186, 208, 215*, 218, 218t
  serratus posterior, 214, 214t
  serratus posterior inferior, *206*, 214, 214t, *216*
  serratus posterior superior, *206*, 214, 214t, *216*
  soleus, 80, *186, 187, 241, 242, 243*, 244, 245t
  sphincter pupillae, 321
  spinalis, 205, 211t
  spinalis capitis, 205, 211t
  spinalis cervicis, 205, 211t
  spinalis thoracis, 205, *206, 207*, 211t, *216*
  splenius, 203t, 205
  splenius capitis, *187, 200, 201, 202*, 203t, 205, *206, 216*
  splenius cervicis, *200*, 203t, 205
  stapedius, 334
  sternocleidomastoideus, 74, *74, 186, 187, 199, 200, 202*, 203t, 204
  sternohyoideus, *191, 196*, 197, 197t, *198*
  sternothyroideus, *196*, 197t, 198, *198*
  styloglossus, 194, *194*, 195t
  stylohyoideus, *194, 196*, 197, 197t, *198*
  subclavius, *196, 215*, 218, 218t
  subscapularis, *215*, 220, 221t, 222
  supinator, 223, 223t, *224*, 225
  supraspinatus, *206, 216*, 220, 221t
  temporalis, 72, *191, 192*, 193, 193t
  temporoparietalis, 190
  tensor fasciae latae, *186, 233*, 235t, 239
  tensor tympani, 333
  teres major, *187, 206, 216*, 219, 221t, *222*
  teres minor, *187, 206, 216*, 202, 221t
  thyroarytenoideus, 411
  thyroepiglotticus, 411
  thyrohyoideus, *194, 196*, 197t, 198, *198*
  tibialis anterior, *186*, 240, *241, 242*, 245t
  tibialis posterior, *243*, 244, 245t
  transverse lingual, 194
  transversospinalis, 203t, 205
  transversus abdominis, 79, *207, 208, 209, 209*, 211t, *215*
  trapezius, *186, 187, 191, 196, 200, 202, 206, 216*, 218, 218t
  triangularis, *188*, 189, 189t, *191*
  triceps brachii, 77, *187*, 219, 222, *222*, 223t
  vastus intermedius, 232, *233*, 234t
  vastus lateralis, *186*, 232, *233*, 234t
  vastus medialis, *186*, 232, *233*, 234t
  vertical lingual, 194
  zygomaticus major, *188*, 189, 190t, *191*
  zygomaticus minor, *188*, 189, 190t, *191*
Muscular tissue, 54, 55, 168-181
  cardiac, 54, *54*
  development, 168
  skeletal, 54, *55*
  smooth, 54, *54*
  types, 169
Myasthenia gravis, 180
Myelencephalon, 249, *249*
  derivatives, 249, *249*
Myelin sheath, 254, 255, *256*, 258

Myeloma, 100
Myoblast, 168
Myocardium, 354, *354*
Myofibrils, 168
Myometrium, 476, *476*
Myopathy, 180
Myopia, 325, *326*
Myosin, *172*, 173, *173*

N
Nail(s), *93*
  bed, 93
  body, 93
  development, 84, *84*
  eponychium, 93
  field, 84, *84*
  fold, *84*, 85
  free edge, 93
  groove, 93
  hyponychium, 93
  root, 93
Naked nerve endings, 310
Nasal cavities, *407*
  functions, 407
Nasal fossa, 104, *104*
Nasolabial fold, 73, *73*
Nasolacrimal duct, 325, *325*
Nearsightedness, 325, *326*
Neoplasm, 18
Nephron(s)
  function, 457, 458
  structure, 454, *455, 456*
  types, 454, *455*
Nerve, 262
  structure, 262, *262*
Nerve(s)
  abducent, 293t, *295*, 296
  accessory, 293t, 299, *299*
  axillary, *301*, 302
  cochlear, 296, *297*
  common peroneal, 303, *303*
  facial, 293t, 296, *296*
  femoral, 302, *303*
  genitofemoral, 302, *303*
  glossopharyngeal, 293t, 297, *297*
  hamstring, 303, *303*
  hypoglossal, 293t, 299, *299*
  iliohypogastric, 302, *303*
  ilioinguinal, 302, *303*
  inferior gluteal, 303
  infraorbital, 73
  lateral femoral cutaneous, 302, *303*
  mandibular, *295*, 296
  maxillary, *295*, 296
  median, *301*, 302
  musculocutaneous, *301*, 302
  obturator, 302, *303*
  oculomotor, 293t, *294*, 295
  olfactory, 293t, *294*, *294*
  ophthalmic, *295*, *295*
  optic, 293t, *294*, *294*
  radial, *301*, 302
  sciatic, 303, *303*

Nerve(s)—cont'd
  spinal, 300
  superior gluteal, 303
  supraorbital, 73
  thoracic, 302, *302*
  tibial, 303, *303*
  trigeminal, 293t, *295*, *295*
  trochlear, 293t, *294*, *295*
  ulnar, *301*, 302
  vagus, 293t, *298*, 299
  vestibular, 296, *297*
  vestibulocochlear, 293t, 296, *297*
Nerve impulse
  conduction, 258
  formation, 258
Nervous system, 248-305; *see also* specific names of parts
  autonomic, 303
  cells and tissues, 253
  central, 263
  development, 248-250
  meninges, 286
  organization, 250, *251*
  peripheral, 252, 292
  physiology, 255
  ventricles and cerebrospinal fluid, 290
Nervous tissue, 55, *55*
Neural crest, 249, *249, 250*
Neural folds, 248, *249*
Neural groove, 248, *249*
Neural plate, 248, *249*
Neural tube, *249*
  wall, *249, 250*
Neurilemma, *254*, 255
Neuroblast, 253
Neurofibril, *254*, 255
Neurohypophysis, 495, *495*, 498
Neuromere, 282
Neuromuscular junction, *176*
  function, 176
  structure, 176
Neuron(s), 253-259
  physiology, 255, 258, 259
  properties, 258, 259
  structure, *254*, 255
  types, 253, *253*, *254*
Neurotendinous spindle, 316, *316*
Neutrophil, *346*, 347, 348t
Nipple, 478, *478*
Nissl body, *254*, 255
Nodal tissue, 358, 359, *359*
Node of Ranvier, *254*, 255
Norepinephrine, 504
Nose, 406, *406*
Notochord, 23
Nuclear fluid, 16
Nuclear membrane, 16, *16*
Nucleoli, 16, *16*
Nucleotides, 16
  structure, 16
Nucleus of cell, *16*
  functions, 16
  structure, 16

O

Obstetrical perineum, 478
Odontoblast, 428
Odors, 340
Olecranon process, 76, 77
Olfactory epithelium, 340, 340
Olfactory pathway, 341, 341
Oligodendroglia, 259, 259
Oligomenorrhea, 479
Olivary nucleus, 277, 277
Optic chiasm, 294, 294, 322, 323
Optic cup, 317, 317
Optic groove, 317, 317
Optic nerve, 322, 323
Optic radiation, 322, 323
Optic tract, 294, 294
Optic vesicle, 317, 317, 328
Oral cavity, 427, 427
Orbit, 73, 104, 104, 105
Organ, 56
Organ of Corti, 329, 336, 337
Organelles, 11
Organism, 63
Osseous labyrinth, 328, 334
Osteoblast, 51, 52
Osteoclast, 52
Osteogenic sarcoma, 100
Osteon, 50, 51
Osteoporosis, 100
Otic pit, 328, 328
Otic placode, 328, 328
Otolith, 334, 335
Otosclerosis, 337
Ovarian follicles
    development, 471
    primary, 470, 472
    primordial, 470, 472
    secondary, 470, 473
    vesicular, 471, 473
Ovary, 469-473, 505
    follicles, 470-473, 470, 471
    ligaments, 469, 469
Ovulation, 473
Oxygen toxicity, 419
Oxytocin, 498

P

Pacinian corpuscle, 311, 311
Pain, 310
    neural pathway, 312, 312
Palmar, 3
Palpation, 2
Pancreas, 426, 446, 446, 505
    ducts, 446, 446
    gross anatomy, 446, 446
    microscopic anatomy, 446, 446
    secretion, 447
Pancreatic juice
    characteristics, 447
    control of secretion, 447
    enzymes, 447
Panniculus adiposus, 88
Papillary duct, 454, 455
Paralysis, 180

Parasympathetic division, 252
Parathyroid glands
    cells, 502, 502, 503
    disorders, 506
    hormone, 503
    location, 502, 502
    size, 502
Parathyroid hormone, 503
    effects, 503
Paresthesia, 313
Parietal eminence, 72, 72
Parkinson's disease, 271
Pars distalis, 495, 495
Pars intermedia, 495, 495, 498, 499
Pars nervosa, 495, 495, 499
Pars tuberalis, 495, 495, 498
Peg cells, 474, 474
Pelvic floor
    female, 475
    male, 489
Pelvis, 136, 136
    sex differences, 136, 136
Penis, 480, 488, 489, 489
Peptic ulcer, 448
Perichondrium, 50
Perikaryon, 255
Perimetrium, 476
Perimysium, 178, 178
Perineurium, 262, 262
Periodontal membrane, 428, 429
Periosteum, 52
Peripheral nervous system, 252
Peritoneum, 434
Peritonitis, 448
Peroxisomes, 15
Peyer's patches, 393, 438
Phantom limb, 313
Pharynx
    divisions, 408, 408
    structure, 408, 408
Philtrum, 73, 73
Phosphoric acid, 16
Pia mater, 286, 288, 289, 291
Pigmented connective tissue, 48
Pineal gland, 274, 274, 505
    cells, 505
    hormone, 505
    location, 505, 505
Pinna, 328, 331, 332
Pituicytes, 498
Pituitary gland, 495-499
    blood supply, 495, 496
    cells, 496, 497, 497
    control of secretion, 496, 498, 498t
    disorders, 506
    hormones, 496, 497t, 498
    location, 495-498
    parts, 495, 495
    size, 495
Pituitary infantilism, 506
Pituitary portal system, 495, 496
Placenta
    circulation, 30
    functions, 30

Placenta—cont'd
    structure, 30, 30
Planes of section
    coronal, 4, 4
    cross-section, 4, 4
    frontal, 4, 4
    horizontal, 4, 4
    longitudinal, 4, 4
    median, 4, 4
    midsagittal, 4, 4
    oblique, 4, 4
    parasagittal, 4, 4
    sagittal, 4, 4
    transverse, 4, 4
Plantar, 3
Plasma (cell) membrane, 10
    carrier molecules, 11
    functions, 10, 11
    receptor sites, 11
    structure, 10, 11
Platelet(s), 346, 347, 348t
Pleurae, 414
Pleural cavities, 414
Pneumonocyte, 416
Podocyte, 454, 456
Poliovirus, 286
Polycystic ovary syndrome, 480
Polyribosomes, 11, 12
Pons
    basal, 278
    external anatomy, 276
    internal anatomy, 278, 278
    nuclei, 278, 278
    tegmentum, 278
Popliteal space, 80
Portal canal, 442, 442
Posterior root, 300, 300
Posterior root ganglion, 282
Pott's fracture, 139
Prepuce, 489
Presbyopia, 326
Pressure
    neural pathway, 313, 313
    receptor, 311, 311
Primary amenorrhea, 479
Primary bronchi, 411, 411, 414
Primary enlargement, 249, 249
Primary sex cord, 463, 464, 465
Primitive groove, 23
Primitive pit, 23
Primitive streak, 22, 23
Prolactin, 497t
Pronephric duct, 460, 461
Pronephros, 460, 461
Prosencephalon, 249, 249
Prostate gland, 480, 486, 487
    cancer, 489
Protein, 9
Protein synthesis, 17
Proximal convoluted tubule, 454, 455
Pudendal plexus, 303, 303
Pudendum, 469, 477, 477
Pulmonary circulation, 364, 365
Pulmonary trunk, 390

Puncta, 325
Pupil, 320
Purkinje's fibers, 359, *359*
Putamen, 270, *270*
Pyelonephritis, 461
Pyramid, *276, 277,* 277

Q

Quadrants of abdomen, *6,* 7

R

Radioactive isotopes, 2
Receptor(s)
  characteristics, 309, 310
  classification, 310
  defined, 309
  poorly adapting, 310
  rapidly adapting, 310
Rectum, *426,* 440, *440*
  wall structure, 440
Red nucleus, 279, *279*
Reference lines
  abdominal lines, *6,* 7
  anterior axillary line, 5, *5*
  midclavicular line, 5, *5*
  midsternal line, 5, *5*
  scapular line, 5, *6*
  vertebral line, 5, *6*
Reflex(es)
  characteristics, 261
  crossed-extension, 286
  flexor, 286
  myotatic, 286
Reflex arc, 261
  components, 261, *261*
Renal column, *453,* 454
Renal corpuscle, 454
Renal fascia, 452, *452*
Renal pelvis, 453, *453*
  wall structure, 458
Renal sinus, 452, *453*
Replication, 16
Reproductive systems, 463-489
  development, 463, *464, 465, 466*
  female, 469-480
  homologies, 467t
  male, 480-489
  malformations, 468t
Respiratory bronchiole, 416
Respiratory system, 401-419, 416t
  development, 401-404, *401-404*
  divisions, 405
  functions, 401
  organs, 405, *405*
  physiology, 418
Rete tubules, *482, 484, 485*
Reticular connective tissue, *47,* 48
Reticular formation, 280, *280*
Retina, *318,* 321, *321*
Retroperitoneal, 434
Rhinitis, 407
Rhombencephalon, 249, *249*

Ribonucleic acid (RNA)
  messenger RNA, 17
    function, 17
  ribosomal RNA, 17
    function, 17
  transfer RNA, 17
    function, 17
Ribosomes, *11*
  functions, 11, 12
Right lymphatic duct, *392,* 393
RNA; *see* Ribonucleic acid
Rod, 322, *322*
Rubrospinal tract, 277, *277*
Ruffini's corpuscle, 311, *311*

S

Sacculus (saccule), *334*
  macula, 334, *335*
Saliva
  characteristics, 432
  enzyme, 432
  functions, 432
Salivary glands
  control of secretion, 432
  location, 430, *431*
  microscopic structure, 432, *432*
  names, 431, *431,* 432
  secretion, 432
Saltatory conduction, 259
Sarcolemma, 169, *172, 173*
Sarcoplasm, 169, 173
Sarcoplasmic reticulum, *172,* 174
Satellite cell, 259
Schlemm's canal, *318,* 320, *320*
Schwann's cell, *254,* 255, *256, 257, 258,* 259
Schwann's sheath, *254,* 255
Sclera, *318,* 318
Scleroderma, 50
Scoliosis, 120, *120*
Scrotum, *480,* 481, *481*
Sebaceous gland, 94, *94*
Sebum, 94
Secondary bronchi, *411,* 414
Secretin, 447, 506
Secretion granule, 13
Semen, 489
Semicircular canal(s), *332,* 334, *334*
  crista, 334, *335*
Seminal vesicle, *480,* 486, *487*
Seminiferous tubule, *482, 482*
  cells, *482, 482*
Sertoli's cells, 483
Sexual precocity, 479
Sharpey's fibers, 178
Shin splints, 181
Sinoatrial node, 358, *359*
Sinus(es)
  dural, 388, *388*
  frontal, 105, *105*
  maxillary, 106, *107*
  sphenoid, *105*
Skeletal muscle, 173-181
  blood supply, 179, *179*

Skeletal muscle—cont'd
  connective tissue, 178, *178*
  contractile mechanisms, 176
  disorders, 180, 181
  energy sources for contraction, 177
  fiber arrangement, 174, *175*
  myofibrils, *172,* 173
  myofilaments, *172, 173, 173*
  physiological properties, 177
  striations, *172,* 173
  structure, *172,* 173
  tendon attachment, 179, *179*
  types of fibers, 174
Skeleton, 96-146
  appendicular, 101
  axial, 101
  bones; *see* individual bone names under Bone(s)
  development, 97, *97, 98-100*
  functions, 96
  ossification times, 142, *142, 143,* 144
  terms to describe features on bones, 102
Skin, 82-94
  blood vessels, 90, *90*
  burns, 91, *91*
  color, 80
  dermis, 88
  development, 82, *83*
  epidermis, 86, *86, 87, 87*
  functions, 90, 91
  glands, 93, *93,* 94, *94*
  gross features, 85, *85*
  hypodermis, 88
  lymphatic vessels, 90
  sensory corpuscles, 88, *88, 89*
  structure, 86, *86*
  thick, 86
  thin, 86, 87
Small intestine, 438, *438, 439*
  absorption, 447
  digestion, 447
  glands, 438
  parts, 438
  villi, 438, *439*
  wall structure, 438, *438, 439*
Smell (sense of), 340
Smoking, 419
Smooth muscle, 169, *169, 170, 171*
  multiunit, 171
  unitary, 171
Somatic fibers, 252
Somites, 24, *24*
Spasm, 181
Specificity of receptors, 310
Sperm
  development, 482, *482,* 483
  morphology, *483*
Spermatogenic cells, 482, *482*
Spinal cord, 281-286
  cell columns, 283, *283*
  disorders, 286
  external anatomy, 281, *282*
  functions, 286
  internal anatomy, 282, *282, 283, 283*

Spinal cord—cont'd
   tracts, *284*, 285t
Spinal lemniscus, 277, *277*
Spinal nerves, 252, 300, *300*
   divisions, 300, *301*
   roots, 300, *301*
   trunks, 300, *301*
Spinal shock, 286
Spinal tap (puncture), 291
Spleen
   functions, 396
   gross structure, *396*
   microscopic structure, *396, 397*
Spongioblast, 253
Stapes, 333, *333*
Sternal angle, 78
Stomach
   control of secretion, 437
   digestion, 437
   gross anatomy, 436, *436, 437*
   microscopic anatomy, *436*, 437
   wall structure, 437
Strabismus, 327
Straight tubule, *482*, 484, *484*
Subarachnoid space, 286, 288, *291*
Subdural hematoma, 291
Subdural space, 286, *288*
Subendocardium, 354
Subthalamic nuclei, 270
Sucrase, 447
Sulcus (sulci)
   cingulate, *265*, 266
   collateral, *265*, 266
   frontal, *265*, 265
   marginal, *265*, 266
   postcentral, *265*, 266
   precentral, *265*, 265
   rhinal, *265*, 266
   temporal, *265*, 266
Superciliary ridge, 73, *73*, 104, *104*
Superior colliculus, 276, *279*, 279
Superior extensor retinaculum, *186*
Supraorbital margin, 104, *104*
Supraorbital notch (foramen), *104*, 105
Surface anatomy, 66-80
   head, 72, *72, 73*, 73
   lower appendage, 79, *79*, 80
   neck, 74, *74, 75*, 75
   thorax, 78, *78*
   upper appendage, 76-78, *76-78*
Surfactant, 416
Suture(s), 102
   cranial vault, 103, *103*
Sweat gland, *84*, 85, 93, *93*
   apocrine, 93
   eccrine, 93
Sympathetic division, 252
Synapse
   properties, 261
   structure, 260, *260*
Synaptic vesicle, 260, *260*
System, 56
   list, 56-63, *56-63*
Systemic circulation, 364, *365*

T

Tabes dorsalis, 286
Taenia coli, *426*, 440
Tanning, 85
Taste
   bud, 338, *338*
   cells, 338, *338*
   neural pathway, 338, *339*
   pore, 338, *338*
   sensations, 338, *339*
Tay-Sachs disease, 14
Teeth, 427, 428
   blood vessels, 430, *430*
   deciduous, 427, 428t
   nerves, 430, *430*
   permanent, 427, 428t
   structure, 428, *429*
   time of eruption, 428t
   types, 427, *427*
Telencephalon, 249, *249*
   derivatives, 249, *249*
Tendo calcaneus, *79*, 80, *243*, 244
Tendon, 178
Tendon sheaths, 179, *179*
Tentorium cerebelli, 286, *287*
Teratology, 2
Terminal bronchiole, *413*, 414
Testes, 505
Testis, 480-486
   descent, 481
   ducts, 484-486
   germinal epithelium, 482, *482*
   tunics, 481, *482*
   structure, *482*
Testosterone, 505
Tetany, 506
Thalamic syndrome, 275
Thalamus
   functions, 275
   nuclei, *274*, 275
Thenar eminence, *77*, 78
Thermal sense
   neural pathway, 312, *312*
   receptors, 311, *311*
Thigh, *79*, 80
Thoracic duct, *392*, 393
Thymine, 16
Thymosin, 505
Thymus, 398, *398*, 505
   functions, 398
   location, *398*
   structure, 398
Thyroid cartilage, 74, *74*
Thyroid gland
   blood supply, *500*, 501
   cells, 501, *501*
   disorders, 506
   follicles, *500*, 501
   hormones, 501
   location, 500, *500*
   structure, 500, *500*
Thyroid-stimulating hormone, 497t
Thyroxin, 501
   actions, 501
Tight junction, 38, *38*

Tissue groups; *see also* individual tissue group names
   connective tissues, 31
   epithelial tissues, 31
   muscular tissues, 31
   nervous tissues, 31
Tongue, 430, *430*
   papillae, *430*, 431
   von Ebner's glands, 431
Tonsil(s)
   lingual, 395, *395*
   palatine, 395, *395*
   pharyngeal, 395, *395*
Torticollis, 181
Touch
   neural pathway, 313, *313*
   receptor, 311, *311*
Trachea, 411, *411, 412*
Tracheotomy, 414
Trachoma, 327
Transcription, 17
Translation, 17
Transversalis fascia, 209
Transverse tubules, *172*, 174
Triangles of neck
   anterior, 75, *75*
   inferior carotid, 75, *75*
   occipital, 75, *75*
   omoclavicular, 75, *75*
   posterior cervical, 74, *75*
   submandibular, 75, *75*
   superior carotid, 75, *75*
   suprahyoid, 75, *75*
Trophoblast, 20, *20*
Tropic hormone, 496
Tropomyosin, 173, *173*
Troponin, 173, *173*
Trypsin, 447
Tubular transport, 458
Tunica albuginea
   ovary, 470, *470*
   penis, *488*, 489
   testis, 481, *482*
Tympanic cavity, *330*, 332, *332*
Tympanic membrane, *330, 332*, 333

U

Ureter, 458
   wall structure, 458, *458*
Urethra, *460*
   sex differences, 460
   wall structure, 460
Urinary bladder, *459*
   disorders, 461
   wall structure, 459
Urinary system, 450-461
   congenital anomalies, 451t
   development, 450, 451, *451*
   disorders, 461
   organs, 452
Urogenital ridge, 460
Uterine tube
   function, 474
   mesentery, 474

Uterine tube—cont'd
  obstruction, 480
  parts, *469*, 474
  wall structure, 474, *474*, 475
Uterus, 475–477
  function, 475
  ligaments, *469*, 475, *476*
  parts, *469*, 475, *476*
  wall structure, 476, *476*
Utriculus (utricle), *334*
  macula, 334, *335*
Uvea, 320

V

Vagina, *469*, 477, *477*
  function, 477
Vaginal cancer, 480
Varicosities, 373, *373*
Vas deferens, *480*, 486, *486*, *487*, *488*
Vasectomy, 486
Vasopressin-ADH, 498
Vein
  anterior cardiac, *360*, 361
  anterior tibial, 385, *385*
  axillary, 384, *384*
  azygos, 382, *382*, 382t
  basilic, 77, 384, *384*
  brachiocephalic, 382, *382*, 382t
  cavernous sinus, 388, *388*
  cephalic, 77, 384, *384*
  common iliac, 383, *383*, 383t, 385, *385*
  coronary, 386, *386*
  cystic, 386, *386*
  digital, 384, *384*, 385, *385*
  diploic, 389, *389*
  dorsal venous network, 78, 384
  dural sinuses, 388, *388*
  emissary, 389
  external iliac, 385, *385*
  external jugular, 382, *382*, 382t
  femoral, 385, *385*
  gastric, 386, *386*
  gonadal, 383, *383*, 383t
  great cardiac, *360*, *361*
  great cerebral, 388, *388*
  great saphenous, 385, *385*
  hepatic, 383, *383*, 383t
  hepatic portal, 386, *386*

Vein—cont'd
  inferior sagittal sinus, 388, *388*
  inferior vena cava, 382, 383, *383*, 383t
    vessels forming, 383, *383*, 383t
  innominate, 382, *382*, 382t
  internal iliac, 385, *385*
  internal jugular, 382, *382*, 382t
  median antebrachial, 384, *384*
  median antecubital, 77
  median cubital, 384, *384*
  metacarpal, 384, *384*
  metatarsal, 385
  occipital sinus, *388*, 389
  peroneal, 385, *385*
  petrosal sinus, 388, *388*
  plantar arch, 385
  plantar venous arch, 385
  popliteal, 385, *385*
  posterior tibial, 385, *385*
  posterior vein of left ventricle, 360, *361*
  pulmonary, 390, *390*
  pyloric, 386, *386*
  radial, 384, *384*
  renal, 383, *383*, 383t
  saphenous, 80
  sigmoid sinus, *388*, 389
  small saphenous, 385, *385*
  splenic, 386, *386*
  straight sinus, *388*, 389
  subclavian, 382, *382*, 382t, 384, *384*
  superior mesenteric, 386, *386*
  superior sagittal sinus, 388, *388*
  superior vena cava, 382, *382*, 382t
    vessels forming, 382, *382*, 382t
  thebesian, 361
  transverse sinus, *388*, 389
  ulnar, 384, *384*
  umbilical, 391, *391*
  vertebral, 382
Veins of
  brain, 388, *388*
  lower appendage, 385, *385*
  lungs, 390, *390*
  upper appendage, 384, *384*
Ventricles of brain
  fourth, 290, *290*
  lateral, 290, *290*
  third, 290, *290*

Ventricular zone, *249*, 250
Vertebra(e)
  cervical, 122, *122*
  lumbar, *122*, 123
  prominens, 73, *78*
  thoracic, *122*, 123
Vertebral column, 118, *119*
Vertebral curvatures, 120, *120*
Vertebral structure, 118, *119*
Vesalius, 1
Vestibule (ear), 334
Visceral sensation
  neural pathway, 314
  receptors, 314, *314*
Visual pathway, 322, *323*
  lesions, 327, *327*
Vitreous body, *318*, 322
Vocal folds
  false, *409*, 410
  true, *409*, 410
Volar, 3
Volkmann's canals, 52, *52*, 100
Volume
  reserve expiratory, 418
  reserve inspiratory, 418
  residual, 418
  tidal, 418
von Ebner's glands, 431
Vulva, *469*, 477, *477*

W

Waldeyer's ring, 395, *395*
Wallerian degeneration, 262
Warmth receptor, 311, *311*
Wilms' tumor, 461
Wryneck, 181

X

X-ray examination, 2

Y

Yolk sac, 30

Z

Zona pellucida, 20, *20*, *471*, 472
Zonula adherens, 38, *38*
Zonula occludens, 38, *38*
Zygomatic arch, 72, *72*